Difference and Differential Equations

FIELDS INSTITUTE COMMUNICATIONS

THE FIELDS INSTITUTE FOR RESEARCH IN MATHEMATICAL SCIENCES

Difference and Differential Equations

Saber Elaydi
Gerry Ladas
Jianhong Wu
Xingfu Zou
Editors

American Mathematical Society
Providence, Rhode Island

The Fields Institute for Research in Mathematical Sciences

The Fields Institute is named in honour of the Canadian mathematician John Charles Fields (1863–1932). Fields was a visionary who received many honours for his scientific work, including election to the Royal Society of Canada in 1909 and to the Royal Society of London in 1913. Among other accomplishments in the service of the international mathematics community, Fields was responsible for establishing the world's most prestigious prize for mathematics research—the Fields Medal.

The Fields Institute for Research in Mathematical Sciences is supported by grants from the Ontario Ministry of Training, Colleges and Universities, and the Natural Sciences and Engineering Research Council of Canada. The Institute is sponsored by Carleton University, McMaster University, the University of Ottawa, the University of Toronto, the University of Waterloo, the University of Western Ontario, and York University. In addition there are several affiliated universities and corporate sponsors, from Canada and the United States.

2000 *Mathematics Subject Classification.* Primary 00B20, 34K11, 34K13, 34K20, 34K28, 34K30, 39A05, 39A11.

Library of Congress Cataloging-in-Publication Data

International Conference on Difference Equations (7th: 2002: Changsha Shi, China)
Difference and differential equations/Saber Elaydi...[et al.].
p. cm. — (Fields Institute Communications, ISSN 1069-5265 ; v. 42)
Includes bibliographical references.
ISBN 0-8218-3354-5 (alk. paper)
1. Difference equations–Congresses. 2. Differential equations–Congresses. I. Elaydi, Saber, 1943– II. Title. III. Series.
QA431.I15145 2002
515′.625–dc22 2004040309

∞ The paper used in this book is acid-free and falls within the guidelines
established to ensure permanence and durability.
This publication was prepared by The Fields Institute.
Visit the AMS home page at http://www.ams.org/

10 9 8 7 6 5 4 3 2 1 09 08 07 06 05 04

Contents

OPEN PROBLEMS AND CONJECTURES

Preface

This special volume is the proceedings of the 7th International Conference on Difference Equations and Applications (ICDEA) that took place in the beautiful "cultural city" of Changsha, the capital of China's Hunan province, between Monday, August 12, and Saturday August 17, 2002. The conference was held in Lushan Hotel near the campus of Hunan University. There were more than 80 participants who came from Nine countries including Canada, China, Germany, Hungary, Italy, Japan, Poland, UAE, USA.

This conference was held under the auspices of the International Society of Difference Equations (ISDE), and it was a continuation of the series of the International Conferences on Difference Equations and Applications. Previous meetings in this series were held in San Antonio, Texas/USA (1994), Veszprém/Hungary (1995), Taipei/Taiwan (1997), Poznan/Poland (1998), Temuco/Chille (2000) and Augsburg/Germany (2001). The 7th ICDEA was also designated as one of the Satellite Conferences of the ICM2002 in Beijing, reflecting the growing recognition of the importance of the research fields in difference equations and allied subjects. The Organizing Committee consists of S. S. Cheng (Taiwan/China), B. X. Dai (China), L. H. Huang (co-chair, China), X. Z. Qian (China), Z. C. Wang, (China), J. S. Yu (China), Z. Zhou (China), X. Zou (co-chair, Canada), and the Scientific Committee consists of B. Aulbach (Germany), H. Brunner (Canada), S. Elaydi (USA), I. Györi (Hungary), L. H. Huang (China), G. Ladas (USA), Z. C. Shi (China), H. Smith (USA), Z. C. Wang (China), J. Wu (Canada), J. S. Yu (China), X. Zou (Canada),

Participants benefitted from plenary talks by well-known researchers in the field of difference equations and discrete dynamical systems, and from many invited and contributed lectures. Other activities included an inspiring "Open Problems and Conjectures" session coordinated by Professor Elaydi. These proceedings reflect the spirit of the conference and include carefully peer-reviewed contributions in three categories: (i) survey papers; (ii) research papers; (iii) open problems and conjectures. The articles cover a wide range of topics such as stability, oscillation, chaos, symmetries, boundary value problems and bifurcations for discrete dynamical systems, and discretization of continuous equations. The reader will find these proceedings very useful in helping him/her grasp the trends and in bringing him/her to the frontiers of the research in difference equations and the related areas.

This successful conference was hosted by Hunan University, an academic institute that has a proud history of over 1000 years, and also one of the Chinese centers in the conference themes led by a few celebrated scholars including professor Zhicheng Wang. It is thus entirely appropriate that some contributing authors dedicated their work to the 70th birthday of Professor Wang.

We would like to take this opportunity to express our gratitudes to Hunan University and local organizing committee led by Professor Lihong Huang and

Professor Jianshe Yu for their great hospitality, to the Ministry of Education of China and Hunan University for their generous financial support, to ICM2002 for including this meeting as one of its satellite conferences, and to the Fields Institute for including these conference proceedings in its prestigious communications series. We would also like to thank all the participants for making the conference a great success. Finally, we are grateful to all the authors for their contributions to these proceedings and to all the referees for their timely and valuable reviews of the manuscripts.

Saber Elaydi
Gerry Ladas
Jianhong Wu
Xingfu Zou

SURVEY ARTICLES

Fields Institute Communications
Volume **42**, 2004

The Discretization of Volterra Functional Integral Equations with Proportional Delays

Hermann Brunner
Department of Mathematics and Statistics
Memorial University of Newfoundland
St. John's, Newfoundland, Canada A1C 5S7
hermann@math.mun.ca

Abstract. Collocation methods in piecewise polynomial spaces for discretizing Volterra functional integral and integro-differential equations with (vanishing) proportional delays lead to systems of difference equations whose convergence and asymptotic stability properties are still not well understood, especially if the underlying meshes are uniform. Hence, the analysis of the optimal order of (local) superconvergence at the mesh points and the asymptotic stability of collocation solutions remains largely open. In the present paper I shall survey some of the history of these problems and then focus on recent and ongoing work.

1 Volterra functional equations with proportional delays

The delay differential equation

$$y'(t) = ay(t) + by(qt), \quad t \in I := [0,T], \quad 0 < q < 1, \tag{1.1}$$

has become known as the *pantograph equation*: it was introduced by Ockendon & Tayler [A2] (see also [A4]) in connection with the mathematical modelling of the dynamics of an overhead current collection system ("pantograph") for an electric locomotive, and subsequently analyzed by Fox, Mayers, Ockendon & Tayler [T7] and by Kato & McLeod [T8] (see also the insightful earlier study [T6], and [T9–T12], [T15], [T18], [T20–T22], [T24] and [T25]). The pantograph equation (1.1) is a simple delay differential equation (DDE) with (linear) variable and vanishing delay $\tau(t) := (1-q)t$ describing the delay function $\theta(t) := qt = t - (1-q)t$. It is

2000 *Mathematics Subject Classification.* Primary 65R20, 34K28; Secondary 34K06.
Key words and phrases. Volterra functional equations, proportional delays, collocation methods .
This research was supported in part by the Natural Sciences and Engineering Research Council of Canada (NSERC). The author also gratefully acknowledges the support and the generous hospitality extended to him by Hunan University during his visit to Changsha in August 2002.

equivalent to the delay Volterra integral equation (DVIE)

$$y(t) = y_0 + \int_0^t ay(s)ds + \int_0^{qt} (b/q)y(s)ds, \quad t \in I. \tag{1.2}$$

This is a special case of the general linear DVIE

$$y(t) = g(t) + (\mathcal{V}y)(t) + (\mathcal{V}_\theta y)(t), \quad t \in I, \tag{1.3}$$

with delay function $\theta(t) := qt$ $(0 < q < 1)$. The Volterra integral operators $\mathcal{V} : C(I) \to C(I)$ and $\mathcal{V}_\theta : C([0, \theta(T)]) \to C(I)$ are given by

$$(\mathcal{V}y)(t) := \int_0^t K_1(t,s)y(s)ds \quad \text{and} \quad (\mathcal{V}_\theta y)(t) := \int_0^{\theta(t)} K_2(t,s)y(s)ds, \tag{1.4}$$

with kernels K_1 and K_2 assumed to be continuous on their respective domains $D := \{(t,s) : 0 \le s \le t \le T\}$ and $D_\theta := \{(t,s) : 0 \le s \le \theta(t), t \in I\}$.

For $K_2 = -K_1 =: -K$ the DVIE (1.3) reduces to

$$y(t) = g(t) + (\mathcal{W}_\theta y)(t), \quad t \in I, \tag{1.5}$$

where

$$(\mathcal{W}_\theta y)(t) := \int_{\theta(t)}^t K(t,s)y(s)ds. \tag{1.6}$$

Although Fox et al. [T7] studied the discretization of (1.1) by means of the trapezoidal method and the τ-method (including iterative corrections) – compare pp. 287-292 for the resulting difference equations and error analysis – , numerical analysts did not recognize for some 20 years the formidable challenge presented by the analysis of, e.g., asymptotic stability or optimal superconvergence of discrete versions of the pantograph equation and analogous Volterra integral or integro-differential equations. Since the early 1990s, however, largely thanks to the work by Iserles, Buhmann, Nørsett, Liu, and their collaborators (see [T27–T35], [N3–N11]), this field of numerical analysis and difference equations has seen considerable progress.

It is the main purpose of this survey paper to exhibit the complex nature of the (systems of) difference equations that arise when such initial-value problems are approximated by collocation in certain piecewise polynomial spaces, and to show that many problems regarding, e.g., the uniform boundedness and the asymptotic stability of solutions remain open, since they are not tractable by the standard tools of the "usual" analysis.

We conclude this introductory section with some remarks on the early history of delay equations with proportional delays. The first detailed analysis of a functional equation with proportional delay is contained in Vito Volterra's 1897 paper [T1] (see also his book [T4] of 1913, and the generalizations by Lalesco in [T2], [T3]). He considered the first-kind integral equation

$$(\mathcal{W}_\theta y)(t) = g(t), \quad t \in I, \tag{1.7}$$

with $\theta(t) := qt$ $(0 < q < 1)$ and $g(0) = 0$. If $g \in C^1(I)$, and if the kernel $K = K(t,s)$ in the integral operator $\mathcal{V}_\theta$ is continuously differentiable on $\bar{D}_\theta := \{(t,s) : \theta(t) \le s \le t, t \in I\}$ and satisfies $|K(t,t)| \ge \kappa_0 > 0$ on I, then this equation is equivalent to a second-kind DVIE more general than (1.5),

$$y(t) = g_1(t) + b(t)y(\theta(t)) + \int_{\theta(t)}^t H(t,s)y(s)ds, \quad t \in I. \tag{1.8}$$

Here, we have defined

$$g_1(t) := g'(t)/K(t,t), \quad b(t) := qK(t,qt), \quad H(t,s) := -[\partial K(t,s)/\partial y]/K(t,t).$$

Volterra showed, by an elegant, rather nontrivial argument, that the Picard iteration approach he employed in his two fundamental papers of 1896 (for (1.7) with $q = 0$) can be extended to the DVIE (1.7) for arbitrary $q \in (0,1)$, to show the existence and uniqueness of a solution $y \in C(I)$.

This same question was studied in 1913/14 by G. Andreoli [T5] for the "pure delay" case of the DVIE (1.3) (corresponding to $K_1 = 0$ in (1.4)),

$$y(t) = g(t) + (\mathcal{V}_\theta y)(t), \quad t \in I. \tag{1.9}$$

Using again Picard iteration he showed (see also the 1990 paper [T23] by Chambers) that the solution of (1.9) can be expressed in the form

$$\begin{aligned} y(t) &= g(t) + \int_0^t \left(\sum_{j=1}^{\infty} q^j K_j(t, q^j s) g(q^j s) \right) ds, \\ &= g(t) + \sum_{j=1}^{\infty} \int_0^{q^j t} K_j(t,s) g(s) ds, \quad t \in I, \end{aligned} \tag{1.10}$$

with uniformly convergent infinite series based on the iterated kernels $\{K_j(t,s)\}$ of K,

$$K_{j+1}(t,s) := \int_{q^{-j}s}^{qt} K(t,v) K_j(v,s) dv, \quad (t,s) \in D_\theta \quad (j \geq 1;\ K_1 := K).$$

For $q = 1$ we obtain of course the classical result of Volterra (1896) which states that the unique solution $y \in C(I)$ of the VIE

$$y(t) = g(t) + \int_0^t K(t,s) y(s) ds, \quad t \in I,$$

is given by

$$y(t) = g(t) + \int_0^t R(t,s) g(s) ds, \quad t \in I, \tag{1.11}$$

where $R(t,s) := \sum_{j=1}^{\infty} K_j(t,s)$ denotes the resolvent kernel defined by the Neumann series corresponding to the given kernel $K(t,s)$. As (1.10) shows, for $0 < q < 1$ this "resolvent representation" (1.11) of y is no longer valid, due to the "overlap" of the values of the iterated kernels K_j with those of the nonhomogeneous term g. As we shall see below, this has dramatic consequences for the analysis of the discrete versions of pantograph VIEs (and for the pantograph equation (1.1) itself).

In addition to the delay integral equations (1.3) and (1.7) we will also consider the nonlocal version of the pantograph DDE (1.1) given by the delay Volterra integro-differential equation (DVIDE)

$$y'(t) = a(t)y(t) + b(t)y(\theta(t)) + (\mathcal{V}y)(t) + (\mathcal{V}_\theta y)(t), \quad t \in I, \tag{1.12}$$

with $\theta(t) = qt$, and its discretization by piecewise polynomial collocation methods. In view of applications, and to illustrate a number of open problems we shall also look at the two special cases given by

$$y'(t) = a(t)y(t) + b(t)y(\theta(t)) + (\mathcal{V}_\theta y)(t), \quad t \in I, \tag{1.13}$$

and

$$y'(t) = a(t)y(t) + b(t)y(\theta(t)) + (\mathcal{W}_\theta y)(t), \quad t \in I. \tag{1.14}$$

In the numerical solution of Volterra functional integral or integro-differential equations of the above forms by, e.g., piecewise polynomial collocation methods, the following questions regarding the quantitative or qualitative behaviour of solutions of the difference equations describing the resulting discrete problems play a central role:

1. If the given functional equation is solved on the compact interval $I := [0, T]$ and if the diameter of the mesh I_h underlying the method (see Section 2) tends to zero, do the solutions of these difference equations always remain (uniformly) bounded? If so, what can be said about the (asymptotic) rate at which they converge?
2. Suppose that the functional equation possesses solutions on $\mathbb{R}_+$ that are asymptotically stable; that is, they tend to zero as $t \to \infty$. Do the solutions of the discrete version of such an equation have the the same asymptotic behaviour; i.e., do they also tend to zero if they correspond to meshes I_h with fixed mesh parameter $h > 0$?

We shall see in the following sections that the answers to these questions – especially to the second one – remain to be found in many cases, due to the very complex nature of the difference equations arising in the discretization of delay problems with vanishing proportional delays.

2 Discretizations based on collocation methods

Let $I_h := \{t_n : 0 = t_0 < t_1 < \cdots < t_N = T\}$ denote a mesh on the interval $I := [0, T]$, and set

$$\begin{aligned} \sigma_n &:= (t_n, t_{n+1}], \quad \bar{\sigma}_n := [t_n, t_{n+1}], \quad h_n := t_{n+1} - t_n \ (0 \le n \le N-1), \\ h &:= \max_{(n)} h_n. \end{aligned}$$

For given integers $m \ge 1$ and $d \ge -1$ we denote by

$$S_{m+d}^{(d)}(I_h) := \{u_h \in C^d(I) : u_h|_{\sigma_n} \in \pi_{m+d} \ (0 \le n \le N-1)\} \tag{2.1}$$

the linear space of (real) piecewise polynomials with respect to the mesh I_h whose degree does not exceed $m + d$. If $d = -1$ then $u_h \in S_{m-1}^{(-1)}(I_h)$ will in general have finite (jump) discontinuities at the interior points of I_h; the space of step functions, $S_0^{(-1)}(I_h)$, is the most obvious example of such a discontinuous piecewise polynomial space.

The dimension of the linear space defined by (2.1) is given by

$$\dim S_{m+d}^{(d)}(I_h) = Nm + (d+1).$$

The choice of d, the degree of regularity, will be governed by the type of functional equation whose solution will be approximated by collocation in $S_{m+d}^{(d)}(I_h)$: for the functional integral equations (1.3), (1.7) and (1.8) the "natural" piecewise polynomial space is $S_{m-1}^{(-1)}(I_h)$ $(d = -1)$, while for functional integro-differential equations containing as the highest derivative of the unknown solution $y^{(k)}$ we choose $d = k - 1$. This means that the collocation space for equation (1.12) will be $S_m^{(0)}(I_h)$ $(d = 0)$.

The desired collocation solution $u_h \in S_{m+d}^{(d)}(I_h)$ will be determined by requiring that it satisfy the given functional equation on the set of collocation points

$$X_h := \{t_{n,i} := t_n + c_i h_n : 0 < c_1 < \cdots < c_m \leq 1 \ (0 \leq n \leq N-1)\}, \tag{2.2}$$

described by given *collocation parameters* $\{c_i\}$. Clearly,

$$|X_h| = Nm = \dim S_{m+d}^{(d)}(I_h) - (d+1).$$

If $d \geq 0$ the collocation solution will also be required to concide, at $t = 0$, with the prescribed initial values; e.g., in the case of the DVIDE (1.12) we will have $u_h(0) = y_0$.

The collocation equations corresponding to the delay Volterra equations introduced in Section 1 will contain the delay integral terms $(\mathcal{V}_\theta u_h)(t)$ and $(\mathcal{W}_\theta u_h)(t)$ where $t = t_{n,i} := t_n + c_i h_n \in X_h \cap \sigma_n$. Thus, the structure of the difference equations corresponding to these collocation points will be governed by the location of the images of these collocation points,

$$\theta(t_{n,i}) = q(t_n + c_i h_n) \quad (i = 1, \ldots, m).$$

For arbitrary non-uniform meshes these difference equations are obviously very complex. Therefore, in order to make the analysis tractable we will for the present assume that the mesh I_h is *uniform:*

$$I_h := \{t_n := nh : \ n = 0, 1, \ldots, N; \ t_N = T\}.$$

(Non-uniform meshes that are *quasi-geometric* will be briefly discussed in Section 5.) For uniform meshes we will write

$$\theta(t_{n,i}) = q(t_n + c_i h) =: h\{q_{n,i} + \gamma_{n,i}\} = t_{q_{n,i}} + \gamma_{n,i} h, \tag{2.3}$$

where

$$q_{n,i} := \lfloor q(n + c_i) \rfloor \in \mathbb{N}_0, \qquad \gamma_{n,i} := q(n + c_i) - q_{n,i} \in [0, 1).$$

For given collocation parameters with $0 < c_1 < \cdots < c_m \leq 1$ and $q \in (0, 1)$ define

$$q^I := \lceil qc_1/(1-q) \rceil, \qquad q^{II} := \lceil qc_m/(1-q) \rceil. \tag{2.4}$$

Here, $\lceil x \rceil$ denotes the least upper integer bound for $x \in \mathbb{R}$.

The validity of the following lemma – which characterizes the *"overlap"* of the images of the collocation points under the mapping θ – is easily verified.

Lemma 2.1 *Let $q \in (0, 1)$ and $0 < c_1 < \cdots < c_m \leq 1$, and assume that I_h is a uniform mesh on $I := [0, T]$ with diameter $h := T/N$. Then:*

(i) *For $n = 0$ we have $q(t_n + c_i h) \in (t_n, t_{n+1})$ for $i = 1, \ldots, m$.*
(ii) *If $n \geq 1$ we have $q(t_n + c_i h) \in (t_n, t_{n+1})$ iff $n < q^I$.*
(iii) *$q(t_n + c_i h) \leq t_n$ for $i = 1, \ldots, m$ iff $q^{II} \leq n \leq N - 1$.*

This result reveals that the recursive computation of the collocation solution for functional equations with (vanishing) proportional delay consists in general of *three phases:*

- *Phase I:* During this "initial phase" we have complete overlap which is described by the values of n satisfying

$$0 \leq n < \left\lceil \frac{q}{1-q} c_1 \right\rceil =: q^I.$$

- *Phase II:* The "transition phase" corresponds to those n where *partial overlap* occurs; they are given by

$$q^I \leq n < \left\lceil \frac{q}{1-q} c_m \right\rceil =: q^{II}.$$

If this set of indices n is non-empty, there exists a $\nu_n \in \{1, \ldots, m-1\}$ so that

$$q(t_n + c_i h) \leq t_n \quad (i \leq \nu_n) \qquad \text{and} \qquad q(t_n + c_i h) > t_n \quad (i > \nu_n).$$

- *Phase III:* In this "pure delay phase", described by

$$q^{II} \leq n \leq N - 1,$$

there is *no* longer any *overlap* of $\theta(t_{n,i})$ and $t_{n,i}$: we have

$$q(t_n + c_i h) \leq t_n \quad \text{for} \quad i = 1, \ldots, m.$$

More precisely, for such a value of n there exist integers $\nu_n \in \{1, \ldots, m\}$ and $q_n < n - 1$ so that

$$q_{n,i} = q_n \quad (i \leq \nu_n) \qquad \text{and} \qquad q_{n,i} = q_{n+1} \quad (i > \nu_n).$$

Note that the integers q^I and q^{II} do not depend on the underlying mesh and are thus independent of N.

Illustration:

We list a selection values of q^I and q^{II} for $m = 2$ and $m = 3$, corresponding to the

- *Gauss points:*
  $m = 2$: $c_1 = (3 - \sqrt{3})/6$, $c_2 = (3 + \sqrt{3})/6$,
  $m = 3$: $c_1 = (5 - \sqrt{15})/10$, $c_2 = 1/2$, $c_3 = (5 + \sqrt{15})/10$;
- *Radau II points:*
  $m = 2$: $c_1 = 1/3$, $c_2 = 1$,
  $m = 3$: $c_1 = (4 - \sqrt{6})/10$, $c_2 = (4 + \sqrt{6})/10$, $c_3 = 1$.

Table 1 $m = 2$

	Gauss points				*Radau II points*			
$q =$	1/2	2/3	0.9	0.99	1/2	2/3	0.9	0.99
$q^I =$	1	1	2	21	1	1	3	33
$q^{II} =$	1	2	8	79	1	2	9	99

Table 2 $m = 3$

	Gauss points				*Radau II points*			
$q =$	1/2	2/3	0.9	0.99	1/2	2/3	0.9	0.99
$q^I =$	1	1	2	12	1	1	2	16
$q^{II} =$	1	2	8	88	1	2	9	99

2.1 Integral equations with proportional delays. The collocation solution $u_h \in S_{m-1}^{(-1)}(I_h)$ for the delay integral equation (1.3) is defined by

$$u_h(t) = g(t) + (\mathcal{V}u_h)(t) + (\mathcal{V}_\theta u_h)(t), \quad t \in X_h, \tag{2.5}$$

while the one, in the same collocation space, for the first-kind DVIE (1.7) is described by

$$(\mathcal{W}_\theta u_h)(t) = g(t), \quad t \in X_h. \tag{2.6}$$

Here, $\theta(t) = qt$ $(0 < q < 1)$, and the set X_h is given by (2.2).

Consider now the contribution of $(\mathcal{V}_\theta u_h)(t_{n,i})$ to the collocation equation (2.5): it follows from the definition of $q_{n,i}$ and $\gamma_{n,i}$ in (2.3) that

$$\begin{aligned}(\mathcal{V}_\theta u_h)(t_{n,i})&\\ = \int_0^{t_{q_{n,i}}} K(t_{n,i}, s)u_h(s)ds - h\int_0^{\gamma_{n,i}} K(t_{n,i}, t_{q_{n,i}} + sh)u_h(t_{q_{n,i}} + sh)ds.\end{aligned} \tag{2.7}$$

For the special DVIEs (1.5) and (1.7) we find – in complete analogy to the above – that $(\mathcal{W}_\theta u_h)(t_{n,i})$ is given by

$$\begin{aligned}&(\mathcal{W}_\theta u_h)(t_{n,i})\\ &= h\int_{\gamma_{n,i}}^1 K(t_{n,i}, t_{q_{n,i}} + sh)u_h(t_{q_{n,i}} + sh)ds + \int_{t_{q_{n,i}+1}}^{t_n} K(t_{n,i}, s)u_h(s)ds\\ &\quad +h\int_0^{c_i} K(t_{n,i}, t_n + sh)u_h(t_n + sh)ds.\end{aligned} \tag{2.8}$$

In order to obtain suitable computational forms of the collocation equations (2.5) and (2.6) (leading to systems of difference equations whose solution describes the collocation solution on I), we express u_h on the subinterval σ_n by the local Lagrange representation,

$$u_h(t_n + vh_n) = \sum_{j=1}^{m} L_j(v)U_{n,j}, \quad v \in (0,1] \quad (0 \le n \le N-1), \tag{2.9}$$

with

$$L_j(v) := \prod_{k\neq j}^{m}(s - c_k)/(c_j - c_k) \qquad \text{and} \qquad U_{n,j} := u_h(t_n + c_jh_n),$$

and where at present we have assumed that $h_n = h = T/N$ for $n = 0, 1, \ldots, N-1$.

The above expressions (2.7) and (2.8) for $(\mathcal{V}_\theta u_h)(t_{n,i})$ and $(\mathcal{W}_\theta u_h)(t_{n,i})$ now assume the forms

$$\begin{aligned}(\mathcal{V}_\theta u_h)(t_{n,i}) &= h\sum_{\ell=0}^{q_{n,i}-1}\sum_{j=1}^{m}\left(\int_0^1 K(t_{n,i}, t_\ell + sh)L_j(s)ds\right)U_{\ell,j}\\ &\quad +h\sum_{j=1}^{m}\left(\int_0^{\gamma_{n,i}} K(t_{n,i}, t_{q_{n,i}} + sh)L_j(s)ds\right)U_{q_{n,i},j}\end{aligned} \tag{2.10}$$

and

$$\begin{aligned}(\mathcal{W}_\theta u_h)(t_{n,i}) &= h\sum_{j=1}^{m}\left(\int_{\gamma_{n,i}}^{1} K(t_{n,i},t_{q_{n,i}}+sh)L_j(s)ds\right)U_{q_{n,i},j}\\ &+h\sum_{\ell=q_{n,i}+1}^{n-1}\left(\int_0^1 K(t_{n,i},t_\ell+sh)L_j(s)ds\right)U_{\ell,j}\\ &+h\sum_{j=1}^{m}\left(\int_0^{c_i} K(t_{n,i},t_n+sh)KL_j(s)ds\right)U_{n,j}.\end{aligned} \tag{2.11}$$

Thus, the collocation equations lead to systems of linear difference equations for the vectors $\mathbf{U}_n \in \mathbb{R}^m$ $(0 \le n \le N-1)$ whose structure will change as we pass from Phase I of *complete overlap* $(0 \le n < q^I)$ via Phase II of *partial overlap* $(q^I \le n < q^{II})$ to the *pure delay* Phase III $(q^{III} \le n \le N-1)$. To be more precise, consider first the collocation equation (2.5) for the general DVIE (1.3) and set $\mathbf{g}_n := (\ g(t_{n,1}), \dots, g(t_{n,m})\ )^T \in \mathbb{R}^m$,

$$B_n := \begin{pmatrix} \int_0^{c_i} K_1(t_{n,i},t_n+sh)L_j(s)ds \\ (i,j=1,\dots,m)\end{pmatrix}, \tag{2.12}$$

$$B_{n,\ell} := \begin{pmatrix} \int_0^{1} K_1(t_{n,i},t_\ell+sh)L_j(s)ds \\ (i,j=1,\dots,m)\end{pmatrix} \quad (\ell < n). \tag{2.13}$$

These matrices correspond to the contribution to the difference equation due to the "classical" (non-delay) Volterra operator $\mathcal{V}$ in (2.3). For the concise formulation of the terms in the difference equation describing Phases I, II, and III and corresponding to the delay operator $\mathcal{V}_\theta$ we introduce the following $m \times m$ matrices:

<u>*Phase I:*</u>

$$B_n^I(q) := \begin{pmatrix} \int_0^{\gamma_{n,i}} K_2(t_{n,i},t_n+sh)L_j(s)ds \\ (i,j=1,\dots,m)\end{pmatrix}, \tag{2.14}$$

$$B_{n,\ell}^I(q) := \begin{pmatrix} \int_0^{1} K_2(t_{n,i},t_\ell+sh)L_j(s)ds \\ (i,j=1,\dots,m)\end{pmatrix} \quad (\ell < n). \tag{2.15}$$

Here, $\mathbf{U}_n$ $(0 \le n < q^I)$ is given by the solution of

$$[I_m - h(B_n + B_n^I(q))]\mathbf{U}_n = \mathbf{g}_n + h\sum_{\ell=0}^{n-1}(B_{n,\ell}+B_{n,\ell}^I(q))\mathbf{U}_\ell. \tag{2.16}$$

Phase II:

$$B_n^{II}(q) := \operatorname{diag}(\underbrace{0,\dots,0}_{\nu_n},1,\dots,1)B_n^I(q), \tag{2.17}$$

$$B_{n-1}^{II}(q) := \left(\begin{array}{c} \int_0^{\gamma_{n,i}} K_2(t_{n,i}, t_{n-1}+sh)L_j(s)ds \\ (i,j=1,\dots,m) \end{array} \right), \tag{2.18}$$

$$S_{n-1}^{II}(q) := \operatorname{diag}(\underbrace{0,\dots,0}_{\nu_n},1,\dots,1)B_{n,n-1}^I(q), \tag{2.19}$$

$$\hat{S}_{n-1}^{II}(q) := \operatorname{diag}(\underbrace{1,\dots,1}_{\nu_n},0,\dots,0)B_{n-1}^{II}(q). \tag{2.20}$$

It is the readily verified that now $\mathbf{U}_n$ $(q^I \le n < q^{II})$ is determined by the solution of

$$\begin{aligned}[I_m - h(B_n + B_n^{II}(q))]\mathbf{U}_n &= \mathbf{g}_n + h\sum_{\ell=0}^{n-1} B_{n,\ell}\mathbf{U}_\ell + h\sum_{\ell=0}^{n-2} B_{n,\ell}^I(q)\mathbf{U}_\ell \\ &\quad + h[\hat{S}_{n-1}^{II}(q) + S_{n-1}^{II}(q)]\mathbf{U}_{n-1}. \end{aligned} \tag{2.21}$$

Phase III:

$$B_{q_n}^{III}(q) := \left(\begin{array}{c} \int_0^{\gamma_{n,i}} K_2(t_{n,i}, t_{q_n}+sh)L_j(s)ds \\ (i,j=1,\dots,m) \end{array} \right), \tag{2.22}$$

$$S_{q_n+1}^{III}(q) := \operatorname{diag}(\underbrace{0,\dots,0}_{\nu_n},1,\dots,1)B_{q_n+1}^{III}(q), \tag{2.23}$$

$$\hat{S}_{q_n}^{III}(q) := \operatorname{diag}(\underbrace{1,\dots,1}_{\nu_n},0,\dots,0)B_{q_n}^{III}(q). \tag{2.24}$$

Once we have reached this pure delay phase the vector $\mathbf{U}_n$ $(q^{II} \le n \le N-1)$ is given by the solution of

$$\begin{aligned}[I_m - hB_n]\mathbf{U}_n &= \mathbf{g}_n + h\sum_{\ell=0}^{n-1} B_{n,\ell}\mathbf{U}_\ell + h\sum_{\ell=0}^{q_n-1} B_{n,\ell}^I(q)\mathbf{U}_\ell \\ &\quad + h[\hat{S}_{q_n}^{III}(q) + B_{n,q_n}^I(q)]\mathbf{U}_{q_n} + hS_{q_n+1}^{III}(q)\mathbf{U}_{q_n+1}. \end{aligned} \tag{2.25}$$

The (different) integers ν_n occurring in Phases II and III, and the integer $q_n < n-1$ in Phase III were defined following Lemma 2.1.

Consider now the collocation equations corresponding to the DVIEs (1.5) and (1.7) with the special delay operator $\mathcal{W}_\theta$. Starting with the second-kind VIE (1.5), it follows from (2.8) that the difference equations for $\mathbf{U}_n$ determining the local representations for $u_h \in S_{m-1}^{(-1)}(I_h)$ become, for:

Phase I:

$$[I_m - h\bar{B}_n^I(q)]\mathbf{U}_n = \mathbf{g}_n \quad (0 \le n < q^I), \tag{2.26}$$

with

$$\bar{B}_n^I(q) = \left(\begin{array}{c} \int_{\gamma_{n,i}}^{c_i} K(t_{n,i}, t_n+sh)L_j(s)ds \\ (i,j=1,\dots,m) \end{array} \right),$$

which is of course equivalent to $B_n + B_n^I(q)$ in (2.16) when $K_2 = -K_1 =: -K$;

Phase II:

$$[I_m - h\bar{B}_n^{II}(q)]\mathbf{U}_n = \mathbf{g}_n + h\bar{S}_{n-1}^{II}(q)\mathbf{U}_{n-1} \quad (q^I \le n < q^{II}), \tag{2.27}$$

where

$$\bar{B}_n^{II}(q) := \mathrm{diag}(\underbrace{1,\dots,1}_{\nu_n},0,\dots,0)B_n + \mathrm{diag}(\underbrace{0,\dots,0}_{\nu_n},1,\dots,1)\bar{B}_n^I(q),$$

and

$$\bar{S}_{n-1}^{II}(q) := \mathrm{diag}(\underbrace{1,\dots,1}_{\nu_n},0,\dots,0)\left(\int_{\gamma_{n,i}}^1 K(t_{n,i},t_{n-1}+sh)L_j(s)ds\right);$$

Phase III:

$$[I_m - hB_n]\mathbf{U}_n = \mathbf{g}_n + h[\bar{S}_{q_n}^{III}(q)\mathbf{U}_{q_n} + \sum_{\ell=q_n+1}^{n-1} B_{n,\ell}\mathbf{U}_\ell + S_{q_n+1}^{III}(q)\mathbf{U}_{q_n+1}], \tag{2.28}$$

with

$$\begin{aligned}\bar{S}_{q_n}^{III}(q) &:= \mathrm{diag}(\underbrace{1,\dots,1}_{\nu_n},0,\dots,0)\left(\int_{\gamma_{n,i}}^1 K(t_{n,i},t_{q_n}+sh)L_j(s)ds\right),\\ S_{q_n+1}^{III}(q) &:= \mathrm{diag}(\underbrace{1,\dots,1}_{\nu_n},0,\dots,0)B_{n,q_n+1} +\\ &\quad +\mathrm{diag}(\underbrace{0,\dots,0}_{\nu_n},1,\dots,1)\left(\int_{\gamma_{n,i}}^1 K(t_{n,i},t_{q_n+1}+sh)L_j(s)ds\right).\end{aligned}$$

The matrices $B_{n,\ell}$ coincide with the ones in (2.16).

Turning to the first-kind VIE (1.7) the above difference equations reduce to

$$\bar{B}_n^I(q)\mathbf{U}_n = h^{-1}\mathbf{g}_n \quad (0 \le n < q^I), \tag{2.29}$$

$$\bar{B}_n^{II}(q)\mathbf{U}_n = h^{-1}\mathbf{g}_n - \bar{S}_{n-1}^{II}(q)\mathbf{U}_{n-1} \quad (q^I \le n < q^{II}), \tag{2.30}$$

and

$$B_n\mathbf{U}_n = h^{-1}\mathbf{g}_n - [\bar{S}_{q_n}^{III}(q)\mathbf{U}_{q_n} + \sum_{\ell=q_n+1}^{n-1} B_{n,\ell}\mathbf{U}_\ell + S_{q_n+1}^{III}(q)\mathbf{U}_{q_n+1} \quad (q^{II} \le n \le N-1), \tag{2.31}$$

respectively.

Example 2.1: Suppose that the second-kind delay VIE (1.9) is solved by collocation in $S_0^{(-1)}(I_h)$, with uniform mesh I_h and collocation points $X_h = \{t_n + c_1 h : 0 < c_1 \le 1\ (0 \le n \le N-1)\}$. Since u_h is constant on each subinterval σ_n we write $y_{n+1} := u_h(t_n + vh)\ (v \in (0,1])$.
For $m = 1$ we have $q^I = q^{II} = \lceil qc_1/(1-q)\rceil$. Hence, the collocation equation (2.5) (with $\mathcal{V} = 0$) becomes

$$\left(1 - h\int_0^{\gamma_{n,1}} K(t_{n,1},t_n+sh)ds\right)y_{n+1} = h\sum_{\ell=0}^{n-1}\left(\int_0^1 K(t_{n,1},t_\ell+sh)ds\right)y_{\ell+1} + g(t_{n,1})$$

when $0 \le n < q^I$. For $q^I \le n \le N-1$ it reads

$$\begin{aligned} y_{n+1} &= g(t_{n,1}) + h\sum_{\ell=0}^{q_{n,1}-1}\left(\int_0^1 K(t_{n,1}, t_\ell + sh)ds\right)y_{\ell+1} \\ &\quad + h\left(\int_0^{\gamma_{n,1}} K(t_{n,1}, t_{q_{n,1}+1} + sh)ds\right)y_{q_{n,1}+1}. \end{aligned}$$

Thus, if $K(t,s) \equiv b/q$ and $g(t) \equiv 1$ the collocation solution to the resulting DVIE

$$y(t) = 1 + \int_0^{qt}(b/q)y(s)ds, \quad t \in I$$

(which is equivalent to the initial-value problem $y'(t) = by(qt),\ y(0) = 1$) is determined by the solution of he difference equation

$$y_{n+1} = 1 + \frac{hb}{q}\sum_{\ell=0}^{q_{n,1}-1} y_{\ell+1} + \frac{hb}{q}\gamma_{n,1}y_{q_{n,1}+1}, \tag{2.32}$$

where $q_{n,1} := \lfloor qc_1/(1-q)\rfloor$ and $\gamma_{n,1} := q(n+c_1) - q_{n,1}$.

We list, also for use in Example 2.3 below, a sample of values of $q_{n,1}$ and $\gamma_{n,1}$ for $c_1 = 1/2$, to provide some insight into the structure of the above difference equations when collocation is at the Gauss point $t_n + h/2$.

Table 3 $q = 1/2,\ c_1 = 1/2\ (q^I = q^{II} = 1)$

$n =$	0	1	2	3	4	5	6
$q_{n,1} =$	0	0	1	1	2	2	3
$\gamma_{n,1} =$	1/4	3/4	1/4	3/4	1/4	3/4	1/4

Table 4 $q = 0.9,\ c_1 = 1/2\ (q^I = q^{II} = 5)$

$n =$	0	1	2	3	4	5	6
$q_{n,1} =$	0	1	2	3	4	4	5
$\gamma_{n,1} =$	0.45	0.35	0.25	0.15	0.05	0.95	0.85

Example 2.2: Consider the first-kind delay VIE (1.7), and assume that its collocation solution is to be in the collocation space of Example 2.1. Thus, for $0 \le n < q^I$ the collocation equation (2.4) may be written in the form

$$\left(\int_{\gamma_{n,1}}^{c_1} K(t_{n,1}, t_n + sh)ds\right)y_{n+1} = h^{-1}g(t_{n,1}).$$

If $n \ge q^I = q^{II}$ (see Example 2.1) we have, by (2.11),

$$(\mathcal{W}_\theta u_h)(t_{n,1}) = \int_{qt_{n,1}}^{t_n} K(t_{n,1}, s)u_h(s)ds + \int_{t_n}^{t_{n,1}} K(t_{n,1}, s)u_h(s)ds,$$

and this can be written as

$$\left(\int_0^{c_1} K(t_{n,1}, t_n + sh)ds\right) y_{n+1}$$
$$= h^{-1}g(t_{n,1}) - \left(\int_{\gamma_{n,1}}^1 K(t_{n,1}, t_{q_{n,1}} + sh)ds\right) y_{q_{n,1}+1}$$
$$- \sum_{\ell=q_{n,1}+1}^{n-1} \left(\int_0^1 K(t_{n,1}, t_\ell + sh)ds\right) y_{\ell+1}.$$

Setting

$$A_n := \int_0^{c_1} K(t_{n,1}, t_n+sh)ds, \quad B_{n,\ell} := \int_0^1 K(t_{n,1}, t_\ell+sh)ds \ (q_{n,1}+1 \le \ell \le n-1),$$

and

$$C_{n,q_n} := \int_{\gamma_{n,1}}^1 K(t_{n,1}, t_{q_{n,1}} + sh)ds,$$

with $q_{n,1} := \lfloor q(n + c_1)\rfloor$ and $\gamma_{n,1} := q(n + c_1) - q_{n,1}$, the above difference equation for $\{y_{n+1}\}$ becomes

$$A_n y_{n+1} + \sum_{\ell=q_{n,1}+1}^{n-1} B_{n,\ell} y_{\ell+1} + C_{n,q_n} y_{q_{n,1}+1} = h^{-1}g(t_{n,1}). \tag{2.33}$$

If $K(t, s) \equiv 1$ the delay integral equation (1.7) reduces to

$$\int_{qt}^t y(s)ds = g(t), \quad t \in I \quad (g(0) = 0),$$

and this is equivalent to the functional equation

$$y(t) - qy(qt) = g'(t), \quad t \in I.$$

The corresponding collocation solution $u_h \in S_0^{(-1)}(I_h)$ is thus determined by the solution of the difference equation

$$c_1 y_{n+1} + \sum_{\ell=q_{n,1}+1}^{n-1} y_{\ell+1} + (1 - \gamma_{n,1})y_{q_{n,1}+1} = h^{-1}g(t_n + c_1 h) \quad (n \ge 0). \tag{2.34}$$

In order to obtain the difference equation corresponding to collocation at the Gauss points ($c_1 = 1/2$) recall Tables 2 and 3 of Example 1 for the values of $q_{n,1}$ and $\gamma_{n,1}$.

Remark: For the more general second-kind DVIE (1.8), which we write now as

$$y(t) = g(t) + b(t)y(\theta(t)) + (\mathcal{W}_\theta y)(t), \quad t \in I \quad (\theta(t) = qt,\ 0 < q < 1),$$

the existence of a unique collocation solution $u_h \in S_{m-1}^{(-1)}(I_h)$ is no longer guaranteed in Phases I and II for all sufficiently small mesh diameters $h > 0$. This is due to the presence of the terms $b(t_{n,i})u_h(\theta(t_{n,i}))$ which, using the local representation of u_h on $\sigma_{q_{n,i}}$ (cf. (2.9)), assume the form $b(t_{n,i}) \sum_{j=1}^m L_j(\gamma_{n,i})U_{q_{n,i},j}$ $(i = 1, \dots, m)$. Thus, the matrix describing the left-hand side of the difference equation (2.16) (Phase I) is now $I_m - \mathcal{B}_n - h(B_n + B_n^I(q))$, with

$$\mathcal{B}_n := \mathrm{diag}(\ b(t_{n,i})\ ) \begin{pmatrix} L_j(\gamma_{n,i}) \\ (i, j = 1, \dots, m) \end{pmatrix},$$

and hence its inverse will no longer exist for all sufficiently small $h > 0$. (Compare also [N10] where this problem is studied for the case $\mathcal{W}_\theta = 0$, $m = 1$, and $c_1 = 1$.)

2.2 Volterra integro-differential equations with proportional delays. The collocation solution $u_h \in S_m^{(0)}(I_h)$ for the DVIDE (1.14) is determined by

$$u_h'(t) = a(t)u_h(t) + b(t)u_h(\theta(t)) + (\mathcal{W}_\theta u_h)(t), \quad t \in X_h, \tag{2.35}$$

with initial condition $u_h(0) = y_0$. Since the collocation space is now a subspace of $C(I)$ the system of difference equations arising in the computation of u_h has now a somewhat more complex structure than the ones we encountered in the previous section. To be somewhat more precise, the local representation of $u_h \in S_m^{(0)}(I_h)$ is now

$$u_h(t_n + vh) = y_n + h\sum_{j=1}^{m} \beta_j(v)Y_{n,j}, \quad v \in [0,1] \quad (0 \le n \le N-1), \tag{2.36}$$

where $y_n := u_h(t_n)$, $Y_{n,j} := u_h'(t_{n,j})$, and $\beta_j(v) := \int_0^1 L_j(s)ds$. Hence, the key ingredients in the derivation of the difference equations are essentially the same as in Section 2.1, that is, the terms of $(\mathcal{W}_\theta u_h)(t_{n,i})$ (see (2.11)) corresponding to Phases I, II, and III and containing the vectors $\mathbf{Y}_n$ are described by matrices similar to those in (2.25) – (2.27), except that instead of $L_j(s)$ their integrands contain the integrated Lagrange polynomials $\beta_j(s)$. However, there are now also additinal terms reflecting the continuity constraint of u_h at the interior mesh points $t_1, \ldots, t_{N-1}$. We leave the details to the reader; compare also Chapter 5 in [N35]. Instead, we present an illustration from which the difference equations for arbitrary $m \ge 2$ can readily be deduced.

Example 2.3: Consider (2.35) and suppose that $u_h \in S_1^{(0)}(I_h)$ $(m = 1)$, with $0 < c_1 \le 1$. The collocationi equation defining this collocation solution u_h,

$$u_h'(t_{n,1}) = a(t_{n,1})u_h(t_{n,1}) + b(t_{n,1})u_h(qt_{n,1}) + (\mathcal{W}_\theta u_h)(t_{n,1}),$$

where, by (2.36), $u_h(t_n + vh) = y_n + hvY_{n,1}$ $(v \in [0,1])$, can be written as

$$Y_{n,1} = a(t_{n,1})\{y_n + hc_1Y_{n,1}\} + b(t_{n,1})\{y_{q_{n,1}} + h\gamma_{n,1}Y_{q_{n,1}}\} + (\mathcal{W}_\theta u_h)(t_{n,1}),$$

with

$$\begin{aligned}(\mathcal{W}_\theta u_h)(t_{n,1}) &= \int_{t_{q_{n,1}}}^{t_{q_{n,1}+1}} K(t_{n,1},s)u_h(s)ds + \int_{t_{q_{n,1}+1}}^{t_n} K(t_{n,1},s)u_h(s)ds \\ &\quad + h\int_0^{c_1} K(t_{n,1},t_n+sh)\{y_n + hsY_{n,1}\}ds.\end{aligned}$$

Hence, the resulting difference equation is

$$\begin{aligned}&\left(1 - ha(t_{n,1})c_1 - h^2\int_0^{c_1} K(t_{n,1},t_n+sh)s\,ds\right)Y_{n,1} \\ &= [hb(t_{n,1})\gamma_{n,1} + h^2\left(\int_{\gamma_{n,1}}^1 K(t_{n,1},t_{q_{n,1}}+sh)s\,ds\right)]Y_{q_{n,1}} \\ &+h^2\sum_{\ell=q_{n,1}+1}^{n-1}\left(\int_0^1 K(t_{n,1},t_\ell+sh)s\,ds\right)Y_{\ell,1} + \rho_n,\end{aligned} \tag{2.37}$$

where

$$\rho_n := a(t_{n,1})y_n + \left(b(t_{n,1}) + h\int_{\gamma_{n,1}}^{1} K(t_{n,1}, t_{q_{n,1}} + sh)ds\right) y_{q_n,1}$$

$$+h \sum_{\ell=q_{n,1}+1}^{n-1} \left(\int_0^1 K(t_{n,1}, t_\ell + sh)ds\right) y_\ell + h\left(\int_0^{c_1} K(t_{n,1}, t_n + sh)ds\right) y_n.$$

The values of $q_{n,1}$ and $\gamma_{n,1}$ can be found in Example 2.1. Observe also that the above difference equation (2.37) can be reformulated as a difference equation for $\{y_n\}$, by setting $Y_{n,1} = (y_{n+1} - y_n)/h$ (cf. (2.36) with $m = 1$ and $v = 1$).

The special case corresponding to $\mathcal{W} = 0$ and constant a, b, i.e. the pantograph equation (1.1), was studied in [N6] and [N7] for $q = 1/2$.

3 Collocation on compact intervals: optimal orders of convergence

Suppose that a given delay integral or integro-differential equation is solved by collocation in $S_{m-1}^{(-1)}(I_h)$ or $S_m^{(0)}(I_h)$, with the underlying mesh I_h being a *uniform* one, and assume that the functions defining the functional equation have arbitrarily high degree of regularity on their respective domains (implying that the exact solution has at least the same regularity properties). What can be said, as $h \to 0$, $N \to \infty$ ($Nh = T$), about the uniform boundedness of the solution of the resulting difference equations (derived in Section 2) and about the optimal values of p and p^* in the estimates

$$||y - u_h||_\infty := \max\{|y(t) - u_h(t)| : t \in I\} \leq Ch^p \tag{3.1}$$

and

$$||y - u_h||_{h,\infty} := \max\{|y(t) - u_h(t)| : t \in I_h \setminus \{0\}\} \leq Ch^{p^*} \ ? \tag{3.2}$$

It turns out that for *second-kind* delay VIEs and for delay VIDEs, these solutions do remain uniformly bounded and $p = m$ in the global estimate (3.1) for any choice of the $\{c_i\}$; i.e. the "classical" results that hold for their non-delay versions remain valid. However, the question regarding the optimal value of p^* in (3.2) (attainable order of *local superconvergence* on $I_h \setminus \{0\}$) remains open. Even more seriously, it is not known under what conditions on the collocation parameters $\{c_i\}$ the collocation solutions u_h to the *first-kind* delay VIE (1.7) remain uniformly bounded, or converge uniformly on I to the exact solution y: we only know from numerical evidence that the condition

$$-1 \leq \rho_m := (-1)^m \prod_{i=1}^{m} \frac{1 - c_i}{c_i} \leq 1$$

(which guarantees uniform convergence when $q = 0$ in $\mathcal{W}_\theta$ with $\theta(t) = qt$) is no longer sufficient for this to be true. In particular, it is not known for which values of $c_1 \in (0, 1]$ the solution of the simple difference equation (2.34) remains uniformly bounded as $N \to \infty$ ($h \to 0$, $Nh = T$) when $q \in (0, 1)$.

Theorem 3.1 *Consider the second-kind delay VIE (1.3) and assume that the given functions g, K_1, K_2 are at least m times continuously differentiable on their respective domains. Then, for all sufficiently small $h > 0$ (so that the difference equations (2.16), (2.21), (2.25) possess unique solutions), we have $p = m$ in (3.1)* *for arbitrary* $\{c_i\}$ *in (2.2) and for all delay functions* $\theta(t) = qt$ *with* $0 < q < 1$. *The constant* C *in (3.1) depends on the* $\{c_i\}$ *and on* q *but not on* h.

The proof of this result is an adaptation of the one for classical second-kind VIEs and delay VIEs with non-vanishing delays: it consists in showing that in each of the Phases I – III the ℓ^1-norms of the vectors $\mathbf{U}_n$ defined by (2.16), (2.21), and (2.25) satisfy a discrete Gronwall inequality. An elementary induction argument then leads to the assertion, observing the local representation (2.9) of u_h and the fact that, by the assumption on the regularity of the given data, $y \in C^m(I)$. The detailed arguments can be found in Chapter 5 (Section 5.3) of [N35].

The next theorem shows that the classical global order of convergence remains also valid for delay VIDEs with vanishing proportional delay. The proof is again based on a "Phase I – III" Gronwall-type inductive argument; the details can be found in [N35] (Section 5.5). See also [N22,N23] for some related results.

Theorem 3.2 *Consider the delay VIDE (1.12) (which includes (1.14) as a special case) and assume that the given functions* a, b, K_1 *and* K_2 *are at least* m *times continuously differentiable on their respective domains. Then for any* $\theta(t) = qt$ *with* $q \in (0,1)$ *and for all sufficiently small* $h > 0$ *the unique collocation solution* $u_h \in S_m^{(0)}(I_h)$ *for (1.12) satisfies*

$$||y^{(\nu)} - u_h^{(\nu)}||_\infty := \sup\{|y^{(\nu)}(t) - u_h^{(\nu)}(t)| : t \in I\} \le C_\nu h^m \quad (\nu = 0, 1),$$

for arbitrary $\{c_i\}$ *in (2.2). The constants* C_ν *depend on the* $\{c_i\}$ *and on* q *but not on* h.

We now turn to the question of global and local *superconvergence* on I and I_h, respectively: for which collocation parameters $\{c_i\}$ in (2.2) is $p^* > p$ possible in (3.2), and how large can p^* become? Consider first the second-kind delay VIE (1.3). We associate with the collocation solution $u_h \in S_{m-1}^{(-1)}(I_h)$ the *iterated collocation solution*,

$$u_h^{it}(t) := g(t) + (\mathcal{V}u_h)(t) + (\mathcal{V}_\theta u_h(t), \quad t \in I. \tag{3.3}$$

Note that $u_h^{it}(t) = u_h(t)$ whenever $t \in X_h$; in particular, if $c_m = 1$ then $t_n \in X_h$ $(n = 1, \ldots, N)$ and hence $u_h^{it}(t_n) = u_h(t_n)$.

It is well known that if $\mathcal{V}_\theta = 0$ and if the $\{c_i\}$ are the *Gauss (-Legendre) points* (given by the zeros of the shifted Legendre polynomial $P_m(2s-1)$), then $p^* = 2m$. For the general delay VIE (1.3) with $\theta(t) = qt$ $(0 < q < 1)$ and for its particular case (1.5) this is no longer true (see [N16,N17,N22,N23,N27,N32]): for *arbitrary* $q \in (0,1)$ we have $p^* < 2m$ when $m \ge 3$. The results of Theorems 3.3 and 3.4 have recently been established in [N36]. Theorem 3.4 disproves a conjecture in [N16,N17,N28].

Theorem 3.3 *Let the collocation parameters* $\{c_i\}$ *satisfy the orthogonality condition*

$$J_0 := \int_0^1 \prod_{i=1}^m (s - c_i)ds = 0. \tag{3.4}$$

Then

$$||y - u_h^{it}||_\infty \le Ch^{m+1},$$

where u_h^{it} is the iterated collocation approximation (3.3) corresponding to the collocation solution $u_h \in S_{m-1}^{(-1)}(I_h)$ for (1.3).

We note that the most prominent set of parameters $\{c_i\}$ satifying the above orthogonality condition are the *Gauss points.*

Proof: We will sketch the principal steps leading to the above global superconvergence result by using an approach different from the one in [N36]. For ease of notation we will do this for (2.5) with $\mathcal{V} = 0$. The corresponding collocation error $e_h := y - u_h$ satisfies the integral equation

$$e_h(t) = \delta_h(t) + (\mathcal{V}_\theta e_h(t), \quad t \in I, \tag{3.5}$$

where the defect δ_h vanishes on X_h and inherits (piecewise, on each subinterval σ_n) the regularity of g and K_2. Moreover, we have $e_h^{it} := y - u_h^{it} = e_h - \delta_h$. Hence, it follows from the representation (1.10) that

$$e_h^{it}(t) = \sum_{j=1}^{\infty} \int_0^{q^j t} K_{2,j}(t,s)\delta_h(s)ds, \quad t \in I. \tag{3.6}$$

Here, $K_{2,j}(t,s)$ denotes the jth iterated kernel of the given kernel $K_2(t,s)$ in $\mathcal{V}_\theta$ (cf. (1.4)), and the infinite series converges uniformly on I.

Let know $t = t_n + vh,\ v \in [0,1]$ be given, and define

$$q_{j,n}(v) := \lfloor q^j(n+v) \rfloor, \qquad \gamma_{j,n}(v) := q^j(n+v) - q_{j,n}(v) \in [0,1) \quad (j \in \mathbb{N}).$$

This allows us to write (3.6) in the form

$$\begin{aligned} e_h^{it}(t) \;&=\; \sum_{j=1}^{\infty} \Bigg(\int_0^{t_{q_{j,n}(v)}} K_{2,j}(t,s)\delta_h(s)ds \\ &\quad + h \int_0^{\gamma_{j,n}(v)} K_{2,j}(t, t_{q_{j,n}}(v) + sh)\delta_h(t_{q_{j,n}}(v) + sh)ds \Bigg). \end{aligned} \tag{3.7}$$

The assertion in Theorem 3.3 now follows from the following observations:

(i) Since $0 < q < 1$, we have $q_{j,n}(v) < N$ for all $n \leq N-1$ and $v \in [0,1]$. Upon writing

$$\int_0^{t_{q_{j,n}(v)}} K_{2,j}(t,s)\delta_h(s)ds = h \sum_{\ell=0}^{q_{j,n}(v)-1} \int_0^1 K_{2,j}(t, t_\ell + sh)\delta_h(t_\ell + sh)ds,$$

we can resort to the classical "quadrature error argument", consisting in replacing each of the above integrals by the sum of the interpolatory m-point quadrature approximation based on the points $\{t_\ell + c_i h\}$ and the corresponding quadrature error. As $\delta_h(t_\ell + c_i h) = 0$, and $Nh = T$, it follows that the absolute values of the integrals are bounded by $C_Q h^{m+1}$, because the orthogonality condition (3.4) implies that the quadrature formulas all possess a degree of precision of (at least) m.

(ii) The global convergence estimate given by Theorem 3.1 can be used in (3.5), to obtain the estimate

$$||\delta_h||_\infty \leq (1 + ||\mathcal{V}_\theta||)Ch^m,$$

where

$$||\mathcal{V}_\theta|| := \max_{t \in D_\theta} \int_0^{\theta(t)} |K_2(t,s)|ds.$$

(iii) The iterated kernels $K_{2,j}(t,s)$ are readily seen to satisfy

$$|K_{2,j}(t,s)| \leq \frac{q^{j(j-1)/2}}{(j-1)!} T^{j-1} \bar{K}_\theta^j, \quad t \in D_\theta^{(j)},$$

where

$$\bar{K}_\theta := \max_{(D_\theta)} |K_2(t,s)| \qquad \text{and} \qquad D_\theta^{(j)} := \{(t,s) : \ 0 \leq s \leq q^j t, \ t \in I\}.$$

This result implies that the infinite series involving the second terms on the right-hand side of (3.7) converges uniformly and is $\mathcal{O}(h^{m+1})$.

In contrast to this result (the analogue of the global superconvergence result for ODEs), collocation at the Gauss points does no longer lead to *local superconvergence* of order $2m$ on I_h when $m \geq 3$, as the following theorem shows. (Observe that for $m = 2$ we have $2m = m + 2$.) Its proof, which is based on interpolatory projection techniques, can be found in [N36].

Theorem 3.4 *Assume that g, K_1 and K_2 in (1.3) are at least $d \geq m+2$ times continuously differentiable on their respective domains. Let $u_h \in S_{m-1}^{(-1)}(I_h)$ be the collocation solution corresponding to the Gauss points $\{c_i\}$, and let the iterated collocation solution be defined by (3.3). Then for any $q \in (0,1)$ the order p^* in the local estimate*

$$||y(t) - u_h^{it}(t)||_{h,\infty} \leq Ch^{p^*}$$

cannot exceed $m + 2$, regardless of the value of d. More precisely, the following is true:

(i) *If $q = 1/2$ then*

$$||y - u_h^{it}||_{h,\infty} \leq C \begin{cases} h^{m+2} & \text{if } m \text{ is even,} \\ h^{m+1} & \text{if } m \text{ is odd.} \end{cases}$$

(ii) *For $q \in (0,1) \setminus \{1/2\}$, we have*

$$||y - u_h^{it}||_{h,\infty} \leq Ch^{m+1},$$

for all $m \geq 2$.

We conclude this section by briefly turning to the delay VIDEs (1.12)–(1.14). While *global* superconvergence of order $m + 1$ is possible for the collocation solution $u_h \in S_m^{(0)}(I_h)$ (Theorem 3.3), the problem regarding the optimal order of *local* superconvergence on I_h is not yet completely solved (Conjecture 3.4). Note, incidentally, that these delay VIDEs include the pantograph equation (1.1) and its counterpart with variable coefficients a and b as special cases.

Theorem 3.5 *Let $u_h \in S_m^{(0)}(I_h)$ be the collocation solution to the proportional delay VIDE (1.12). If the given functions possess continuous derivatives of at least order $m + 1$ on their respective domains, and if the $\{c_i\}$ satisfy the orthogonality condition of Theorem 3.3, then*

$$||y - u_h||_\infty \leq Ch^{m+1}$$

holds for all $q \in (0,1)$.

Proof: We write the error equation for (1.12),

$$e_h'(t) = a(t)e_h(t) + b(t)e_h(qt) + \delta_h(t) + (\mathcal{V}e_h)(t) + (\mathcal{V}_\theta e_h)(t), \quad t \in I, \quad e_h(0) = 0,$$

in integrated form. The analysis in [N36] (or, if $a(t) \equiv 0$, $\mathcal{V} = 0$, the one employed in the proof of Theorem 3.3) can the be applied to the resulting delay integral equation for e_h. We leave the details to the reader.

Conjecture 3.6 *The order of local superconvergence of $u_h \in S_m^{(0)}(I_h)$ on I_h cannot exceed $p^* = m + 2$. If the $\{c_i\}$ defining the collocation points X_h are the Gauss points, then we have $p^* = m + 2$ for any $q \in (0, 1)$ and all $m \geq 3$.*

4 Collocation on $\mathbb{R}^+$: asymptotic stability

Suppose that the pantograph equation (1.1) is solved by collocation in $S_m^{(0)}(I_h)$ where now $I_h := \{t_n := nh : n = 0, 1, ...; h > 0 \text{ fixed.}\}$ It has been shown in, e.g., [T7,T8,T10,T27,T34] that for $q \in (0, 1)$, the solutions of (1.1)

(a) remain *bounded* as $t \to \infty$ if $\mathrm{Re}(a) < 0$ and $|b| = |a|$;
(b) satisfy $\lim_{t\to\infty} y(t) = 0$ if $\mathrm{Re}(a) < 0$ and $|b| < |a|$;
(c) converge *monotonically* to zero if $\mathrm{Re}(a) < 0$ and $|b| < -\mathrm{Re}(a)$.

There are at present very few analogous results for the solutions of the difference equations resulting from collocation applied to (1.1) (see [N2–N9], [N11–N14], [N18–N21], and [N25] for partial results, especially on special non-uniform (geometric or quasi-geometric) meshes; compare also the monograph [N33] by Bellen & Zennaro). In [N6] an asymptotic stability result is derived for $u_h \in S_1^{(0)}(I_h)$ when $q = 1/2$; the analysis does not seem to be extendible to values $q \neq 1/2$ except when $q = 1/r$, $r \in \mathbb{N}\,(r \geq 2)$ (see [N9]).

The same situation exists for collocation solutions $u_h \in S_{m-1}^{(-1)}(I_h)$ for the second-kind delay integral equations (1.3) and (1.5). As we have said before, the pantograph equation (1.1) is equivalent to the special delay VIE (1.2): it can be shown that collocation in $S_m^{(0)}(I_h)$ for (1.1) and iterated collocation, based on the space $S_{m-1}^{(-1)}(I_h)$, with the same collocation points X_h, yield identical approximations. Hence, the results of [N6] carry over to delay VIEs of the form (1.2). However, the asymptotic stability analysis of collocation solutions for more general delay VIEs with vanishing proportional delays is completely open.

This is also true for the delay VIDEs (1.13) and (1.14): there are as yet no general boundedness/asymptotic stability results for collocation solutions in $S_m^{(0)}(I_h)$. (Compare, however, [T29,T40] and [N27,N32].)

5 Concluding remarks

While the extension of the global convergence results of Section 3 to nonlinear integral and integro-differential equations with proportional delays is fairly straightforward, the analysis of the optimal local superconvergence and asymptotic stability properties of collocation solutions for such problems appears at present rather hopeless; it is not clear what the appropriate mathematical tools are to be or whether they exist.

The quantitative and qualitative theory of nonlinear versions of the pantograph equation, such as the proprtional delay Riccati equation,

$$y'(t) = ay(t) + by(qt)[1 - cy(qt)],$$

and rational DDE

$$y'(t) = \frac{ay(t)}{1 + by(qt)},$$

(see [T28,T31,T35]) is now fairly advanced, but much work remains to be done on the behaviour of solutions to analogous nonlinear VIEs and VIDEs with proportional delays, such as "Hammerstein-type" delay VIDEs of the form

$$y'(t) = a(t)y(t) + b(t)y(qt)[1 - c(t)y(qt)] + \int_{qt}^{t} k(t-s)G(y(s))ds, \qquad (5.1)$$

where $k(t-s)$ is a "memory kernel" and G is a smooth function. Results for the more general version of (5.1) in which $G(y(s))$ is replaced by $G(y(t), y(s))$ will also be of interest.

In order to circumvent the difficulties arising in the analysis of collocation methods on *uniform meshes* for pantograph-type functional integral and integro-differential, a number of authors have studied the asymptotic stability and optimal local superconvergence properties of collocation solutions on *geometric* or *quasi-geometric meshes* (see, e.g., [N11,N18,N31,N34]). The idea underlying the latter approach is to generate a suitable approximation $y_0(t)$ on some "small" initial interval $[0, t_0]$ with feasible $t_0 > 0$, and then to solve the given delay problem by collocation on $[t_0, T]$ with $y(t_0) := y_0(t_0)$. The delay $\tau(t) = (1-q)t$ does no longer vanish on $[t_0, T]$, and hence standard analysis can be used, for the quasi-geometric mesh given by

$$I_h := \bigcup_{\mu=0}^{M} I_h^{(\mu)}, \quad I_h^{(\mu)} := \{t_n^{(\mu)} : \xi_\mu = t_0^{(\mu)} < t_1^{(\mu)} < \cdots < t_N^{(\mu)} = \xi_{\mu+1}\},$$

where the points $\{\xi_\mu\}$ are defined by $\xi_{\mu+1} =: \theta(\xi_\mu)$, $\mu = 0, 1, \ldots, M$ (setting, without loss of generality, $T := \xi_{M+1}$). If, in addition, it is assumed that this mesh I_h is *θ-invariant*, that is, $\theta(I_h^{(\mu)}) = I_h^{(\mu-1)}$ $(\mu = 1, \ldots, M)$, then the structure of the resulting difference equations is such that the well-known standard analysis leads to optimal superconvergence and asymptotic stability results on such non-uniform meshes.

6 A (chronological) selection of references

We conclude this survey with a representative list of references on the theory, application, and numerical analysis of functional equations with proportional (and related vanishing) delays. This list uses chronological order and is divided into three parts: the first part (references labelled **(T#)**) contains contributions to the theory of equations with proportional delays and related equations, while the second part (**(A#)**) and the third part (**(N#)**) list papers and books on applications and numerical analysis of such problems. It is hoped that it will provide an up-to-date and reasonably complete picture of the current "state of the art" in this field of research.

1. Theory:

(T1) V. VOLTERRA, *Sopra alcune questioni di inversione di integrali definite*, Ann. Mat. Pura Appl., 25 (1897), 139-178.
[$\int_{qx}^{x} K(x,\xi)u(\xi)d\xi = \phi(x)$, $|q| < 1$. Compare also (T2),(T3),(T4), the monograph (T16), and the generalization in (T26), (T33).]

(T2) T. LALESCO, *Sur l'équation de Volterra*, Thèse, Gauthier-Villars, Paris, 1908; J. de Math. (6), 4 (1908), 309-317.

(T3) T. LALESCO, *Sur une équation intégrale du type Volterra*, C.R. Acad. Sci. Paris, 152 (1911), 579-580.
[Generalization of result in (T1): qx replaced by $\theta(x)$ with $\theta(0) = 0$, $\theta'(0) = 1$. See also (T26).]

(T4) V. VOLTERRA, *Leçons sur les équations intégrales*, Gauthier-Villars, Paris, 1913.
[VIEs with proportional limits are treated on pp. 92-101.]

(T5) G. ANDREOLI, *Sulle equazioni integrali*, Rend. Circ. Mat. Palermo, 37 (1914), 76-112.
[Compare also (T23).]

(T6) A. FELDSTEIN and C.P. KOLB, *Retarded exponentials*, Tech. Report No. A.M.-28, Brown University, Providence, RI, 1967.

(T7) L. FOX, D.F. MAYERS, J.R. OCKENDON and A.B. TAYLER, *On a functional differential equation*, J. Inst. Math. Appl., 8 (1971), 271-307.
[This is the first paper on the theory and numerical solution of the "pantograph equation". See also (A2) for its physical background.]

(T8) T. KATO and J.B. McLEOD, *The functional-differential equation $y'(x) = ay(\lambda x) + by(x)$*, Bull. Amer. Math. Soc., 77 (1971), 891-937.

(T9) P.O. FREDERICKSON, *Dirichlet solutions for certain functional differential equations*, in: Japan-United States Seminar on Ordinary Differential and Functional Equations (M. Urabe, ed.), pp. 249-254, Lecture Notes in Math., Vol. 243, Springer-Verlag, Berlin-Heidelberg-New York, 1971.
[See also for references on work of the 1920s on DDEs with proportional delays.]

(T10) T. KATO, *Asymptotic behavior of solutions of the functional differential equation $y'(x) = ay\lambda x) + by(x)$*, in: Delay and Functional Differential Equations and Their Applications (K. Schmitt, ed.), pp. 197-217, Academic Press, New York, 1972.
[Extension of asymptotic results in (T8) to complex b.]

(T11) R.D. NUSSBAUM, *Existence and uniqueness theorems for functional differential equations of neutral type*, J. Differential Equations, 11 (1972), 607-623.
[Results for $x'(t) = k(t)x'(\phi(t)) + b(t)x(\theta(t)) + g(t)$, motivated by $x'(t) = kx'(mt) + g(t)$ $(0 < m < 1)$.]

(T12) G.R. MORRIS, A. FELDSTEIN and E.W. BOWEN, *The Phragmén-Lindelöf principle and a class of functional differential equations*, in: Ordinary Differential Equations (L. Weiss, ed.), pp. 513-540, Academic Press, New York, 1972.

(T13) J. CERHA, *On some linear Volterra delay equations*, Časopis Pěst. Mat., 101 (1976), 111-123.
[Resolvent equations and solution representation for $y(t) = g(t) + \int_0^t K(t,s)y(\theta(s))ds$, $\theta(s) \leq s$.]

(T14) G.A. DERFEL, *Asymptotic behavior of the solutions of a class of functional-differential equations* (in Russian), in: The Asymptotic Behaviour of Solutions of Differential-Functional Equations, pp. 58-65, Akad. Nauk Ukrain. SSR, Inst. Mat., Kiev, 1978.

[Growth of solutions of $y^{(m)}(x) = \sum_{j=1}^{\ell} \sum_{k=0}^{m} a_{ij} y^{(k)}(\alpha_j x)$, $0 < |\alpha_j| < 1$.]

(T15) J. BÉLAIR, *Sur une équation différentielle fonctionnelle analytique*, Canad. Math. Bull., 24 (1981), 43-46.

[This paper deals with proportional delay DEs of order 2. See also (T21) and references [3] and [6] in (T25).]

(T16) S. FENYÖ and H.W. STOLLE, *Theorie und Praxis der linearen Integralgleichungen*, Band 3, Birkhäuser Verlag, Basel-Boston-Stuttgart, 1984.

[See pp. 324-328 for the theory of VIEs introduced in (T1).]

(T17) O.J. STAFFANS, *Some well-posed functional equations which generate semigroups*, J. Differential Equations, 58 (1985), 157-191.

(T18) A. FELDSTEIN and Z. JACKIEWICZ, *Unstable neutral functional differential equations*, Canad. Math. Bull., 33 (1990), 428-433.

(T19) T.P. PUKHNACHEVA, *A functional equation with contracting argument*, Siberian Math. J., 31 (1990), 365-367.

[$u(x) - a(x)u(\omega(x)) = \int_0^x K(x,\xi)u(x-\xi)d\xi + f(x)$, $|\omega(x)| \leq x/k_1$ $(k_1 > 1)$.]

(T20) Y. KUANG and A. FELDSTEIN, *Monotonic and oscillatory solutions of a linear neutral differential equation with infinite lag*, SIAM J. Math. Anal., 21 (1990), 1633-1641.

(T21) G.A. DERFEL and S.A. MOLCHANOV, *Spectral methods in the theory of differ-
ential-functional equations*, Math. Notes Acad. Sci. USSR, 47 (1990), 254-260.

[Existence of bounded (almost periodic) solutions to $y''(t) = \sum_{j=-\ell}^{\ell} a_j y(\alpha_j t) + \lambda y(t)$.]

(T22) G.A. DERFEL, *Kato problem for functional differential equations and difference Schrödinger operators*, Operator Theory: Adv. Appl., 46 (1990), 319-321.

(T23) Ll.G. CHAMBERS, *Some properties of the functional equation $\phi(x) = f(x) + \int_0^{\lambda x} g(x,y,\phi(y))dy$*, Internat. J. Math. Math. Sci., 14 (1990), 27-44.

(T24) G.A. DERFEL, *Functional differential equations with linearly transformed arguments and their applications*, in: Proceedings of EQUADIFF 91 (Barcelona) (C. Perelló et al., eds.), pp. 421-423, World Scientific Publ. Co., River Edge, NJ, 1993.

(T25) Á. ELBERT, *Asymptotic behaviour of the analytic solution of the differential equation $y'(t) + y(qt) = 0$ as $q \to 1^-$*, J. Comput. Appl. Math., 41 (1992), 5-22.

[Asymptotic distribution of zeros; sequel to (T6).]

(T26) A.M. DENISOV and S.V. KOROVIN, *On Volterra's integral equation of the first kind*, Moscow Univ. Comput. Math. Cybernet., 3 (1992), 19-24.

(T27) A. ISERLES, *On the generalized pantograph functional differential equation*, Europ. J. Appl. Math., 4 (1993), 1-38.

[Illuminating survey; contains also numerous references on applications.]

(T28) A. ISERLES, *On nonlinear delay-differential equations*, Trans. Amer. Math. Soc., 344 (1994), 441-477.

[Equations of the types $y'(t) = ay(t) + by(qt)(1 - y(qt))$ (Riccati) and $y'(t) = ay(t)/(1 + by(qt))$.]

(T29) A. ISERLES and Y. LIU, *On pantograph integro-differential equations*, J. Integral Equations Appl., 6 (1994), 213-237.

[See also the paper (N26) by Ishiwata.]

(T30) A. FELDSTEIN, A. ISERLES and D. LEVIN, *Embedding of delay equations into an infinite-dimensional ODE system*, J. Differential Equations, 117 (1995), 127-150.

(T31) A. ISERLES and J. TERJÉKI, *Stability and asymptotic stability of functional-differential equations*, J. London Math. Soc. (2), 51 (1995), 559-572.

(T32) J. TERJÉKI, *Representation of the solutions to linear pantograph equations*, Acta Sci. Math. (Szeged), 60 (1995), 705-713.

(T33) A.M. DENISOV and A. LORENZI, *On a special Volterra integral equation of the first kind*, Boll. Un. Mat. Ital. B (7), 9 (1995), 443-457.

(T34) Y. LIU, *Asymptotic behaviour of functional-differential equations with proportional time delays*, Europ. J. Appl. Math., 7 (1996), 11-30.

(T35) Y. LIU, *On a functional Riccati equation*, Numerical Analysis Report NA19-1996, DAMTP, University of Cambridge, 1996.

[Contains extensive list of references on functional differential equations with proportional delays. See also (T28).]

(T36) J. PIILA and J. PITKÄRANTA, *On the integral equation* $f(x) - (c/L(x)) \int_0^{L(x)} f(y)dy = g(x)$, *where* $L(x) = \min\{ax,\ 1\}$, $a > 1$, J. Integral Equations Appl., 8 (1996), 363-378.

(T37) G.A. DERFEL and F. VOGL, *On the asymptotics of solutions of a class of linear functional-differential equations*, Europ. J. Appl. Math., 7 (1996), 511-518.

[Systems of the form $u'(t) = \sum_{j=1}^{\ell} C_j(t)u(\lambda_j t)$, $0 < \lambda_j < 1$.]

(T38) C.T.H. BAKER and C.A.H. PAUL, *Pitfalls in parameter estimation for delay differential equations*, SIAM J. Sci. Comput., 18 (1997), 305-314.

[See p. 307 and p. 310.]

(T39) H. BRUNNER, *1896-1996: One hundred years of Volterra integral equations of the first kind*, Appl. Numer. Math., 24 (1997), 83-93.

(T40) A.M. DENISOV and A. LORENZI, *Existence results and regularization techniques for severely ill-posed integrofunctional equations*, Boll. Un. Mat. Ital. B (7), 11 (1997), 713-732.

[$z(t) = k(t)z(\alpha(t)) + \int_{\beta(t)}^{t} b(t,s)z(s)ds$, with $\alpha(t) \leq qt$ $(0 < q < 1)$, $0 \leq \beta(t) < t$.]

(T41) A. ISERLES and Y. LIU, *On functional-differential equations with proportional delays*, J. Math. Anal. Appl., 207 (1997), 73-95.

(T42) A. FELDSTEIN and Y. LIU, *On neutral functional-differential equations with variable time delays*, Math. Proc. Cambridge Phil. Soc., 124 (1998), 371-384.

[Compare also for many additional references on linear and nonlinear pantograph equations.]

(T43) L. BERG and M. KRÜPPEL, *On the solution of an integral-functional equation with a parameter*, Z. Anal. Anwend., 17 (1998), 159-181.

[VIEs of the form $y(t) = b\int_{qt-q+1}^{qt} y(s)ds$, $q > 1$, $b = q/(q-1)$.]

(T44) L. BERG and M. KRÜPPEL, *Cantor sets and integral-functional equations*, Z. Anal. Anwend., 17 (1998), 997-1020.
[Sequel to (T43); includes a spline approximation method.]

(T45) V. MUREŞAN, *On a class of Volterra integral equations with deviating argument*, Studia Univ. Babeş-Bolyai Math., XLIV (1999), 47-54.
[$x(t) = x(0) + \int_0^t f(s, x(\lambda s))ds$, $0 < \lambda < 1$.]

2. Applications:

(Compare also (T24), (T27) and (T35) for additional references.)

(A1) K. MAHLER, *On a special functional equation*, J. London Math. Soc., 15 (1940), 115-123.
[$(f(z+\omega) - f(z))/\omega = f(qz)$, $\omega \neq 0$, $0 < q < 1$.]

(A2) J.R. OCKENDON and A.B. TAYLER. *The dynamics of a current collection system for an electric locomotive*, Proc. Roy. Soc. London Ser. A, 322 (1971), 447-468.
[The numerical solution of the resulting "pantograph equation" is studied in (N1).]

(A3) R. RODEMAN, D.B. LONGCOPE and L.F. SHAMPINE, *Response of a string to an accelerating mass*, J. Appl. Mech. Ser. E, 43 (1976), 675-680.

(A4) A.B. TAYLER, *Mathematical Models in Applied Mechanics*, Clarendon Press, Oxford, 1986.
[See pp. 40-45 and 50-53 for a derivation of the pantograph equation.]

(A5) J. PIILA, *Characterization of the membrane theory of a clamped shell. The hyperbolic case*, Math. Methods Appl. Sci., 6 (1996), 169-194.
[See also (T36).]

3. Numerical analysis: (vanishing) proportional delay qt $(0 < q < 1)$

(N1) L. FOX, D.F. MAYERS, J.R. OCKENDON and A.B. TAYLER, *On a functional differential equation*, J. Inst. Math. Appl., 8 (1971), 271-307.

(N2) Z. JACKIEWICZ, *Asymptotic stability analysis of θ-methods for functional differential equations*, Numer. Math., 43 (1984), 389-396.

(N3) M.D. BUHMANN and A. ISERLES, *Numerical analysis of functional differential equations with a variable delay*, in: Numerical Analysis (Dundee 1991) (D.F. Griffiths and G.A. Watson, eds.), pp. 17-33. Pitman Research Notes in Math., Vol. 260, Longman, Harlow, 1992.

(N4) M.D. BUHMANN and A. ISERLES, *On the dynamics of a discretized neutral equation*, IMA J. Numer. Anal., 12 (1992), 339-363.

(N5) M.D. BUHMANN and A. ISERLES, *Stability of the discretized pantograph differential equation*, Math. Comp., 60 (1993), 575-589.

(N6) M.D. BUHMANN, A. ISERLES and S.P. NØRSETT, *Runge-Kutta methods for neutral differential equations*, in: Contributions in Numerical Analysis (R.P. Agarwal, ed.), pp. 85-98, World Scientific Series in Applicable Analysis, Vol. 2, World Scientific Publ. Co., Singapore, 1993.
[Asymptotic stability of one-point collocation at Gauss point $c_1 = 1/2$; $q = 1/2$.]

(N7) A. ISERLES, *Numerical analysis of delay differential equations with variable delays*, Ann. Numer. Math., 1 (1994), 133-152.
[Survey paper, including pantograph equation; see also (T25).]

(N8) A. ISERLES, *The asymptotic behaviour of certain difference equations with proportional delay*, Comput. Math. Appl., 28 (1994), 141-152.

(N9) Y. LIU, *Stability analysis of θ-methods for neutral functional-differential equations*, Numer. Math., 70 (1995), 473-485.

(N10) Y. LIU, *The linear q-difference equation*, Appl. Math. Lett., 8 (1995), 15-18.
[Analyzes (systems of) functional equations of the form $y(x) = ay(qx) + f(x)$, $q \in (0,1)$ and convergence of one-point collocation methods.]

(N11) Y. LIU, *On the θ-method for delay differential equations with infinite lag*, J. Comput. Appl. Anal., 71 (1996), 177-190.

(N12) J. LIANG and M. LIU, *Stability of numerical solutions to pantograph delay systems* (in Chinese), J. Harbin Inst. Tech., 28 (1996), 21-26.

(N13) J. LIANG and M. LIU, *Numerical stability of θ-methods for pantograph delay differential equations* (in Chinese), J. Numer. Methods Comput. Appl., 12 (1996), 271-278.

(N14) J. LIANG, S. QIU and M. LIU, *The stability of θ-methods for pantograph delay differential equations*, Numer. Math. (Engl. Ser.), 5 (1996), 80-85.
[(N12),(N13), and (N14) discuss, respectively, the asymptotic stability of the linear θ-method and the corresponding one-leg method.]

(N15) A. ISERLES, *Beyond the classical theory of computational ordinary differential equations*, in: *The State of the Art in Numerical Analysis (York 1996)* (I.S. Duff and G.A. Watson, eds.), pp. 171-192, Clarendon Press, Oxford, 1997.

(N16) H. BRUNNER, *On the discretization of differential and Volterra integral equations with variable delay*, BIT, 37 (1997), 1-12.
[Local superconvergence results for DDEs and DVIEs with proportional delay.]

(N17) N. TAKAMA, Y. MUROYA and E. ISHIWATA, *On the attainable order of collocation methods for the delay differential equation with proportional delay*, Tech. Report No. 97-7, Advanced Research Insitute for Science and Engineering, Waseda University, Tokyo, 1997; BIT, 40 (2000), 374-394.
[Extension of results in (N16).]

(N18) A. BELLEN, N. GUGLIELMI and L. TORELLI, *Asymptotic stability properties of θ-methods for the pantograph equation*, Appl. Numer. Math., 24 (1997), 275-293.
[Use of quasi-geometric meshes depending on q in the delay qt. Compare also (N11), (N24), (N27), (N30), (N32) and (N33).]

(N19) A. ISERLES, *Exact and discretized stability of the pantograph equation*, Appl. Numer. Math., 24 (1997), 295-308.

(N20) Y. LIU, *Numerical investigation of the pantograph equation*, Appl. Numer. Math., 24 (1997), 309-317.

(N21) L.A.V. CARVALHO and K.L. COOKE, *Collapsible backward continuation and numerical approximations in a functional differential equation*, J. Differential Equations, 143 (1998), 96-109.
[Behaviour of discrete analogues of oscillatory solutions of $y'(t) = -ay(t) + ay(qt)$, $q < 1$.]

(N22) W. ZHANG, *Numerical Analysis of Delay Differential and Integro-Differential Equations*, Ph.D. Thesis, Dept. of Mathematics & Statistics, Memorial University of Newfoundland, St. John's, NF, 1998.

(N23) H. BRUNNER and W. ZHANG, *Collocation approximations for second-order differential equations and Volterra integro-differential equations with variable delays*, Canad. Appl. Math. Quart., 6 (1998), 269-285.
[Extension of analysis in (N16).]

(N24) D. LI and M. LIU, *Asymptotic stability of numerical solution of pantograph delay differential equations* (in Chinese), J. Harbin Inst. Tech., 31 (1999), 57-59.

[Proof of a conjecture in (N7).]

(N25) T. KOTO, *Stability of Runge-Kutta methods for the generalized pantograph equation*, Numer. Math., 84 (1999), 233-247.

(N26) L. QIU, T. MITSUI and J.-X. KUANG, *The numerical stability of the θ-method for delay differential equations with many variable delays*, J. Comput. Math., 17 (1999), 523-532.

[$y'(t) = ay(t) + \sum_{j=1}^{m} b_j y(\lambda_j t)$, $0 < \lambda_m \leq \ldots \leq \lambda_1 < 1$.]

(N27) E. ISHIWATA, *On the attainable order of collocation methods for the neutral functional-differential equations with proportional delays*, Computing, 64 (2000), 207-222.

(N28) H. BRUNNER, Q. HU and Q. LIN, *Geometric meshes in collocation methods for Volterra integral equations with proportional delays*, IMA J. Numer. Anal., 21 (2001), 783-798.

[Quasi-optimal order of local superconvergence; in contrast to (N18,N31,N34), no initial approximation is needed.]

(N29) A. BELLEN, S. MASET and L. TORELLI, *Contractive initializing methods for the pantograph equation of neutral type*, in: *Recent Trends in Numerical Analysis* (D. Trigiante, ed.), pp. 35-41, Nova Science Publ., Huntington, N.Y., 2000.

[Extension of approach in (N18).]

(N30) C.T.H. BAKER and E. BUCKWAR, *Continuous θ-methods for the stochastic pantograph equation*, Tech. Report 361, Department of Mathematics, University of Manchester, May 2000; Electr. Trans. Numer. Anal., 11 (2000), 131-151.

(N31) A. BELLEN, *Preservation of superconvergence in the numerical integration of delay differential equations with proportional delay*, IMA J. Numer. Anal., 22 (2002), 529-536.

(N32) Y. MUROYA, E. ISHIWATA and H. BRUNNER, *On the attainable order of collocation methods for pantograph-integro-differential equations*, J. Comput. Appl. Math., 152 (2003), 347-366.

(N33) A. BELLEN and M. ZENNARO, *Numerical Methods for Delay Differential Equations*, Oxford University Press, Oxford, 2003.

[Compare Ch. 6 and Ch. 8.]

(N34) A. BELLEN, H. BRUNNER, S. MASET and L. TORELLI, *Superconvergence of collocation solutions on quasi-geometric meshes for Volterra integro-differential equations with vanishing delays*, Preprint, Dip. di Scienze Matematiche, University of Trieste, November 2002.

(N35) H. BRUNNER, *Collocation Methods for Volterra Integral and Related Functional Differential Equations*, Cambridge Monographs on Applied and Computational Mathematics, Cambridge University Press, Cambridge, 2004 (to appear).

[Compare Ch. 5.]

(N36) H. BRUNNER and Q.-Y. HU, *Superconvergence orders of iterated collocation solutions for Volterra integral equations with variable delays*, Preprint, June 2003.

Fields Institute Communications
Volume **42**, 2004

The LPA Model

J. M. Cushing
Department of Mathematics
617 N Santa Rita
University of Arizona
Tucson, AZ 85721 U.S.A.
cushing@math.arizona.edu

1 Introduction

Since the seminal papers of Lewis and Leslie in the 1940's [28, 29, 30], difference equations have played an important role in population dynamics and ecology. The utilization of models based on difference equations ranges from investigations of a very theoretical nature (a notable example being their seminal role in the development of chaos theory [31, 32, 33]) to studies of a very applied nature (see [2] for many examples). A difference equation model that has been particularly successful in applications to population dynamics and ecology is the so-called "LPA model". The success of this model is due to the fact that it has been rigorously connected to real data by statistical parameterization and validation methods and shown capable of quantitatively accurate descriptions and predictions by means of controlled experiments. The model has played a key role in many projects designed to document and study the predictions of nonlinear theory (including chaos) [2, 11]. While this kind of success is common, even expected, in many other scientific disciplines (the so-called "hard" sciences), it is rather unprecedented in ecology where mathematical models are all too commonly used only for theoretical speculations and expected to provide no more than qualitative accuracy, "guesstimates", or verbal metaphors. Because of the role it has played – and is continuing to play – in population dynamics and ecology, the LPA model is a worthy object of study for mathematicians interested in difference equations. In this paper I summarize the mathematical results known about the LPA model. I will also discuss many open questions. In addition to the LPA model itself, there are also a number of variants of the model that have also played important roles in applications. One example is a periodically forced LPA model which I discuss in Section 4. In Section 5 I mention several other variations of the LPA model that have received less mathematical attention.

2000 *Mathematics Subject Classification.* Primary 39A11; Secondary 92D25.

2 Basic properties of the LPA model

The LPA model is the nonlinear (Leslie) matrix model

$$\begin{pmatrix} L(t+1) \\ P(t+1) \\ A(t+1) \end{pmatrix}$$
$$= \begin{pmatrix} 0 & 0 & b\exp\left(-\frac{c_{el}}{V}L(t)-\frac{c_{ea}}{V}A(t)\right) \\ 1-\mu_l & 0 & 0 \\ 0 & \exp\left(-\frac{c_{pa}}{V}A(t)\right) & 1-\mu_a \end{pmatrix}\begin{pmatrix} L(t) \\ P(t) \\ A(t) \end{pmatrix}.$$

This matrix equation is equivalent to the system

$$L(t+1) = bA(t)\exp\left(-\frac{c_{el}}{V}L(t)-\frac{c_{ea}}{V}A(t)\right) \tag{2.1}$$
$$P(t+1) = (1-\mu_l)\,L(t) \tag{2.2}$$
$$A(t+1) = P(t)\exp\left(-\frac{c_{pa}}{V}A(t)\right)+(1-\mu_a)\,A(t). \tag{2.3}$$

of three difference equations.

It is useful as a guide to mathematical investigations of the system (2.1)-(2.3) to understand some of the biological interpretations for the equations. The name of the model derives from the biological interpretation of three state variables used in the applications from which the model arose. The laboratory experiments utilized species of insects from the genus *Tribolium* (flour beetles) which have three life cycle stages. The state variables $L(t)$, $P(t)$, $A(t)$ denote the number of individuals, at time t, present in the larval, pupal, and adult stages respectively. The exponential terms appearing in the equations arise from contacts among individuals that decrease survival probabilities. The exponential form of the nonlinearities derive from a Poisson process involving random encounters among individual organisms [6]. These encounters are assumed inversely proportional to habitat size V, which accounts for the parameter ratios appearing in the exponential terms. (This law has been validated for *Tribolium* by laboratory experiments.) The parameter $b > 0$ is the *inherent* per adult recruitment of larvae ("inherent" means when the negative effects of encounters among individuals are negligible). The two parameters μ_l and μ_a satisfy $0 < \mu_l < 1$, $0 < \mu_a \leq 1$ and represent death rates (the fractions of the larvae and adult stages that die each unit of time). For notational simplicity one can assume, without loss of generality, that $V = 1$. There are, however, applications in which it is advantageous to keep V in the model equations. In applications involving *Tribolium* the intra-stage contacts are cannibalistic encounters, and in that context the parameters c_{el}, c_{ea}, and c_{pa} are called "cannibalism coefficients".

Although developed with a particular organism in mind, system (2.1)-(2.3) can be an appropriate model for other organisms possessing three life cycle stages. In this regard, it should be noted that the model assumes that the unit of time is equal to that spent by individuals in the larvae stage. This assumption accounts for the lack of an additive term (such as, for example, $(1-\mu)L(t)$) on the right hand side of (2.1) that would account for larvae at time t which are still larvae at time $t+1$. Another assumption of the model is that the time spent in the pupal stage is identical to that spent in the larvae stage. Hence, there appears no additive term (such as $(1-\mu)P(t)$) on the right hand side of (2.2) that accounts for pupae at time t which are still pupae at time $t+1$. Finally, in equation (2.3) the model assumes that all pupae successfully emerge as adults (in the absence of encounters with adults). This accounts for the absence of a multiplicative factor $1-\mu_p$ in the first term on the right hand side of (2.3). A more general three life cycle stage model would contain some or all of these additional terms. However, these biological assumptions present in the LPA model (2.1)-(2.3) are appropriate for *Tribolium* populations. In this paper we restrict our attention to this model.

In applications to population dynamics, one is interested only in nonnegative values of the three state variables. It is clear from the difference equations (2.1)-(2.3) that

$$\begin{pmatrix} L(0) \\ P(0) \\ A(0) \end{pmatrix} \geq \begin{pmatrix} 0 \\ 0 \\ 0 \end{pmatrix} \implies \begin{pmatrix} L(t) \\ P(t) \\ A(t) \end{pmatrix} \geq \begin{pmatrix} 0 \\ 0 \\ 0 \end{pmatrix}$$

and

$$\begin{pmatrix} L(0) \\ P(0) \\ A(0) \end{pmatrix} \neq \begin{pmatrix} 0 \\ 0 \\ 0 \end{pmatrix} \implies \begin{pmatrix} L(t) \\ P(t) \\ A(t) \end{pmatrix} \neq \begin{pmatrix} 0 \\ 0 \\ 0 \end{pmatrix} \tag{2.4}$$

for all $t = 1, 2, 3, \ldots$. That is to say, the nonnegative cone is forward invariant.

Moreover, all orbits in the nonnegative cone are (forward) bounded. To see this note from (2.1) and (2.2) that

$$0 \leq L(t+1) \leq bA(t)\exp\left(-\frac{c_{pa}}{V}A(t)\right) \leq b\frac{V}{c_{pa}e}$$

for $t = 0,\ 1,\ 2,\ldots$ and consequently

$$0 \leq P(t+1) \leq (1-\mu_l)\, b\frac{V}{c_{pa}e}$$

for $t = 1,\ 2,\ 3,\ldots$. In other words

$$0 \leq L(t) \leq b\frac{V}{c_{pa}e}, \qquad t = 1,\ 2,\ 3,\ldots$$

$$0 \leq P(t) \leq (1-\mu_l)\, b\frac{V}{c_{pa}e} \qquad t = 2,\ 3,\ 4,\ldots\ .$$

For $t \geq 2$ we find from (2.3) that

$$0 \leq A(t+1) \leq (1-\mu_l)\, b\frac{V}{c_{pa}e} + (1-\mu_a)\, A(t)$$

from which the inequality

$$0 \leq A(t) \leq \frac{1}{\mu_a}(1-\mu_l)\, b\frac{V}{c_{pa}e} + (1-\mu_a)^{t-2} A(2) \qquad t = 3,\ 4,\ 5,\ldots$$

follows by an induction. After a finite number of iterations

$$0 \leq A(t) \leq 2\frac{1-\mu_l}{\mu_a} b\frac{V}{c_{pa}e}.$$

Thus, there is a compact set in the nonnegative cone inside of which all nonnegative orbits lie after a finite number of steps.

From (2.1)-(2.3) we have, for orbits in the nonnegative cone, the inequalities

$$\begin{aligned} &0 \leq L(t-1) \leq bA(t) \\ &0 \leq P(t+1) \leq (1-\mu_l)\, L(t) \\ &0 \leq A(t+1) \leq P(t) + (1-\mu_a)\, A(t) \end{aligned} \tag{2.5}$$

for all $t = 0, 1, 2, \ldots$. An induction shows

$$\begin{aligned} 0 &\le L(t) \le x(t) \\ 0 &\le P(t) \le y(t) \\ 0 &\le A(t) \le z(t) \end{aligned} \tag{2.6}$$

where $x(t)$, $y(t)$, $z(t)$ are defined by the linear system

$$\begin{aligned} x(t+1) &= bz(t) \\ y(t+1) &= (1-\mu_l)\,y(t) \\ z(t+1) &= y(t) + (1-\mu_a)\,z(t) \end{aligned} \tag{2.7}$$

with initial conditions $x(0) = L(0) \ge 0$, $y(0) = P(0) \ge 0$, and $z(0) = A(0) \ge 0$. If $\mu_a < 1$ the coefficient matrix

$$\begin{pmatrix} 0 & 0 & b \\ 1-\mu_l & 0 & 0 \\ 0 & 1 & 1-\mu_a \end{pmatrix} \tag{2.8}$$

associated with the linear system (2.7) is irreducible and primitive. (Its fourth power is positive.) Therefore, (2.8) has a real, positive, simple, and strictly dominant r eigenvalue [1]. The dominant eigenvalue is less than 1 if

$$n \mathrel{\hat{=}} \frac{1-\mu_l}{\mu_a} b < 1$$

and greater than 1 if $n > 1$ [6, 9].

Consequently, if $n < 1$ it follows that all solutions of (2.7) tend to the origin as $t \to +\infty$. From (2.6) we see $n < 1$ also implies that all orbits of the LPA model (lying in the nonnegative cone) tend to the origin as $t \to +\infty$. Furthermore, the matrix (2.8) is the Jacobian of the LPA model at the origin and, as a result, $n < 1$ also implies the origin is locally asymptotically stable (LAS) [6, 16].

The origin is a "trivial" equilibrium of the LPA model (2.1)-(2.3). Its global asymptotic stability when $n < 1$ means, in the biological context, that the population goes extinct for all initial conditions. The quantity n has an important biological interpretation, namely, it is the expected number of offspring per individual per life time [2, 6]. The inequality $n < 1$ means, then, that on average individuals do not replace themselves by reproduction. Given this biological interpretation of n, we expect $n > 1$ to imply that the population will not go extinct. Since the dominant eigenvalue of the Jacobian matrix (2.8) is greater than 1, the origin is in fact unstable [6, 16]. However, the origin might have a stable manifold and some nonnegative orbits might approach it. If $\mu_a < 1$ we will show that the origin cannot have a stable manifold that intersects the nonnegative cone (except at the origin).

In addition to the (positive) dominant eigenvalue, the Jacobian matrix (2.8) has the complex conjugate pair of eigenvalues

$$\begin{aligned} \lambda_\pm &= \alpha \pm \beta i \\ \alpha &= \frac{1}{3}s_a - \frac{1}{2}\gamma - \frac{1}{18}\frac{s_a^2}{\gamma} \\ \beta &= \frac{\sqrt{3}}{2}\frac{1}{\gamma}\left(\gamma^2 - \frac{1}{9}s_a^2\right) > 0 \end{aligned}$$

where

$$\gamma \doteq \left(\frac{1}{2}bs_l + \frac{1}{27}s_a^3 + \left(\frac{1}{27}bs_a^3 s_l + \frac{1}{4}b^2 s_l^2\right)^{1/2}\right)^{1/3} > 0$$

$$s_l \doteq 1 - \mu_l > 0, \quad s_a \doteq 1 - \mu_a > 0.$$

These complex conjugate eigenvalues (calculated with the aid of an algebraic computer program) satisfy

$$\begin{aligned} &\left|\lambda^{\pm}\right| < 1 \quad \text{when} \quad \mu_a < 1 \\ &\left|\lambda^{\pm}\right| = (bs_l)^{2/3} = n^{2/3} \quad \text{when} \quad \mu_a = 1. \end{aligned} \tag{2.9}$$

The stable manifold at the origin is tangent to the plane spanned by the real and imaginary parts of a complex eigenvector associated with λ^- (or its conjugate λ^+) [6, 16]. An eigenvector associated with eigenvalue λ^- is the vector

$$\begin{pmatrix} \lambda^-(\lambda^- - s_a) \\ (\lambda^- - s_a)\, s_l \\ s_l \end{pmatrix}$$

whose real and imaginary parts are

$$\begin{pmatrix} -\frac{1}{2}\left(\alpha^2 + \beta^2\right) \\ \frac{1}{2\alpha}\left(\beta^2 - \alpha^2\right) s_l \\ s_l \end{pmatrix}, \quad \begin{pmatrix} -\frac{\beta}{2\alpha}\left(\alpha^2 + \beta^2\right) \\ -\beta s_l \\ s_l \end{pmatrix}.$$

Since both of these real vectors have components of opposite signs, the plane they span intersects the nonnegative cone only at the origin. It follows that in a neighborhood of the origin the stable manifold intersects the nonnegative cone only at the origin. From this fact, and from (2.4), we see that an orbit in the nonnegative cone cannot approach the origin. In the terminology of [25], the origin is an isolated invariant set in the nonnegative cone that is its own stable set. By Theorem 4.1 of [25], the origin is uniformly persistent. This means that there exists a positive constant $\mu > 0$ such that

$$\lim_{t \to +\infty} \inf \left(L(t) + P(t) + A(t)\right) \geq \mu > 0$$

for all orbits in the nonnegative cone (except the origin).

We have seen that when $n > 1$ and $\mu_a < 1$ all nonnegative orbits are (forward) bounded and cannot approach the origin. A fundamental problem is to describe the omega limit set of such orbits. First we point out that when $n > 1$ there exists a positive equilibrium (i.e., an equilibrium all of whose components are positive) and this equilibrium is the only nontrivial, nonnegative equilibrium. To see this consider the equilibrium equations

$$L = bA\exp\left(-\frac{c_{el}}{V}L - \frac{c_{ea}}{V}A\right) \tag{2.10}$$

$$P = (1 - \mu_l)\, L \tag{2.11}$$

$$A = P\exp\left(-\frac{c_{pa}}{V}A\right) + (1 - \mu_a)\, A \tag{2.12}$$

of (2.1)-(2.3). Equations (2.12) and (2.11) respectively imply

$$P = \mu_a A \exp\left(\frac{c_{pa}}{V}A\right), \quad L = \frac{\mu_a}{1 - \mu_l} A \exp\left(\frac{c_{pa}}{V}A\right). \tag{2.13}$$

If $A = 0$ then $L = P = 0$. Suppose $A > 0$. A substitution of (2.13) into equation (2.10) yields (after some simplification) the equation

$$\frac{c_{el}}{V}\frac{\mu_a}{1-\mu_l}A\exp\left(\frac{c_{pa}}{V}A\right) + \left(\frac{c_{ea}}{V}+\frac{c_{pa}}{V}\right)A = \ln n \tag{2.14}$$

for A. The left hand side of this equation, as a function of A, vanishes for $A = 0$ and is strictly increasing and unbounded for $A > 0$. We conclude that for $\ln n > 0$ there is a unique, positive solution $A = A_n > 0$ of equation (2.14) which, when used in equations (2.13), yields a positive equilibrium of the LPA model. Note that

$$A_1 = 0, \quad \frac{dA_n}{dn} > 0 \quad \text{for } n > 1 \quad \text{and} \lim_{n\to+\infty} A_n = +\infty. \tag{2.15}$$

(For $\ln n < 0$ there is a unique, negative solution of (2.14) which yields a negative equilibrium of the LPA model. Thus, a transcritical bifurcation of two equilibrium branches occurs at the origin when $n = 1$.)

When $n > 1$ the positive equilibrium becomes a candidate for an attractor in the nonnegative cone. However, little is known about the stability properties of this equilibrium. When $\mu_a < 1$ it is known that when $n > 1$ is sufficiently close to 1 the positive equilibrium is LAS. This is the familiar "exchange of stability" phenomenon that generically occurs at a transcritical bifurcation. A proof using Liapunov-Schmidt methods appears in [6]. Another way to prove this fact is by considering the characteristic polynomial p of the Jacobian

$$J = \begin{pmatrix} -\frac{c_{el}}{V}bA\exp\left(-\frac{c_{ea}}{V}A - \frac{c_{el}}{V}L\right) & 0 & \left(1-\frac{c_{ea}}{V}A\right)b\exp\left(-\frac{c_{ea}}{V}A-\frac{c_{el}}{V}L\right) \\ 1-\mu_l & 0 & 0 \\ 0 & \exp\left(-\frac{c_{pa}}{V}A\right) & -\frac{c_{pa}}{V}P\exp\left(-\frac{c_{pa}}{V}A\right)+1-\mu_a \end{pmatrix}$$

of the LPA model. An evaluation of J at the positive equilibrium (using the equilibrium equations (2.10)-(2.12)) yields

$$J(n) = \begin{pmatrix} -\frac{c_{el}}{V}\frac{\mu_a}{1-\mu_l}A\exp\left(\frac{c_{pa}}{V}A\right) & 0 & \frac{\mu_a}{1-\mu_l}\left(1-\frac{c_{ea}}{V}A\right)\exp\left(\frac{c_{pa}}{V}A\right) \\ 1-\mu_l & 0 & 0 \\ 0 & \exp\left(-\frac{c_{pa}}{V}A\right) & -\frac{c_{pa}}{V}\mu_a A+1-\mu_a \end{pmatrix}$$

and

$$\begin{aligned} p(\lambda, A) = {} & \lambda^3 + \left(\frac{c_{el}}{V}\frac{\mu_a}{1-\mu_l}A\exp\left(\frac{c_{pa}}{V}A\right)+\frac{c_{pa}}{V}\mu_a A - 1 + \mu_a\right)\lambda^2 \\ & + \frac{c_{el}}{V}\frac{\mu_a}{1-\mu_l}\left(\frac{c_{pa}}{V}\mu_a A - 1 + \mu_a\right)A\exp\left(\frac{c_{pa}}{V}A\right)\lambda \\ & + \left(\frac{c_{ea}}{V}A - 1\right)\mu_a. \end{aligned} \tag{2.16}$$

The positive equilibrium is a function of n and therefore so are the eigenvalues $\lambda = \lambda_n$.

When $n = 1$, the positive equilibrium collapses to the origin and the Jacobian reduces to the Leslie matrix (2.8). Therefore, the eigenvalues of J at $n = 1$ are $\lambda = 1$ and $\lambda^{\pm}$. Let $\lambda = \lambda_n$ denote the eigenvalue of J that satisfies $\lambda_1 = 1$. An implicit differentiation of $p(\lambda_n, A_n) = 0$ with respect to n yields

$$\frac{\partial}{\partial\lambda}p(\lambda_n, A_n)\frac{d\lambda_n}{dn} + \frac{\partial}{\partial A}p(\lambda_n, A_n)\frac{dA_n}{dn} = 0$$

and

$$\frac{d\lambda_n}{dn} = -\frac{\frac{\partial}{\partial A}p(\lambda_n, A_n)}{\frac{\partial}{d\lambda}p(\lambda_n, A_n)}\frac{dA_n}{dn}.$$

A calculation shows of the partial derivatives of p are

$$\left.\frac{\partial p(\lambda,A)}{\partial A}\right|_{(\lambda,A)=(1,0)} = \frac{\mu_a}{1-\mu_l}\left(\frac{c_{pa}}{V}(1-\mu_l)+\frac{c_{el}}{V}\mu_a+\frac{c_{ea}}{V}(1-\mu_l)\right) > 0$$

$$\left.\frac{\partial p(\lambda,A)}{d\lambda}\right|_{(\lambda,A)=(1,0)} = 1+2\mu_a > 0.$$

It follows (from (2.15)) that

$$\left.\frac{d\lambda_n}{dn}\right|_{n=1} < 0$$

and hence $\lambda_1(n) < 1$ for $n > 1$ sufficiently close to 1.

If $\mu_a < 1$ then, for n close to 1, the remaining two eigenvalues are close to the complex eigenvalues $\lambda^{\pm}$ which have magnitude less than 1 (see (2.9)). It follows in this case that all three eigenvalues of J are inside the unit circle when $n > 1$ is close to 1, and the positive equilibrium is LAS.

However, if $\mu_a = 1$ then the stability of the positive equilibrium is determined by the complex eigenvalues $\lambda^{\pm}(n)$ for $n > 1$. These eigenvalues have magnitude 1 when $n = 1$ (see (2.9)) and the problem is to determine their magnitude for $n > 1$. It is shown in [8] that $|\lambda^{\pm}(n)| > 1$ for $n > 1$ close to 1 and therefore that the positive equilibrium is unstable in this case.

In the following theorem we summarize the facts we have obtained so far about the LPA model.

Theorem 2.1 *The following hold for the LPA model (2.1)-(2.3).*

(a) Orbits associated with nonnegative initial conditions remain (forward) bounded and in the nonnegative cone for all $t = 0, 1, 2, \ldots$.

(b) If $n < 1$ *then the origin is locally asymptotically stable and globally attracting in the nonnegative cone.*

(c) If $n > 1$ *there exists a unique nonnegative, nontrivial equilibrium and this equilibrium is positive.*

(d) If $\mu_a < 1$ *and* $n > 1$ *then the LPA model is uniformly persistent on the nonnegative cone with respect to the origin.*

(e) If $\mu_a < 1$ *then the positive equilibrium is locally asymptotically stable for* $n > 1$ *sufficiently close to* 1.

(f) If $\mu_a = 1$ *then the positive equilibrium is unstable for* $n > 1$ *sufficiently close to* 1.

For $n < 1$ the the origin is globally asymptotically stable. Little is known about the global asymptotic stability of the positive equilibrium when $n > 1$. I conjecture that the positive equilibrium is globally asymptotically stability for $n > 1$ sufficiently close to 1, but I have no proof of this assertion. One known global stability result concerns the special case when

$$c_{el} = 0. \tag{2.17}$$

In this case, it is shown in [42] that the positive equilibrium is globally asymptotically stable if

$$1 < n < \min\left\{e, \frac{1-\mu_a}{\mu_a}\frac{c_{ea}}{c_{pa}}e\right\}.$$

This condition requires $\mu_a < 1$ and the inequality

$$r \triangleq \frac{c_{pa}}{c_{ea}} < \frac{1-\mu_a}{\mu_a} e$$

for the ratio r of cannibalism coefficients. The case (2.17), about which more is said in the next section, is an interesting mathematical case, but it is not relevant to the applications that have been made to *Tribolium* populations. (Larval on egg cannibalism is very significant in *Tribolium* populations.) Global stability results when $c_{el} > 0$ would be of interest.

The local stability results in Theorem 2.1 are valid for $n > 1$ close to 1. In the next section we show that the positive equilibrium is not LAS for all $n > 1$.

3 Non-equilibrium dynamics

We begin by examining the positive equilibrium of the LPA model (2.1)-(2.3) when $\mu_a < 1$. In the proof of Theorem 2.1 we saw that all three eigenvalues λ of the Jacobian J evaluated at the positive equilibrium lie inside the unit circle for n sufficiently close to 1. In order for the positive equilibrium to lose stability as n increases, at least one eigenvalue must migrate outside the unit circle. Eigenvalues λ are the roots of the characteristic polynomial (2.16) where $A = A_n$ is the third component of the positive equilibrium. From the constant term of this polynomial we see that the product of the three eigenvalues of J is larger than 1 if $A_n > (1+\mu_a) V/c_{ea}\mu_a$. From (2.15) we have the following result.

Theorem 3.1 *There exists a number $n^* > 1$ such that positive equilibrium of the LPA model (2.1)-(2.3) is unstable for $n > n^*$.*

Let $n_1 \leq n^*$ be the smallest value of $n > 1$ for which a root of the characteristic polynomial $p(\lambda, A_n)$ lies on the unit circle. Then $1 < n < n_1$ implies the positive equilibrium is LAS. The number n_1 is the smallest candidate for a local bifurcation point at which the positive equilibrium destabilizes and a bifurcation to another attractor occurs. Local bifurcations are classified according to the location where a root leaves the unit circle [17]. Note that $\lambda = 1$ cannot be a root of $p(\lambda, A(n))$ for any $n > 1$, as the calculation

$$p(1, A_n) = \frac{c_{pa}}{V}\mu_a A_n + \frac{c_{el}}{V}\frac{\mu_a}{1-\mu_l}\left(\frac{c_{pa}}{V}\mu_a A_n + \mu_a\right) A_n \exp\left(\frac{c_{pa}}{V}A_n\right) + \frac{c_{ea}}{V}\mu_a A_n > 0$$

shows. This implies that certain types of local bifurcations of positive equilibria, such as transcritical, saddle-node, and pitchfork bifurcations, never occur in the LPA model for $n > 1$ [17]. (This is, of course, consistent with the fact that there is a unique positive equilibrium for $n > 1$.) There are only two ways the positive equilibrium can lose stability at n_1 (or any other value of $n > 1$): a root of $p(\lambda, A_n)$ at $n = n_1$ equals either -1 or $e^{i\theta}$ for some θ satisfying $0 < \theta < \pi$. In the first case a period doubling bifurcation "generically" occurs in which a 2-cycle comes into existence. In the second case an invariant loop bifurcation "generically" occurs (sometimes called a Neimark-Sacker or discrete Hopf bifurcation) in which an aperiodic attractor comes into existence. (Certain technical sufficiency conditions are needed to guarantee that these bifurcations actually occur [17].) It turns out that both of these types of local bifurcations are possible, depending on the values of the model parameters. See Figure 1.

There are few rigorous bifurcation results for the model (2.1)-(2.3). One special case for which period doubling and invariant loop bifurcation points n_1 have been calculated is when $c_{el} = 0$. In [13] it is shown in this case that $p(\lambda, A_n)$ has root $\lambda = -1$ when

$$n = n_1 = \exp\left(\frac{2}{\mu_a}\right)$$

and a pair of complex conjugate roots of magnitude equal to 1 when

$$n = n_1 = \exp\left(\frac{K^{\pm}(\mu_a, r)}{\mu_a}\right)$$

where

$$K^{\pm}(\mu_a, r) \mathrel{\hat{=}} -\frac{1}{2}\frac{1+r}{1-r}\left[(r-3)\,\mu_a + 1\right]$$
$$\pm\frac{1}{2}\frac{1+r}{1-r}\left(\left[(r-3)\,\mu_a + 1\right]^2 + 4\,(1-r)\,(1-\mu_a)\,(2\mu_a+1)\right)^{1/2}.$$

While these formulae allow us to draw "parameter maps" that show the boundaries in parameter space where equilibrium destabilization occurs, rigorous proofs that local bifurcations occur at these points are currently lacking, as is the determination of the bifurcation characteristics (e.g., whether the bifurcation is supercritical or subcritical and whether the non-equilibrium attractors that result are stable or unstable).

Bifurcations have played a central role in the applications of the LPA model (2.1)-(2.3). Although the quantity n is a mathematically and biologically useful bifurcation parameter, other model parameters can be used as a bifurcation parameter. For example, the parameters μ_a and c_{pa} were used in the bifurcation experiments involving *Tribolium* described in [11]. An interesting special case arises from the "route-to-chaos" experiment in those studies. In that experiment μ_a was experimentally manipulated to equal 0.96 and a complicated sequence of bifurcations, that included chaotic dynamics, was observed as c_{pa} increases. These bifurcations have only been studied numerically. No rigorous mathematical analysis has been made of the bifurcation sequence and the chaotic attractors it contains. The fact that $\mu_a = 0.96$ in that well-known experiment motivates at study of the LPA model for $\mu_a = 1$. Although this case is not covered by some of the basic results presented in Sec. 2, it turns out to be a mathematically tractable case. A first study of this case appears in [8].

With $\mu_a = 1$ in (2.1)-(2.3) the coordinate axes are invariant. In fact, orbits on the axes move sequentially from one to the next, re-visiting each axis every third step. We call these orbits *fully synchronized* (a name that derives from the biological application in which the three life cycle stages are never mutually present). An orbit is fully synchronized if and only if its initial condition is fully synchronized (i.e., contains two zeroes). If we denote every third L component by x

$$x(t) = L(3t), \quad t = 0, 1, 2, \ldots$$

we find that $x(t)$ satisfies the so-called Ricker map

$$x(t+1) = nx(t)\exp\left(-cx(t)\right)$$

with

$$n = b\,(1-\mu_l)\,, \quad c \mathrel{\hat{=}} c_{ea}\,(1-\mu_l) > 0.$$

In so far as fully synchronous orbits are concerned, it follows that the LPA model with $\mu_a = 1$ exhibits the famous period doubling route-to-chaos with respect to the dynamics at every third step. (A similar result follows if either of the other two coordinate axes is considered.) For example, if $1 < n$ the Ricker equation has a unique positive equilibrium $x = c^{-1}\ln n$. This equilibrium corresponds to a fully synchronous 3-cycle of the LPA model. The Ricker equilibrium is (asymptotically) stable and globally attracting (on the positive axis) if $1 < n < e^2$ and consequently the 3-cycle is (asymptotically) stable and globally attracting with respect to fully synchronous, positive orbits of the LPA model. For $n > e^2$ the Ricker equation has a 2-cycle and this 2-cycle corresponds to a fully synchronous 6-cycle of the LPA model. A cycle of period p of the Ricker equation produces a fully synchronous $3p$-cycle of the LPA model. A quasi-periodic or chaotic orbit of the Ricker map produces a similar type of fully synchronous orbit of the LPA model. The stability properties of Ricker orbits is inherited by the LPA orbits when one restricts attention to fully synchronous orbits (motion on the coordinate axes).

With $\mu_a = 1$ it is also true that the coordinate planes are invariant and that an orbit lying in the coordinate planes moves sequentially from one to the other. Let

$$\begin{pmatrix} x(t) \\ y(t) \\ z(t) \end{pmatrix}$$

denote every third point on an orbit of the LPA model (2.1)-(2.3). An orbit lying in a coordinate plane (but not on a coordinate axis) always has one zero and is called *partially synchronized.* After three steps a partially synchronized orbit returns to the same coordinate plane. Consider, for example, the (L, A) coordinate plane. From (2.1)-(2.3) we find that an initial condition in this plane results, at every third step, in an orbit satisfying the two dimensional system

$$x(t+1) = nx(t)\exp\left(-cx(t)\right) \tag{3.1}$$
$$z(t+1) = \left[n\exp\left(-\alpha x(t)\right)\right] z(t)\exp\left(-\beta z(t)\right) \tag{3.2}$$

where

$$\alpha \mathrel{\mathring{=}} c_{el} + c_{pa}\left(1 - \mu_l\right)$$
$$\beta \mathrel{\mathring{=}} c_{ea}.$$

This system has nonnegative equilibria

$$\begin{pmatrix} x \\ z \end{pmatrix} = \begin{cases} (0,0) & \\ \left(\frac{1}{c}\ln n, 0\right) & \text{if } n > 1 \\ \left(\frac{1}{c}\ln n, \frac{\gamma}{\beta}\ln n\right) & \text{if } n > 1 \text{ and } \gamma > 0 \end{cases}$$

where

$$\gamma \mathrel{\mathring{=}} 1 - \frac{\alpha}{c} = \frac{\left(c_{ea} - c_{pa}\right)\left(1 - \mu_l\right) - c_{el}}{c_{ea}\left(1 - \mu_l\right)}.$$

A linearization analysis shows the following results:

$$\begin{pmatrix} 0 \\ 0 \end{pmatrix} \quad \text{is a repeller if } n > 1$$

$$\begin{pmatrix} \frac{1}{c}\ln n \\ 0 \end{pmatrix} \quad \begin{cases} \text{is LAS stable if } 1 < n < e^2 \text{ and } \gamma < 0 \\ \text{is a saddle if } 1 < n < e^2,\ \gamma > 0 \\ \text{is a saddle if } n > e^2,\ \gamma < 0 \\ \text{is a repeller if } n > e^2,\ \gamma > 0 \end{cases}$$

$$\begin{pmatrix} \frac{1}{c}\ln n \\ \frac{\gamma}{\beta}\ln n \end{pmatrix} \quad \begin{cases} \text{is LAS stable if } 1 < n < \min\left\{e^2, e^{2/\gamma}\right\} \text{ and } \gamma > 0 \\ \text{is a saddle if } \gamma > 1 \text{ and } e^{2/\gamma} < n < e^2 \\ \text{is a saddle if } 0 < \gamma < 1 \text{ and } e^2 < n < e^{2/\gamma} \\ \text{is a repeller if } n > \max\left\{e^2, e^{2/\gamma}\right\}. \end{cases}$$

Of these many cases, only the case

$$1 < n < e^2 \text{ and } \gamma < 0 \tag{3.3}$$

has been analyzed in detail.

In the system (3.1)-(3.2) the first equation shows that the x-component satisfies a Ricker equation which is uncoupled from the second equation. Thus, an initial condition $x(0) > 0$ equation (3.1) determines a sequence $x(t)$, which when substituted into equation (3.2) produces a non-autonomous Ricker equation for the dynamics of the z-component. Under the assumption (3.3) it follows that $\lim_{t\to+\infty} x(t) = c^{-1}\ln n$ and equation (3.2) becomes "asymptotically autonomous" with a "limiting equation" obtained by replacing $x(t)$ with its limit $c^{-1}\ln n$. Let $w(t)$ denote the solutions of the limiting equation

$$w(t+1) = n^{\gamma} w(t) \exp\left(-\beta w(t)\right).$$

This is a Ricker equation and from (3.3) it follows that $lim_{t\to+\infty} w(t) = 0$. Known theorems concerning asymptotically autonomous difference equations [5] relate the dynamics of this limiting equation to those of (3.2). In [8] these theorems are used to show that the equilibrium $(x,z) = \left(c^{-1}\ln n, 0\right)$ is globally attracting in the positive (x,z) quadrant.

A completely analogous analysis can also be carried out in the other two coordinate planes with regard to the equilibrium points $(x,y) = (0, \beta^{-1}\ln n)$ and $(y,z) = (0, \beta^{-1}\ln n)$. It follows, under assumption (3.3), that all orbits $(x(t), y(t), z(t))$ starting in a coordinate plane asymptotically equilibrate. What this result means for the LPA model is that all orbits $(L(t), P(t), A(t))$ lying in the coordinate planes asymptotically approach a 3-cycle located on the coordinate axes, namely the cycle defined by the three axis points

$$\begin{pmatrix} L \\ P \\ A \end{pmatrix} = \begin{pmatrix} \frac{1}{c}\ln \\ 0 \\ 0 \end{pmatrix} \longrightarrow \begin{pmatrix} 0 \\ \frac{1}{\beta}\ln n \\ 0 \end{pmatrix} \longrightarrow \begin{pmatrix} 0 \\ 0 \\ \frac{1}{\beta}\ln n \end{pmatrix}. \tag{3.4}$$

These three points are also fixed points of the three step (second composite) LPA map, which we denote by

$$\begin{pmatrix} x_1 \\ y_1 \\ z_1 \end{pmatrix} \doteq \begin{pmatrix} \frac{1}{c}\ln \\ 0 \\ 0 \end{pmatrix}, \quad \begin{pmatrix} x_2 \\ y_2 \\ z_2 \end{pmatrix} \doteq \begin{pmatrix} 0 \\ \frac{1}{\beta}\ln n \\ 0 \end{pmatrix}, \quad \begin{pmatrix} x_3 \\ y_3 \\ z_3 \end{pmatrix} \doteq \begin{pmatrix} 0 \\ 0 \\ \frac{1}{\beta}\ln n \end{pmatrix} \tag{3.5}$$

The cases other than (3.3) provide open problems concerning the asymptotic dynamics of synchronous orbits of the LPA model when $\mu_a = 1$. It is worth noting, however, that in terms of the original model parameters, the condition $\gamma < 0$ (i.e., $(c_{ea} - c_{pa})(1 - \mu_l) - c_{el} < 0$) holds for the experimentally estimated values of the parameters in the *Tribolium* experiments.

Theorem 3.2 [8] *Under assumptions*

$$\mu_a = 1, \quad 1 < n < e^2, \quad (c_{ea} - c_{pa})(1 - \mu_l) - c_{el} < 0 \tag{3.6}$$

all synchronous orbits of the LPA model (2.1)-(2.3) tend asymptotically to the fully synchronous 3-cycle (3.4).

More can be said about the case (3.3). In the full, three dimensional phase space of the LPA model, the fully synchronous 3-cycle (3.4) is unstable. A calculation of the products of the Jacobian evaluated at the three points of the 3-cycle shows, in fact, that the 3-cycle is a saddle, with a two dimensional stable manifold (the coordinate planes) and a one dimensional unstable manifold [8]. What is the attractor of nonsynchronous orbits in this case?

We return, for a moment, to the general case of the LPA model when $\mu_a = 1$. It was shown in Sec. 2 that there exists a (unique) positive equilibrium when $n > 1$. It was also shown that when, in addition, $\mu_a < 1$ the positive equilibrium is stable at least for n near 1. It turns out, however, that when $\mu_a = 1$ this is no longer true. A tedious perturbation analysis carried out in [8] shows that for all $n > 1$ near 1 the positive equilibrium is a saddle (with a one dimensional stable manifold and a two dimension stable manifold on which the equilibrium is a stable spiral). Therefore, in case (3.3) when $n > 1$ is close to 1, neither the origin, the positive equilibrium, nor the fully synchronous 3-cycle are stable attractors.

The reason for the failure of the exchange of stability principle for the transcritical bifurcation at the origin when $n = 1$ is seen from the Jacobian (2.8) of the LPA model evaluated at the origin and with $\mu_a = 1$. The eigenvalues of that matrix are the cube roots of $n = b(1 - \mu_l)$ and hence are the cube roots of unity when $n = 1$. Thus, when $\mu_a = 1$ the origin loses stability as n increases through 1 because *all three* eigenvalues leave the unit circle simultaneously and, as a result, the bifurcation is "nongeneric". The eigenvalue that crosses at $+1$ is associated with the transcritical bifurcation of two equilibrium branches at $n = 1$ and the existence of the positive equilibrium for $n > 1$. The two complex eigenvalues that also occur at $n = 1$ would be associated, in a generic situation (when they were the only eigenvalues crossing the unit circle at $n = 1$), with the bifurcation of an invariant loop. This suggests the bifurcation of an invariant loop from the origin at $n = 1$.

Although the general invariant loop bifurcation theorem (Neimark/Sacker or discrete Hopf Theorem) cannot be used for the nongeneric case bifurcation at $n = 1$, the existence of an invariant loop can be proved in case (3.3) by a straightforward analysis as follows. With regard to the dynamics of every third step (the second composite of the LPA model), the first equilibrium point in (3.5) lies in both the (x, z) and the (x, y) coordinate planes. We saw above that in the (x, z) plane all positive orbits tend to this equilibrium while in the (x, y) all orbits tend to the second equilibrium point in (3.5). Therefore, the (one dimensional) unstable manifold of the first equilibrium point [6] tends to the second equilibrium point. That is to say, there a heteroclinic orbit lying in the (x, y) coordinate plane that connects the first two equilibrium points in (3.5). Similarly, there are heteroclinic orbits lying in the (y, z) and (x, z) coordinate planes that connect the second to the third and the third to the first equilibrium points respectively. This set of three equilibria and their connecting heteroclinic orbits form an invariant loop. In terms of the original (one step) LPA model, each of the three equilibria (3.5) correspond, as initial points, to the three different phases of the 3-cycle (3.4). The heteroclinic orbits are partially synchronized orbits connecting the three phases of the 3-cycle.

Theorem 3.3 [8] *Under assumptions (3.6) there exists an invariant loop for the LPA model (2.1)-(2.3) having the form of a cycle chain of synchronous orbits. The cycle chain consists of the fully synchronous 3-cycle (3.4) and partially synchronous orbits that heteroclinically connect the three temporal phases of the 3-cycle.*

The 3-cycle is unstable in the three dimensional phase space of the LPA model. When (3.6) holds, numerical simulations suggest, however, that the synchronous cycle chain on which the 3-cycle lies is an attractor (for nonnegative orbits) at least for $n > 1$ close to 1 [8]. A rigorous proof of this is currently unavailable. If this is true, then we see in this case that the nongeneric bifurcation at $n = 1$ results in both a positive equilibrium and an invariant loop, but it is the invariant loop that inherits the stability from the origin.

Many open questions remain concerning the case $\mu_a = 1$. When $\gamma > 0$ replaces $\gamma < 0$ in (3.6), the equilibrium $(x, z) = \left(c^{-1} \ln n, \gamma\beta^{-1} \ln n\right)$ gives rise to a partially synchronous 3-cycle. Does this produce a cycle chain invariant loop for the LPA model? If so, when is it stable? When $n > e^2$ the Ricker equation (3.1) has non-equilibrium attractors, including chaos. What kind of attractors do these produce for the planar system (3.1)-(3.2)? What kind of synchronous orbits of the LPA model do they yield?

Recall that $\mu_a = 0.96$ has been an important case in applications. Another open mathematical question is how the dynamics in the case $\mu_a = 1$ relate to those when μ_a is near 1.

4 The periodically forced LPA model

The LPA model (2.1)-(2.3) is an autonomous model; all parameters are assumed constant in time. The majority of models used in the mathematical ecology literature are autonomous, even though real world ecological systems are likely to be significantly influenced by external forces. If random external forces are taken into account (i.e., perturbations in the model system that are described probabilistically rather than deterministically), then a model determines a stochastic process. At the other extreme, if the external forces have strong pattern of regularity, which can be reasonably modeled deterministically, then the dynamic model becomes non-autonomous. An example of external fluctuations that can show such regularity are those related to natural cycles such as daily, monthly, or seasonal oscillations in physical and/or biological parameters. In laboratory experiments, such as those carried out using *Tribolium* and the LPA model, environmental fluctuations can be held to a minimum (temperature, humidity, nutrient availability, habitat size, etc.). On the other hand, in controlled laboratory experiments one can also deliberately fluctuate an environmental parameter, in order to study the resulting effects on the population dynamics. In the case of laboratory cultures of *Tribolium* such an experiment was reported by Jillson [26]. In this experiment the habitat of the beetle cultures (the volume of flour medium) was changed according to a periodic schedule. Jillson noted in one of his experiments (he performed experiments using several different periodic patterns of volume changes) that the total biomass of *Tribolium* was significantly higher in the oscillating habitat than it was in a habitat held constant at the average volume. This observation was noteworthy for at least two reasons. First, it suggested that a population manager might increase yield by periodically oscillating resource availability, rather than holding it constant. Second, Jillson's experimental result contradicts what at the time was a commonly held tenet that fluctuating environments were deleterious to biological populations [35, 34, 36]. Jillson made no mathematical study of his experiment, but a mathematical explanation of his result has been obtained by using the LPA model, suitably modified to account for periodically oscillating volumes V [4, 7, 18, 19, 22].

If a sequence $V(t) > 0$ of period p is substituted for V in (2.1)-(2.3), the result is the periodically forced LPA model

$$L(t+1) = bA(t)\exp\left(-\frac{c_{el}}{V(t)}L(t) - \frac{c_{ea}}{V(t)}A(t)\right) \tag{4.1}$$

$$P(t+1) = (1-\mu_l)\,L(t) \tag{4.2}$$

$$A(t+1) = P(t)\exp\left(-\frac{c_{pa}}{V(t)}A(t)\right) + (1-\mu_a)\,A(t). \tag{4.3}$$

An important mathematical problem is to determine the long term dynamics of solutions with nonnegative initial conditions. It is clear from the equations that nonnegative (positive) initial conditions produce nonnegative (positive) solutions.

The origin is an equilibrium point of the periodic LPA model (4.1)-(4.3). As for the autonomous LPA model, *the origin is LAS and it globally attracts all nonnegative solutions if* $n < 1$. The proof of this statement follows closely that of the autonomous case since the inequalities (2.5) remain valid for the periodic model (4.1)-(4.3).

Suppose $n > 1$. The linearization of (4.1)-(4.3) at the origin has coefficient matrix (2.8) which, as we noted in Sec. 2, has a real eigenvalue greater than 1 (if also $\mu_a < 1$). Thus, as in the autonomous case, *the origin is unstable when* $n > 1$ *and* $\mu_a < 1$.

Furthermore, *when* $n > 1$ *solutions of (4.1)-(4.3) are (forward) bounded.* The proof of this fact follows from the same comparison argument given for the autonomous case in Section 2 once the inequalities

$$0 \leq L(t+1) \leq bA(t)\exp\left(-\frac{c_{pa}}{V_m}A(t)\right) \leq b\frac{V_m}{c_{pa}e}, \quad \text{for } t = 0,1,2,\ldots$$

$$0 \leq P(t+1) \leq (1-\mu_l)\,b\frac{V_m}{c_{pa}e}, \quad \text{for } t = 1,2,3,\ldots$$

$$0 \leq A(t+1) = (1-\mu_l)\,b\frac{V_m}{c_{pa}e} + (1-\mu_a)\,A(t), \quad \text{for } t = 2,3,4,\ldots$$

are noted. Here $V_m = \max V(t) > 0$.

The challenge is to determine the asymptotic properties of solutions when $n > 1$. A first approach is to obtain basic results for the periodic LPA model that parallel those of the autonomous LPA model in Theorem 2.1. For example, is the periodic LPA model (4.1)-(4.3) uniformly persistent with respect to the origin (and nonnegative orbits) when $n > 1$? This result has been established for period $p = 2$ in [22], but has not been proved for periods $p > 2$. (Perhaps the proof in [22] can be generalized to the case $p > 2$.). It is shown in [7] (using degree theory) that when $\mu_a < 1$ the periodic LPA model has an unbounded continuum of positive p-cycles for values of $n > 1$ that bifurcates from the origin at $n = 1$. This continuum is the counterpart to the continuum of (unique) positive equilibria that exists for all $n > 1$ in the autonomous case. Aside from its unboundedness, little is known about the continuum and the properties of the p-cycles lying on it. One case that has been studied in some detail is period $p = 2$ case.

When $p = 2$. we can write $V(t) = \langle V \rangle \left(1 + \alpha(-1)^t\right)$ where $\langle V \rangle > 0$ is the average of the cycle and $|\alpha|$ is the relative amplitude. Absorbing $\langle V \rangle$ into the coefficients c_{el}, c_{ea}, c_{pa} we write

$$V(t) = 1 + \alpha(-1)^t \tag{4.4}$$

in the periodic LPA model equations (4.1)-(4.3). In [18] it is shown, for a general class of periodic maps, that the bifurcating continuum of positive p-cycles has a parameterization (Liapunov-Schmidt expansion) near the origin for n near 1, and formulas for calculating the lowest order terms are given. When applied to the $p = 2$ periodic LPA model, these formulas yield the expansions

$$\begin{aligned}
L_\alpha(t) &= \frac{\mu_a}{1-\mu_l}\varepsilon_\alpha + O\left(\left|\varepsilon_\alpha^2\right|\right)(t) \\
P_\alpha(t) &= \mu_a \varepsilon_\alpha + O\left(\left|\varepsilon_\alpha^2\right|\right)(t) \\
A_\alpha(t) &= \varepsilon_\alpha + O\left(\left|\varepsilon_\alpha^2\right|\right)(t) \\
n &= 1 + \left(c_{ea} + \frac{\mu_a}{1-\mu_l}c_{el} + c_{pa}\right)\frac{1}{1-\alpha^2}\varepsilon_\alpha + O\left(\left|\varepsilon_\alpha^2\right|\right)
\end{aligned} \tag{4.5}$$

for the positive 2-cycles near the bifurcation point (i.e., for small $\varepsilon_\alpha > 0$) [22]. In each equation $O\left(\left|\varepsilon_\alpha^2\right|\right)(t)$ represents terms that are 2-periodic in t and of order two or higher in ε_a for small $|\varepsilon_\alpha$. The subscript α denotes dependence on the (amplitude/phase) parameter α.

The expansions (4.5) can be used to approximate the dominant eigenvalue associated with the linearization of the model taken at the 2-cycle in order to determine the cycle's stability. It turns out that the dominant eigenvalue has the expansion

$$1 - 2\frac{\mu_a}{1+2\mu_l}\left(c_{ea} + \frac{\mu_a}{1-\mu_l}c_{el} + c_{pa}\right)\frac{1}{1-\alpha^2}\varepsilon_\alpha + O\left(\left|\varepsilon_\alpha^2\right|\right)$$

and hence is less than 1 for $\varepsilon_\alpha > 0$ small. This shows that the bifurcating 2-cycles are LAS for $n > 1$ close to 1.

These facts about the periodic LPA model are summarized in the theorem below. More details can be found in [7, 18, 22].

Theorem 4.1 *Assume $V(t) > 0$ is a p-periodic sequence. The following hold for the periodically forced LPA model (4.1)-(4.3).*

(a) Orbits associated with nonnegative initial conditions are (forward) bounded and remain in the nonnegative cone for all $t = 0, 1, 2, \ldots$.

(b) If $n < 1$ then the origin is locally asymptotically stable and global attracts all nonnegative solutions.

(c) If $\mu_a < 1$ there exists an unbounded continuum of positive p-cycles for positive values of $n > 1$ that bifurcates from the origin at $n = 1$.

(d) If $\mu_a < 1$ and $p = 2$ then the LPA model is uniformly persistent on the nonnegative cone with respect to the origin for $n > 1$.

(e) If $\mu_a < 1$ and $p = 2$ then the positive equilibrium is locally asymptotically stable for $n > 1$ sufficiently close to 1.

There are several interesting, unsolved problems left open by this theorem. Is the range of n values associated with the continuum in (c) the entire half line $n > 1$ (i.e., is there a positive p-cycle for every $n > 1$)? For a given value of n is the p-cycle from the continuum the only positive cycle? Under what conditions are the positive p-cycles from the continuum LAS? Unstable? Globally attracting (for positive orbits)? What kinds of bifurcations do they undergo? Do (d) and (e) hold for periods $p > 2$? For $n > 1$ the case $\mu_a = 1$ has not been studied.

The existence and stability results in Theorem 4.1 do not address the issue of interest in Jillson's experiment, namely the unexpected increase in population numbers observed in the periodically fluctuating volumes. Although Jillson conducted experiments using several different periods, he observed an increase in population numbers only when the period of the flour volume oscillation was $p = 2$. For this reason studies of this problem using (4.1)-(4.3) have concentrated on the period $p = 2$ case.

To address Jillson's observation by using the periodic LPA model, we need to relate the attractor of the periodic model ($\alpha \neq 0$) to that of the autonomous model ($\alpha = 0$). One way to do this, under the assumption that the attractors in both cases are cycles (an equilibrium is a 1-cycle), is to compare averages. This can be done component-wise (comparing the cycle average of the L-stage components and so on) or, more in keeping with Jillson's result, using the cycle averages of total population numbers $L + P + A$. If the cycle average is greater in the periodic case (all other model parameters equal), then we say that the periodic forcing has caused a *resonance*; if it is less, we say an *attenuation* occurs as a result of the periodic forcing.

In general, whether resonance or attenuation occurs in periodically forced difference equations depends on the type of nonlinearities and on the values of parameters that appear in the equations. Resonance never occurs in certain types of one dimensional monotone maps [10]. On the other hand, for a general class of period $p = 2$ forced difference equations that exhibit a bifurcation of positive equilibria from the origin, Henson [19] gives conditions under which attenuation or resonance occurs near the bifurcation point [19].

Both attenuation and resonance is possible in the LPA model, depending on the parameter values. For example, it is shown in [22] that attenuation always occurs for $n > 1$ near 1, but that for n is sufficiently large resonance occurs for some values of the model parameters.

For $n > 1$ near 1 we can see why attenuation occurs by making use of the expansions (4.5), from which follow the expansions

$$\begin{aligned}
\langle L_\alpha(t) \rangle &= \frac{\mu_a}{1-\mu_l}\varepsilon_\alpha + O\left(\left|\varepsilon_\alpha^2\right|\right) \\
\langle P_\alpha(t) \rangle &= \mu_a \varepsilon_\alpha + O\left(\left|\varepsilon_\alpha^2\right|\right)
\end{aligned}$$

$$\begin{aligned}\langle A_\alpha(t)\rangle &= \varepsilon_\alpha + O\left(\left|\varepsilon_\alpha^2\right|\right)\\ n &= 1+\left(c_{ea}+\frac{\mu_a}{1-\mu_l}c_{el}+c_{pa}\right)\frac{1}{1-\alpha^2}\varepsilon_\alpha + O\left(\left|\varepsilon_\alpha^2\right|\right)\end{aligned}$$

for the cycle averages of each 2-cycle component. For autonomous case $\alpha = 0$ we have the expansions

$$\begin{aligned}L_0 &= \frac{\mu_a}{1-\mu_l}\varepsilon_0 + O\left(\left|\varepsilon_0^2\right|\right)\\ P_0 &= \mu_a\varepsilon_0 + O\left(\left|\varepsilon_0^2\right|\right)\\ A_0 &= \varepsilon_0 + O\left(\left|\varepsilon_0^2\right|\right)\\ n &= 1-\left(c_{ea}+\frac{\mu_a}{1-\mu_l}c_{el}+c_{pa}\right)\varepsilon_0 + O\left(\left|\varepsilon_0^2\right|\right)\end{aligned}$$

for the positive equilibria for $n > 1$ near 1. In order to make a comparison between the equilibria and the cycle averages for the same value of n, we need to choose the parameter ε_α so that

$$\frac{1}{1-\alpha^2}\varepsilon_\alpha = \varepsilon_0.$$

Thus,

$$\begin{aligned}L_0 - \langle L_\alpha(t)\rangle &= \alpha^2\frac{\mu_a}{1-\mu_l}\varepsilon_0 + O\left(\left|\varepsilon_0^2\right|\right)\\ P_0 - \langle P_\alpha(t)\rangle &= \alpha^2\mu_a\varepsilon_0 + O\left(\left|\varepsilon_0^2\right|\right)\\ A_0 - \langle A_\alpha(t)\rangle &= \alpha^2\varepsilon_0 + O\left(\left|\varepsilon_0^2\right|\right).\end{aligned}$$

and for $\varepsilon_0 > 0$ sufficiently small it follows for all $\alpha \neq 0$ that

$$L_0 - \langle L_\alpha(t)\rangle > 0, \quad P_0 - \langle P_\alpha(t)\rangle > 0, \quad A_0 - \langle A_\alpha(t)\rangle > 0.$$

We see that there is a component-wise attenuation for $\varepsilon_0 > 0$ small (i.e., $n > 1$ close to 1). It follows, of course, that there is also an attenuation in total population size

$$L_0 + P_0 + A_0 > \langle L_\alpha(t)\rangle + \langle P_\alpha(t)\rangle + \langle A_\alpha(t)\rangle.$$

as well.

Theorem 4.2 *Consider the 2-periodically forced LPA model (4.1)-(4.3) with $V(t)$ given by (4.4) and $\mu_a < 1$. For $n > 1$ sufficiently close to 1 the locally asymptotically stable, positive 2-cycles that bifurcate from the origin at $n = 1$(see Theorem 4.1) are (component-wise) attenuant.*

This Theorem does not provide an explanation of Jillson's experimental observation. However, resonance can occur in the periodic LPA model when $n > 1$ is sufficiently large. A numerical example appears in Figure 3, in which the positive 2-cycles change from attenuant to resonant as n increases. In [22] sufficient conditions for resonance when $n > 1$ is large were obtained using regular perturbation techniques.

Implicit function based proofs show, if $n > 1$ and $\mu_a < 1$, that there exist positive 2-cycles for $|\alpha|$ sufficiently small, that these cycles are infinitely differentiable functions of α, and that they approach the equilibrium as $\alpha \to 0$. These 2-cycles have Taylor expansions

$$\begin{aligned} L_\alpha(t) &= L_0 + L_1(t)\alpha + L_2(t)\alpha^2 + O\left(\alpha^3\right)(t) \\ P_\alpha(t) &= P_0 + P_1(t)\alpha + P_2(t)\alpha^2 + O\left(\alpha^3\right)(t) \\ A_\alpha(t) &= A_0 + A_1(t)\alpha + A_2(t)\alpha^2 + O\left(\alpha^3\right)(t) \end{aligned} \tag{4.6}$$

where $O\left(\alpha^3\right)(t)$ represents terms that are 2-periodic in t and of order three or higher in α for small $|\alpha|$ and (L_0, P_0, P_0) is the positive equilibrium of the autonomous LPA model (which is assumed hyperbolic). It turns out that $\langle L_1(t)\rangle = \langle P_1(t)\rangle = \langle A_1(t)\rangle = 0$ and some tedious calculations using (4.6) show

$$\lim_{n\to+\infty} \langle L_\alpha(t)\rangle - L_0 = 2\frac{c_{pa}\mu_a^2 + 2c_{ea}\mu_a - 2c_{ea}}{\mu_a^2 c_{pa} c_{el}}\alpha^2 + O\left(\alpha^3\right) \tag{4.7}$$

$$\lim_{n\to+\infty} \langle P_\alpha(t)\rangle - P_0 = 2\frac{c_{pa}\mu_a^2 + 2c_{ea}\mu_a - 2c_{ea}}{\mu_a^2 c_{pa} c_{el}}\left(1-\mu_l\right)\alpha^2 + O\left(\alpha^3\right) \tag{4.8}$$

$$\lim_{n\to+\infty} \langle A_\alpha(t)\rangle - A_0 = 4\frac{1-\mu_a}{\mu_a^2 c_{pa}}\alpha^2 + O\left(\alpha^3\right). \tag{4.9}$$

From the coefficients of α^2 in these expressions, one can determine conditions on the model parameters under which these limits are positive or negative for small $|\alpha|$ and hence when the components of the 2-cycle are resonant or attenuant for large n.

For example, for small amplitudes α and sufficiently large n we see from (4.9) that the A component of the 2-cycles is resonant. On the other hand, the L and P components are resonant (attenuant) if $c_{pa}\mu_a^2 + 2c_{ea}\mu_a - 2c_{ea} > 0$ (< 0). Thus, we find that it is possible for some components to be resonant while others are attenuant. By summing the three limits (4.7)-(4.9) we can determine conditions under with the total population size is resonant or attenuant from the α^2 coefficient in

$$\begin{aligned} &\lim_{n\to+\infty} \langle L_\alpha(t) + P_\alpha(t) + A_\alpha(t)\rangle \\ &= 2\left(\left(2-\mu_l\right)\frac{c_{pa}\mu_a^2 + 2c_{ea}\mu_a - 2c_{ea}}{\mu_a^2 c_{pa} c_{el}} + 2\frac{1-\mu_a}{\mu_a^2 c_{pa}}\right)\alpha^2 + O\left(\alpha^3\right). \end{aligned}$$

These results show that the question of resonance posed by Jillson's experiment is a complicated one. Moreover, there is yet another issue. In order to be meaningful in applications the 2-cycles (4.6) must be stable. If the equilibrium (L_0, P_0, A_0) is LAS then, for $|\alpha|$ small, so are the 2-cycles. It turns out, however, that this is not the case for parameter values estimated from Jillson's data (see the caption of Figure 3). Instead, for those parameters the autonomous LPA model has an unstable equilibrium and the attractor is a 2-cycle.

Consequently, to address Jillson's resonance question we have to ask what happens when the LPA model with a stable 2-cycle is periodically forced? A perturbation analysis in α can also be carried out in this case. The result is that for α small there exist two stable 2-cycles. These cycles result from the two different phases of the 2-cycle of the autonomous case, but they themselves are not (in general) has shifts of one another. This result is a corollary of a general theorem of Henson [20] in which a perturbation analysis is done for a general class of difference equations when the autonomous case has a stable q-cycle and the system is periodically forced with period r (also see [41]). Moreover, Henson shows that of all the cycles of the forced system that perturb from the phase shifts of the q-cycle at least one is attenuant and at least one is resonant.

When applied to the LPA model (4.1)-(4.3) and (4.4), Henson's results show for the Jillson estimated parameters that there exist two LAS 2-cycles, one resonant and one attenuant, for small amplitude forcing. Thus, for small α there exist in this case two stable 2-cycles (perturbations of the phases of the stable 2-cycle of the autonomous model $\alpha = 0$) and one unstable 2-cycle (a perturbation of the unstable equilibrium of the autonomous model $\alpha = 0$). This multiple attractor scenario has been experimentally verified by laboratory experiments using *Tribolium* [21].

In Jillson's experiment, however, $\alpha = 0.6$ and numerical simulations show there are not multiple attractors, but instead a globally attracting 2-cycle. Numerical simulations suggest an explanation for this. As α is increased from 0, the unstable 2-cycle and the stable attenuant 2-cycle collide and annihilate one another in a saddle node bifurcation at approximately $a = 0.45$ [4], leaving a globally attracting resonance 2-cycle for $\alpha > 0.45$. There are at this time no mathematical proofs of these assertions.

As we see, many challenging mathematical questions concerning periodically forced difference equations have arisen from applications involving the LPA model. While some results have been obtained, there remain numerous open questions concerning the existence, stability, and resonance of cycles induced by periodic forcing (including a rigorous explanation of the Jillson experiment).

5 Concluding remarks

In this paper I have discussed the LPA model and one of its variations (the periodically forced LPA model). Other variations of the model have arisen in applications, most of which have been little studied from a rigorous mathematical point of view. In this final section, I briefly discuss some of these variations.

Animals come in whole numbers. Even so, most mathematical models in population dynamics utilize state space variables that are continuous. This is the case of the LPA models in the previous sections. In recent studies it has been shown how restriction of state variables in the LPA model to a finite lattice of values (such as whole integers or to a finite number of densities obtained from integers divided by volume) explains distinctive patterns observed in data that find no explanation from the continuous state space version of the model [11, 23, 27]. An example of an "integerized" LPA model results from simply rounding the right hand sides of each equation (2.1)-(2.3) to the nearest integer. This rather ad hoc mathematical way of obtaining an integer model can be replaced by a more sophisticated derivation from an integer valued probabilistic model [11]. The continuous state space LPA model (2.1)-(2.3) derives from that probabilistic model by taking the expectations (means) of the random state variables. On the other hand, a deterministic model that remains integer valued results from the modes of the random state variables. Thus, the "mean" LPA models treated in this paper can be derived from a stochastic "master" model, but so can an integer valued "mode" LPA model, which it turns out is given by the difference equations

$$\begin{aligned} L(t+1) &= floor\left[bA(t)\exp\left(-\frac{c_{el}}{V}L(t) - \frac{c_{ea}}{V}A(t)\right)\right] \\ P(t+1) &= floor\left[(1-\mu_l)\,L(t)\right] \\ A(t+1) &= floor\left[P(t)\exp\left(-\frac{c_{pa}}{V}A(t)\right)\right] + floor\left[(1-\mu_a)\,A(t)\right]. \end{aligned}$$

(*floor* $[x]$ is the largest integer less than or equal to x). The mathematical properties of such integerized versions of the LPA model have not been studied in a systematic or rigorous way. It is not difficult to show that the nonnegative cone (on the lattice) is forward invariant for these integerized LPA models, that all nonnegative orbits are bounded, and, as a result, that all orbits become periodic after a finite number of steps. However, a study remains to be made of the set of final periodic states, their uniqueness or lack of uniqueness, their regions of attraction, and so on. Other interesting questions concern the relationship among the dynamics of the mean, integer, and random LPA models; see [11, 27, 24].

A genetics variation of the LPA model has recently played a role in experimental studies of how genetic adaptation can alter a population's dynamics. For the case of 2 alleles at a single locus the model takes the form [3]

$$\begin{aligned} L_i(t+1) &= \frac{2}{i!\,(2-i)!} p^{2-i}(t) q^i(t) b_i A(t) \exp\left(-c_{el} L(t) - c_{ea} A(t)\right) \\ P_i(t+1) &= \left(1-\mu_{l,i}\right) L_i(t) \\ A_i(t+1) &= P_i(t) \exp\left(-c_{pa} A(t)\right) + \left(1-\mu_a\right) A_i(t) \end{aligned} \tag{5.1}$$

where $i = 0, 1, 2$ and

$$L(t) = \sum_{i=0}^{2} L_i(t), \qquad P(t) = \sum_{i=0}^{2} P_i(t), \qquad A(t) = \sum_{i=0}^{2} A_i(t)$$

$$p(t) = \frac{A_0(t) + \frac{1}{2} A_1(t)}{A(t)}, \qquad q(t) = 1 - p(t).$$

(A number of simplifying biological assumptions have been made in order to attain this model, including random mating, a Hardy-Weinberg distribution of newly formed zygotes, and the restriction that all three genotypes have the same cannibalism coefficients and adult stage survival rates.) In this model L_0 and L_2 are the larval numbers of the homozygote population and L_1 is the larval numbers of the heterozygote population. The notation is similar for the pupal and adult stages P_i and A_i.

If initially only the homozygote population $i = 0$ is present, then the nine dimensional genetics LPA model (5.1) reduces to the three dimensional LPA model (2.1)-(2.3) with $L(t) = L_0(t)$, $P(t) = P_0(t)$, $A(t) = A_0(t)$.

With other genetic strains present, some algebraic manipulations show that the total numbers L, P, and A satisfy the equations

$$\begin{aligned} L(t+1) &= b(t) A(t) \exp\left(-c_{el} L(t) - c_{ea} A(t)\right) \\ P(t+1) &= \left(1-\mu_l(t-1)\right) L(t) \\ A(t+1) &= P(t) \exp\left(-c_{pa} A(t)\right) + \left(1-\mu_a\right) A(t) \end{aligned} \tag{5.2}$$

where

$$b(t) \mathrel{\mathring{=}} p^2(t) b_0 + 2p(t)q(t) b_1 + q^2(t) b_2$$

$$\mu_l(t) \mathrel{\mathring{=}} \frac{p^2(t) b_0 \mu_{l,0} + 2p(t)q(t) b_1 \mu_{l,1} + q^2(t) b_2 \mu_{l,2}}{b(t)}.$$

From the similarity of the equations (5.2) and the LPA model (2.1)-(2.3) one can see how the genetics (embodied in $b(t)$ and $\mu_l(t)$) cause the parameters b and μ_l in the LPA model to change with time. From this point of view, genetic change and adaptation in a population is expressed as parameter changes in the LPA model. As we have seen, parameter changes in the LPA model can result in dynamic bifurcations. The experiment reported in [3] (involving the adaptation of species of *Tribolium* to the pesticide malathion) documented such dynamic changes as predicted by the genetic LPA model (5.1).

A mixture of population dynamics and genetics is unusual in biomathematical modeling. This, together with its role in the genetics experiments involving flour beetles, make the genetics LPA model (5.1) an interesting model to investigate mathematically. There has been virtually no analysis made of the model.

Another interesting variation of the LPA model arises from the historical importance of *Tribolium* in the theory of ecological competition. During the formulation of competition theory in the 1940-1950's, laboratory experiments utilizing two species of *Tribolium* were influential is establishing some fundamental principles relating to the concepts of ecological niche, limiting similarity of species, and competitive exclusion or coexistence [37, 38, 39, 40]. A basic tenet was formulated that an ecological community could have no more species than the number of available (limited) resources and, in particular, that two similar species relying on one resource could not coexist indefinitely. This principle is supported by many mathematical models (including the famous two species model of Lotka/Volterra). Given this history and the recent establishment of the LPA model as an accurate model for the dynamics of *Tribolium*, it is interesting to consider a competition version of the LPA model.

In particular, it is interesting to compare the predictions of a LPA competition model to those of the classical Lotka/Volterra model (and similar competition models). Such a study was initiated in [14]. The competition LPA model results from two copies of the system (2.1)-(2.3) which have been coupled as follows:

$$\begin{aligned}
L(t+1) &= BA(t)\exp\left(-\frac{c_{EL}}{V}L(t)-\frac{c_{EA}}{V}A(t)\right)\exp\left(-\frac{c_{El}}{V}l(t)-\frac{c_{Ea}}{V}a(t)\right)\\
P(t+1) &= (1-\mu_L)\,L(t)\\
A(t+1) &= P(t)\exp\left(-\frac{c_{PA}}{V}A(t)\right)\exp\left(-\frac{c_{Pa}}{V}a(t)\right)+(1-\mu_A)\,A(t)\\
l(t+1) &= bAa(t)\exp\left(-\frac{c_{el}}{V}l(t)-\frac{c_{ea}}{V}a(t)\right)\exp\left(-\frac{c_{eL}}{V}L(t)-\frac{c_{eA}}{V}A(t)\right)\\
p(t+1) &= (1-\mu_l)\,l(t)\\
a(t+1) &= p(t)\exp\left(-\frac{c_{pa}}{V}a(t)\right)\exp\left(-\frac{c_{pA}}{V}A(t)\right)+(1-\mu_a)\,a(t).
\end{aligned}$$

Some results concerning asymptotic dynamics appear in [14], including analyses of the extinction equilibrium (in which both species are absent) and the competitive exclusion equilibria (in which one species is absent), a persistence result when both $n_1 = B(1-\mu_L)/\mu_A > 1$ and $n_2 = b(1-\mu_l)/\mu_a > 1$, and some existence and bifurcation results for coexistence (positive) equilibria. However, a great deal is unknown about the asymptotic dynamics of this six dimensional system (the properties of attractors, their uniqueness or lack of uniqueness, their bifurcations as parameters change, and so on). Given the model's high dimensionality and large number of parameters, it perhaps comes as no surprise that complicated non-equilibrium dynamics are possible, including chaotic dynamics [14], unlike the famous Lotka/Volterra equations.

Interestingly, recent numerical studies have also shown that the model can possess scenarios that are not found in classical competition theory [14, 15]. For example, it is possible for the system to possess two exclusion attractors (equilibria) and a coexistence attractor (a 2-cycle in which both species are present) and, as a result, whether or not one species goes extinct or both coexist depends on the initial conditions. (There is even some evidence for this scenario in the historical experiments involving *Tribolium* carried out by T. Park [15].) Currently there is no mathematical proof of this dynamic scenario, or for any of the many other complicated scenarios of multiple attractors that have been discovered by numerical simulations.

Difference equations have a long history of application to population dynamics. The LPA model has a distinctive place in that history because of its close ties with data and of its key role in a long list of experimental studies. For a mathematician the model, and its many variations, provide many interesting problems to study and abundant opportunities to make meaningful contributions to both theoretical and applied population biology.

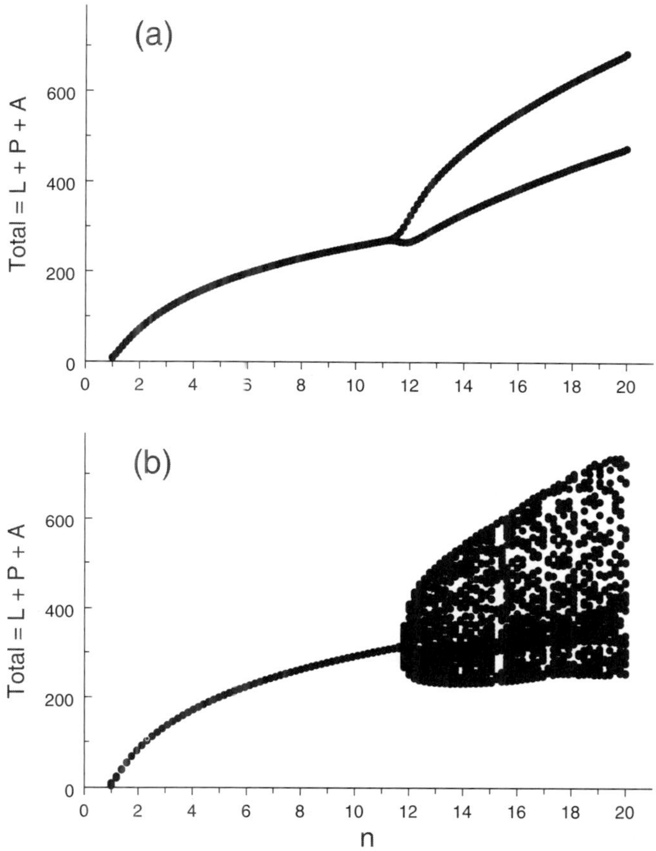

FIGURE 1. Bifurcation diagrams are shown in which the total "population size" $L+P+A$ of the attractor of the LPA model (2.1)-(2.3) is plotted against n. In both plots $\mu_l = 0.2$, $c_{el} = 0.01$, $c_{ea} = 0.01$, $c_{pa} = 0.005$, and $V = 1$. In (a) $\mu_a = 0.5$ and in (b) $\mu_a = 0.96$.

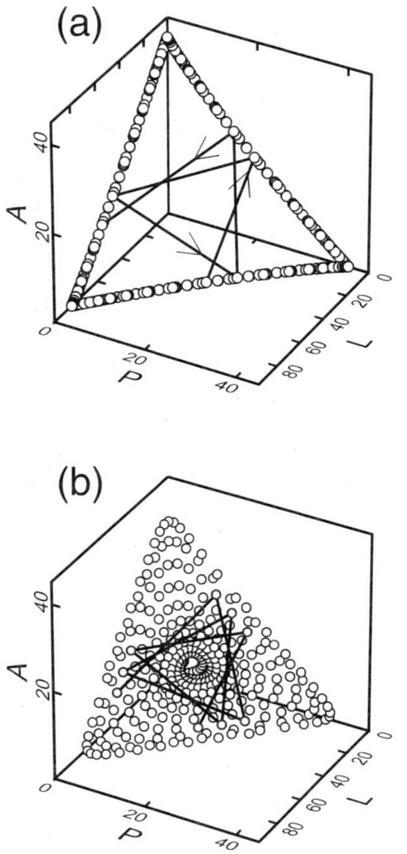

FIGURE 2. (a) The cycle chain in Theorem 3.3 is shown for parameter values $\mu_a = 1$, $\mu_l = 0.5$, $c_{el} = 0.007$, $c_{ea} = 0.01$, $c_{pa} = 0.01$, and $V = 1$. (b) A positive orbit is shown approaching the cycle chain in (a). Initial conditions are $(L(0), P(0), A(0)) = (4, 2, 2)$. Several consecutive points are connected by straight lines in order to show the temporal sequence.

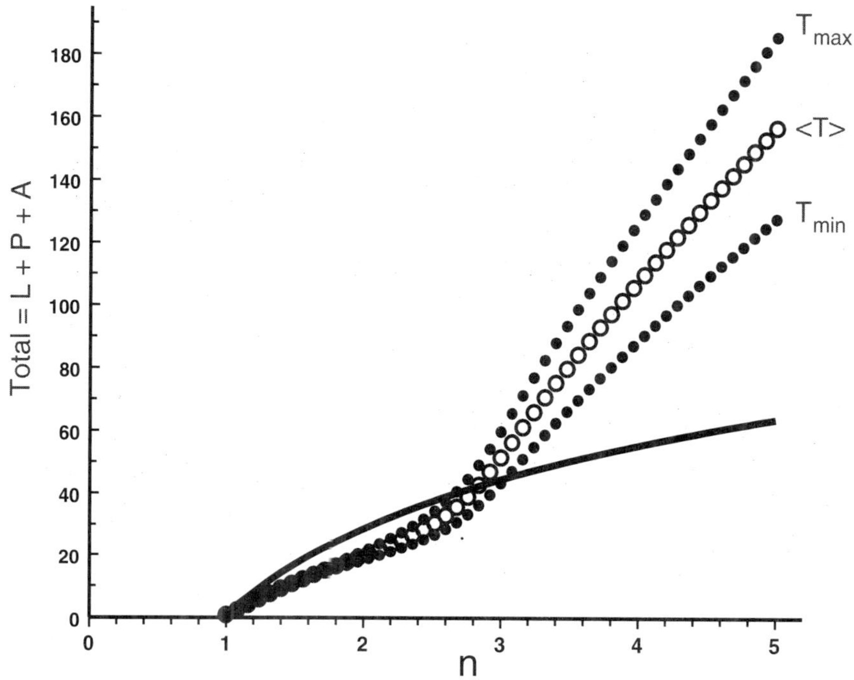

FIGURE 3. The solid line shows the equilibrium total population size $T = L + P + A$ of the autonomous LPA model as a function of $n = b(1-\mu_l)/\mu_a$. Parameter values are those estimated from the control cultures of Jillson's experiment [12]: $\mu_a = 0.1542$, $\mu_l = 0.4794$, $c_{el} = 0.0584$, $c_{ea} = 0.0058$, $c_{pa} = 0.0105$, $b = 4.445$ with $V = 1$ (representing the volume occupied by the standard laboratory volume of $20g$ of flour medium). The lines of solid circles show the maxima $T_{\max}$ and minima $T_{\min}$ of the positive 2-cycles from the periodically forced LPA model with the same parameter values and $\alpha = 0.6$ (which is the relative amplitude of Jillson's resonance experiment). The line of open circles shows the cycle averages $\langle T \rangle = (T_{\max} + T_{\min})/2$ of the 2-cycles. For n less than 2.8 (approximately) the cycle averages are less than the equilibria they are greater for n greater than 2.8. For n greater than 3 (approximately) even the cycle minima are greater than the equilibria. This diagram is consistent with the attenuation assertion of Theorem 4.2, but shows resonance for n sufficiently large. Note that the estimated value of $n = 15.18$ for Jillson's experiment does not appear in this graph.

References

[1] A. Berman and R. J. Plemmons, *Nonnegative Matrices in the Mathematical Sciences,* Classics in Applied Mathematics, SIAM, Philadelphia, 1994

[2] H. Caswell, *Matrix Population Models: Construction, Analysis and Interpretation*, Second edition, Sinauer Associates, Inc. Publishers, Sunderland, Massachusetts, 2001

[3] W. Cheung, *The Effects of Natural Selection on Nonlinear Population Dynamics*, Masters of Science thesis, California State University at Los Angeles, 2002

[4] R. F. Costantino, J. M. Cushing, B. Dennis, R. A. Desharnais, and S. M. Henson, Resonant population cycles in temporally fluctuating habitats, *Bulletin of Mathematical Biology* 60 (1998), 247-273

[5] J. M. Cushing, A strong ergodic theorem for some nonlinear matrix models for the dynamics of structured populations, *Natural Resource Modeling* 3 (1989), 331-357

[6] J. M. Cushing, *An Introduction to Structured Population Dynamics*, CBMS-NSF Regional Conference Series in Applied Mathematics, Vol. 71, SIAM, Philadelphia, 1998

[7] J. M. Cushing, Periodically forced nonlinear systems of difference equations, *J. Difference Equations and Applications*, *Journal of Difference Equations and Applications* 3 (1998), 547-561

[8] J. M. Cushing, Cycle chains and the LPA model, *J. Difference Equations and Applications* 9, No. 7 (2003), 655-670

[9] J. M. Cushing and Zhou Yicang, The net reproductive value and stability in structured population models, *Natural Resource Modeling* 8, No. 4 (1994), 1-37

[10] J. M. Cushing and S. M. Henson, Global dynamics of some periodically forced, monotone difference equations), *Journal of Difference Equations and Applications* 7 (2001), 859-872

[11] J. M. Cushing, R. F. Costantino, B. Dennis, R. A. Desharnais, and S. M. Henson, *Chaos in Ecology: Experimental Nonlinear Dynamics*, Theoretical Ecology Series, Academic Press, San Diego, 2003

[12] S. M. Henson, R. F. Costantino, J. M. Cushing, B. Dennis, and R. A. Desharnais, Multiple attractors, saddles, and population dynamics in periodic habitats, *Bulletin of Mathematical Biology* (1999) 61, 1121-1149

[13] B. Dennis, R. A. Desharnais, J. M. Cushing, and R. F. Costantino, Nonlinear demographic dynamics: mathematical models, statistical methods, and biological experiments, *Ecological Monographs* 65, No. 3 (1995), 261-281

[14] J. Edmunds, *A Study of a Stage-Structured Model of Two Competing Species*, Ph.D. dissertation, University of Arizona, Tucson, 2001

[15] J. Edmunds, J. M. Cushing, R. F. Costantino, S. M. Henson, B. Dennis, and R. A. Desharnais, Park's *Tribolium* competition experiments: a non-equilibrium species coexistence hypothesis, *Journal of Animal Econolgy* 72 (2003), 703-712

[16] S. N. Elaydi, *An Introduction to Difference Equations*, Springer-Verlag, New York, Second edition, 1999

[17] J. Guckenheimer and P. Holmes, *Nonlinear Oscillations, Dynamical Systems, and Bifurcations of Vector Fields*, Springer-Verlag, Berlin, 1983

[18] S. M. Henson, Existence and stability of nontrivial periodic solutions of periodically forced discrete dynamical systems, *Journal of Difference Equations and Applications* 2 (1996), 314-331

[19] S. M. Henson, The effect of periodicity in maps, *Journal of Difference equations and Applications* 5 (1999), 31-56

[20] S. M. Henson, Multiple attractors and resonance in periodically forced population models, *Physica D* 140 (2000), 33-49

[21] S. M. Henson, R. F. Costantino, J. M. Cushing, B. Dennis, and R. A. Desharnais, Multiple attractors, saddles, and population dynamics in periodic habitats, *Bulletin of Mathematical Biology* 61 (1999), 1121-1149

[22] S. M. Henson, and J. M. Cushing, The effect of periodic habitat fluctuations on a nonlinear insect population model, *Journal of Mathematical Biology* 36 (1997), 201-226

[23] S. M. Henson, R. F. Costantino, J. M. Cushing, R. A. Desharnais, B. Dennis, and A. King, Lattice effects observed in chaotic dynamics of experimental populations, *Science* 294 (19 Oct. 2001), 602-605

[24] S. M. Henson, R. F. Costantino, J. M. Cushing, Brian Dennis, Robert A. Desharnais, and A. A. King, Explaining and predicting patterns in stochastic population systems, *The Proceedings of the Royal Society, London B* 270 (2003), 1549-1553

[25] Josef Hofbauer and Joseph W.-H. So, Uniform persistence and repellers for maps, *Proceedings of the American Mathematical Society* 107, No. 4 (1989), 1137-1142

[26] D. Jillson, Insect populations respond to fluctuating environments, *Nature* 288 (1980), 699-700

[27] A. King, J. M. Cushing, R. F. Costantino, S. M. Henson, R. A. Desharnais, and B. Dennis, Quantification of temporal patterns in population data: revealing order in chaos, to appear in the *Proceedings of the National Academy of Sciences*

[28] P. H. Leslie, On the use of matrices in certain population mathematics, *Biometrika* 33 (1945), 183-212

[29] P. H. Leslie, Some further notes on the use of matrices in population mathematics, *Biometrika* 35 (1948), 213-245

[30] E. G. Lewis, On the generation and growth of a population, *Sankhya* 6 (1942), 93-96

[31] T. Y. Li and J. A. Yorke, Period three implies chaos, *American Mathematical Monthly* 82 (1975), 985-992

[32] R. M. May, Biological populations with nonoverlapping generations: stable points, stable cycles and chaos, *Science* 186 (1974), 645-647

[33] R. M. May, Simple mathematical models with very complicated dynamics, *Nature* 261 (1976), 459-467

[34] R. M. May, *Theoretical Ecology: Principles and Applications*, W. B. Saunders, 1976

[35] R. M. May, *Stability and Complexity in Model Ecosystems*, Princeton Landmarks in Biology, Princeton University Press, Princeton, New Jersey, 2001

[36] R. M. Nisbet and W. S. C. Gurney, Population dynamics in a periodically varying environment, *Journal of Theoretical Biology* 56 (1976), 499-475

[37] T. Park, Experimental studies of interspecies competition. I. Competition between populations of the flour beetles *Tribolium confusum* Duval and *Tribolium castaneum* Herbst, *Ecological Monographs* 18 (1948), 265-308

[38] T. Park, Experimental studies of interspecies competition. II. Temperature, humidity and competition in two species of *Tribolium*, *Physiological Zoology* 27 (1954), 177-238

[39] T. Park, Experimental studies of interspecies competition. III. Relation of initial species proportion to the competitive outcome in populations of *Tribolium*, *Physiological Zoology* 30 (1957), 22-40

[40] R. E. Ricklefs and G. Miller, *Ecology*, fourth edition, W. H. Freeman, New York, 2000

[41] J. F. Selgrade and J. H. Roberds, On the structure of attractors for discrete, periodically forced systems with applications to population models. *Physica D* 158 (2001), 69-82

[42] Yang Kuang and J. M. Cushing, Global stability in a nonlinear difference-delay equation model of flour beetle population growth, *Journal of Difference Equations and Applications* 2 (1995), 31-37

Fields Institute Communications
Volume **42**, 2004

A Survey on Oscillation of Linear Delay Difference Equations

X. Z. Qian
College of Mathematics and Econometrics
Hunan University
Changsha 410082, P. R. China
qian4704@sina.com

Z. C. Wang
College of Mathematics and Econometrics
Hunan University
Changsha 410082, P. R. China
zcwang@hnu.net.cn

Abstract. The aim of the paper is to survey the most basic results for the oscillation of solutions of the following linear difference equation

$$x_{n+1} - x_n + p_n x_{n-k} = 0, \quad n = 0, 1, 2, \cdots,$$

where $\{p_n\}$ is a sequence of real numbers and k is a positive integer.

1 Introduction

Recently there has been a lot of activities concerning the oscillation of delay difference equations. See, for example, [1-30]. For the general theory of difference equations, we refer to [1, 2, 4, 6, 8-10, 13, 15]. We should note that the oscillatory behavior of ordinary differential equations and its discrete analogue can be quite different. For example, it is well known that every solution of the logistic equation

$$x'(t) = rx(t)\left[1 - \frac{x(t)}{K}\right]$$

is monotonic. But its discrete analogue

$$x_{n+1} = ax_n(1 - x_n)$$

exhibits chaotic behavior when $a = 4$(see [7] or [9]).

In this paper, our aim is to survey the recent results for the oscillation of the following linear difference equation

$$x_{n+1} - x_n + p_n x_{n-k} = 0, \quad n = 0, 1, 2, \cdots, \tag{1.1}$$

where $\{p_n\}$ is a sequence of real numbers and k is a positive integer.

2000 *Mathematics Subject Classification.* Primary 39A10; Secondary 39A11.
Project Supported by NNSF of China (19971026; 10271044).

By a solution of Eq. (1.1) we mean a sequence $\{x_n\}$ which is defined for $n \geq -k$ and satisfies Eq. (1.1) for $n \geq 0$. A solution $\{x_n\}$ of Eq. (1.1) is said to be oscillatory if the terms x_n of the sequence are neither eventually positive nor eventually negative. Otherwise, the solution is called nonoscillatory.

2 Oscillation of Eq. (1.1) with nonnegative coefficients

In this Section, we always assume that $p_n \geq 0, \quad n = 0, 1, 2, \cdots$.

2.1 Oscillation criteria of inferior limit. The first result listed here for the oscillation of Eq. (1.1) was obtained by Erbe and Zhang [5] in 1989.

Theorem 2.1 ([5]) *If*

$$\liminf_{n\to\infty} p_n > \frac{k^k}{(k+1)^{k+1}}, \tag{2.1}$$

then every solution of Eq. (1.1) oscillates. If for large n

$$p_n \leq \frac{k^k}{(k+1)^{k+1}}, \tag{2.2}$$

then Eq. (1.1) has a nonoscillatory solution.

In the same year, Ladas, Philos and Sficas [12] proved that the condition (2.1) can be replaced by the weaker condition

$$\liminf_{n\to\infty} \sum_{i=n-k}^{n-1} p_i > \left(\frac{k}{k+1}\right)^{k+1}. \tag{2.3}$$

In 1998, Tang [18] obtained a better sufficient condition than (2.3) for the oscillation of Eq. (1.1).

Theorem 2.2 ([18]) *Assume that*

$$\sum_{i=n-k}^{n-1} p_i \geq \left(\frac{k}{k+1}\right)^{k+1} \quad \textit{for large } n \tag{2.4}$$

and

$$\sum_{n=k}^{\infty} p_n \left[\sum_{i=n-k}^{n-1} p_i - \left(\frac{k}{k+1}\right)^{k+1}\right] = \infty. \tag{2.5}$$

Then every solution of Eq. (1.1) oscillates.

When the sum in (2.5) is convergent, Tang and Yu [24] proved the following theorem.

Theorem 2.3 ([24]) *Assume that (2.4) holds and*

$$\limsup_{m\to\infty} m \sum_{n=mk}^{\infty} p_n \left[\sum_{i=n-k}^{n-1} p_i - \left(\frac{k}{k+1}\right)^{k+1}\right] > \left(\frac{k}{k+1}\right)^{2k+1} + \frac{k+1}{k} L^2, \tag{2.6}$$

where $L^2 = \limsup_{n\to\infty} \sum_{i=n-k}^{n-1} p_i^2$. *Then every solution of Eq. (1.1) oscillates.*

The aforementioned theorems always assume

$$\liminf_{n\to\infty}\sum_{i=n-k}^{n-1} p_i \geq \left(\frac{k}{k+1}\right)^{k+1}. \tag{2.7}$$

Is (2.7) a necessary condition for oscillation of all solutions of Eq. (1.1)? The following three theorems obtained by Tang and Yu in a recent paper [20] show that the answer to this problem is negative.

Theorem 2.4 ([20]) *Suppose that*

$$\sum_{n=0}^{\infty} p_n \left[\frac{k+1}{k}\left(\sum_{i=n+1}^{n+k} p_i\right)^{\frac{1}{k+1}} - 1\right] = \infty. \tag{2.8}$$

Then every solution of Eq. (1.1) oscillates.

Let

$$P_n^1 = \sum_{i=n+1}^{n+k} p_i, \quad P_n^{j+1} = \sum_{i=n+1}^{n+k} p_i P_i^j, \quad j = 1, 2, \cdots.$$

Theorem 2.5 ([20]) *Assume that there exists an integer $l \geq 1$ such that*

$$\sum_{n=0}^{\infty} p_n \left[\left(\frac{k+1}{k}\right)^l (P_n^l)^{1/(k+1)} - 1\right] = \infty. \tag{2.9}$$

Then every solution of Eq. (1.1) oscillates.

Theorem 2.6 ([20]) *Assume that there exists an integer $l \geq 1$ such that*

$$\liminf_{n\to\infty} P_n^l > \left(\frac{k}{k+1}\right)^{(k+1)l}. \tag{2.10}$$

Then every solution of Eq. (1.1) oscillates.

Example 2.7 Consider the delay difference equation

$$x_{n+1} - x_n + p_n x_{n-3} = 0, \quad n = 0, 1, 2, \cdots, \tag{2.11}$$

where $k = 3$ and for $n = j \bmod (12)$,

$$p_n = \begin{cases} 0, & 0 \leq j \leq 2, \\ 0.24, & 3 \leq j \leq 11. \end{cases}$$

By a simple calculation, we see that

$$P_n^1 = \sum_{i=n+1}^{n+3} p_i = \begin{cases} 0.24(j+1), & 0 \leq j \leq 2, \\ 0.72, & 3 \leq j \leq 8, \\ 0.24(11-j), & 9 \leq j \leq 11. \end{cases}$$

Clearly, $P_n^1 = 0$ for $n = 11 \bmod (12)$, and so

$$\liminf_{n\to\infty}\sum_{i=n-3}^{n-1} p_i = 0 \quad \text{and} \quad \limsup_{n\to\infty}\sum_{i=n-3}^{n} p_i = 0.96 < 1.$$

But, it is easy to see that

$$\sum_{n=0}^{11} p_n \left[\frac{4}{3}\left(\sum_{i=n+1}^{n+3} p_i\right)^{\frac{1}{4}} - 1\right] = 0.32\sum_{n=3}^{11}\left(\sum_{i=n+1}^{n+3} p_i\right)^{\frac{1}{4}} - 2.16$$
$$= 0.32\left(6\sqrt[4]{0.72} + \sqrt[4]{0.48} + \sqrt[4]{0.24}\right) - 2.16$$
$$\approx 0.09895 > 0.$$

It follows that

$$\sum_{n=0}^{\infty} p_n \left[\frac{4}{3}\left(\sum_{i=n+1}^{n+3} p_i\right)^{1/4} - 1\right] = \infty.$$

By Theorem 2.4, every solution of Eq.(2.11) oscillates.

Example 2.8 Consider the delay difference equation

$$x_{n+1} - x_n + p_n x_{n-2} = 0, \quad n = 0, 1, 2\cdots, \tag{2.12}$$

where $k = 2$, and $p_{4n} = \frac{3}{8}$, $p_{4n+1} = \frac{1}{4}$, $p_{4n+2} = 0$, $p_{4n+3} = \frac{1}{4}$, $n = 0, 1, 2, \cdots$. By direct calculation, one can obtain

$$P^1_{4n} = \frac{1}{4}, \quad P^1_{4n+1} = \frac{1}{4}, \quad P^1_{4n+2} = \frac{5}{8}, \quad P^1_{4n+3} = \frac{5}{8}, \quad n = 0, 1, 2, \cdots;$$
$$P^2_{4n} = \frac{1}{4^2}, \quad P^2_{4n+1} = \frac{1}{2\times 4}, \quad P^2_{4n+2} = \frac{1}{4^2}, \quad P^1_{4n+3} = \frac{5}{2\times 4^2}, \quad n = 0, 1, 2, \cdots;$$
$$P^3_{4n} = \frac{5}{2\times 3}, \quad P^3_{4n+1} = \frac{5}{2\times 4}, \quad P^2_{4n+2} = \frac{8}{2\times 4^3}, \quad P^3_{4n+3} = \frac{8}{2\times 4^3}, \quad n \in N;$$

$$\liminf_{n\to\infty} P^3_n = \frac{5}{2\times 4^3} > \left(\frac{2}{3}\right)^9.$$

Therefore, we may conclude from Theorem 2.6 that every solution of Eq. (2.12) oscillates. The same conclusion, however, cannot be inferred from Theorems 2.1–2.3. Since

$$\liminf_{n\to\infty} \sum_{i=n-2}^{n-1} p_i = \frac{1}{4} < \left(\frac{2}{3}\right)^3.$$

In 1990, Ladas [11] presented six open problems on the oscillation of difference equations, in the sequel, we will give three of them in this paper.

Open Problem 2.9 ([11]) Assume that

$$\sum_{i=n-k}^{n-1} p_i \le \left(\frac{k}{k+1}\right)^{k+1}. \tag{2.13}$$

Does Eq. (1.1) have an eventually positive solution?

Ladas' reason is that condition (2.3) can be looked upon as a discrete analogue of the corresponding oscillation criteria for the following first order delay differential equation

$$x'(t) + p(t)x(t-\tau) = 0, \quad t \ge 0,$$

where $p(t) \in C([0,\infty), [0,\infty))$, $\tau \in (0,\infty)$. And so condition (2.13) should be a discrete analogue of the corresponding nonoscillation criteria of the above differential equation.

We also remark that the above theorems all fail to the case when (2.13) holds.

In 1992, Qian and Yan [16] gave an example which shows that (2.1) can not guarantee that Eq. (1.1) has an eventually positive solution.

In 1994, Yu, Zhang and Wang [30] obtained a better sufficient condition than (2.3) for the oscillation of all solutions of Eq. (1.1) and used the result to construct an example which also shows that the answer to Problem 2.9 is negative.

Theorem 2.10 ([30]) *Let* $E = \{\lambda > 0 : 1 - \lambda p_n > 0\}$.

1. *Assume that for some integer* $N \geq 1$.

$$\sup_{\lambda \in E, n \geq N} \left[\lambda \prod_{i=n-k}^{n-1} (1 - \lambda p_i) \right] < 1. \tag{2.14}$$

Then every solution of Eq. (1.1) oscillates.

2. *Assume that there exists a* $\lambda \in E$ *such that for large* n

$$\lambda \prod_{i=n-k}^{n-1} (1 - \lambda p_i) \geq 1. \tag{2.15}$$

Then Eq. (1.1) has an eventually positive solution.

Conditions (2.14) and (2.15) are difficult to verify. In paper [21], Tang and Yu established the following explicit oscillation criteria which fit to the case when (2.13) holds.

Theorem 2.11 ([21]) *If*

$$\sum_{n=0}^{\infty} \left[\sum_{i=n}^{n+k} p_i \ln \left(\sum_{i=n}^{n-k} p_i + 1 - sign \sum_{i=n}^{n+k} p_i \right) \right. \tag{2.16}$$

$$\left. - \sum_{i=n+1}^{n+k} p_i \ln \left(\sum_{i=n+1}^{n+k} p_i + 1 - sign \sum_{i=n+1}^{n+k} p_i \right) \right] = \infty, \tag{2.17}$$

then every solution of Eq. (1.1) oscillates.

Corollary 2.12 ([21]) *Assume that there exists an integer* $N \geq 0$ *such that*

$$\sum_{i=n+1}^{n+k} p_i > 0 \quad \textit{for} \quad n \geq N,$$

and

$$\sum_{n=N}^{\infty} \left[\sum_{i=n}^{n+k} p_i \ln \left(\sum_{i=n}^{n+k} p_i \right) - \sum_{i=n+1}^{n+k} p_i \ln \left(\sum_{i=n+1}^{n+k} p_i \right) \right] = \infty. \tag{2.18}$$

Then every solution of Eq. (1.1) oscillates.

Corollary 2.13 ([21]) *Assume that (2.4) holds and*

$$\limsup_{n \to \infty} p_n > \frac{k^k}{(k+1)^{k+1}}. \tag{2.19}$$

Then every solution of Eq. (1.1) oscillates.

Remark 2.14 Condition (2.18) substantially improves (2.3). In fact, if (2.3) holds, then there exist $\theta > 1$ and an integer $s > 0$ such that

$$\sum_{i=n+1}^{n+k} p_i \geq \theta \left(\frac{k}{k+1} \right)^{k+1} = \theta_1 \quad \text{for} \quad n \geq sk. \tag{2.20}$$

For $d \geq 0$, the function $(d+x)\ln(d+x) - x\ln x$ is nondecreasing on $(0, +\infty)$, it follows that for $n \geq sk$

$$\sum_{i=n}^{n+k} p_i \ln \left(\sum_{i=n}^{n+k} p_i \right) - \sum_{i=n+1}^{n+k} p_i \ln \left(\sum_{i=n+1}^{n+k} p_i \right) \geq (p_n + \theta_1)\ln(p_n + \theta_1) - \theta_1 \ln \theta_1,$$

and so for $j = s, s+1, s+2, \cdots$

$$\begin{aligned} &\sum_{n=jk}^{(j+1)k-1} \left[\sum_{i=n}^{n+k} p_i \ln \left(\sum_{i=n}^{n+k} p_i \right) - \sum_{i=n+1}^{n+k} p_i \ln \left(\sum_{i=n+1}^{n+k} p_i \right) \right] \\ &\quad \geq \sum_{n=jk}^{(j+1)k-1} [(p_n + \theta_1)\ln(p_n + \theta_1) - \theta_1 \ln \theta_1]. \end{aligned}$$

By noting that for $a > 0$ and $x_i \geq 0$,

$$\sum_{i=1}^{k} (a + x_i)\ln(a + x_i) \geq k \left(a + \frac{1}{k}\sum_{i=1}^{k} x_i \right) \ln \left(a + \frac{1}{k}\sum_{i=1}^{k} x_i \right),$$

we see that

$$\begin{aligned} &\sum_{n=jk}^{(j+1)k-1} \left[\sum_{i=n}^{n+k} p_i \ln \left(\sum_{i=n}^{n+k} p_i \right) - \sum_{i=n+1}^{n+k} p_i \ln \left(\sum_{i=n+1}^{n+k} p_i \right) \right] \\ &\quad \geq k \left[\left(\theta_1 + \frac{1}{k} \sum_{n=jk}^{(j+1)k-1} p_n \right) \ln \left(\theta_1 + \frac{1}{k} \sum_{n=jk}^{(j+1)k-1} p_n \right) - \theta_1 \ln \theta_1 \right]. \end{aligned} \tag{2.21}$$

Since the function $x \ln x$ is increasing on $[\frac{1}{e}, \infty)$, and by (2.20)

$$\theta_1 + \frac{1}{k} \sum_{n=jk}^{(j+1)k-1} p_n \geq \frac{k+1}{k}\theta_1 = \theta \left(\frac{k}{k+1} \right)^k > \frac{1}{e},$$

it follows from (2.21) that

$$\begin{aligned} &\sum_{n=jk}^{(j+1)k-1} \left[\sum_{i=n}^{n+k} p_i \ln \left(\sum_{i=n}^{n+k} p_i \right) - \sum_{i=n+1}^{n+k} p_i \ln \left(\sum_{i=n+k}^{n+k} p_i \right) \right] \\ &\quad \geq k \left[\theta \left(\frac{k}{k+1} \right)^k \ln \left(\theta \left(\frac{k}{k+1} \right)^k \right) - \theta \left(\frac{k}{k+1} \right)^{k+1} \ln \left(\theta \left(\frac{k}{k+1} \right)^{k+1} \right) \right] \\ &\quad = \left(\frac{k}{k+1} \right)^{k+1} \theta \ln \theta > 0. \end{aligned}$$

This implies that (2.18) holds.

The following example shows that Theorem 2.11 and Corollary 2.12 fit the case when (2.13) holds.

Example 2.15 Consider the delay difference equation

$$x_{n+1} - x_n + p_n x_{n-3} = 0, \quad n = 0, 1, 2, \dots, \tag{2.22}$$

where $p_{3n} = 0$, $p_{3n+1} = p_{3n+2} = d$, $n = 0, 1, 2, \dots, d \in (4/27, 81/512]$. Observe that

$$\sum_{i=n-3}^{n-1} p_i = 2d \le \left(\frac{3}{4}\right)^4,$$

which implies that condition (2.13) is satisfied. But, for $j = 0, 1, 2, \cdots$

$$\begin{aligned}
&\sum_{n=3j}^{3(j+1)-1} \left[\sum_{i=n}^{n+3} p_i \ln \left(\sum_{i=n}^{n+3} p_i \right) - \sum_{i=n+1}^{n+3} p_i \ln \left(\sum_{i=n+1}^{n+3} p_i \right) \right] \\
&\quad = \sum_{n=0}^{2} \left[\sum_{i=n}^{n+3} p_i \ln \left(\sum_{i=n}^{n+3} p_i \right) - \sum_{i=n+1}^{n+3} p_i \ln \left(\sum_{i=n+1}^{n+3} p_i \right) \right] \\
&\quad = 2(3d \ln 3d - 2d \ln 2d) \\
&\quad = 2d \ln(27d/4) > 0.
\end{aligned}$$

It follows that

$$\sum_{n=0}^{\infty} \left[\sum_{i=n}^{n+3} p_i \ln \left(\sum_{i=n}^{n+3} p_i \ln \left(\sum_{i=n+1}^{n+3} p_i \right) \right) \right] = \infty.$$

In view of Corollary 2.12, every solution of Eq. (2.22) oscillates.

The following example further shows that condition (2.16) or (2.18) is better than condition (2.14).

Example 2.16 Consider the delay difference equation

$$x_{n+1} - x_n + p_n x_{n-2} = 0, \quad n = 0, 1, 2, \dots, \tag{2.23}$$

where $p_{8n} = p_{8n+1} = 0, p_{8n+2} = p_{8n+3} = \cdots = p_{8n+7} = d, \quad d \in (2/9, 1/3)$.

By a simple calculation, one can obtain for $j = 0, 1, 2, \cdots$

$$\begin{aligned}
&\sum_{n=8j}^{8j+7} \left[\sum_{i=n}^{n+2} p_i \ln \left(\sum_{i=n}^{n+2} p_i + 1 - \operatorname{sign} \sum_{i=n}^{n+2} p_i \right) \right. \\
&\qquad \left. - \sum_{i=n+1}^{n+2} p_i \ln \left(\sum_{i=n+1}^{n+2} p_i + 1 - \operatorname{sign} \sum_{i=n+1}^{n+2} p_i \right) \right] \\
&\quad = \sum_{n=0}^{7} \left[\sum_{i=n}^{n+2} p_i \ln \left(\sum_{i=n}^{n+2} p_i + 1 - \operatorname{sign} \sum_{i=n}^{n+2} p_i \right) \right. \\
&\qquad \left. - \sum_{i=n+1}^{n-2} p_i \ln \left(\sum_{i=n+1}^{n+2} p_i + 1 - \operatorname{sign} \sum_{i=n+1}^{n+2} p_i \right) \right] \\
&\quad = 4(3d \ln 3d - 2d \ln 2d) + (2d \ln 2d - d \ln d) + d \ln d \\
&\quad = 6d \ln(9d/2) > 0.
\end{aligned}$$

It follows that

$$\sum_{n=0}^{\infty}\Bigg[\sum_{i=n}^{n+2} p_i \ln\left(\sum_{i=n}^{n+2} p_i + 1 - \operatorname{sign}\sum_{i=n}^{n+2} p_i\right) - \sum_{i=n+1}^{n+2} p_i \ln\left(\sum_{i=n+1}^{n+2} p_i + 1 - \operatorname{sign}\sum_{i=n+1}^{n+2} p_i\right)\Bigg] = \infty.$$

In view of Theorem 2.11, every solution of Eq. (2.23) oscillates. The same conclusion, however, cannot be inferred from Theorems 2.1–2.10 since

$$\liminf_{n\to\infty} \sum_{i=n-2}^{n-1} p_i = 0, \quad \text{and} \quad \limsup_{n\to\infty} \sum_{i=n-2}^{n} p_i = 3d < 1,$$

and for $\lambda > 1$

$$\lambda \prod_{i=8n}^{8n+1} (1 - \lambda p_i) = \lambda > 1, \quad n = 0, 1, 2, \cdots,$$

namely, conditions (2.3) and (2.14) do not hold.

In [21], the authors also presented two nonoscillation results for Eq. (1.1).

Theorem 2.17 ([21]) *Assume that for large n*

$$\sum_{i=n-k}^{n} p_i \le \frac{1}{e}. \tag{2.24}$$

Then Eq. (1.1) has an eventually positive solution.

Corollary 2.18 ([21]) *Assume that for large n*

$$\sum_{i=n-k}^{n} p_i \le \left(\frac{k}{k+1}\right)^{k+1}. \tag{2.25}$$

Then Eq. (1.1) has an eventually positive solution.

Remark 2.19 Condition (2.25) may be looked upon as a correction for (2.13).

2.2 Oscillation in a critical state. In recent papers [22] and [25], Tang and Yu and Tang and Zhang studied the oscillation and nonoscillation of Eq. (1.1) in the critical case when

$$\liminf_{n\to\infty} p_n = \frac{k^k}{(k+1)^{k+1}}. \tag{2.26}$$

Theorem 2.20 ([25]) *Assume that*

$$\limsup_{n\to\infty} p_n > \frac{k^k}{(k+1)^{k+1}} \lim \inf_{n\to\infty} \frac{1}{k} \sum_{i=n-k}^{n} p_i. \tag{2.27}$$

Then every solution of Eq. (1.1) oscillates.

Remark 2.21 Theorem 2.20 shows that when

$$\lim \inf_{n\to\infty} \frac{1}{k} \sum_{i=n-k}^{n} p_i = \frac{k^k}{(k+1)^{k+1}},$$

if the limit $\lim_{n\to\infty} p_n$ does not exist, then every solutions of (1.1) oscillates; if the limit $\lim_{n\to\infty} p_n$ exists, then $\lim_{n\to\infty} p_n = \frac{k^k}{(k+1)^{k+1}}$. In this case, following Theorems 2.22–2.24 provide very effective tools to determine the oscillation of (1.1).

Theorem 2.22 ([22]) *Assume that for sufficiently large n*

$$p_n - \frac{k^k}{(k+1)^{k+1}} \geq 0. \tag{2.28}$$

Then every solution of Eq. (1.1) oscillates if and only if every solution of the following difference equation

$$\Delta^2 y_{n-1} + \frac{2(k+1)^k}{k^{k+1}} \left(p_n - \frac{k^k}{(k+1)^{k+1}} \right) y_n = 0, \quad n \in N \tag{2.29}$$

oscillates.

Theorem 2.23 ([22]) *The following two statements are valid:*

1. *If*

$$\liminf_{n\to\infty} \left[\left(p_n - \frac{k^k}{(k+1)^{k+1}} \right) n^2 \right] > \frac{k^{k+1}}{8(k+1)^k}, \tag{2.30}$$

then every solution of Eq. (1.1) oscillates.

2. *If eventually*

$$\left(p_n - \frac{k^k}{(k+1)^{k+1}} \right) n^2 \leq \frac{k^{k+1}}{8(k+1)^k}, \tag{2.31}$$

then Eq. (1.1) has an eventually positive solution.

Theorem 2.24 ([22]) *Assume that (2.27) holds. Then the following two statements are valid:*

1. *If*

$$\liminf_{n\to\infty} \left[n \sum_{i=n+1}^{\infty} \left(p_i - \frac{k^k}{(k+1)^{k+1}} \right) \right] > \frac{k^{k+1}}{8(k+1)^k}, \tag{2.32}$$

then every solution of Eq. (1.1) oscillates.

2. *If eventually*

$$n \sum_{i=n+1}^{\infty} \left(p_i - \frac{k^k}{(k+1)^{k+1}} \right) \leq \frac{k^{k+1}}{8(k+1)^k}, \tag{2.33}$$

then Eq. (1.1) has an eventually positive solution.

Example 2.25 By Theorem 2.23, every solution of the following delay difference equation

$$x_{n+1} - x_n + \left(\frac{k^k}{(k+1)^{k+1}} + c(1+n)^{-\alpha} \right) x_{n-k} = 0, \quad n \in N.$$

oscillates if and only if $\alpha < 2$ and $c > 0$ or $\alpha = 2$ and $c > k^{k+1}/[8(k+1)^k]$.

2.3 Oscillation criteria of superior limit. Set

$$\alpha = \liminf_{n\to\infty} \sum_{i=n-k}^{n-1} p_i \quad \text{and} \quad \Lambda = \limsup_{n\to\infty} \sum_{i=n-k}^{n} p_i.$$

In 1989, Erbe and Zhang [5] proved that every solution of Eq. (1.1) oscillates provided $\Lambda > 1$. By (2.25), if $\Lambda < \left(\frac{k}{k+1}\right)^{k+1}$, then Eq. (1.1) has an eventually positive solution. Obviously, there is a gap between $\left(\frac{k}{k+1}\right)^{k+1}$ and 1, therefore, Ladas [11] presented the following open problem in 1990.

Open Problem 2.26 ([11]) Obtain sufficient conditions for the oscillation of Eq. (1.1) when $\alpha \le [k/(k+1)]^{k+1}$ and $\Lambda \le 1$.

Remark 2.27 Theorems 2.2–2.22 all fit the case when $\alpha \le [k/(k+1)]^{k+1}$ and $\Lambda \le 1$.

In 1992, Yu, Zhang and Qian [29] proved that all solutions of Eq. (1.1) oscillates if

$$\alpha \le \left(\frac{k}{k+1}\right)^{k+1} \quad \text{and} \quad \Lambda > 1 - \frac{\alpha^2}{4}. \tag{2.34}$$

In 1994, Chen and Yu proved that (2.34) can be replaced by the weaker condition

$$\alpha \le \left(\frac{k}{k+1}\right)^{k+1} \quad \text{and} \quad \Lambda > 1 - \frac{1-\alpha-\sqrt{1-2\alpha-\alpha^2}}{2}. \tag{2.35}$$

For $0 \le \alpha \le \left(\frac{k}{k+1}\right)^{k+1}$, let $f(\alpha) \in [1, k/(k+1)\alpha]$ satisfy the following equation

$$f(\alpha)\left[1 - \frac{\alpha}{k} f(\alpha)\right]^k = 1. \tag{2.36}$$

Then $f(\alpha)$ is strictly increasing in $\left[0, \left(\frac{k}{k+1}\right)^{k+1}\right]$ from 1 to $\left(\frac{k+1}{k}\right)^k$.

In a recent paper [23], Tang and Yu obtained the following result.

Theorem 2.28 ([23]) *Assume that* $0 \le \alpha \le \left(\frac{k}{k+1}\right)^{k+1}$. *If*

$$\limsup_{n\to\infty} p_n > \frac{1}{f(\alpha)} - \frac{1-\alpha-\sqrt{1-2\alpha-\alpha^2}}{2}, \tag{2.37}$$

then every solution of Eq. (1.1) oscillates.

Remark 2.29 Condition (2.37) is independent of (2.34), but it improves the following condition obtained by Stavroulakis in [17],

$$\limsup_{n\to\infty} p_n > 1 - \frac{\alpha^2}{4}.$$

Theorem 2.30 ([23]) *Assume that* $0 \le \alpha \le \left(\frac{k}{k+1}\right)^{k+1}$. *If*

$$\limsup_{n\to\infty} \sum_{i=n-k}^{n} p_i \prod_{j=i-k}^{n-k-1} (1 - p_j f(\alpha))^{-1} > 1 - \frac{1-\alpha-\sqrt{1-2\alpha-\alpha^2}}{2}, \tag{2.38}$$

then every solution of Eq. (1.1) oscillates.

Theorem 2.31 ([23]) *Assume that* $0 \le \alpha \le \left(\frac{k}{k+1}\right)^{k+1}$. *If*

$$\limsup_{n\to\infty} \sum_{i=n-k}^{n} p_i \left[\min\left\{\prod_{j=i-k}^{n-k-1}(1-p_j f(\alpha))^{-1}, \frac{1}{f(\alpha)}\prod_{j=i-k}^{i-1}(1-p_j f(\alpha))^{-1}\right\}\right]$$
$$> \frac{1+\ln f(\alpha)}{f(\alpha)} - \frac{1-\alpha-\sqrt{1-2\alpha-\alpha^2}}{2}, \qquad (2.39)$$

then every solution of Eq. (1.1) oscillates.

Corollary 2.32 ([23]) *Assume that* $0 \le \alpha \le \left(\frac{k}{k+1}\right)^{k+1}$. *If*

$$\limsup_{n\to\infty} \sum_{i=n-k}^{n} p_i \prod_{j=i-k}^{n-k-1}(1-p_j)^{-1} > 1 - \frac{1-\alpha-\sqrt{1-2\alpha-\alpha^2}}{2}, \qquad (2.40)$$

then every solution of Eq. (1.1) oscillates.

Corollary 2.33 ([23]) *Assume that* $0 \le \alpha \le \left(\frac{k}{k+1}\right)^{k+1}$. *If*

$$\Lambda = \limsup_{n\to\infty} \sum_{i=n-k}^{n} p_i > \frac{1+\ln f(\alpha)}{f(\alpha)} - \frac{1-\alpha-\sqrt{1-2\alpha-\alpha^2}}{2}, \qquad (2.41)$$

then every solution of Eq. (1.1) oscillates.

Remark 2.34 Obviously, (2.41) improves (2.35) when $0 < \alpha \le \left(\frac{k}{k+1}\right)^{k+1}$, as $\alpha \to 0$, (2.41), together with (2.34), (2.35), reduces to $\Lambda > 1$. Nevertheless, the following Example 2.36 illustrates that (2.40) still possible improves condition $\Lambda > 1$ for the case $\alpha = 0$.

Example 2.35 Consider the difference equation

$$x_{n+1} - x_n + p_n x_{n-2} = 0, \quad n = 0, 1, 2, \cdots, \qquad (2.42)$$

where $p_{10n} = p_{10n+1} = \cdots = p_{10n+8} = 0.1, p_{10n+9} = 0,73$, $n = 0, 1, 2, \cdots$. It is easy to observe that

$$\alpha = \liminf_{n\to\infty} \sum_{i=n-2}^{n-1} p_i = 0.2 < \left(\frac{2}{3}\right)^3$$

and

$$\Lambda = \limsup_{n\to\infty} \sum_{i=n-2}^{n} p_i = 0.93 < 1.$$

In addition, we find

$$f(\alpha) = 1.336 \quad \text{and} \quad A(\alpha) = \frac{1}{2}\left(1-\alpha-\sqrt{1-2\alpha-\alpha^2}\right) = 0.026.$$

By these, one can easy verify that

$$\Lambda < \frac{1+\ln f(\alpha)}{f(\alpha)} - A(\alpha).$$

These show that conditions (2.34), (2.35) and (2.41) are not satisfied. But

$$\limsup_{n\to\infty} p_n = 0.73 > \frac{1}{f(\alpha)} - A(\alpha) = 0.7225.$$

Hence, the conditions of Theorem 2.24 are satisfied and therefore every solution of (2.42) is oscillatory.

Example 2.36 Consider the difference equation

$$x_{x+1} - x_n + p_n x_{n-3} = 0, \quad n = 0, 1, 2, \cdots, \tag{2.43}$$

where $k = 3$ and $p_{15n} = p_{15n+1} = \cdots = p_{15n+7} = 0$, $p_{15n+8} = p_{15n+9} = \cdots = p_{15+14} = 0.2$, $n = 0, 1, 2 \cdots$. It is easy to observe that

$$\alpha = \liminf_{n\to\infty} \sum_{i=n-3}^{n-1} p_i = 0 \quad \text{and} \quad \Lambda = \limsup_{n\to\infty} \sum_{i=n-3}^{n} p_i = 0.8 < 1,$$

which show that conditions (2.34) and (2.35) are not satisfied. In addition, it is easy to verify that (2.7) is not satisfied either. But

$$\sum_{i=15n+11}^{15n+14} p_i \prod_{j=i-3}^{15n+10} (1 - p_j)^{-1} = \frac{369}{320},$$

and so

$$\limsup_{n\to\infty} \sum_{i=n-3}^{n} p_i \prod_{j=i-3}^{n-3-1} (1 - p_j)^{-1} > 1.$$

Hence, the conditions of Corollary 2.32 are satisfied and therefore all solutions of (2.42) oscillate.

Example 2.37 Consider the difference equation

$$x_{n+1} - x_n + p_n x_{n-3} = 0, \quad n = 0, 1, 2, \cdots, \tag{2.44}$$

where $p_{15n} = p_{15n+1} = \cdots = p_{15n+7} = 0.1$, $p_{15n+8} = p_{15n+9} = \cdots = p_{15n+14} = 0.16$, $n \in N$. It is easy to observe that

$$\alpha = \liminf_{n\to\infty} \sum_{i=n-3}^{n-1} p_i = 0.3 < \left(\frac{3}{4}\right)^4, \quad \Lambda = \limsup_{n\to\infty} \sum_{i=n-3}^{n} p_i = 0.64 < 1,$$

$$f(\alpha) = f(0.3) = 1.842, \quad A(\alpha) = \left(1 - \alpha - \sqrt{1 - 2\alpha - \alpha^2}\right)/2 = 0.0716.$$

Hence

$$\frac{1 + \ln f(\alpha)}{f(\alpha)} - A(\alpha) = 0.802 > \Lambda,$$

which shows that condition (2.40) is not satisfied. But

$$\begin{aligned}
&\sum_{i=15n+11}^{15n+14} p_i \left[\min\left\{\prod_{j=i-3}^{15n+10} (1 - p_j f(\alpha))^{-1}, \frac{1}{f(\alpha)} \prod_{j=i-3}^{i-1} (1 - r_j f(\alpha))^{-1}\right\}\right] \\
= \;& 0.16 \sum_{i=15n+11}^{15n+14} \left[\min\left\{(1 - 0.16 \times 1.842)^{-(15n+14-i)}, \frac{1}{1.842}(1 - 016 \times 1.842)^{-3}\right\}\right] \\
= \;& 0.16 \left(\frac{1.418^3}{1.842} + \frac{1.418^3}{1.842} + 1.418 + 1\right) = 0.882 \\
> \;& 0.802 = \frac{1 + \ln f(\alpha)}{f(\alpha)} - A(\alpha).
\end{aligned}$$

Theses show that the conditions of Theorem 2.30 are satisfied and therefore every solution of (2.44) oscillates.

3 Oscillation of Eq. (1.1) with oscillating coefficients

In this section, the coefficients p_n themselves may be allowed to be oscillatory. We define

$$N(a) = \{a, a+1, a+2, \cdots\} \quad \text{and} \quad N(a,b) = \{a, a+1, \cdots, b\},$$

where a and b are integers with $a \le b$.

Open Problem 3.1 ([11]) Extend condition (2.3) to Eq. (1.1) with oscillating coefficients.

In 1993, Yu, Zhang and Qian [29] investigated the oscillation of all solutions of Eq. (1.1) with oscillating coefficients and obtained several sufficient conditions which partly solved Open Problem 3.1.

Theorem 3.2 ([29]) *Assume that there exists a sequence $\{n_i\}$ such that $n_i \to \infty$ as $i \to \infty$ and $p_n \ge 0$ on $N(n_i - 2k, n_i)$, and*

$$\sum_{j=n_i-k}^{n_i} p_j \ge 1, \quad i = 1, 2, \cdots. \tag{3.1}$$

Then every solution of Eq. (1.1) oscillates.

Theorem 3.3 ([14]) *Assume that there exists a positive integer sequence $\{n_i\}$ such that $n_i \to \infty$ as $i \to \infty$, $p_n \ge 0$ for $n \in N(n_i - 3k, n_i)$, and*

$$\sum_{j=n-k}^{n-1} p_j \ge \alpha \quad \textit{for} \quad n \in N(n_i - 2k, n_i). \tag{3.2}$$

Assume also that

$$\limsup_{i\to\infty} \sum_{j=n_i-2k}^{n_i-k-1} p_j > 1 - \frac{\alpha^2}{4}. \tag{3.3}$$

Then every solution of Eq. (1.1) oscillates.

In 1995, Stavroulakis [17] proved the following theorem which answered Open Problem 3.1.

Theorem 3.4 ([17]) *Assume that there exists a sequence $n_i \to \infty$ such that $p_n \ge 0$ for $n \in N(n_i - (N^*+1)k, n_i)$ and*

$$\sum_{j=n-k}^{n-1} p_j \ge c > \left(\frac{k}{k+1}\right)^{k+1} \quad \textit{for} \quad n \in \bigcup_{i=1}^{\infty} N(n_i - N^*k, n_i), \tag{3.4}$$

where

$$N^* = 1 + \left[\frac{\ln 4 - 2\ln c}{\ln c + (k+1)(\ln(k+1) - \ln k)}\right] \tag{3.5}$$

and $[\cdot]$ denotes the greatest integer function. Then every solution of Eq. (1.1) oscillates.

In a recent paper [28], Yu and Tang proved the following theorem.

Theorem 3.5 ([28]) *Assume that*

1. *there exist two sequences* $\{a_j\}_{j=1}^{\infty}$ *and* $\{b_j\}_{j=1}^{\infty}$ *of nonnegative integers such that* $b_j \le a_{j+1} - 2$ *and* $b_j - a_j \ge 2k$ *for* $j = 1, 2, \cdots$, *and*

$$p_n \ge 0 \quad \text{for} \quad n \in \cup_{j=1}^{\infty} N(a_j, b_j); \tag{3.6}$$

2.

$$\sum_{n=0}^{\infty} \overline{p}_n \left[\frac{k+1}{k} \left(\sum_{i=n+1}^{n+k} \overline{p}_i \right)^{\frac{1}{k+1}} - 1 \right] = \infty, \tag{3.7}$$

where

$$\overline{p}_n = \begin{cases} p_n, & n \in \cup_{j=1}^{\infty} N(a_j + k, b_j), \\ 0, & n \in N(0, a_1 + k - 1) \cup \cup_{j=1}^{\infty} (b_j + 1, a_{j+1} + k - 1). \end{cases} \tag{3.8}$$

Then every solution of Eq. (1.1) oscillates.

In [19], Tang and Cheng proved another theorem which is independent of Theorem 3.5.

Theorem 3.6 ([19]) *Assume that*

1. *is same as (1) in Theorem 3.5;*
2.

$$\sum_{n=0}^{\infty} \left[\sum_{i=n}^{n+k} \overline{p}_i \ln \left(\sum_{i=n}^{n+k} \overline{p}_i + 1 - sign \sum_{i=n}^{n+k} \overline{p}_i \right) \right.$$
$$\left. - \sum_{i=n+1}^{n+k} \overline{p}_i \ln \left(\sum_{i=n+1}^{n+k} \overline{p}_i + 1 - sign \sum_{i=n+1}^{n+k} \overline{p}_i \right) \right] = \infty, \tag{3.9}$$

where $\overline{p}_n$ *is defined by (3.8).*

Then every solution of Eq. (1.1) oscillates.

From Theorem 3.6, we can prove the following corollary.

Corollary 3.7 ([19]) *Assume that (3.4) holds, and that there exists a sequence* $n_i \to \infty$ *such that* $p_n \ge 0$ *for* $n \in N(n_i - (N^* + 1)k, n_i)$, *where*

$$N^* = 1 + \left[\frac{1/(ec)}{\ln c + (k+1)(\ln(k+1) - \ln k)} \right]$$

and $[\cdot]$ *denotes the greatest integer function. Then every solution of Eq. (1.1) oscillates.*

Remark 3.8 Clearly, $1/(ec)$ is much less that $\ln \frac{4}{4c^2}$ for $\left(\frac{k}{k+1}\right)^{k+1} < c \le 1$. Therefore, Corollary 3.7 improves Theorem 3.4 greatly and answers Open Problem 3.1 better.

Example 3.9 Consider the delay difference equation

$$x_{n+1} - x_n + p_n x_{n-3} = 0, \quad n = 0, 1, 2 \cdots, \tag{3.10}$$

where $p_{12n} = -1$, $p_{12n+1} = p_{12n+2} = p_{12n+3} = 0$, $p_{12n+4} = \cdots = p_{12n+11} = 0.245$, $n \in N$.

Observe that

$$\liminf_{n\to\infty} \sum_{i=n-3}^{n-1} p_i = -1 \quad \text{and} \quad \limsup_{n\to\infty} \sum_{i=n-3}^{n} p_i = 0.98 < 1.$$

But if we let

$$a_j = 12j+1, b_j = 12j+11, \quad j = 0,1,2,\cdots,$$

then $p_n \geq 0$ for $n \in \cup_{j=0}^{\infty} N(a_j, b_j)$ and

$$\overline{p}_n = \begin{cases} 0.245, & n \in \cup_{j=0}^{\infty} N(12j+4, 12j+11), \\ 0, & n \in N(0,3) \cup \cup_{j=1}^{\infty} N(12j, 12j+3). \end{cases}$$

By a simple calculation, we have

$$\begin{aligned} \sum_{n=0}^{11} \overline{p}_n \left[\frac{4}{3} \left(\sum_{i=n+1}^{n+3} \overline{p}_i \right)^{1/4} - 1 \right] &= \frac{0.98}{3} \sum_{n=4}^{10} \left(\sum_{i=n+1}^{n+3} \overline{p}_i \right)^{\frac{1}{4}} - 1.96 \\ &= \frac{0.98}{3} \left(5\sqrt[4]{0.735} + \sqrt[4]{0.49} + \sqrt[4]{0.245} \right) - 1.96 \\ &= 0.0630 > 0. \end{aligned}$$

Thus

$$\sum_{n=0}^{\infty} \overline{p}_n \left[\frac{4}{3} \left(\sum_{i=n+1}^{n+3} \overline{p}_i \right)^{\frac{1}{4}} - 1 \right] = \infty.$$

By Theorem 3.5, every solution of Eq. (3.10) oscillates.

Example 3.10 Consider the following delay difference equation

$$x_{n+1} - x_n + p_n x_{n-3} = 0, \quad n \in N, \tag{3.11}$$

where $p_{12n} = -1$, $p_{12n+1} = p_{12n+2} = p_{12n+3} = 0$, $p_{12n+4} = \cdots = p_{12n+11} = d$, $n \in N$, $d \in (3\sqrt{3}/32, 1/5]$.

Observe that

$$\liminf_{n\to\infty} \sum_{i=n-3}^{n-1} p_i = -1 \quad \text{and} \quad \limsup_{n\to\infty} \sum_{i=n-3}^{n} p_i = 4d \leq 0.8.$$

But if we let

$$a_j = 12j+1, b_j = 12j+11, \quad j = 0,1,2,\cdots,$$

then $p_n \geq 0$ for $n \in \cup_{j=0}^{\infty} N(a_j\ b_j)$ and

$$\overline{p}_n = \begin{cases} d, & n \in \cup_{j=0}^{\infty} N(12j+4, 12j+11), \\ 0, & n \in N(0,3) \cup \cup_{j=1}^{\infty} N(12j, 12j+3). \end{cases}$$

By a simple calculation, we have

$$\begin{aligned} &\sum_{n=12j}^{12j+11} \left[\sum_{i=n}^{n+3} \overline{p}_i \ln \left(\sum_{i=n}^{n+3} \overline{p}_i + 1 - \operatorname{sign} \sum_{i=n}^{n+3} \overline{p}_i \right) \right. \\ &\quad \left. - \sum_{i=n+1}^{n+3} \overline{p}_i \ln \left(\sum_{i=n+1}^{n+3} \overline{p}_i + 1 - \operatorname{sign} \sum_{i=n+1}^{n+3} p_i \right) \right] = 4d \ln(4^5 d^2/3^3) > 0. \end{aligned}$$

Thus

$$\begin{aligned} &\sum_{n=0}^{\infty} \left[\sum_{i=n}^{n+3} \overline{p}_i \ln \left(\sum_{i=n}^{n+3} \overline{p}_i + 1 - \operatorname{sign} \sum_{i=n}^{n+3} \overline{p}_i \right) \right. \\ &\quad \left. - \sum_{i=n+1}^{n-3} \overline{p}_i \ln \left(\sum_{i=n+1}^{n+3} \overline{p}_i + 1 - \operatorname{sign} \sum_{i=n+1}^{n+3} \overline{p}_i \right) \right] = \infty. \end{aligned}$$

By Theorem 3.5, every solution of Eq. (3.11) oscillates.

References

[1] Agarwal, R. P. [1992] *Difference Equations and Inequalities*, Marcel Dekker, Inc., New York.

[2] Agarwal, R. P. and Wong, P. J. Y. [1991] *Advanced Topics in Difference Equations*, Kluwer Academic Publishers, Dordrecht.

[3] Chen, M. P. and Yu, J. S. [1995] *Oscillations for delay difference equations with variable coefficients*, Proceedings of the first International Conference on Difference Equations, Edited by S. N. Elaydi, et al., Gordon and Breach Publishers, pp. 105-114.

[4] Elaydi, N. [1996] *An Introduction to Difference Equations*, Springer-Verlag, New York.

[5] Erbe, L. H. and Zhang, B. G. [1989] *Oscillation of discrete analogue of delay equations*, Differential Integral Equations **2**, 300-309.

[6] Gyori, I. and Ladas, G. [1991] *Oscillatory Theory of Delay Differential Equations with Applications*, Oxford University Press, Oxford.

[7] Hale, J. K. [1977] *Theory of functional Differential Equations*, Springer-Verlag, New York.

[8] Jerri, A. J. [1996] *Linear Difference Equations with Discrete Transform Methods*, Kluwer Academic Publishers, Dordrecht.

[9] Kelley, W. G. and Peterson, A. C. [1991] *Difference Equations with Applications*, Acad. Press, New York.

[10] Kocic, V. L. and Ladas, G. [1993] *Global Behassior of Nonlinear Difference Equations of Higher Order with Applications*, Kluwer Academse Publishers, Dordrecht.

[11] Ladas, G. [1990] *Recent developments in the oscillation of delay difference equations*, Differential Equations: Stability and Control, Marcel Dekker, New York, pp. 321-332.

[12] Ladas, G., Philos, Ch. G. and Sficas, Y. G. [1989] *Sharp condition for the oscillation of delay difference equations*, J. Appl. Math. **2**, 101-112.

[13] Lakshmikantham, V. and Trigiante, D. [1988] *Difference Equations with Application to Numerical Analysis*, Academic Press, New York.

[14] Li, B. [1996] *Discrete oscillations*, J. Difference Equations and Appl. **2**, 389-399.

[15] Mickens, R. E. [1990] *Difference Equations: Theory and Applications*, 2nd ed., Van Nostrand Reinhold, New York.

[16] Qian, C. and Yan, J. [1992] *Existence and comparsion results of positive solutions of difference equations*, Funk. Ekvac. **35**, 581-595.

[17] Stavroulakis, I. P. [1995] *Oscillations of delay difference equations*, Computers Math. Appl. **29(7)**, 83-88.

[18] Tang, X. H. [1998] *r Oscillations of delay difference equations with variable coefficients*, J. Central South University of Technology **29(3)**, 287-288(Chinese).

[19] Tang, X. H. and Cheng, S. S. [2001] *An oscillation criterion for linear difference equations with oscillating coefficients*, J. Computational and Applied Mathematics **132**, 319-329.

[20] Tang, X. H. and Yu, J. S. [1999] *Oscillation of delay difference equations*, Computers Math. Appl. **37**, 11-20.

[21] Tang, X. H. and Yu, J. S. [1999] *Some further results on oscillation of delay difference equations*, Computers Math. Applic. **38**, 229-237.

[22] Tang, X. H. and Yu, J. S. [2000] *Oscillation of delay difference equation in a critical state*, Appl. Math. Lett. **13**, 9-15.

[23] Tang, X. H. and Yu, J. S. [2000] *Oscillation for delay difference equations*, Hokkaido Math. J. **29**, 213-228.

[24] Tang, X. H. and Yu, J. S. [2000] *A new oscillation criterion for linear delay difference equations*, Dynamics of Continuous Discrete and Impulsive Systems **7**, 525-531.

[25] Tang, X. H. and Zhang, R. Y. [2001] *New oscillation criterion for delay difference equations*, Computers Math. Applic. **42**, 1319-1330.

[26] Tang, X. H. and Yu, J. S. [2000] *Oscillation of nonlinear delay difference equations*, J. Math. Anal. Appl. **249**, 476-490.

[27] Yan, J. and Qian, C. [1992] *Oscillation and comparison result for delay difference equations*, J. Math. Anal. Appl. **165**, 346-360.

[28] Yu, J. S. and Tang, X. H. [2000] *Sufficient conditions for the oscillation of linear delay difference equations with oscillating coefficients*, J. Math. Anal. Appl. **250**, 735-742.

[29] Yu, J. S., Zhang, B. G. and Qian, X. Z. [1993] *Oscillations of delay difference equations with deviating coefficients*, J. Math. Anal. Appl. **177**, 432-444.

[30] Yu, J. S., Zhang, B. G. and Wang, Z. C. [1994] *Oscillations of delay difference equations*, Appl. Anal. **53**, 117-124.

Fields Institute Communications
Volume **42**, 2004

On Almost Automorphic Oscillations

Yingfei Yi
School of Mathematics
Georgia Institute of Technology
Atlanta, GA 30332 U.S.A.
yi@math.gatech.edu

Abstract. Almost automorphic dynamics have been given a notable amount of attention in recent years with respect to the study of almost periodically forced monotone systems. There are solid evidences that these types of dynamics should also be of fundamental importance in non-monotone especially conservative systems due to the interaction of multi-frequencies. This article will give a preliminary discussion in this regard by reviewing certain known cases and raising some problems for potential future studies.

1 Introduction

Almost automorphic functions, generating the (Bohr) almost periodic ones, were first introduced by Bochner ([12]) in 1955 in a differential geometry context, and, with a minor modification, they were shown in [27, 94] to coincide with the Levitan class of N-almost periodic functions ([56]). Let T be a locally compact, σ -compact, Abelian, first countable topological group and let E be a complete metric space. A function $f \in C(T, E)$ is said to be *almost automorphic* if whenever $\{t_n\}$ is a sequence such that $f(t_n + t) \to g(t) \in C(T, E)$ uniformly on compact sets, then $g(t - t_n) \to f(t)$ uniformly on compact sets, as $n \to \infty$. f becomes *almost periodic* if any sequence $\{t_n\}$ admits a subsequence $\{t_{n'}\}$ such that $f(t_{n'} + t)$ converges uniformly on T, as $n \to \infty$. In the literature, the above notion of almost automorphy is referred to as *sequential* or *continuous* almost automorphy. In fact, almost automorphic functions (and flows) can be defined in an abstract fashion on any topological group with respect to pointwise net convergence (see [27, 94, 98] for details). A function is sequential almost automorphic iff it is net almost automorphic and uniformly continuous ([98]). We choose the sequential

2000 *Mathematics Subject Classification.* Primary 34C27, 34C28, 37B55; Secondary 54H20, 58F14, 58F05, 58F30.

Key words and phrases. Almost automorphic dynamics, multi-frequency, nonlinear oscillations.

Partially supported by NSF grant DMS0204119.

notion in this paper because of its convenience in the applications to differential equations.

An almost periodic function is necessarily almost automorphic, but not vice versa. One can define Fourier series for both almost periodic and almost automorphic functions valued in a Banach space but the one for an almost periodic function is unique and converges uniformly in terms of Bochner-Fejer summation, while the one for an almost automorphic function is in general non-unique and its Bochner-Fejer sum only converges pointwise ([98]). Although there can be many Fourier series associated with a given almost automorphic function, one can define the *frequency module* of an almost automorphic function in the usual way as the smallest Abelian group containing a Fourier spectrum - the set of Fourier exponents associated with a Fourier series, and, it has been shown that such a frequency module is uniquely defined ([106]). In the above sense, both almost periodic and almost automorphic functions can be viewed as natural generalizations to the periodic ones in the strongest and the weakest sense respectively.

Almost automorphic minimal flows were first introduced and studied by Veech ([97]-[100]). A flow (E, T) is called *almost automorphic* (*almost periodic*) *minimal* if E is the closure of an almost automorphic (almost periodic) orbit. An almost automorphic minimal flow contains residually many almost automorphic points and becomes almost periodic only if every point in E is almost automorphic ([98, 100]). Typical examples of almost automorphic minimal sets include the well known Toeplitz minimal sets in symbolic dynamics ([19, 61]), the Denjoy set ([17]) on the circle, and the Aubry-Mather sets ([5, 62]) on an annulus. Unlike an almost periodic minimal flow, an almost automorphic one can be non-uniquely ergodic and can admit positive topological entropy ([61]), and its general measure theoretical characterization is completely random ([31]). Topologically, while an almost periodic minimal set is always a compact topological group ([22]), a non-almost periodic, almost automorphic one is only an almost 1-cover of a topological group ([98]) and can be topologically complicated ([46]). Hence, on one hand, an almost automorphic flow resembles an almost periodic one harmonically, but on the other hand, it presents certain complicated dynamical, topological and measure theoretical features which are significantly different from an almost periodic one.

Systematic studies of almost automorphic dynamics in differential equations were made in a series of recent works of the author with Shen ([89]-[93]) with respect to almost periodically forced monotone systems which are roughly those admitting no internal frequencies. Consider a skew-product semi-flow π_t over an almost periodic minimal base flow. Loosely speaking, it has been shown that if π_t is fiber-wise totally monotone (e.g., skew-product flows and semi-flows generated by almost periodically forced scalar ODEs and almost periodically forced parabolic PDEs in one space dimension, respectively), then all its minimal sets are almost automorphic with frequency modules responding to that of the base flow harmonically, and, if it is fiber-wise strongly monotone (e.g., skew-product flows and semi-flows generated by cooperative almost periodic systems of ODEs and almost periodically forced parabolic PDEs in higher space dimensions, respectively), then each linearly stable minimal set is almost automorphic with frequency module responding to that of the base flow sub-harmonically, and moreover, a minimal set in the fiber-wise strongly monotone skew-product semi-flow π_t over an almost periodic minimal base flow becomes almost periodic if it is uniformly stable. It is well known that even in the simplest almost periodically forced monotone system such as a quasi-periodically forced scalar ODE with only two frequencies and a linear scalar ODE with limiting periodic coefficients, almost periodic motions need not exist ([26, 46, 56, 75, 93]). Thus, the finding of almost automorphic dynamics in [89]-[93] actually shows a fundamental phenomenon in almost periodically forced monotone systems, i.e., almost automorphic solutions largely exist but almost periodic ones need not. We refer the readers to [85, 93] for references on the study of almost automorphic dynamics in almost periodically forced differential systems and to [3, 41, 42, 74, 76, 86, 88] for recent developments in the subject.

Comparing with periodically forced monotone systems in which periodic solutions are generically expected, the existence of almost automorphic solutions in almost periodically forced monotone systems reflects a general harmonic nature of the systems due to the interactions of several frequencies especially when they are close to the resonance. It is therefore well expected that almost automorphic dynamics should also largely exist in almost periodically forced non-monotone systems (i.e., systems with self-excitations or containing internal frequencies) and even in autonomous conservative systems. Some examples of periodically forced nonlinear oscillators and Hamiltonian systems have already supported this assertion but no systematic investigation is made to the existence of almost automorphic dynamics in general almost periodically forced non-monotone systems yet. In this article, we will review some known cases of the existence of almost automorphic dynamics in non-monotone systems and formulate some general open problems as potential starting points in the subject. We are also attempting to link the study of almost automorphic oscillations with other active areas in dynamical systems such as the almost periodic Floquet theory, Aubry-Mather theory, chaos, Hamiltonian systems, non-chaotic strange attractors, nonlinear oscillations, and toral flows. Problems arising in multi-frequency, non-monotone systems are far more complicated and challenging than those in monotone systems. At this stage, we have more questions than answers. Therefore, instead of an expository article as it was supposed to be, the present article is rather a research note, aiming at making some preliminary discussions in this interesting subject, which is by no means complete or well thought out.

Throughout the paper, we let (Y, R) be an almost periodic minimal flow and denote $y \cdot t$ as the orbit of the flow passing through a point $y \in Y$. We will mainly consider ODEs with almost periodic forcing of the form $f(x, y \cdot t)$, where $x \in R^n$ is a state variable (the global existence of solutions of the ODEs is always assumed). This is without loss of generality. Suppose that an ODE

$$x' = F(x,t), \; x \in R^n, \tag{1.1}$$

is considered with F being uniformly Lipschitz in x and almost periodic in t uniformly with respect to x. Let $Y = H(F) = cl\{F_\tau | \tau \in R\}$ be the *hull* of $y_0 = F$ under the compact open topology (hence Y is compact metric), where $F_\tau(x,t) \equiv F(x, t+\tau)$. Define $f : R^n \times Y \to R^n$: $f(x,y) = y(x,0)$, $y \in Y$. Then the time translation $y \cdot t = y_t$ defines a natural almost periodic minimal flow (Y, R) and $F(x,t) \equiv f(x, y_0 \cdot t)$. Thus instead of studying the single equation (1.1), we will consider a family of equations

$$x' = f(x, y \cdot t), \; x \in R^n, \tag{1.2}$$

which consist of the translated and limiting equations of (1.1). To study the dynamical behaviors of solutions of (1.1), we note that the equations (1.2) give rise to a skew-product flow $(R^n \times Y, \pi_t)$:

$$\pi_t(x, y) = (X(x, y, t), y \cdot t),$$

where $X(x, y, t)$ denotes the solution of (1.2) with the initial value x. Such a formulation was originated in [67, 84] in studying dynamics of non-autonomous differential equations (see also [48, 105]). We note that if F is quasi-periodic in t with k frequencies, then $Y \simeq T^k$ and (Y, R) is topologically conjugated to the standard quasi-periodic flow on the k-torus.

Part of the material in this article has been lectured at the 7th International Conference on Difference Equations and Applications (ICDEA), Changsha, China, 2002, the International Conference on Dynamical Methods for Differential Equations, Medina, Spain, 2002, and the International Workshop on Global Analysis of Dynamical Systems, Leiden, The Netherlands, 2001. The author would like to thank the organizers of these conferences for opportunities to present part of the results contained in this article, in particular the organizers for the 7th ICDEA for their invitation to write this article.

2 Almost automorphic functions and flows

In this section, unless specified otherwise, we let T be a locally compact, σ-compact, Abelian, first countable group, and E be a complete metric space.

2.1 Harmonic properties. Fourier theory was first developed in [98] for complex valued almost automorphic functions. Parallel Fourier theory can be also found in [56] for N-almost periodic functions.

Theorem 2.1 *Let f be an almost automorphic function on T valued in a Banach space. Then f admits Fourier series (not necessarily unique) whose Bochner-Fejer sums converge to f pointwise (in fact, uniformly on compact sets).*

The theorem was first proved in [98] for complex valued functions and was generalized in [106] for almost automorphic functions on R valued in a Banach space. The same proof can be carried over for general T.

Let f be an almost automorphic function on T. Then the topological hull $H(f)$ is almost automorphic minimal under the 'time' translation flow which is an almost 1-cover of its maximal almost periodic factor (Y, T) (see Theorem 2.3 below). The almost periodic factor is unique up to flow isomorphism and admits a natural compact Abelian topological group structure inherited from the flow ([22]), e.g., when $T = R$, Y is a solenoidal group ([2]). Hence the dual group Y' is discrete. We define the *frequency module* $\mathcal{M}(f)$ of f as the subgroup of Y' generated by the Fourier spectrum associated with a Fourier series of f. This is well defined because of the following theorem which was originally shown in [106] for $T = R$ but also holds in general.

Theorem 2.2 *With respect to any Fourier series of f, $\mathcal{M}(f) \simeq Y'$.*

The following module containment result can be shown similarly as in [106] for functions defined on R.

Proposition 2.1 *For two almost automorphic functions $f, g \in C(T, E)$, $\mathcal{M}(g) \subset \mathcal{M}(f)$ iff whenever $f(t + t_n) \to f(t)$ for some sequence $\{t_n\} \subset T$ then $g(t + t_n) \to g(t)$, uniformly on compact sets.*

2.2 Structural properties. The following result shown in [98] is known as the Veech *almost automorphic structure theorem.*

Theorem 2.3 *A compact minimal flow* (E, T) *is almost automorphic iff it is an almost* 1-1 *extension of its maximal almost periodic factor* (Y, T), *i.e., there is a residual subset* $Y_0 \subset Y$ *such that each fiber over* Y_0 *is a singleton.*

Let (E, T), (Y, T) be as in the above. It is shown in [100] that as long as there is a $y_0 \in Y$ corresponding to a singleton fiber, then there are residually many of them. The set of points in E lying in the singleton fibers consists of precisely the almost automorphic points, and, E becomes almost periodic iff all points in E are almost automorphic. Again, as (Y, T) is almost periodic minimal, Y admits a compact Abelian group structure inherited from the flow whose dual group Y' is discrete. We thus define the *frequency module* $\mathcal{M}(E)$ of E as the dual group Y'. This definition is in consistent with that for an almost automorphic function which we introduced above.

Theorem 2.4 1) *A function* $f \in C(T, E)$ *is almost automorphic iff it is a pointwise limit of a sequence of jointly almost automorphic, almost periodic functions.*

2) $f \in C(T, E)$ *is almost automorphic iff there exists a dense subgroup* G *of a compact group* G_*, *a continuous homomorphism* $\phi : T \to G$, *and a function* $g : G_* \to E$ *which is continuous on* G *such that* $f = g \circ \phi$ *is uniformly continuous.*

3) *If* $\phi : T \to D$, *where* D *is complete metric, is almost automorphic,* $g : D \to E$ *is continuous on* $R(\phi)$, *then* $f = g \circ \phi \in C(T, E)$ *is almost automorphic if it is uniformly continuous.*

Part 1) of the theorem was shown in [98] for complex valued (sequential) almost automorphic functions. Part 2) of the theorem was stated and shown in [96] for complex valued function f as follows: f is (net) almost automorphic iff there is a continuous homomorphism $\phi : T \to G$, where G is a totally bounded group, and a complex valued function g which is continuous on G such that $f = g \circ \phi$. Part 3) of the theorem was stated and shown in [99] for complex valued functions f, g as follows: if $\phi : T \to C$ is (net) almost automorphic and $g : R(\phi) \to C$ is continuous, then $f = g \circ \phi$ is (net) almost automorphic. With proper modifications as the above, all these results can be extended to the ones stated in the theorem by using similar arguments as originally used in [96, 98, 99].

Parts 2) 3) of the above theorem is particularly useful in constructing non-quasi-periodic, almost automorphic functions with finite many frequencies. Let $\omega = (\omega_1, \omega_2, \cdots, \omega_k) \in R^k$ be a given non-resonant vector and $f : T^k \to R^n$ be a measurable function which is continuous on an orbit $\{\theta_0 + \omega t : t \in T\}$ of the quasi-periodic flow on T^k with the frequency ω, here $T = R$ or Z. As $\{\omega t : t \in R\}$ is embedded into T^k as a dense subgroup (or the map $R \to T^k : t \mapsto \theta_0 + \omega t$ is almost periodic hence almost automorphic), an immediate application of part 2) or 3) of Theorem 2.4 and Proposition 2.1 yields the following.

Corollary 2.1 *Let* f, ω, θ_0 *be as in the above. Then* $F(t) = f(\theta_0 + \omega t)$, $t \in R$ *or* Z, *is almost automorphic whose frequency module* $\mathcal{M}(F)$ *is contained in* $\mathcal{M}(\omega)$ *- the additive subgroup of* R *generated by* $\omega_1, \omega_2, \cdots, \omega_k$.

Remark 2.1 For each $\theta \in T^k$, the function $f(\theta + \omega t)$, where $t \in R$ or Z, may be regarded as a *weak quasi-periodic* function. Weak quasi-periodic orbits are known to existence in monotone twist maps on an annulus as orbits lying in the Aubry-Mather sets ([62]) and in scalar ODEs with quasi-periodic time dependence ([76]). The existence of such orbits in these dynamical systems were shown under certain monotonicity and properties that $f : T^k \to R^1$ is semi-continuous and admits bounded directional derivative along ω. Indeed, the semi-continuity implies that f admits a residual set Y_0 of points of continuity, and the directional differentiability together with the continuous dependence on initial conditions imply the continuity of f on each orbit $\{\theta_0 + \omega t\}$ if $\theta_0 \in Y_0$. It is clear that a weak quasi-periodic orbit $f(\theta + \omega t)$ becomes almost automorphic precisely when θ is a point of continuity of f, and becomes quasi-periodic only if f itself is a continuous function on the torus.

It is important to note that the family of weak quasi-periodic orbits such defined always contains almost automorphic ones. This is because of the harmonic similarity between almost automorphic functions with finite many frequencies and quasi-periodic ones. Indeed, if θ is not a point of continuity of f, then the function $f(\theta+\omega t)$ needs not even be recurrent and needs not admit any Fourier series whose Bochner-Fejer sum is pointwise convergent, due to the lack of continuity of f on the set $\{\theta + \omega t\}$.

2.3 Inherit properties. Similar to almost periodic minimal flows ([22]), it was shown in [9] that an almost automorphic minimal flow (X, T) also admits an inheritance property as follows.

Theorem 2.5 *Let S be a syndetic subgroup of T. Then (X, T) is almost automorphic minimal with maximal almost periodic factor (Y, T) iff (X, S) is almost automorphic minimal with maximal almost periodic factor (Y, S).*

The above inherit property can be easily extended to a Poincaré map associated with a real compact flow (X, R). Recall that a closed subset $Z \subset X$ is a *global cross section of* (X, R) if i) all orbits meet Z; ii) there is a positive continuous function $T : Z \to R$, called *first return time*, such that $z \cdot T(z) \in Z$ and $z \cdot t \notin Z$ for all $z \in Z$ and $0 < t < T(z)$. In case a global cross section Z exists, one can define the *Poincaré map* $P : Z \to Z : \ z \mapsto z \cdot T(z)$, which is a homeomorphism on Z. The following can be shown similarly to a special case considered in [9].

Proposition 2.2 *Suppose that (X, R) admits a global cross section Z and let $P : Z \to Z$ be the associated Poincaré map. Then (X, R) is almost automorphic minimal iff (Z, P) is.*

Let $P : S^1 \to S^1$ be an orientation preserving homeomorphism with irrational rotation number. Then P has a unique minimal set E which is either topologically conjugated to a pure rotation ([17]) or is an almost 1-1 extension of a pure rotation on S^1 ([60]), both with zero topological entropy. In any case, E is (discrete) almost automorphic.

Now consider a C^1 fixed-point-free flow π_t on the 2-torus T^2. Then π_t admits a Poincaré section and its *rotation vector*

$$\omega = \lim_{t \to \infty} \frac{\tilde{\pi}_t(y)}{t}$$

exists and is independent of $y \in T^2$, where $\tilde{\pi}_t : R^2 \to R^2$ denotes a continuous lift of π_t. We assume further that ω is non-resonant, or equivalently, π_t admits no periodic orbits. An easy application of the Denjoy theory implies that π_t admits a unique minimal set E which is topologically equivalent to a suspension of either a pure rotation or a Denjoy Cantor set on the circle (hence E is uniquely ergodic with zero topological entropy). Applying Proposition 2.2, we further conclude that E is actually almost automorphic

Similarly, let f be an area preserving, orientation preserving, boundary components preserving, twist homeomorphism of an annulus A. Let $\rho_0 < \rho_1$ denote the rotation numbers of f restricted to the lower and upper boundaries of A, respectively. Then it has been shown in [62] that for any $\rho \in [\rho_0, \rho_1]$ irrational, f admits a minimal set M_ρ which is topologically conjugated to either a pure rotation of the circle or a Denjoy set of the circle, with the rotation number ρ. Hence, M_ρ is an almost automorphic minimal set. In fact, Fourier series and their pointwise convergence for the almost automorphic orbits (referred to as quasi-periodic orbits by Mather) in M_ρ were also discussed in [62] which fit in the harmonic nature of general almost automorphic functions described in Theorem 2.1.

To summarize, we have the following.

Theorem 2.6 *A Denjoy minimal set of a circle homeomorphism, a Denjoy minimal set of a toral flow, and Aubry-Mather sets of an area preserving monotone twist map on an annulus are all almost automorphic minimal sets with zero topological entropy.*

3 Almost periodically forced circle flows

Consider an almost periodically forced circle flow generated by the following almost periodically forced scalar ODE:

$$\phi' = f(\phi, y \cdot t), \tag{3.1}$$

where $\phi \in R^1$, $f : R^1 \times Y \to R^1$ is Lipschitz continuous, periodic in the first variable with period 1. Using the identification $\phi \equiv \phi' \pmod 1$, (3.1) clearly generates a skew-product flow $\Lambda_t : S^1 \times Y \to S^1 \times Y$:

$$\Lambda_t(\phi_0, y_0) = (\phi(\phi_0, y_0, t), y_0 \cdot t) \tag{3.2}$$

when both $\phi(\phi_0, y_0, t)$ and ϕ_0 are identified modulo 1, where $\phi(t) = \phi(\phi_0, y_0, t)$ is the solution of (3.1) with $y = y_0$, $\phi(0) = \phi_0$.

3.1 Rotation number and mean motion. Using arguments of [49] and the Birkhoff ergodic theorem, it is easily seen that the equation (3.1), or equivalently, the skew-product flow (3.2) admits a well defined *rotation number*

$$\rho = \lim_{t\to\infty} \frac{\phi(\phi_0, y_0, t)}{t}$$

i.e., the limit exists and is independent of initial values $(\phi_0, y_0) \in R^1 \times Y$.

We say that the equation (3.1), or equivalently, the skew-product flow (3.2) admits *mean motion* if

$$\sup_{t\in R^1} |\phi(\phi_0, y_0, t) - \phi_0 - \rho t| < \infty$$

for all $(\phi_0, y_0) \in R^1 \times Y$.

Remark 3.1 1) It is well known that if either (3.1) is periodically dependent on time ([38]), i.e., $(Y, R) = (S^1, R)$ is a pure rotation of the circle, or almost periodically dependent on time but admits an almost periodic solution ([25]), then it always admits mean motion. This is however not the case for general almost periodic time dependence. As an extreme example, let $Y = T^k$, $f \equiv f(\omega t)$ be quasi-periodic in (3.1), where $\omega \in R^k$ is a non-resonant frequency vector. Then $\rho = [f]$ - the mean value of f, and, it is well known that $\phi(\phi_0, t) - \phi_0 - \rho t = \int_0^t (f(\omega s) - [f]) \mathrm{d}s$ can be unbounded if $k > 1$ unless some additional conditions are assumed on f and ω (e.g., f is sufficiently smooth and ω is Diophantine).

2) A natural question here is that under what conditions (3.1) can admit mean motion. Another question is that whether the mean motion property is generic, e.g., among the class of real analytic functions $f : S^1 \times T^k \to R^1$ endowed with sup-norm, with $(Y, R) = (T^k, R)$ being a fixed Diophantine, quasi-periodic flow.

We do not have answers to these questions but conjecture that the answer to the second question should be affirmative. Below, we give some equivalent conditions for mean motion to hold in (3.1).

Proposition 3.1 *The equation* (3.1) *admits mean motion iff there is a* $(\phi_0, y_0) \in R^1 \times Y$ *such that*

$$|\phi(\phi_0, y_0, t) - \phi_0 - \rho t|$$

is bounded for either $t \geq 0$ *or* $t \leq 0$.

Proof We note that, by the periodicity of f in ϕ, for any $t \in R^1$, $y \in Y$, if $|\phi_1^* - \phi_2^*| < l$ for some positive integer l, then also

$$|\phi(\phi_1^*, y, t) - \phi(\phi_2^*, y, t)| < l. \tag{3.3}$$

Without loss of generality, assume that $(\phi_0, y_0) \in R^1 \times Y$ is such that

$$\sup_{t \geq 0} |\phi(\phi_0, y_0, t) - \phi_0 - \rho t| < \infty.$$

Then it follows from the flow property that

$$\sup_{t \in R^1} |\phi(\phi_*, y_*, t) - \phi_* - \rho t| < \infty$$

for any $(\phi_*, y_*) \in \omega(\phi_0, y_0)$ with respect to the flow (3.2). Hence (3.1) admits mean motion by (3.3). □

Lemma 3.1 *Consider* (3.1). *Then*

$$|\int_Y \phi(\phi_0, y, t) \mathrm{d}y - \phi_0 - \rho t| \leq 4$$

for any $\phi_0 \in R^1$.

Proof Again, by the periodicity of f in ϕ, for any $t \in R^1$, $y \in Y$, if $\phi_1^* \equiv \phi_2^*$ (mod 1), then

$$\phi(\phi_1^*, y, t) - \phi_1^* = \phi(\phi_2^*, y, t) - \phi_2^*, \tag{3.4}$$

and, if $|\phi_1^* - \phi_2^*| < l$ for some positive integer l, then (3.3) holds.

Now, fix $t, \phi_0 \in R^1$. For any $y \in Y$, $0 \leq |s| \leq |t|$, let $0 \leq \phi_1, \phi_2 < 1$ be such that

$$\phi_1 \equiv \phi_0, \ \ \phi_2 \equiv \phi(\phi_0, y, s), \ \ (\mathrm{mod}\ 1).$$

It follows from (3.4) that

$$\phi(\phi_0, y \cdot s, t) - \phi_0 = \phi(\phi_1, y \cdot s, t) - \phi_1, \tag{3.5}$$

$$\phi(\phi_0, y, t + s) - \phi(\phi_0, y, s) = \phi(\phi_2, y \cdot s, t) - \phi_2. \tag{3.6}$$

Hence, by (3.3)-(3.6),

$$\begin{aligned}&|\phi(\phi_0, y, t+s) - \phi(\phi_0, y, s) - \phi(\phi_0, y \cdot s, t) + \phi_0|\\&= |\phi(\phi_2, y \cdot s, t) - \phi(\phi_1, y \cdot s, t) + \phi_2 - \phi_1|\\&\le |\phi(\phi_2, y \cdot s, t) - \phi(\phi_1, y \cdot s, t)| + |\phi_2 - \phi_1| \le 4,\end{aligned}$$

i.e.,

$$|\int_s^{t+s} \phi'(\phi_0, y, \lambda)\mathrm{d}\lambda - \phi(\phi_0, y \cdot s, t) + \phi_0| \le 4. \tag{3.7}$$

Let $T \neq 0$. An application of Frobine's theorem yields

$$\begin{aligned}&|\phi_0 - \int_0^t \frac{1}{T}\phi(\phi_0, y, \lambda)\mathrm{d}\lambda + \int_0^t \frac{\phi(\phi_0, y, T+s)}{T}\mathrm{d}s - \frac{1}{T}\int_0^t \phi(\phi_0, y \cdot s, t)\mathrm{d}s|\\&= |\frac{1}{T}\int_0^T \int_s^{t+s} \phi'(\phi_0, y, \lambda)\mathrm{d}\lambda\mathrm{d}s - \frac{1}{T}\int_0^T \phi(\phi_0, y \cdot s, t)\mathrm{d}s + \phi_0| \le 4. \qquad (3.8)\end{aligned}$$

Since

$$\lim_{T\to\infty} \frac{\phi(\phi_0, y, T+s)}{T} = \rho$$

uniformly in $y \in Y$, $s \in [0, t]$, and, by the Birkhoff ergodic theorem,

$$\lim_{T\to\infty} \frac{1}{T}\int_0^T \phi(\phi_0, y \cdot s, t)\mathrm{d}s = \int_Y \phi(\phi_0, y, t)\mathrm{d}y,$$

the lemma is proved by passing limit $T \to \infty$ in (3.8). □

Theorem 3.1 (3.1) *admits mean motion iff there is a* $(\phi_0, y_0) \in R^1 \times Y$ *such that*

$$|\phi(\phi_0, y_0, t) - \int_Y \phi(\phi_0, y, t)dy|$$

is bounded for $t \ge 0$ *or* $t \le 0$.

Proof This follows immediately from Proposition 3.1 and Lemma 3.1. □

3.2 Mean motion and almost automorphic dynamics. One significance for having mean motion is that it actually implies the existence of almost automorphic dynamics in the skew-product flow (3.2).

First, we consider the following almost periodically forced scalar ODE

$$x' = f(x, y \cdot t), \tag{3.9}$$

where $x \in R^1$, $f : R^1 \times Y \to R^1$ is Lipschitz continuous. Let $\pi_t : R^1 \times Y \to R^1 \times Y$:

$$\pi_t(x_0, y_0) = (x(x_0, y_0, t), y_0 \cdot t) \tag{3.10}$$

be the skew-product flow generated by (3.9), where $x(t) = x(x_0, y_0, t)$ denotes the solution of (3.9) with $y = y_0$ and $x(0) = x_0$. As (3.9) forms a special class of almost periodically forced, 1-dimensional, scalar parabolic PDEs with the Neumann boundary condition, the following lemma follows from the corresponding result in [91, 93]. As the proofs for the scalar ODE case is much less involved than the parabolic PDE case, we include them here for the readers' convenience.

Lemma 3.2 *Let* E *be a minimal set of* (3.10). *Then the following holds.*

1) (E, R) *is almost automorphic and in fact an almost* 1-1 *extension of* (Y, R);
2) E *is uniquely ergodic iff* Y_0 *admits full Haar measure, where* $Y_0 \subset Y$ *is the residual set corresponding to the singelton fibers.*

3) *Let* $x(x_0, y_0, t)$ *be an almost automorphic solution of* (3.9) *for some* $y_0 \in Y$. *Then* $\mathcal{M}(x) \subset \mathcal{M}(f)$.

Proof Let $p : R^1 \times Y \to Y$ be the natural projection and define

$$a_M(y) = \max\{x : (x, y) \in E \cap p^{-1}(y)\}, \; a_m(y) = \min\{x : (x, y) \in E \cap p^{-1}(y)\}.$$

1) Let 2^E be furnished with Hausdorff metric. Since the map $h : Y \to 2^E$: $y \mapsto E \cap p^{-1}(y)$ is upper semi-continuous, the points of continuity of h form a residual subset $Y_0 \subset Y$. Fix a $y \in Y_0$ and let $t_n \to \infty$ be such that $\pi_{t_n}(a_M(y), y) \to (a_m(y), y)$. By the continuity of h at y, there is a sequence of points $(x_n, y) \in E \cap p^{-1}(y)$ such that $\pi_{t_n}(x_n, y) \to (a_M(y), y)$. Now, by taking limits to both side of the inequalities

$$x(x_n, y, t_n) \le x(a_M(y), y, t_n),$$

we obtain that $a_M(y) \le a_m(y)$. Hence $a_M(y) = a_m(y)$ and $E \cap p^{-1}(y)$ is a singleton.

2) Let μ denote the Haar measure on Y. If $\mu(Y_0) = 1$, then E is clearly uniquely ergodic. Now suppose that $\mu(Y_0) = 0$ (note that Y_0 is invariant). Then the functionals $l_{M,m} : C(E) \to R$,

$$l_{M,m}(f) = \int_Y f(a_{M,m}(y), y) d\mu$$

would define two distinct invariant measures on E.

3) follows immediately from 1) and Proposition 2.1. □

Remark 3.2 1) If (3.9) is periodically dependent on time, then each minimal set of its associated skew-product flow becomes periodic (hence is uniquely ergodic and admits zero topological entropy). We conjecture that in the almost periodic time dependent case, the skew-product flow associated with (3.9) can have almost automorphic minimal sets which are non-uniquely ergodic and admit positive topological entropy.

2) Assume that (3.9) is quasi-periodically dependent on time, i.e., $(Y, R) = (T^k, R) : y \cdot t = y + \omega t$ is a quasi-periodic flow on the k-torus for some k with the toral frequency $\omega \in R^k$. It was shown in [76] that if (3.9) admits a bounded solution $x(t)$ for some $y = y_0$, then it has two families of weak quasi-periodic solutions $x_-(y + \omega t), x_+(y + \omega t)$, $y \in T^k$, in the sense of Remark 2.1, such that $x_-, x_+ : T^k \to R^1$ are lower and upper semi-continuous respectively, and,

$$\inf x(t) \le x_-(y) \le x_+(y) \le \sup x(t), \quad y \in T^k, \tag{3.11}$$

$$x_-(y_0 + \omega t) \le \phi(t) \le x_+(y_0 + \omega t), \quad t \in R. \tag{3.12}$$

Moreover, if $y_0 \in T^k$ is a point of continuity for x_- (x_+ resp.), then $x_-(y_0 + \omega t)$ ($x_+(y_0 + \omega t)$ resp.) becomes almost automorphic.

Theorem 3.2 *Assume that* (3.1) *admits mean motion and let* $E \subset S^1 \times Y$ *be a minimal set of the skew-product flow* (3.2). *Then the following holds.*

1) E *is almost automorphic with* $\mathcal{M}(E)$ *being the additive subgroup of* R *generated by* ρ *and* $\mathcal{M}(Y)$.
2) E *is uniquely ergodic iff its set of almost automorphic points admits full measure with respect to an invariant measure on* E.

Proof 1) Consider the change of variable $x = \phi - \rho t$. This transforms the equation (3.1) into

$$x' = f(x + \rho t, y \cdot t) - \rho \tag{3.13}$$

which generates a skew-product flow π_t on $R^1 \times Y$ or $R^1 \times (S^1 \times Y)$, in the same way as (3.10), depending on whether $\rho \in \mathcal{M}(Y)$.

Since (3.1) admits mean motion,

$$\int_0^t (f(\Lambda_t(\phi_0, y_0)) - \rho)\mathrm{d}t$$

is bounded for all $(\phi_0, y_0) \in E$. It follows from [33] that there is a continuous function $H : E \to R^1$ such that

$$\phi(\phi_0, y_0, t) - \phi_0 - \rho t = H(\Lambda_t(\phi_0, y_0)) - H(\phi_0, y_0) = \int_0^t (f(\Lambda_t(\phi_0, y_0)) - \rho)\mathrm{d}t,$$

i.e.,

$$\phi(\phi_0, y_0, t) - H(\Lambda_t(\phi_0, y_0)) = \phi_0 - H(\phi_0, y_0) + \rho t \tag{3.14}$$

for all $t \in R^1$ and all $(\phi_0, y_0) \in E$. We note that either $H(E) \times Y$ or $H(E) \times (S^1 \times Y)$ is a minimal set of π_t, which by Lemma 3.2 is almost automorphic. It follows from (3.14) and the definition of almost automorphic minimal set that E is almost automorphic.

Let $\mathcal{M}$ be the additive subgroup of R generated by ρ and $\mathcal{M}(Y)$. If $(\phi_0, y_0) \in E$ is an almost automorphic point, then $x(t) = H(\Lambda_t(\phi_0, y_0))$ is an almost automorphic solution of (3.13) for $y = y_0$. It follows from Lemma 3.2 that $\mathcal{M}(x) \subset \mathcal{M}$. Since $\Lambda_t(\phi_0, y_0) = (x(t) + \rho t + \phi_0 - x(0), y_0 \cdot t)$, we have by Proposition 2.1 that $\mathcal{M}(E) = \mathcal{M}(\Lambda_t(\phi_0, y_0)) = \mathcal{M}$.

2) follows from 1) and Lemma 3.2 2). □

We now consider a special case of (3.1) in which $Y = T^k$ and (Y, R) is a quasi-periodic minimal flow with non-resonant toral frequency $\omega = (\omega_1, \omega_2, \cdots, \omega_k)$. In this case, more can be said about the structure of a minimal set of (3.2).

Corollary 3.1 *Consider (3.1) with $(Y, R) = (T^k, R)$ being the quasi-periodic minimal flow on T^k with the frequency ω. Assume that (3.1) admits mean motion and let E be a minimal set of (3.2). Then there is an almost periodic minimal flow (X, R), a quasi-periodic minimal flow $(\hat{Y}, R)$, and flow epimorphisms $p : (E, R) \to (X, R)$, $q : (X, R) \to (\hat{Y}, R)$ for which the following holds.*

1) *p is almost 1-1 and q is open.*
2) *$\mathcal{M}(E) = \mathcal{M}(X) = \mathcal{M}(\hat{Y})$ and equals*

$$\{n_0\rho + n_1\omega_1 + \cdots + n_k\omega_k : n_0, n_1, \cdots, n_k \in Z\}.$$

3) *If ρ is rationally independent of ω, then X is a $(k+1)$-solenoid and $\hat{Y}$ is a $(k+1)$-torus. If ρ is rationally dependent of ω, then X is a k-solenoid and $\hat{Y}$ is a k-torus. In fact, if n_0 is the smallest positive number such that $n_0\rho + n_1\omega_1 + \cdots + n_k\omega_k = 0$, for some relatively prime numbers $n_1, \cdots, n_k$, then $\hat{Y}$ is an n_0-fold cover of $Y = T^k$.*

Proof 1) Let X be a maximal almost periodic factor of E. The existence of $p : (E, R) \to (X, R)$ follows from Theorem 2.3. In fact, consider the proximal relation

$$P(E) = \{(e_1, e_2) \in E \times E : \inf_{t \in R^1} d(\Lambda_t(e_1), \Lambda_t(e_2)) = 0\},$$

where d denotes the standard metric on $S^1 \times Y$. Then $P(E)$ is a closed (in particular, an equivalence), equivariance relation, and, X can be chosen to be $E/P(E)$ with flow being induced by Λ_t.

Define $q : X \to S^1 \times Y$,

$$q(p(\phi_0, y_0)) = (\phi_0 - H(\phi_0, y_0), y_0) \pmod 1, \ (\phi_0, y_0) \in E.$$

We need to verify that q is well defined. Let $(e_1, e_2) \in P(E)$. Then it is clear that there is a sequence $t_n \to \infty$ and $\phi_1, \phi_2 \in S^1$, $y_0 \in Y$ such that $e_1 = (\phi_1, y_0)$, $e_2 = (\phi_2, y_0)$, and, $d(\Lambda_{t_n}(e_1), \Lambda_{t_n}(e_2)) \to 0$. It follows from (3.14) that $\phi_1 - H(\phi_1, y_0) = \phi_2 - H(\phi_2, y_0)$ $(mod\, 1)$.

Define $\hat{Y} = q(X)$ and the flow on $(\hat{Y}, R)$ by

$$(q \circ p(\phi_0, y_0)) \cdot t = (\phi(\phi_0, y_0, t) - H(\Lambda_t(\phi_0, y_0), y_0 \cdot t) \pmod 1, \ (\phi_0, y_0) \in E.$$

It follows from (3.14) that $(\hat{Y}, R)$ is a parallel flow with the frequency vector (ρ, ω), which further shows that $\hat{Y}$ is either a $k+1$ or a k-torus and $(\hat{Y}, R)$ is quasi-periodic.

As an epimorphism between two almost periodic flows, q is clearly an open map.

2) By definition, $\mathcal{M}(E) = \mathcal{M}(X)$. Since $(\hat{Y}, R)$ is quasi-periodic with the frequency (ρ, ω), we have

$$\mathcal{M}(\hat{Y}) = \mathcal{M} \equiv \{n_0\rho + n_1\omega_1 + \cdots + n_k\omega_k : n_0, n_1, \cdots, n_k \in Z\}.$$

But by Theorem 3.2 1), $\mathcal{M}(E) = \mathcal{M}$. We thus have $\mathcal{M}(E) = \mathcal{M}(X) = \mathcal{M}(\hat{Y}) = \mathcal{M}$.

3) follows from 2) and the fact that

$$\dim X = \operatorname{rank} X' = \operatorname{rank}\mathcal{M}(X) = \operatorname{rank}\mathcal{M} = \dim Y.$$

□

Remark 3.3 1) Consider (3.1) with periodic time dependence, i.e., $(Y, R) = (S^1, R)$ is a pure rotation of the circle. Then the conclusions of both Theorem 3.2 and Corollary 3.1 hold without any condition, because the existence of mean motion for (3.1) is automatic in the case of periodic time dependence (Remark 3.1 1)). In fact, since the Poincaré map in this case is a circle diffeomorphism, this fact also follows from the classical Denjoy theory and Theorem 2.6. As each Denjoy minimal set admits zero topological entropy, so does each almost automorphic minimal set of (3.10) in the periodic time dependent case.

2) We note that the flow Λ_t can well have almost automorphic dynamics without admitting mean motion. An example was given in [47] in which an equation of type (3.1) with limit periodic time dependence was constructed so that the respective skew-product flow admits an almost automorphic minimal set as an almost 1-1 extension of the base flow but the corresponding rotation number is not contained in the frequency module of the base (compare to Theorem 3.2 1)). However, this should be a non-generic case as we conjectured earlier.

3) In the case of Corollary 3.1, the skew-product flow Λ_t is a toral flow on T^{k+1} whose rotation set is however a single point (ρ, ω). Toral flows and maps have been extensively studied in cases that rotation sets have non-empty interiors in which having positive topological entropy was shown to be typical (see [28, 55, 59] and references therein) but little has been known for cases of 'thin' or singleton rotation sets, except for certain cases of nearly quasi-periodic, 'twist' toral flows in which the existence of quasi-periodic orbits was shown in Moser's twist theorem ([69]).

In this regard, Corollary 3.1 seems to suggest that almost automorphic dynamics should be an important subject to look into in toral flows with 'thin' or singleton rotation sets.

4) Assume the conditions of Corollary 3.1. The result of [76] (see Remark 3.2 2)), when applying to (3.13), implies that any solution $\phi(t)$ of (3.1) corresponds to two weak quasi-periodic solutions of the form

$$\phi_{\pm}(t) = \rho t + x_{\pm}(\phi_0 + \rho t, y + \omega t), \quad \phi_0 \in S^1, \ y \in T^k,$$

where $x_{\pm} : T^{k+1} \to R^1$ are semi-continuous functions lying in between $\inf(\phi(t) - \rho t - \phi_0)$ and $\sup(\phi(t) - \rho t - \phi_0)$. Moreover, a such weak quasi-periodic solution becomes almost automorphic if $(\phi_0, y) \in T^{k+1}$ is a point of continuity of $x_{\pm}$.

5) The study of almost automorphic dynamics for the almost periodically forced skew-product circle flow (3.2) is also closely related to the Floquet theory for a two dimensional, almost periodic linear system of ODEs of the form

$$x' = a(y \cdot t)x, \ x \in R^2, \ tr(a) \equiv 0. \tag{3.15}$$

It is well known that Floquet theory does not hold for such a system in general because the linear skew-product flow (cocycle) generated by (3.15) can well have interval Sacker-Sell spectrum ([82]). On one hand, there are cases where Floquet theory does hold (see [45, 93] for certain almost periodic cases and [21, 39, 51, 53, 73, 104] for recent develop in quasi-periodic reducibility), and, on the other hand, as proposed by Johnson ([45]) one can seek for weak Floquet forms concerning finding an almost automorphic (not almost periodic) strong Perron transformation which transforms (3.15) into a canonical (upper triangular or diagonal) form with almost periodic coefficients. In the later case, an almost automorphic strong Perron transformation is determined by an almost automorphic minimal set (if it exists) of the skew-product circle flow of the form (3.2) which is generated by the angular equation associated with the polar-angle reduction of (3.15). We refer the readers to [45, 85] for more discussions in this regard.

4 Almost automorphic oscillations

Multi-frequency oscillations arise naturally in many electrical and mechanical systems. Physical examples include electric networks, power systems, quasi-periodic velocity field of fluid flows, plasma dynamics, mechanical vibrations and coupled biological oscillators, etc., among which almost periodically forced second order oscillators form an important class. In engineering applications, periodic oscillations are referred to as harmonic oscillations and almost periodic ones are interpreted as harmonic oscillations covered with small 'noise'. Accordingly, almost automorphic oscillations can be regarded as harmonic ones covered with big 'noise'.

There have been extensive studies on periodically forced second order oscillators on issues such as strange attractors, period doubling cascades, mode locking, quasi-periodicity, intermittency, etc. (see [15, 16, 34, 36] and references therein). Concerning almost periodically forced second order oscillators, attention has been mainly on the existence of almost periodic oscillations (e.g., [4, 10, 16, 69]) and chaotic dynamics (e.g., [8, 7, 14, 101]). In general, as suggested by recent numerical works, (e.g., [13, 14, 18, 35, 80, 81]), typical non-chaotic dynamics of almost periodically (in particular, quasi-periodically) forced nonlinear, second order oscillators can be far more complicated than almost periodic ones. Here we would like to argue that almost automorphic dynamics should be a main (if not the only) factor largely responsible for such non-chaotic dynamical complexities arising in almost periodically (even periodically) forced second order oscillators.

4.1 Linear and quasi-linear oscillators. Consider an almost periodic linear oscillator

$$\ddot{x} + a(y \cdot t)\dot{x} + b(y \cdot t)x = f(y \cdot t),\ y \in Y, \tag{4.1}$$

where $a, b, f \in C(Y)$. In terms of the phase variable $u = (x, \dot{x})$, the equation (4.1) becomes the linear planar system

$$\dot{u} = A(y \cdot t)u + B(y \cdot t), \tag{4.2}$$

where

$$A(y) = \begin{pmatrix} 0 & 1 \\ -b(y) & -a(y) \end{pmatrix},\quad B(y) = \begin{pmatrix} 0 \\ -f(y) \end{pmatrix}.$$

The system (4.2) clearly generates a skew-product flow π_t on $R^2 \times Y$. The following was stated in [93], Remark 4.3.

Proposition 4.1 *Assume that for some $y_0 \in Y$, $\int_0^t a(y_0 \cdot s)ds$ is bounded and the equation* (4.1) *corresponding to y_0 has a bounded solution with bounded derivative, then there is an almost automorphic minimal set $E \subset R^2 \times Y$ of the flow π_t satisfying $\mathcal{M}(E) \subset \mathcal{M}(Y)$.*

The proposition implies the existence of almost automorphic solutions in (4.1) for residually many $y \in Y$ which harmonically respond to the external frequencies (the frequency module of Y). The existence of almost automorphic dynamics in the case described in the proposition is particularly significant when the homogeneous part of (4.2) admits a nontrivial bounded solution and an unbounded solution for all y (especially when its Sacker-Sell spectrum is a non-degenerate interval) since the existence of almost periodic ones are not generally expected in this case.

Consider the quasi-linear case

$$\ddot{x} + \alpha\dot{x} = f(x, y \cdot t), \tag{4.3}$$

where $\alpha \neq 0$ is a constant. The equations generate a planar skew-product flow, again denoted by π_t, on $R^2 \times Y$, the same way as above. It is shown in [77] that such equations need not admit any almost periodic solutions and the usual upper-lower solutions techniques for periodic time dependence are not applicable to (4.3) to yield the existence of almost periodic solutions. Nevertheless, using the approach in [83], one can obtain almost automorphic solutions of (4.3) for some y (or equivalently, an almost automorphic minimal set of π_t) in between upper and lower solutions.

In the case that f is quasi-periodic time dependent in (4.3), i.e., $(Y, R) = (T^k, R) : y \cdot t = y + \omega t$, for some k, is a quasi-periodic flow on the torus T^k with toral frequency $\omega \in R^k$, the following was shown in [76].

Theorem 4.1 *Assume that for some $y_0 \in T^k$ the equation* (4.3) *has a bounded solution $x(t)$ such that*

$$\frac{f(x_1, y) - f(x_2, y)}{x_1 - x_2} \geq -\frac{\alpha^2}{4}$$

for all $\inf x(t) \leq x_2 < x_1 \leq \sup x(t)$ and $y \in T^k$. Then (4.3) *admits two families of weak quasi-periodic solutions $x_-(y + \omega t)$, $x_+(y + \omega t)$, $y \in T^k$, satisfying* (3.11), (3.12), *where $x_-, x_+ : T^k \to R^1$ are lower and upper semi-continuous, respectively.*

As remarked in Remark 2.1, each continuity point $y_0 \in T^k$ of $x_\pm$ yields an almost automorphic solution of (4.3) for $y = y_0$, which then gives rise to an almost automorphic minimal set of the skew-product flow $(R^2 \times T^k, R)$ generated by (4.3).

4.2 Damped nonlinear oscillators. Below, we let $\varepsilon \in [-\varepsilon_0, \varepsilon_0]$ be a small parameter.

There are two types of almost periodically forced nonlinear oscillators for which the results in the previous section can be applied naturally. One type concerns a damped, perturbative, nonlinear oscillator of the form:

$$x'' + a(x, x') = \varepsilon b(x, x', y \cdot t), \ \ x \in R^1, \tag{4.4}$$

where $a : R^2 \to R^1$ is C^1, $b : R^2 \times Y \to R^1$ are uniformly Lipschitz in $y \in Y$ and C^1 in other variables whose derivatives are also continuous in $y \in Y$.

Proposition 4.2 *Suppose that as $\varepsilon = 0$ (4.4) admits a hyperbolic periodic solution. Then for ε sufficiently small it has an integral manifold on which the equation reduces to the form*

$$\phi' = 1 + \varepsilon f(\phi, y \cdot t, \varepsilon), \tag{4.5}$$

where $f : S^1 \times Y \times [-\varepsilon_0, \varepsilon_0] \to R^1$ is C^1 in ϕ, ε and Lipschitz in y.

Proof The condition implies that the unperturbed part of the system

$$\begin{cases} u' &= v, \\ v' &= -a(u, v) + \varepsilon b(u, v, y \cdot t) \end{cases} \tag{4.6}$$

admits an invariant normally hyperbolic cycle. Let $\phi \in S^1$ be an angular coordinate parameterizing the cycle and I be an associated normal coordinate. Then in the vicinity of the cycle there is a change of coordinate of the form

$$\begin{pmatrix} u \\ v \end{pmatrix} = h(\phi) + g(I, \phi, y \cdot t),$$

where $h : S^1 \to R^2, g : R^1 \times S^1 \times Y \to R^2$ are C^1 in ϕ, Lipschitz in y, and $g(0, \phi, y) \equiv 0$, under which (4.6) has the form

$$\begin{cases} I' &= \lambda I + \tilde{g}(I, \phi) + \varepsilon \tilde{g}(I, \phi, y \cdot t, \varepsilon), \\ \phi' &= 1 + \tilde{f}(I, \phi) + \varepsilon \tilde{f}(I, \phi, y \cdot t, \varepsilon), \end{cases} \tag{4.7}$$

where $\lambda \neq 0$, $\tilde{g} = O(|I|^2)$, $\tilde{f} = O(|I|)$, and, for fixed I, ε, all terms are multiple periodic with periods 1 ([105]). The proposition now follows from the general integral manifolds theorem ([37, 105]). □

Example (*van der Pol equation*): A special case of (4.4) is the following quasi-periodically forced van der Pol equation

$$x'' - \alpha(1 - x^2)x' + x = \varepsilon b(x, x', y \cdot t), \qquad x \in R^1, \tag{4.8}$$

which models an electric circuit with a triode valve, the resistive properties of which change with current, where $\alpha > 0$ is a constant and b is as in (4.4). It is well known that the skew-product system associated with (4.8) admits a normally hyperbolic, asymptotically stable limit cycle for $\varepsilon = 0$, hence admits an integral manifold with flow governed by an equation of form (4.5). Thus, by Theorems 3.2, if for ε sufficiently small (4.5) admits mean motion, then almost automorphic oscillations will occur in the skew-product flow on $R^2 \times Y$ generated by the system associated with (4.8). □

Another type of application of the Theorem 3.2 concerns damped, almost periodically forced non-linear oscillators which admit certain periodic structures. Consider the following system

$$x'' + \beta x' + a(x, y\cdot t) = \varepsilon b(x, x', y\cdot t), \ x \in R^1, \tag{4.9}$$

where $\beta > 0$ is a constant, $a : S^1 \times Y \to R^1, b : S^1 \times R^1 \times Y \to R^1$ are bounded, uniform Lipschitz in y and C^1 in other variables whose derivatives are uniformly bounded. We re-write (4.9) into the following equivalent system

$$\begin{cases} \phi' &= \psi, \\ \psi' &= -\beta\psi - a(\phi, y\cdot t) + \varepsilon b(\phi, \psi, y\cdot t), \end{cases} \tag{4.10}$$

where $\phi \in S^1$.

The following result is a slight generalization of a similar result contained in [87] for the quasi-periodic case.

Proposition 4.3 *Let* $\alpha = \sup_{S^1\times Y} |\dfrac{\partial a}{\partial \phi}|$. *If* $\beta > 2\sqrt{\alpha}$, *then as* ε *sufficiently small,* (4.12) *admits an attracting integral manifold on which the equation reduces to the form*

$$\phi' = f(\phi, y\cdot t, \varepsilon), \tag{4.11}$$

where $f : S^1 \times Y \times [-\varepsilon_0, \varepsilon_0] \to R^1$ *is* C^1 *in* ϕ, ε *and Lipschitz in* y.

Proof We only show the existence and Lipschitz continuity of the integral manifold.

After reversing time $t \to -t$, (4.10) becomes

$$\begin{cases} \phi' &= -\psi, \\ \psi' &= \beta\psi + a(\phi, y\cdot(-t)) - \varepsilon b(\phi, \psi, y\cdot(-t)). \end{cases} \tag{4.12}$$

Fix $\eta > 0, 0 < \alpha' < \alpha$ such that

$$\frac{\beta - \sqrt{\beta^2 - 4(\alpha-\alpha')}}{2} < \eta < \min\{\beta - \sqrt{\alpha}, \frac{\beta + \sqrt{\beta^2 - 4(\alpha-\alpha')}}{2}\}.$$

One can certainly make ε sufficiently small so that

$$\begin{aligned} &\frac{\alpha + \varepsilon\gamma((\eta + \frac{\alpha'}{\beta} + 2) + \varepsilon(\eta + \frac{\alpha'}{\beta}))}{\beta - \eta} \le \eta, \\ &\frac{\alpha + \varepsilon\gamma}{(\beta-\eta)^2} + \frac{\varepsilon\gamma}{\beta} < 1, \end{aligned} \tag{4.13}$$

where $\gamma = \text{Lip}\, b$.

Let X be the set of Lipschitz continuous functions h on $S^1 \times Y$ with $\max|h| \le K \equiv (\max|a| + \varepsilon \max|b|)/\beta$ and $\text{Lip}_\phi h \le \eta, \text{Lip}_y h \le \eta + \alpha'/\beta$. With the sup-norm $\|\cdot\|$, X is a complete metric space.

Define mapping $T : X \to X$:

$$Th(\phi_0, y_0) = -\int_0^\infty e^{-\beta s}(a(\phi(s), y_0\cdot(-s)) - \varepsilon b(\phi(s), h(\phi(s), y_0\cdot(-s)), y_0\cdot(-s)))\text{d}s, \tag{4.14}$$

where $\phi(t) = \phi(\phi_0, y_0, h, t)$ is the solution of

$$\begin{cases} \phi' &= -h(\phi, y_0\cdot(-t)), \\ \phi(0) &= \phi_0. \end{cases} \tag{4.15}$$

We first check that T is well defined. Let $h \in X$. It is easy to see that $\|Th\| \leq K$ and

$$\mathrm{Lip}_{\phi_0}(Th) \leq \int_0^\infty e^{-\beta s}(\alpha+\varepsilon\gamma\eta+\varepsilon\gamma)\mathrm{Lip}_{\phi_0}\phi(s)\mathrm{d}s, \tag{4.16}$$

$$\mathrm{Lip}_{y_0}(Th) \leq \int_0^\infty e^{-\beta s}((\alpha+\varepsilon\gamma\eta)\mathrm{Lip}_{y_0}\phi(s)+\varepsilon(\alpha'+\eta+2\gamma))\mathrm{d}s. \tag{4.17}$$

Using (4.15) and applying Gronwall's inequality ([38]), we have $\mathrm{Lip}_{\phi_0,y_0}\phi(t) \leq e^{\eta t}$, which, when substituting into (4.16), (4.17) yields that

$$\mathrm{Lip}_{\phi_0}(Th) \leq \eta, \ \ \mathrm{Lip}_{y_0}(Th) \leq \eta + \frac{\alpha'}{\beta}.$$

Next, we show that T is a uniform contraction mapping. Let $h_j \in X$ and $\phi_j(t) = \phi(\phi_0, y_0, h_j, t)$ be the solution of (4.15) with $h := h_j$ for $j = 1, 2$ respectively. An application of Gronwall's inequality yields that

$$|\phi_1(t) - \phi_2(t)| \leq \|h_1 - h_2\| t e^{\eta t}.$$

It then follows from (4.14) that

$$\begin{aligned}\|Th_1 - Th_2\| &\leq \int_0^\infty e^{-\beta s}((\alpha+\varepsilon\gamma)|\phi_1(s)-\phi_2(s)| + \varepsilon\gamma\|h_1-h_2\|)\mathrm{d}s \\ &\leq \int_0^\infty e^{-\beta s}((\alpha+\varepsilon\gamma)se^{\eta s}+\varepsilon\gamma)\mathrm{d}s\|h_1-h_2\| \\ &= (\frac{\alpha+\varepsilon\gamma}{(\beta-\eta)^2}+\frac{\varepsilon\gamma}{\beta})\|h_1-h_2\|.\end{aligned}$$

Thus, by (4.13), T is a uniform contraction mapping.

Now, let h_* be the fixed point of T. It is easy to see from (4.14) that

$$M = \{(h_*(\phi, y), \phi, y) : \phi \in S^1, y \in Y\}$$

is an (topological) invariant manifold of the skew-product flow generated by (4.12), i.e., $h_*(\phi, y\cdot(-t))$ defines an integral manifold of (4.12) with flow governed by the equation

$$\phi' = -h_*(\phi, y\cdot(-t)).$$

Changing t back to $-t$, we then have

$$\phi' = h_*(\phi, y\cdot t)$$

which is invariant to the system (4.10).

To show that M is attracting, we let $x = \psi - h(\phi, y\cdot t)$. By (4.10),

$$x' = -\beta x + \varepsilon b(\phi, h(\phi, y\cdot t), y\cdot t).$$

It follows from Gronwall's inequality that

$$|x(t)| \leq x(0)e^{-(\beta-O(\varepsilon))t} \to 0, \ \text{ as } \ t\to\infty.$$

□

The following example was considered in [87] with respect to quasi-periodic forcing, which also holds in the case of almost periodic forcing.

Example (*Josephson junction*): A particular case of (4.7) is the almost periodically forced Josephson junction equation, or, Damped pendulum:

$$x'' + \beta x' + \sin x = F(y\cdot t) \tag{4.18}$$

which arises in many applications ranging from supersensitive detectors to superfast computers, where $F : Y \longrightarrow R^1$ is a Lipschitz function. By considering the change of variable $x \longrightarrow 2\pi x$, the equation (4.18) becomes

$$x'' + \beta x' + \frac{\sin 2\pi x}{2\pi} = \frac{F(y \cdot t)}{2\pi}$$

which is in the form of (4.9) with $b \equiv 0$ and

$$a(x, y) \equiv \frac{1}{2\pi}(\sin 2\pi x - F(y)).$$

Applying Proposition 4.3 to (4.18) for $\alpha = 1$, we then obtain a desired integral manifold for $\beta > 2$, with flow governed by an equation in the form of (4.11). Thus, by Theorems 3.2 the skew-product flow generated by the system associated with (4.18) will admit almost automorphic oscillations as long as it admits mean motion.

As observed in [87], mean motion holds on the integral manifold above when $\|F\|$ is sufficiently small. To see this, we note that when $F \equiv 0$, the integral manifold clearly contains a relative equilibrium $\phi = 0$ which is also hyperbolic. It follows that as long as $\|F\|$ is sufficiently small the skew-product flow on the integral manifold admits an almost periodic minimal set which is an 1-cover of the base. Hence by Remark 3.2 1) the skew-product flow on the integral manifold admits mean motion with rotation number $\rho = 0$. □

Remark 4.1 Equations (4.8), (4.18) have been used as models to numerically study the existence of so-called non-chaotic strange attractors which typically arise in damped, quasi-periodically forced oscillators such as the Josephson junction and van der Pol (e.g., [14, 35, 81]). Roughly speaking, a non-chaotic strange attractor is one which is geometrically strange (it is neither a finite set, a closed curve, a smooth surface, nor a volume bounded by a piecewise smooth closed surface) but not dynamically chaotic (i.e., the Lyapunov exponents on the attractor are non-positive).

In the case that mean motion exists the above discussions suggest that basic dynamics on a non-chaotic strange attractor of almost perdocially forced Josephson junction or van der Pol are almost automorphic (i.e., such an attractor is formed by almost automorphic minimal sets along with 'connecting orbits'). The topological structure of such an attractor will depend on whether the respective rotation number is in resonant with the forcing frequencies. As an almost 1-1 extension of a solenoid as Corollary 3.1 suggests, an almost automorphic minimal set on such a non-chaotic strange attractor can be a fractal (geometric complexity) and be almost everywhere non-locally connected (topological complexity). Therefore, the structures of almost automorphic minimal sets should be responsible for part of the complexity of a non-chaotic strange attractor.

To be able to capture the full complexity of a non-chaotic strange attractor, one should also consider the existence of weak quasi-periodic orbits lying in the attractor (see Remark 3.3 4), because, as shown in [76], an quasi-periodically forced system can have a complicated attractor with 'simple' almost automorphic orbits.

4.3 Chaotic almost automorphic oscillations. Consider a planar system of the form

$$\dot{u} = f(u) + \varepsilon g(u, y \cdot t), \; u \in R^2. \tag{4.19}$$

Assume that for $\varepsilon = 0$ (4.19) is a Hamiltonian system admitting a hyperbolic equilibrium u_0 along with a homoclinic orbit $u = \gamma(t)$. It follows that the skew-product flow $(R^2 \times Y, \pi_t)$ generated by (4.19) admits a smooth family of almost periodic minimal sets Λ_ε which are 1-covers of Y and skew-hyperbolic (i.e., hyperbolic in the u-direction, see e.g., [105]).

Define the Melnikov functional

$$M(y)(t) = \int_{-\infty}^{\infty} f(\gamma(s)) \wedge g(\gamma(s), y \cdot (t+s))ds.$$

Suppose that the zero set

$$Z = \{y \in Y : M(y)(0) = 0\}$$

is *simple*, i.e.,

$$\frac{d}{dt} M(y)(t)|_{t=0} \neq 0$$

for all $y \in Z$. Then Z is a global cross section for the flow on Y ([66]). Hence $\Sigma = R^2 \times Z$ is a global cross section for the skew-product flow π_t on $R^2 \times Y$. Let $\Psi : R^2 \times Z \to R^2 \times Z$ be the induced Poincaré map.

Theorem 4.2 ([66]) *Suppose that Z is simple. Then there exists an $\varepsilon_0 > 0$ such that for every $0 < |\varepsilon| < \varepsilon_0$ there is a positive integer n_0 such that for all $n \geq n_0$ there is a compact invariant set $\Omega_n \subset R^2 \times Z$ of Ψ whose dynamics is topologically conjugated to the skew-shift map $A \otimes \eta$ on the Bernoulli bundle $\Sigma \times Z$, where A is the full shift on the space of Σ of bi-infinite sequences on n symbols and η is the Poincaré map on Y based on the global cross section Z.*

Chaotic almost automorphic minimal sets have been known to largely exist in symbolic flows. It was first observed in [30, 60] that almost automorphic symbolic flows can exhibit positive topological entropy and lack unique ergodicity. Recently, Toeplitz sequences as a special class of almost automorphic sequences have received considerable attention (see [1, 19, 20, 32, 43, 44, 60, 78, 103] and references therein). While regular Toeplitz sequences are uniquely ergodic with zero topological entropy, it was shown that irregular Toeplitz sequences are typically not uniquely ergodic and exhibit positive topological entropy, though they may also be uniquely ergodic with positive entropy ([32, 43]).

In [9], chaotic almost automorphic symbolic arrays were also shown to be mainly responsible for the complexity (e.g., spatial chaos) of certain lattice dynamical systems. As a particular case, we have the following result for symbolic sequences.

Theorem 4.3 *Consider the full shift dynamics A on the space Σ of bi-infinite sequences of n-symbols. Let Σ_+ denote the space of half infinite sequences of the same symbols. Then the following holds.*

1) *There is a dense set $\Lambda \subset R$ of irrational numbers such that for any given $\rho \in (0, \log n)$ and $\gamma \in \Lambda$ there exists a residual set $\mathcal{R} \subset \Sigma_+$ such that each $\omega \in \mathcal{R}$ corresponds to an almost automorphic minimal set $M(\omega)$ of (Σ, A) which has the (discrete) circle rotation (S^1, γ) as its maximal almost periodic factor, is not uniquely ergodic and has topological entropy $h(M(\omega)) > \rho$.*
2) *Given a prime number p there is a residual set $\mathcal{R} \subset \Sigma_+$ such that each $\omega \in \mathcal{R}$ corresponds to an almost automorphic minimal set $M(\omega)$ of (Σ, A) which has the p-adic odometer group $(\Delta, \oplus)$ as its maximal almost periodic factor, is not uniquely ergodic and has topological entropy $h(M(\omega)) = \log n$.*

The theorem says that the response of chaos in symbolic dynamics to almost automorphic sequences can be made optimal: there are almost automorphic sequences with one frequency whose orbit closures admit topological entropy arbitrarily close to the maximal entropy of the system and there are almost automorphic sequences with infinitely many frequencies whose orbit closures attain the maximal entropy. Recall that having infinitely many periodic sequences is a main feature of chaos in symbolic dynamics. A chaotic almost automorphic sequence attaining maximal entropy should be the (non-uniform) limit of these periodic ones. This links the two commonly adopted definitions of chaos for symbolic dynamics, i.e., the one defined by positive topological entropy and the one requires sensitivity dependence on initial conditions, topological transitivity, and a dense set of periodic orbits.

In the periodically forced case, the Melnikov functional above reduces to the classical Melnikov function. It is known that if the Melnikov function admits a simple zero then the Poincaré map associated with (4.19) admits Smale horseshoe which is topologically conjugated to the full shift on two symbols ([36]).

Using Proposition 2.2 and Theorem 4.3, we then have the following.

Theorem 4.4 *Consider the skew-product flow $(R^2 \times Y, \pi_t)$ generated by (4.19) with periodic time dependence, i.e., $(Y, R) = (S^1, R)$ is a pure rotation of the circle. If Z is simple, then there is an $\varepsilon_0 >$ such that the following holds for all $0 < |\varepsilon| \leq \varepsilon_0$.*

1) *There is a dense set $\Lambda \subset R$ of irrational numbers such that for any given $\rho \in (0, \log 2)$, $\gamma \in \Lambda$ there is an almost automorphic minimal set E_ε, with $\mathcal{M}(E_\varepsilon)$ being the additive subgroup of R generated by $\{\gamma, 1\}$, which is an almost 1-cover of the 2-torus, is not uniquely ergodic and has the topological entropy $h(E_\varepsilon) > \rho$.*
2) *Given a prime number p, there is an almost automorphic minimal set E_ε, with $\mathcal{M}(E_\varepsilon)$ being the additive subgroup of R generated by $\mathcal{P} = \{1/p^l : l = 0, 1, \cdots\}$, which is an almost 1-cover of the 2-solenoid, is not uniquely ergodic and has the topological entropy $h(E_\varepsilon) = \log 2$.*

Remark 4.2 The above result shows the existence of rather chaotic almost automorphic oscillations in periodically forced nonlinear oscillators of form (4.19) if the Melnikov function admits a simple zero. A natural problem is to extend Theorem 4.3 to Bernoulli bundles then use Theorem 4.2 to obtain chaotic almost automorphic oscillations in the almost periodically forced case.

4.4 Hamiltonian systems. Hamiltonian systems is an area in which the existence of almost automorphic dynamics is much less understood, especially for higher degrees of freedom. The well known Denjoy and Aubry-Mather theories have already given strong evidence for such existence in the case of lower degrees of freedom.

As shown in [70], given an area preserving, orientation preserving, boundary components preserving, twist homeomorphism ϕ of the annulus $A = \{(x, u) \in S^1 \times R^1, a \leq u \leq b\}$, there exists a time periodic Hamiltonian $H(x, u, t)$, $(x, u) \in S^1 \times R^1$ of period 1, with $H_x|_{u=a,b} = 0$, $H_{uu} > 0$, whose Poincaré map of the associated motion agrees with ϕ. This, when coupled with Proposition 2.2 and Theorem 2.6, provides an evidence for the existence of almost automorphic solutions of two frequencies in Hamiltonian systems of one and one-half degrees of freedom.

Remark 4.3 Given a time quasi-periodic Hamiltonian system $H(x, u, \omega t)$, $(x, u) \in S^1 \times R^1$ with k (> 1) frequencies $\omega \in R^k$, a natural problem is to develop a variational technique to show the existence of almost automorphic minimal sets (higher dimensional analogue to the Aubry-Mather sets) of $k+1$ frequencies (ρ, ω), where ρ is a rotation number, in the associated Hamiltonian skew-product flow under suitable conditions.

This problem is already significant when $H(x, u, \omega t)$ takes the special form $u^2/2 + V(x, \omega t)$, corresponding to the second order undamped, quasi-periodic time dependent, nonlinear oscillator

$$\ddot{x} + V_x(x, \omega t) = 0, \; x \in S^1, \; V \in C^\infty(T^{k+1}, R^1), \tag{4.20}$$

which can be viewed as a degenerate elliptic equation on the $k+1$-torus. The existence of Aubry-Mather (in fact, almost automorphic) action minimizing solutions for an elliptic equation on $k+1$-torus has been shown in [71, 72] using a variational approach. Similar to [71], a crucial step in tackling the above problem with respect to (4.20) would be to obtain the mean motion property, i.e., for any lifted solution $x(t) \in R^1$ of (4.20) or equivalently any minimizer of the respective Lagrangian there is a unique rotation number ρ such that $|x(t) - \rho t|$ is bounded. This is also closely related to the action-minimizing problems in Hamiltonian systems of higher degrees of freedom (see [63, 64, 65]).

An easier problem is to consider a variational problem for (4.20) without assuming the periodic dependence of V on x. It seems that under reasonable conditions one can obtain almost automorphic dynamics of the respective skew-product flow which respond to ω harmonically. We will discuss this problem in more detail elsewhere.

Almost automorphic dynamics can also exist in Poisson-Hamilton systems defined on a Poisson manifold rather than a symplectic one. In action-angle coordinates, such a system has the form

$$\begin{pmatrix} \dot{y} \\ \dot{x} \end{pmatrix} = J(y, x) \nabla H(y, x), \;\; y \in R^l, \, x \in T^n, \tag{4.21}$$

where J is called a *structure matrix* - a skew-symmetric matrix-valued function satisfying the Jacobi identity (see [57, 58] and references therein). Now let us consider the action-angle-angle case (i.e. $l = 1, n = 2$) with $H_{yy} \neq 0$ for all (x, y) and

$$J = \begin{pmatrix} 0 & -\alpha a(y) & -\beta a(y) \\ \alpha a(y) & 0 & -\gamma \\ \beta a(y) & \gamma & 0 \end{pmatrix},$$

where α, β, γ are constants, $|\alpha| + |\beta| + |\gamma| \neq 0$, and $a(y) \neq 0$ for all y. Then it is clear that each energy surface $M_E = \{H(y, x) = E\}$ is an invariant 2-torus. Moreover, if (α, β) is non-resonant, then the flow on any such invariant 2-torus M_E admits neither fix points nor periodic orbits. By Theorem 2.6, M_E admits a unique 2-frequency almost automorphic minimal set which is not necessarily almost periodic in general. However, for certain nearly integrable case, it was shown in [58] that if (α, β) is Diophantine, then the majority of these 2-tori will be quasi-periodic.

Remark 4.4 1) A related problem is to develop a variational framework for Poisson-Hamilton systems and use it to study the existence of almost automorphic dynamics.

2) Almost automorphic dynamics may exist even in quasi-periodically forced or autonomous, nearly integrable Hamiltonian systems, as possible intermediate orbits in between KAM tori and stochastic layers. Such intermittency has been investigated numerically as the so-called *Cantori* whose topological natures are quite similar to that of almost automorphic minimal sets. The existence of non-KAM intermediate orbits has been theoretically justified in [40] based on an idea of J. C. Yoccoz as the so-called *Lagrangian tori* using a topological approach for fixed small non-integrable Hamiltonian perturbations and in [11] using a variational approach when the perturbations increase in sizes and the KAM theory becomes invalid. Indeed, the approach of taking weak limits to KAM tori adopted in [40] is in spirit quite similar to Theorem 2.4 1) in obtaining almost automorphic functions (and minimal sets). It would be interesting to know whether some of these immediate orbits can actually be almost automorphic. Linking this problem with Theorem 2.4 1), an immediate question is whether some KAM tori form a joint almost automorphic sequence.

References

[1] J.-P. Allouche and R. Bacher, Toeplitz sequences, paperfolding, towers of hanoi and progression-free sequences of integers, *L'Enseignement Mathématique* **38** (1992).

[2] G. Allud and E. S. Thomas, Jr., Almost periodic minimal sets, *J. Differential Equations* **15** (1974).

[3] A. I. Alonso and R. Obaya, The structure of the bounded trajectories set of a scalar convex differential equations, preprint, 2002.

[4] H. Amann, Non-linear almost periodic oscillations, Halsted Press, NY (1973).

[5] S. Aubry, The devil's staircase transformation in incommensurate lattices, *The Riemann problem, complete integrability and arithmetic applications*, Lecture Notes in Mathematics **925** Springer Verlag, 1982.

[6] J. Auslander, Minimal flows and their extensions, North-Holand, Amsterdam - New York - Oxford (1985).

[7] D. Beigie, A. Leonard and S. Wiggins, Chaotic transport in the homoclinic and heteroclinic tangle of quasi-periodically forced two-dimensional dynamical systems, *Nonlinearity* **4** (1991).

[8] A. Belogortsev, Bifurcations of tori and chaos in the quasi-periodically forced Duffing oscillator, *Nonlinearity* **5** (1992).

[9] A. Berger, S. Siegmund and Y. Yi, On almost automorphic dynamics in symbolic lattices, submitted.

[10] M. S. Berger and Y. Y. Chen, Forced quasi-periodic and almost periodic oscillations of nonlinear Duffing equations, *Nonl. Anal. TMA* **19** (1992).

[11] U. Bessi and V. Semijopuva, Minimal orbits close to periodic frequencies, *Comment. Math. Helv.* **73** (1998).

[12] S. Bochner, Curvature and Betti numbers in real and complex vector bundles, *Universitá e Politecnico de Toorino, Rendiconti del Seminario Matematico* **15** (1955-1956).

[13] A. Bondeson, E. Ott and T. M. Antonsen, Quasi-periodically forced damped pendular and Schrödinger equations with quasiperiodic potentials: Implications of their equivalence, *Phy. Rev. Lett.* **55** (1985).

[14] J. Brindley, T. Kaptaniak, and M.S. Naschie, Analytical conditions for strange chaotic and non-chaotic attractors of the quasi-periodically forced van der Pol equation, *Phys. D* **51** (1991).

[15] C. Chicone, Invariant tori for periodically perturbed oscillators, *Publ. Mat.* **41** (1997).

[16] C. Corduneanu, Periodic and almost periodic oscillations in non-linear systems, *Lecture notes in control and information science* **97**, Springer, Berlin and New York (1987).

[17] A. Denjoy, Sur les courbes définies par les équations différentielles à la surface du tore, *J. de Mathé Pures et Appli* **11** (1932).

[18] M. Ding, C. Grebogi, and E. Ott, Evolution of attractors in quasi-periodically forced systems: From quasiperiodic to strange non-chaotic to chaotic, *Phy. Rev. A* **39** (1989).

[19] T. Downarowicz and Y. Lacroix, Almost 1-1 extensions of Furstenberg-Weiss type and applications to Toeplitz flows, *Studia Math.* **130** (1998).

[20] T. Downarowicz, J. Kwiatkowski and Y. Lacroix, A criterion for Toeplitz flows to be topologically isomorphic and applications, *Colloq. Math.* **68** (1995).

[21] L. H. Eliasson, Floquet solutions for the 1-dimensional quasi-periodic Schrödinger equation, *Comm. Math. Phy.* **146** (1992).

[22] R. Ellis, Lectures on topological dynamics, W.A. Benjamin, Inc., NY (1969).

[23] R. Fabbri, R. A. Johnson and R. Pavani, On the nature of the spectrum of the quasi-periodic Schrödinger operator, *Nonlinear Anal. Real World Appl.* **3** (2002).

[24] A. M. Fink, Almost automorphic and almost periodic solutions which minimize functionals, *Tohoku Math. J.* **20** (1968).

[25] A. M. Fink, Almost periodic differential equations, *Lecture Notes in Mathematics,* **377**, Springer-Verlag, Berlin, Heidelberg, NY (1974).

[26] A. M. Fink and P. Frederickson, Ultimate boundedness does not imply almost periodicity, *J. Differential Equations* **9** (1971).

[27] P. Flor, Rhythmische abbildungen abelscher gruppen II, *Zeit. Wahrsch.* **7** (1967).

[28] J. Franks, Realizing rotation vectors for torus homeomorphism, *Trans. Amer. Math. Soc.* **311** No 1 (1989).

[29] J. Franks and M. Misiurewicz, Rotation sets of toral flows, *Proc. Amer. Math. Soc.* **109** No 1 (1990).

[30] H. Furstenberg, Disjointness in ergodic theory, minimal sets, and a problem in Diophantine approximation, *Math. Syst. Theory* **1** (1967).

[31] H. Furstenberg and B. Weiss, On almost 1 - 1 extensions, *Israel J. Math.* **65** (1989).

[32] R. Gjerde and O. Johansen, Bratteli-Vershik models for Cantor minimal systems: applications to Toeplitz flows, *Erg. Th. Dyn. Syst.* **20** (2000).

[33] W. Gottschalk and G. Hedlund, Topological dynamics, Amer. Math. Soc. Colloq. **36** (1955).

[34] J. Grasman, Asymptotic methods for relaxation oscillations and applications, *Appl. Math. Sci.* **63**, Springer-Verlag (1987).

[35] C. Grebogi, E. Ott, S. Pelikan, and J. A. Yorke, Strange attractors that are not chaotic, *Physica D* **13** (1984).

[36] J. Guckenheimer and P. Holmes, Nonlinear oscillations, dynamical systems and bifurcations of vector fields, *Appl. Math. Sci.* **42**, Springer-Verlag (1983).

[37] J. K. Hale, Integral manifolds of perturbed differential systems, *Ann. of Math.* **73** (1961).

[38] J. K. Hale, Ordinary differential equations, 2nd ed, Krieger Publishing Company (1980).

[39] H. He and J. You, On Eliasson's full measure reducibility problem, Preprint 2002.

[40] M. R. Herman, On the dynamics of Lagrangian tori invariant by symplectic diffeomorphisms. Progress in variational methods in Hamiltonian systems and elliptic equations, Pitman Res. Notes Math. Ser., 243, Longman Sci. Tech., Harlow, 1992.

[41] G. Hetzer and W. Shen, Convergence in almost periodic competition diffusion systems *J. Math. Anal. Appl.* **262** (2001).

[42] G. Hetzer and W. Shen, Uniform persistence, coexistence, and extinction in almost periodic/nonautonomous competition diffusion systems, *SIAM J. Math. Anal.* **34** (2002).

[43] A. Iwanik and Y. Lacroix, Some constructions of strictly ergodic non-regular Toeplitz flows, *Studia Math.* **110** (1994).

[44] K. Jacobs and M. Keane, 0-1 sequences of Toeplitz type, *Z. Wahrsch. Verw. Gebiete* **13** (1969).

[45] R. A. Johnson, On a Floquet theory for almost periodic two-dimensional linear systems, *J. Differential Equations* **37** (1980).

[46] R. A. Johnson, Bounded solutions of scalar almost periodic linear equations, *Illinois J. Math.* **25** (1981).

[47] R. A. Johnson, An example concerning the geometric significance of the rotation numbers-integrated density of states, *Proc. Breman Conf., Lect. Notes Math.* **1186**, Springer-Verlag (1984).

[48] R. A. Johnson, M-functions and Floquet exponents for linear differential systems, *Ann. Mat. Pura Appl. IV* **147** (1987).

[49] R. A. Johnson and J. Moser, The rotation number for almost periodic potentials, *Comm. Math. Phys.* **84** (1982).

[50] R. A. Johnson, K. J. Palmer and G. R. Sell, Ergodic properties of linear dynamical systems, *SIAM J. Math. Anal.* **18** (1987).
[51] A. Jorba and C. Simo, On the reducibility of linear differential equations with quasiperiodic coefficients, *J. Differential Equations* **98** (1992).
[52] A. Katok and B. Hasselblatt, Introduction to the modern theory of dynamical systems, Cambridge University Press (1995).
[53] R. Krikorian, Global density of reducible quasi-periodic cocycles on $T^1 \times SU(2)$, *Ann. of Math.* **154** (2001).
[54] J. Kwapisz, An estimate of entropy for toroidal chaos, *Erg. Th. Dyn. Syst.* **13** (1993).
[55] J. Kwapisz, Rotation sets and entropy, Ph.D thesis, SUNY, Stony Brook (1995).
[56] B. Levitan and V. Zhikov, Almost periodic functions and differential equations, Cambridge University Press (1982).
[57] Y. Li and Y. Yi, Persistence of invariant tori in generalized Hamiltonian systems, *Erg. Th. Dyn. Syst.*, **22** (2002).
[58] Y. Li and Y. Yi, On effective stability of generalized Hamiltonian systems, submitted.
[59] J. Llibre and R. Mackay, Rotation vectors and entropy for homeomorphisms of the torus isotopic to the identity, *Erg. Th. Dyn. Syst.* **11** (1991).
[60] N. G. Markely, Homeomorphisms of the circle without periodic points, *Proc. London Math. Soc.* **20** (1970).
[61] N. G. Markley and M. E. Paul, Almost automorphic symbolic minimal sets without unique ergodicity, *Israel J. Math.* **34** (1979).
[62] J. Mather, Existence of quasi-periodic orbits for twist homeomorphisms of the annulus, *Topology* **21** (1982).
[63] J. Mather, Minimal measures, *Comment. Math. Helv.* **64** (1989).
[64] J. Mather, Action minimizing invariant measures for positive definite Lagrangian systems, *Math. Z.* **207** (1991).
[65] J. Mather and G. Forni, Action minimizing orbits in Hamiltonian systems, *Transition to chaos in classical and quantum mechanics* (Montecatini Terme, 1991), Lecture Notes in Math., **1589**, Springer, Berlin, 1994.
[66] K. Meyer and G. R. Sell, Melnikov transforms, Bernoulli bundles, and almost periodic perturbations, *Trans. Amer. Math. Soc.* **314** (1989).
[67] R. K. Miller, Almost periodic differential equations as dynamical systems with applications to the existence of a. p. solutions, *J. Differential Equations* **1** (1965)
[68] M. Misiurewicz and K. Zieman, Rotation sets of tori, *J. London Math. Soc.* **40** No 2 (1989).
[69] J. Moser, On the theory of quasiperiodic motions, *SIAM Rev.* **8** (1966).
[70] J. Moser, Monotone twist mappings and the calculus of variations, *Erg. Th. Dyn. Sys.* **6** (1986).
[71] J. Moser, Minimal solutions of variational problems on a torus, *Ann. Inst. H. Poincaré Anal. Non Linéaire* **3** (1986).
[72] J. Moser, Quasi-periodic solutions of nonlinear elliptic partial differential equations, *Bol. Soc. Brasil. Mat.* **20** (1989).
[73] J. Moser and J. Pöschel, An extension of a result by Dinaburg and Sinai on quasi-periodic potentials, *Comment. Math. Helv.* **59** (1984).
[74] S. Novo, R. Obaya, and M. Sanz, Almost periodic and almost automorphic dynamics for scalar convex differential equations, preprint, 2002.
[75] Z. Opial, Sur une équation différentielle presque-périodique sans solution presque-périodique, *Bulletin de l'Académie Polon. de Sci., Sér. Math. Astr. Phys.* **IX** (1961).
[76] R. Ortega and M. Tarallo, Massera's theorem for quasi-periodic differential equations, *Topol. Methods Nonlinear Anal.* **19** (2002).
[77] R. Ortega and M. Tarallo, Almost periodic upper and lower solutions, to appear in *J. Differential Equations.*
[78] M. E. Paul, Constructions of almost automorphic symbolic minimal flows, *General Topology and its Appl.* **6** (1976).
[79] A. Reich, Prakompakte gruppen und fastperiodizitat, *Math. Z.* **116** (1970).
[80] F. J. Romeiras, A. Bondeson, E. Ott, T. M. Antonsen Jr, and C. Grebogi, Quasiperiodically forced dynamical systems with strange non-chaotic attractors, *Phy. D* **26** (1987).
[81] F. J. Romeiras and E. Ott, Strange non-chaotic attractors of the damped pendulum with quasiperiodic forcing, *Phy. Rev. A* **35** (1987).

[82] R. Sacker and G. R. Sell, A spectral theory for linear differential systems. *J. Differential Equations* **27** (1978).

[83] K. Schmitt and J. R. Ward, Jr., Almost periodic solutions of nonlinear second order differential equations, *Results Math.* **21** (1992).

[84] G. R. Sell, Non-autonomous differential equations and topological dynamics I, II, *Trans. Amer. Math. Soc.* **127** (1967).

[85] G. R. Sell, W. Shen and Y. Yi, Topological dynamics and differential equations, *Contemp. Math.* **215** (1997).

[86] W. Shen, Dynamical systems and traveling waves in almost periodic structures, *J. Differential Equations* **169** (2001).

[87] W. Shen, Global attractor in quasi-periodically forced Josephson junctions, *Far East J. Dyn. Syst.* **3** (2001).

[88] W. Shen, Dynamics in coupled oscillators with recurrent/random forcing, a PDE approach, preprint 2003.

[89] W. Shen and Y. Yi, Dynamic of almost periodic scalar parabolic equations, *J. Differential Equations* **121** (1995).

[90] W. Shen and Y. Yi, Asymptotic almost periodicity of scalar parabolic equations with almost periodic time dependence, *J. Differential Equations* **121** (1995).

[91] W. Shen and Y. Yi, On minimal sets of scalar parabolic equations with skew-product structure, *Trans. Amer. Math. Soc.* **347** No 11 (1995).

[92] W. Shen and Y. Yi, Ergodicity of minimal sets in scalar parabolic equations, *J. Dyns. Diff. Eqs.* **8** No 2 (1996).

[93] W. Shen and Y, Yi, Almost automorphy and skew-product semi-flow, Part II: Skew-product semi-flows, Part III: Applications to differential equations, *Mem. Amer. Math. Soc.* **136** No. 647 (1998).

[94] A. Reich, Prakompakte gruppen und fastperiodizitat, *Math. Z.* **116** (1970).

[95] R. Terras, Almost automorphic functions on topological groups, *Indiana U. Math. J.* **21** (1972).

[96] R. Terras, On almost periodic and almost automorphic differences of functions, *Duke Math. J.* **40** (1973).

[97] W. A. Veech, Almost automorphy and a theorem of loomis, *Archiv. der Math.* **XVIII** (1963).

[98] W. A. Veech, Almost automorphic functions on groups, *Amer. J. Math.* **87** (1965).

[99] W. A. Veech, On a Theorem of Bochner, *Ann. of Math.* **86** (1967).

[100] W. A. Veech, Point-distal flows, *Amer. J. Math.* **92** (1970).

[101] S. Wiggins, Chaos in the quasi-periodically forced Duffing oscillator, *Phys. Lett.* **124A** (1987).

[102] S. Wiggins, Global bifurcations and chaos, *Applied Math Sci.* **73**, Springer-Verlag, New York, Berlin, London, 1988.

[103] S. Williams, Toeplitz minimal flows which are not uniqely ergodic, *Z. Wahrsch. Verw. Gebiete* **67** (1984).

[104] J. Xu and J. You, On reducibility of linear differential equations with almost periodic coefficients, *Chinese J. Contemp. Math.* **17** (1996).

[105] Y. Yi, Generalized integral manifold theorem, *J. Differential Equations* **102** (1993).

[106] Y. Yi, Almost automorphy and skew-product semi-flow, Part I: Almost automorphy and almost periodicity, *Mem. Amer. Math. Soc.* **136** No. 647 (1998).

RESEARCH ARTICLES

Fields Institute Communications
Volume **42**, 2004

Complex Exponential Vandermonde Determinant with Application to Discrete Boundary Value Problem

Raghib M. Abu-Saris
Department of Basic Sciences
University of Sharjah
P.O. Box 27272, Sharjah, U.A.E.
rabusaris@sharjah.ac.ae

Khalil I. Al-Dosary
Department of Basic Sciences
University of Sharjah
P.O. Box 27272, Sharjah, U.A.E.
dosary@sharjah.ac.ae

Abstract. In this research, we develop explicit and recursive formulas for a general complex exponential Vandermonde determinant (CEVD). We employ (CEVD) to establish necessary and / or sufficient conditions for the existence of a unique solution to boundary value problems in difference equations.

1 Introduction

The determinant

$$V_k(z_1, z_2, ..., z_k) = \begin{vmatrix} 1 & 1 & \cdots & 1 \\ z_1 & z_2 & \cdots & z_k \\ \vdots & \vdots & & \vdots \\ z_1^{k-1} & z_2^{k-1} & \cdots & z_k^{k-1} \end{vmatrix} = \prod_{j=2}^{k} \prod_{i=1}^{j-1} (z_j - z_i) \tag{1.1}$$

is called *Vandermonde Determinant (VD)*. It plays an important role in approximation theory and numerical analysis [3]. Recently, several research papers elaborated its importance in difference equations, and the elegance of its hidden structure [6]. Generalizations of (VD) to more general exponential form was developed by Robbin and Salamon [7] and later by Yang, Wu and Zhang [8].

In the present work we consider complex exponential Vandermonde determinant (CEVD). For definiteness, we define

---

2000 *Mathematics Subject Classification.* Primary 39A10, 15A15; Secondary 65F40.

This research is supported by a grant from University of Sharjah.

$$D_k(z_1, ..., z_k; n_2, ..., n_k) = \begin{vmatrix} 1 & \cdots & 1 \\ z_1^{n_2} & \cdots & z_k^{n_2} \\ \vdots & & \vdots \\ z_1^{n_k} & \cdots & z_k^{n_k} \end{vmatrix} \tag{1.2}$$

where the positive integers n_i, $i = 2, ..., k$ are in ascending order with respect to the index, and the complex numbers z_1, z_2, ..., z_k are assumed to be distinct.

Observe that if $n_i = i - 1$, $i = 2, ..., k$, then D_k reduces to V_k. Hence, D_k is a further generalization of V_k. Therefore, from now and on, we assume that $n_i \geq i$ for some $i \in \{2, ..., k\}$.

The main results of the present article are Theorems 2.1, 3.1, and 4.1. Theorems 2.1 and 3.1 are established in Sections 3 and 4, respectively. They establish explicit and recursive formulas for CEVD. They are, also, the keystone in the proof of Theorem 4.1 which provides necessary and / or sufficient conditions for the existence of a unique solution of multi-point boundary value problem in difference equation.

For the convenience of the reader, we list below the adopted notations in this paper.

- $Z_s = \{0, 1, ..., s\}$,
- If $z = (z_1, ..., z_k)$ and $\alpha = (\alpha_1, ..., \alpha_k)$, then

$$|\alpha| = \sum_{i=1}^{k} \alpha_i$$

$$z^\alpha = \prod_{i=1}^{k} z_i^{\alpha_i}$$

- $\sigma_r(z_1, ..., z_k) = \sigma(k, r) = \sum_{|\alpha|=r} z^\alpha; \qquad \alpha_i \in Z_1.$

2 An explicit formula for D_k

Our main result in this section is the following theorem.

Theorem 2.1 *Let $n = \sum_{i=2}^{k} n_i$ and $N = n - k(k-1)/2$, then*

$$D_k(z_1, ..., z_k; n_2, ..., n_k) = \left(\sum_{|\alpha|=N} a_\alpha \, z^\alpha \right) V_k(z_1, ..., z_k); \qquad \alpha_i \in Z_{n_k-k+1} \tag{2.1}$$

where $a_\alpha > 0$.

Proof Since D_k is homogeneous of degree $n = \sum_{j=2}^{k} n_j$ and V_k is homogeneous of degree $k(k-1)/2$, we have

$$D_k(z_1, ..., z_k; n_2, ..., n_k) = \Phi_k(z_1, ..., z_k; n_2, ..., n_k) \; V_k(z_1, ..., z_k)$$

where Φ_k is homogeneous of degree N. Furthermore, since D_k is a polynomial of degree n_k in each z_i and V_k is a polynomial of degree $k-1$ in z_i, we have

$$\Phi_k(z_1, ..., z_k; n_2, ..., n_k) = \sum_{|\alpha|=N} a_\alpha z^\alpha \qquad \alpha_i \in Z_{n_k-k+1}.$$

where $a_\alpha \in \Re$

To this end, we claim that $a_\alpha > 0$ for all α.

Observe that

$$D_2(z_1, z_2; n_2) = \left(\sum_{\ell=0}^{n_2-1} z_1^{\ell} z_2^{n_2-\ell-1} \right) V_2(z_1, z_2),$$

i.e., the claim is true in case of two variables regardless of the exponent. Furthermore, setting $s_j = z_j/z_1$ and applying the well-known properties of determinant, we have

$$\begin{aligned}
& D_k(z_1, z_2, ..., z_k; n_2, ..., n_k) \\
= \; & z_1^n D_k(1, s_2, ..., s_k; n_2, ..., n_k) \\
= \; & z_1^n \begin{vmatrix} 1 & 0 & \cdots & 0 \\ 1 & s_2^{n_2} - 1 & \cdots & s_k^{n_2} - 1 \\ \vdots & \vdots & & \vdots \\ 1 & s_2^{n_k} - 1 & \cdots & s_k^{n_k} - 1 \end{vmatrix} \\
= \; & z_1^n \prod_{j=2}^{k} (s_j - 1) \begin{vmatrix} \sum_{\ell_2=0}^{n_2-1} s_2^{\ell_2} & \cdots & \sum_{\ell_2=0}^{n_2-1} s_k^{\ell_2} \\ \vdots & & \vdots \\ \sum_{\ell_k=n_{k-1}}^{n_k-1} s_2^{\ell_k} & \cdots & \sum_{\ell_k=n_{k-1}}^{n_k-1} s_k^{\ell_k} \end{vmatrix} \\
= \; & z_1^n \prod_{j=2}^{k} (s_j - 1) \\
\cdot \; & \sum_{\ell_2=0}^{n_2-1} \cdots \sum_{\ell_k=n_{k-1}}^{n_k-1} s_2^{\ell_2} \cdots s_k^{\ell_2} D_{k-1}(s_2, ..., s_k; \ell_3 - \ell_2, ..., \ell_k - \ell_2) \\
= \; & z_1^n V_k(1, s_2, ..., s_k) \\
\cdot \; & \sum_{\ell_2=0}^{n_2-1} \cdots \sum_{\ell_k=n_{k-1}}^{n_k-1} s_2^{\ell_2} \cdots s_k^{\ell_2} \Phi_{k-1}(s_2, ..., s_k; \ell_3 - \ell_2, ..., \ell_k - \ell_2)
\end{aligned}$$

Thus

$$\begin{aligned}
\Phi_k(z_1, ..., z_k; n_2, ..., n_k) = & \sum_{\ell_2=0}^{n_2-1} \cdots \sum_{\ell_k=n_{k-1}}^{n_k-1} \Big(z_1^{n-k(k-1)/2-(k-1)\ell_2} z_2^{\ell_2} \cdots z_k^{\ell_2} \\
& \cdot \; \Phi_{k-1}(s_2, ..., s_k; \ell_3 - \ell_2, ..., \ell_k - \ell_2) \Big)
\end{aligned}$$

and so, if the claim holds for D_{k-1}, it holds for D_k. Hence the proof is established by Principle of Mathematical Induction. □

Remark 2.2 It is worth to mention that $\Phi_k(z_1, ..., z_k)$ mentioned above is absolutely invariant under any permutation of $z_i's$. This is due to the fact that interchanges in the columns of D_k correspond to the same interchanges in the columns of V_k.

The following result is an immediate corollary of Theorem 2.1 (see Also [7] and [8]).

Corollary 2.3 *If $z_i = x_i$, $i = 1, ..., k$ are positive real numbers such that $x_i \neq x_j$ whenever $i \neq j$, then $D_k(x_1, ..., x_k; n_2, ..., n_k) \neq 0$. Furthermore, if in addition $x_j > x_i$, $j > i$, then $D_k(x_1, ..., x_k; n_2, ..., n_k) > 0$.*

3 A recursive formula for D_k

The following theorem is our main result in this section.

Theorem 3.1 *Let*

$$\Phi_k(z_1, ..., z_k; n_2, ..., n_k) = \sum_{i=0}^{n_k-k+1} \phi_i(z_1, .., z_{k-1})\ z_k^i.$$

Then

$$\phi_0 = (\sigma(k-1, k-1))^{n_2-1}\ \Phi_{k-1}(z_1, ..., z_{k-1}; n_3 - n_2, ..., n_k - n_2)$$

$$\phi_m = \frac{1}{\sigma(k-1,k-1)} \left[\sum_{\ell=1}^{L} (-1)^{\ell+1}\ \sigma(k-1, k-\ell-1)\ \phi_{m-\ell} - \sum_{\ell=2}^{L+1} (-1)^{\ell}\ \Phi_{k-1}(z_1, ..., z_{k-1}; ..., n_{\ell-1}, n_{\ell+1}, ...)\ \delta_{n_\ell m} \right]$$

where $1 \le m \le n_k - k,\ L = \min\{m, k-1\},$ *&* δ_{ij} *is the Kronecker delta, and*

$$\phi_{n_k-k+1} = \Phi_{k-1}(z_1, ..., z_{k-1}; n_2, ..., n_{k-1}).$$

To prove the above theorem we need the following two lemmas:

Lemma 3.2 [1, Lemma 2.3]
If $0 \le \ell < k$*, then*

$$\begin{vmatrix} 1 & \cdots & 1 \\ \vdots & & \vdots \\ z_1^{\ell-1} & \cdots & z_k^{\ell-1} \\ z_1^{\ell+1} & \cdots & z_k^{\ell+1} \\ \vdots & & \vdots \\ z_1^k & \cdots & z_k^k \end{vmatrix} = \sigma(k, k-\ell)\ V_k(z_1, ..., z_k)$$

Applying Lemma 3.2 and expanding through the kth column, we obtain:

Lemma 3.3 *At* $z_k = 0$,

$$\frac{\partial^\ell V_k}{\partial z_k^\ell} = \begin{cases} (-1)^{k+\ell+1}\ \ell!\ \sigma(k-1, k-\ell-1)\ V_{k-1}(z_1, ..., z_{k-1}) & \text{if } 0 \le \ell < k \\ 0 & \text{if } \ell \ge k \end{cases}$$

Proof of Theorem 3.1 First, evaluating D_k at $z_k = 0$, expanding D_k through the kth column, and using Theorem 2.1, we obtain the first formula. The third formula can be established by realizing the coefficient of $z_k^{n_k}$ in D_k and applying Theorem 2.1.

Now, by Leibniz rule of differentiation:

$$\frac{\partial^m D_k}{\partial z_k^m} = \sum_{\ell=0}^{m} \binom{m}{\ell} \frac{\partial^{m-\ell}\Phi_k}{\partial z_k^{m-\ell}} \frac{\partial^\ell V_k}{\partial z_k^\ell}$$

Differentiating Φ_k and V_k and applying Lemma 3.3 we have at $z_k = 0$

$$\frac{\partial^m D_k}{\partial z_k^m} = m!\; V_{k-1} \sum_{\ell=0}^{L} (-1)^{k+\ell+1}\; \sigma(k-1, k-\ell-1)\; \phi_{m-\ell}.$$

But at $z_k = 0$,

$$\frac{\partial^m D_k}{\partial z_k^m} = \sum_{\ell=2}^{L+1} (-1)^{\ell+k}\; m!\; D_{k-1}(z_1, ..., z_{k-1}; ..., n_{\ell-1}, n_{\ell+1}, ...)\; \delta_{n_\ell m}.$$

Applying Theorem 2.1, simplifying, and rearranging terms we obtain the second formula. □

Remark 3.4 It is interesting to notice that when $k \leq m \leq n_k - k$, then ϕ_m satisfies a "constant coefficient" difference equation of order $k-1$:

$$\begin{aligned}\sum_{\ell=0}^{k-1} &(-1)^\ell \sigma(k-1, k-\ell-1)\phi_{m-\ell} \\ &= -\sum_{\ell=2}^{k} (-1)^\ell\; \Phi_{k-1}(z_1, ..., z_{k-1}; ..., n_{\ell-1}, n_{\ell+1}, ...)\; \delta_{n_\ell m}.\end{aligned}$$

This difference equation can be solved explicitly:

$$\begin{aligned}\phi_m &= \sum_{\ell=1}^{k-1} A_\ell z_\ell^{-m} \\ &\quad - \left[\sum_{\ell=2}^{k} (-1)^\ell\; \Phi_{k-1}(z_1, ..., z_{k-1}; ..., n_{\ell-1}, n_{\ell+1}, ...)\; \delta_{n_\ell m}\right] \prod_{\ell=1}^{k-1} (z_\ell - 1)^{-1}\end{aligned}$$

where $A'_\ell s$ are constants.

Corollary 3.5

$$\Phi_3(z_1, z_2, z_3; n_2, n_3) = \sum_{m=0}^{n_3-2} \phi_m z_3^m$$

where

$$\phi_m = \left(\sum_{|\beta|=m_2} z^\beta\right)\left(\sum_{|\alpha|=n_3-m-2} z^\alpha\right), \quad \beta_i \in Z_{m_2}, \;\; \alpha_i \in Z_{n_3-m-2},$$

and $m_2 = \min\{m, n_2 - 1\}$.

Proof If $k = 3$, then

$$\sigma_0(z_1, z_2) = 1, \quad \sigma_1(z_1, z_2) = z_1 + z_2, \quad \sigma_2(z_1, z_2) = z_1 z_2,$$

and

$$\Phi_2(z_1, z_2; s) = \sum_{\ell=0}^{s-1} z_1^\ell z_2^{s-\ell-1}.$$

On the one hand, if $n_2 = 1$, then by Theorem 3.1 we have:

$$\phi_0 = \Phi_2(z_1, z_2; n_3 - 1) = \sum_{\ell=0}^{n_3-2} z_1^\ell z_2^{n_3-\ell-2},$$

and

$$\begin{aligned} \phi_1 &= \frac{1}{\sigma_2(z_1, z_2)} \left[\sigma_1(z_1, z_2)\phi_0 - \Phi_2(z_1, z_2; n_3)\right] \\ &= \sum_{\ell=0}^{n_3-3} z_1^\ell z_2^{n_3-\ell-3}, \end{aligned}$$

and for $2 \leq m \leq n_3 - 2$, ϕ_m satisfies the second order linear difference equation:

$$\sigma_2(z_1, z_2)\phi_m - \sigma_1(z_1, z_2)\phi_{m-1} + \sigma_0(z_1, z_2)\phi_{m-2} = 0 \tag{3.1}$$

which has the solution

$$\phi_m = \sum_{\ell=0}^{n_3-m-2} z_1^\ell z_2^{n_3-m-\ell-2}.$$

On the other hand, if $(n_2 \geq 2)$, then

$$\begin{aligned} \phi_0 &= \left[\sigma_2(z_1, z_2)\right]^{n_2-1} \Phi_2(z_1, z_2; n_3 - n_2) \\ &= (z_1 z_2)^{n_2-1} \sum_{\ell=0}^{n_3-n_2-1} z_1^\ell z_2^{n_3-n_2-\ell-1} \\ \phi_1 &= \frac{\sigma_1(z_1, z_2)}{\sigma_2(z_1, z_2)} \phi_0 \\ &= \sigma_1(z_1, z_2) \left[\sigma_2(z_1, z_2)\right]^{n_2-2} \phi_0 \\ &= (z_1 + z_2)(z_1 z_2)^{n_2-2} \sum_{\ell=0}^{n_3-n_2-1} z_1^\ell z_2^{n_3-n_2-\ell-1}, \end{aligned}$$

and, by solving Eq.(3.1),

$$\phi_m = (z_1 z_2)^{n_2-m-1} \left(\sum_{\ell=0}^{m} z_1^\ell z_2^{m-\ell}\right) \left(\sum_{\ell=0}^{n_3-n_2-1} z_1^\ell z_2^{n_3-n_2-\ell-1}\right)$$

where $2 \leq m < n_2$. Furthermore, if $m = n_2$, then

$$\begin{aligned} \phi_{n_2} &= \frac{1}{\sigma_2(z_1, z_2)} \left[\sigma_1(z_1, z_2)\phi_{n_2-1} - \sigma_0(z_1, z_2)\phi_{n_2-2} - \Phi_2(z_1, z_2; n_3)\right] \\ &= \left(\sum_{\ell=0}^{n_2-1} z_1^\ell z_2^{n_2-\ell-1}\right) \left(\sum_{\ell=0}^{n_3-n_2-2} z_1^\ell z_2^{n_3-n_2-\ell-2}\right). \end{aligned}$$

Finally, if $n_2 < m \leq n_3 - 2$, then, once more, solving Eq. (3.1) we get

$$\phi_m = \left(\sum_{\ell=0}^{n_2-1} z_1^\ell z_2^{n_2-\ell-1}\right) \left(\sum_{\ell=0}^{n_3-m-2} z_1^\ell z_2^{n_3-m-\ell-2}\right).$$

□

4 Discrete boundary value problems

Consider the kth order linear difference equation with constant coefficients:

$$\sum_{j=0}^{k} a_j y(t+j) = 0, \ t = 0, 1, 2, ..., \qquad a_0 a_k \neq 0 \tag{4.1}$$

subject to boundary conditions of the form:

$$y(n_i) = y_i, \ i = 1, 2, ..., k \tag{4.2}$$

where n_i, $i = 1, 2, ..., k$ are positive integers such that $0 \leq n_1 < n_2 < \cdots n_k < \infty$.

Difference equation (4.1) together with boundary conditions (4.2) will be called discrete boundary value problem (DBVP). They can be used to model dynamical systems governed by the difference equations with available measurements of the system state clocked at times n_i, $i = 1, 2, ..., k$. In general, the fundamental question of existence and uniqueness of solutions of discrete boundary value problems is difficult; there is no simple criterion like that for discrete initial value problems [2]. This fact together with open problem in [5] due to Trigiante initiated our interest in DBVP. It is our intention in this article to characterize distributions of the characteristic roots, i.e., the zeros of the characteristic polynomial:

$$p(\lambda) = \sum_{j=0}^{k} a_j \lambda^j \tag{4.3}$$

that ensures the existence of a unique solution of DBVP (4.1)-(4.2). For this purpose, we assume that the characteristic roots: z_1, z_2, ..., z_k are distinct.

Now, since the characteristic roots are assumed to be distinct, the general solution of Eq. (4.1) is given by

$$y(t) = \sum_{i=1}^{k} c_j z_j^t$$

where c_j are constants [4, page 64]. Applying the boundary conditions 4.2, we obtain a system of linear equations in which the coefficient matrix is given by *complex exponential Vandermonde matrix:*

$$VM = \begin{pmatrix} z_1^{n_1} & z_2^{n_1} & \cdots & z_k^{n_1} \\ z_1^{n_2} & z_2^{n_2} & \cdots & z_k^{n_2} \\ \vdots & \vdots & & \vdots \\ z_1^{n_k} & z_2^{n_k} & \cdots & z_k^{n_k} \end{pmatrix}. \tag{4.4}$$

Hence, the existence of a unique solution of boundary value problem (4.1) and (4.2) is equivalent to the nonsingularity of the matrix VM. But none of the $z_i's$ is zero. Therefore, DBVP (4.1)-(4.2) has a unique solution if and only if $D_k \neq 0$. Consequently, we have the following result:

Theorem 4.1 **(a)** *If all characteristic roots lie on the same ray, then DBVP (4.1)-(4.2) has a unique solution.*

(b) *If $n_2 \geq 2$ such that n_2 divides n_j, $j = 3, ..., k$, and if $z_j = r_j e^{i\phi}$, $r_j > 0$, $\phi \in \{\theta + 2\ell\pi/n_2, \ \ell = 0, 1, n_2 - 1\}$, $j = 1, 2, ..., k$ such that $r_i \neq r_j$ whenever $i \neq j$, then DBVP (4.1)-(4.2) has a unique solution.*

(c) *If r is a positive integer and $n_j = (j-1)r$, $j = 1, 2, ..., k$, then DBVP (4.1)-(4.2) has a unique solution if and only if $z_i^r \neq z_j^r$ whenever $i \neq j$.*

(d) *If* $z_j = re^{i\theta_j}, j = 1, 2, ..., k$ *and DBVP (4.1)-(4.2) has a unique solution, then* $(z_1, ..., z_k)$ *can't be proportional to* $(w_1, ..., w_k)$ *where* $w_j^{n_i} = 1,\ j = 1, 2, ..., k$ *for any* $n_i \geq k$. *Also,* $(z_1, ..., z_k) \notin \{(w_1, ..., w_k),\ w_j^{n_l - n_i} = 1\}$ *whenever* $n_l - n_i \geq k$

Proof Parts (a) and (b) follow from Corollary 2.3, Part (c) follows from the fact that

$$D_k(z_1, ..., z_k; r, 2r, ..., (k-1)r) = V_k(z_1^r, ..., z_k^r),$$

and Part (d) follows from the fact that:

$D_k(z_1, ..., z_k; n_2, ..., n_k) = 0$ *if* $(z_1, ..., z_k)$ *are proportional to* $(w_1, ..., w_k)$ *where* $w_j^{n_i} = 1,\ j = 1, 2, ..., k$ *for some* $n_i \geq k$, *or if* $(z_1, ..., z_k) \in \{(w_1, ..., w_k),$ $w_j^{n_\ell - n_i} = 1$ *for some* $n_\ell - n_i \geq k,\ j = 1, 2, ..., k\}$ □

Remark 4.2 It is worth to mention that parts (a) and (d) are extensions of Theorem 3.3 and Theorem 3.1 in [1], respectively.

References

[1] R. Abu-Saris and H. Yousef, *Existence Theorems for Boundary Value Problems in Difference Equations*, J. Difference Eqs. Appl., **7** (2001), 255-263.

[2] C. Bender and S. Orszag, *Advanced Mathematical Methods for Scientist and Engineers*, Springer, New York, 1999.

[3] P. Davis, *Interpolation and Approximation*, Dover: New York, 1975.

[4] S. Elaydi, *An Introduction to Difference Equations (2nd ed.)*, Springer, New York, 1999.

[5] G. Ladas, *Open Problems and Conjectures*, J. Difference Eqs. Appl., **4** (1998), 395-396.

[6] H. Oruc and G. Phillips, *Explicit Factorization of the Vandermonde Matrix*, Lin Alg. Appl., **315** (2000), 113-123.

[7] J. Robbin and D. Salamon, *The Exponential Vandermonde Matrix*, Lin. Alg. Appl., **317** (2000), 225-226.

[8] S. Yang, H. Wu, Q. Zhang, *Generalization of Vandermonde Determinants*, Lin. Alg. Appl., **336** (2001), 201-204.

Fields Institute Communications
Volume **42**, 2004

A Convergence Theorem Related to Iterative Mean Clustering

Yongqiang Cao
Department of Mathematics and Statistics
York University
4700 Keele Street
Toronto ON M3J 1P3, Canada
Current address: Cognitive and Neural Systems Department
Boston University
677 Beacon Street
Boston, MA 02215 U.S.A.
yqcao@mathstat.yorku.ca, yqcao@bu.edu

Abstract. The iterative mean clustering (IMC) is an algorithm for discovering clusters of data points in a given data set. The convergence of IMC can be reduced to the convergence of a discrete dynamical system. This paper provides a generalized convergence theorem for IMC algorithm in multidimensional spaces.

1 Introduction

For a given data set, a cluster is a subset of points similar to each other according to a certain similarity measurement. The problem of data clustering is to find a partition of the data set into clusters such that those points which are assigned to the same cluster are similar but points from different clusters are dissimilar. Data clustering has extensive applications in many areas such as data mining, information retrieval, on-line commerce, pattern recognition and image segmentation. Clustering problem has also been studied extensively and various methods have been developed such as partition-based or optimization-based algorithms, density-based algorithms, neural network approaches, subspace clustering algorithms [1, 2, 3, 5, 6, 7, 8, 9, 10, 11, 12, 13, 14, 15].

In [4], we proposed a simple clustering algorithm - Iterative Mean Clustering (IMC), which can be simply described as follows:

1. Choose an initial data point x_0.
2. For $i = 1, 2, \cdots$, set
   x_i to be the mean of all data points whose distances to x_{i-1} are less than or equal to σ.

2000 *Mathematics Subject Classification.* Primary 37N99, 37N40; Secondary 91C20, 65K10.

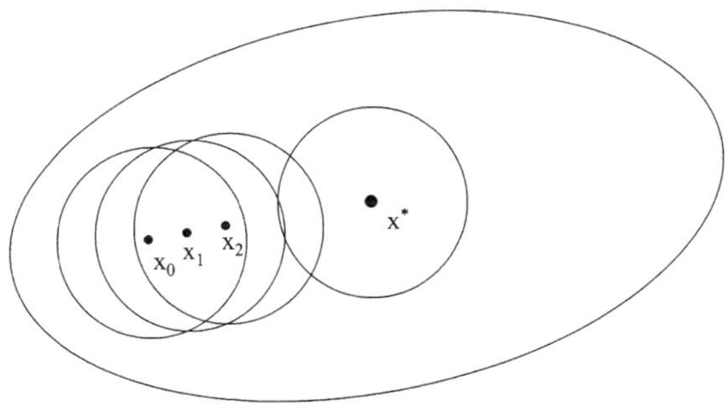

Figure 1 Iterative Mean clustering algorithm.

Here σ is a predefined distance parameter. This algorithm is based on a simple observation that if points in a cluster are distributed around the center, then during the above iteration each new mean x_i should be closer to the center of the cluster than its predecessors, so that the clustering process ultimately converges to the center of the cluster, as shown in Fig. 1. The neural network implementation of IMC was also developed. It is important to point out that it has been shown that the clustering results via IMC are independent of the order at which the data points are input and this represents a significant improvement over many neural network clustering architecture. The convergence of the algorithm can be reduced to the convergence of a discrete dynamical system in the space of data points. This enabled us to obtain the convergence of IMC algorithm in 1-dimensional case and to extend the convergence result to multidimensional case when the data is independently distributed in each dimension. However, in many applications, the data distributions are dependent in general. The purpose of this paper is to provide a generalized convergence theorem for the discrete dynamical system in multidimensional spaces when the independence of data distribution in [4] is not required.

The remaining part of this paper is organized as follows. Section 2 describes the related notations and concepts. Section 3 states and proves our main theorem. Section 4 provides a rough estimate for the rate of convergence. Simulations and applications will be provided elsewhere.

2 Notations and concepts

Let $\mathbb{R}^m$ be m-dimensional Euclidean space with the norm

$$||x|| = (\sum_{i=1}^{m} x_i^2)^{1/2}, \quad x \in \mathbb{R}^m.$$

Definition 2.1 Let $S \subset \mathbb{R}^m$ be a given data set. For a given $x \in \mathbb{R}^m$, let

$$A_{S,\sigma}(x) = \{y \in S : ||y - x|| \leq \sigma\} \tag{2.1}$$

and define an operator $T_{S,\sigma} : \mathcal{D}(T_{S,\sigma}) \subset \mathbb{R}^m \to \mathbb{R}^m$ by

$$T_{S,\sigma}(x) = \frac{1}{|A_{S,\sigma}(x)|} \sum_{y \in A_{S,\sigma}(x)} y, \tag{2.2}$$

$$\mathcal{D}(T_{S,\sigma}) = \{x \in \mathbb{R}^m : A_{S,\sigma}(x) \neq \varnothing\}, \tag{2.3}$$

where $|A_{S,\sigma}(x)|$ denotes the number of elements in the set $A_{S,\sigma}(x)$.

Note that $T_{S,\sigma}(x)$ is nothing but the mean of all elements in the set $A_{S,\sigma}(x)$. The IMC algorithm can be formulated as follows:

1. Choose an initial data point x_0.
2. For $i = 1, 2, \cdots$, iteratively compute x_i by

$$x_i = T_{S,\sigma}(x_{i-1}).$$

Definition 2.2 Let $f : \mathbb{R}^m \to \mathbb{R}$ denote the distribution density function of the data set S. Define an operator $T_{f,\sigma} : \mathcal{D}(T_{f,\sigma}) \subset \mathbb{R}^m \to \mathbb{R}^m$ by

$$T_{f,\sigma}(x) = \frac{\int_{||y-x||\leq\sigma} yf(y)dy}{\int_{||y-x||\leq\sigma} f(y)dy}, \tag{2.4}$$

$$\mathcal{D}(T_{f,\sigma}) = \{x \in \mathbb{R}^m : \int_{||y-x||\leq\sigma} f(y)dy \neq 0\}. \tag{2.5}$$

In what follows, we will consider $T_{f,\sigma}$ with a continuous density function f. This is a very good approximation if the cluster involved has a large amount of points. Therefore, we regard $T_{f,\sigma}$ as a continuous counterpart of the operator $T_{S,\sigma}$, and write $T_{f,\sigma}$ as T in short.

If IMC algorithm converges, it must converge to a fixed point of T. Here $x^* \in \mathbb{R}^m$ is a fixed point of T if $T(x^*) = x^*$. What we are interested in are the location and stability of the fixed points of T.

Definition 2.3 A fixed point x^* of T is said to be stable if for every $\varepsilon > 0$ there exits a $\delta > 0$ such that for every $x_0 \in B(x^*, \delta) \cap \mathcal{D}(T)$ and every integer $n \geq 0$, the sequence $\{x_n\}_{n=0}^{\infty}$ defined by $x_n = T(x_{n-1})$ satisfies

$$x_n \in B(x^*, \varepsilon) \cap \mathcal{D}(T),$$

where $B(x^*, \tau) = \{x \in \mathbb{R}^m : \|x - x^*\| < \tau\}$ for any $\tau > 0$. The fixed point x^* is said to be unstable if it is not stable. The fixed point x^* is asymptotically stable if there exists a $\delta_0 > 0$ such that for every $x_0 \in B(x^*, \delta_0) \cap \mathcal{D}(T)$, the sequence $\{x_n\}_{n=0}^{\infty}$ defined by $x_n = T(x_{n-1})$ satisfies

$$\lim_{n\to\infty} x_n = x^*.$$

Let g be a function defined on $\mathbb{R}^m$. In what follows, we will use Dg or ∇g to denote its gradient, and use $D^2 g$ to denote its Hessian matrix.

Definition 2.4 x^* is said to be a stationary point of g if $\nabla g(x^*) = 0$. The stationary point x^* is said to be a strongly strict minimum of g if the Hessian matrix of g at x^* is positive definite. The stationary point x^* is said to be a strongly strict maximum of g if the Hessian matrix of $-g$ at x^* is positive definite.

From Eq. (2.4) we have

$$T(x) = \frac{\int_{\|y-x\|\leq\sigma} yf(y)dy}{\int_{\|y-x\|\leq\sigma} f(y)dy} = \frac{\int_{\|y\|\leq\sigma}(x+y)f(x+y)dy}{\int_{\|y\|\leq\sigma} f(x+y)dy}.$$

So

$$T(x)\int_{\|y\|\leq\sigma} f(x+y)dy = \int_{\|y\|\leq\sigma}(x+y)f(x+y)dy.$$

Therefore,

$$(T(x)-x)\int_{\|y\|\le\sigma} f(x+y)dy=\int_{\|y\|\le\sigma} yf(x+y)dy. \tag{2.6}$$

It is natural to define

$$F_\sigma(x)=\int_{\|y\|\le\sigma} yf(x+y)dy, \tag{2.7}$$

$$G_\sigma(x)=\int_{\|y\|\le\sigma}(\sigma^2-\|y\|^2)f(x+y)dy. \tag{2.8}$$

3 Main theorem

Lemma 3.1 *x is a fixed point of T if and only if x is a zero of F_σ.*

Proof This result is a direct consequence of Eq. (2.6) and Eq. (2.7). □

Theorem 3.2 *Assume that $f\in C^1(\mathbb{R}^m)$. Then*

(i) *x^* is a fixed point of T if and only if x^* is a stationary point of G_σ.*

(ii) *Every strongly strict local maximum of G_σ is an asymptotically stable fixed point of T, and every strongly strict local minimum of G_σ is a unstable fixed point of T.*

Proof Note that

$$\begin{aligned}\nabla G_\sigma(x) &= \textstyle\int_{\|y\|\le\sigma}(\sigma^2-\|y\|^2)D_xf(x+y)dy\\ &= \textstyle\int_{\|y\|\le\sigma}(\sigma^2-\|y\|^2)D_yf(x+y)dy\\ &= \textstyle\int_{\|y\|\le\sigma}D_y[(\sigma^2-\|y\|^2)f(x+y)]dy+\int_{\|y\|\le\sigma}2yf(x+y)dy\\ &= \textstyle\int_{\|y\|\le\sigma}D_y[(\sigma^2-\|y\|^2)f(x+y)]dy+2F_\sigma(x).\end{aligned}$$

By Stokes' Theorem, we get

$$\int_{\|y\|\le\sigma}D_{y_i}[(\sigma^2-\|y\|^2)f(x+y)]dy=\int_{\|y\|=\sigma}(\sigma^2-\|y\|^2)f(x+y)n_i\,dA=0,$$

where n_i is the ith component of the unit outer normal vector n on the sphere $\|y\|=\sigma$. Therefore,

$$\nabla G_\sigma(x)=2F_\sigma(x). \tag{3.1}$$

Then Lemma 3.1 implies (i).

We now prove (ii). Let x^* be a strongly strict local maximum of G_σ. Then it follows from (i) that $T(x^*)=x^*$. Differentiating both sides of Eq. (2.6), we get

$$(DT(x)-I)\int_{\|y\|\le\sigma} f(x+y)dy+(T(x)-x)D\int_{\|y\|\le\sigma} f(x+y)dy=DF_\sigma(x),$$

where I is the $m\times m$ identity matrix. Let

$$M=\int_{\|y\|\le\sigma} f(x^*+y)dy,$$

and note that $T(x^*)=x^*$ we have

$$DT(x^*)=I+\frac{1}{M}DF_\sigma(x^*)=I-\frac{1}{M}(-DF_\sigma(x^*)). \tag{3.2}$$

Eq. (3.1) implies that

$$-DF_\sigma(x^*)=\frac{1}{2}D^2(-G_\sigma(x^*)),$$

where $D^2(-G_\sigma(x^*))$ is the Hessian matrix of $-G_\sigma$ at x^*. Since x^* is a strongly strict local maximum of G_σ, $-DF_\sigma(x^*)$ is positive definite. Therefore, there exists a orthogonal matrix P with $P^TP = I$ such that

$$P(-DF_\sigma(x^*))P^T = \begin{pmatrix} b_{11} & & & \\ & b_{22} & & 0 \\ & 0 & \ddots & \\ & & & b_{mm} \end{pmatrix} \tag{3.3}$$

is diagonal matrix with $b_{ii} > 0$, $i = 1, \ldots, m$. But

$$P(DF_\sigma(x))P^T = P(D_x \int_{\|y\|\leq\sigma} yf(x+y)dy)P^T = (D_x \int_{\|y\|\leq\sigma} Pyf(x+y)dy)P^T.$$

Let $v = Py$ and note that the orthogonal matrix P has $|\det(P)| = 1$, we get

$$\begin{aligned} P(DF_\sigma(x))P^T &= (D_x \textstyle\int_{\|v\|\leq\sigma} vf(x+P^Tv)|\det(P)|dv)P^T \\ &= (D_x \textstyle\int_{\|v\|\leq\sigma} vf(x+P^Tv)dv)P^T. \end{aligned}$$

Let $u = Px$, we then have

$$\begin{aligned} P(DF_\sigma(x))P^T &= (D_u \textstyle\int_{\|v\|\leq\sigma} vf(P^Tu+P^Tv)dv)P^T \cdot D_x u(x) \\ &= (D_u \textstyle\int_{\|v\|\leq\sigma} vf(P^Tu+P^Tv)dv)P^TP = D_u \textstyle\int_{\|v\|\leq\sigma} vf(P^Tu+P^Tv)dv. \end{aligned}$$

Let $\hat{f}(x) = f(P^Tx)$, we get

$$\begin{aligned} P(DF_\sigma(x))P^T &= D_u \textstyle\int_{\|v\|\leq\sigma} v\hat{f}(u+v)dv \\ &= \textstyle\int_{\|v\|\leq\sigma} D_u\hat{f}(u+v)vdv = \left(\textstyle\int_{\|v\|\leq\sigma} D_{u_i}\hat{f}(u+v)v_j dv\right)_{m\times m}. \end{aligned}$$

Therefore, form Eq. (3.3) we have

$$\begin{aligned} &P(DF_\sigma(x^*))P^T \\ &= \left(\textstyle\int_{\|v\|\leq\sigma} D_{u_i}\hat{f}(u^*+v)v_j dv\right)_{m\times m} = \begin{pmatrix} -b_{11} & & & \\ & -b_{22} & & 0 \\ & 0 & \ddots & \\ & & & -b_{mm} \end{pmatrix}, \end{aligned}$$

where $u^* = Px^*$. In other words,

$$\int_{\|v\|\leq\sigma} D_{u_i}\hat{f}(u^*+v)v_i dv = -b_{ii} < 0,$$

and

$$\int_{\|v\|\leq\sigma} D_{u_i}\hat{f}(u^*+v)v_j dv = 0 \quad \text{if } i \neq j.$$

Similarly,

$$\begin{aligned} &\textstyle\int_{\|y\|\leq\sigma} f(x+y)dy \\ &= \textstyle\int_{\|v\|\leq\sigma} f(P^Tu+P^Tv)dv = \textstyle\int_{\|v\|\leq\sigma} \hat{f}(u+v)dv. \end{aligned}$$

This implies that

$$M = \int_{\|y\|\leq\sigma} f(x^*+y)dy = \int_{\|v\|\leq\sigma} \hat{f}(u^*+v)dv. \tag{3.4}$$

Note that

$$\begin{aligned}&\int_{\|v\|\le\sigma} D_{u_i}\hat f(u^*+v)v_i dv = \int_{\|v\|\le\sigma}(D_{v_i}\hat f(u^*+v))v_i dv\\&= \int_{\|v\|\le\sigma} D_{v_i}(\hat f(u^*+v)v_i)dv - \int_{\|v\|\le\sigma}\hat f(u^*+v)dv\\&= -b_{ii} < 0.\end{aligned}$$

Therefore,

$$0 < b_{ii} = \int_{\|v\|\le\sigma}\hat f(u^*+v)dv - \int_{\|v\|\le\sigma} D_{v_i}(\hat f(u^*+v)v_i)dv. \tag{3.5}$$

According to Stokes' Theorem, we have

$$\int_{\|v\|\le\sigma} D_{v_i}(\hat f(u^*+v)v_i)dv = \int_{\|v\|=\sigma}\hat f(u^*+v)v_i n_i\, dA, \tag{3.6}$$

where n_i is the ith component of the unit outer normal vector n on the sphere $\|v\| = \sigma$. Therefore, n_i and v_i have the same sign and hence $v_i n_i \ge 0$. Note that $\hat f(u^*+v)$ is always nonnegative, we have

$$\int_{\|v\|\le\sigma} D_{v_i}(\hat f(u^*+v)v_i)dv = \int_{\|v\|=\sigma}\hat f(u^*+v)v_i n_i\, dA \ge 0.$$

Therefore, Eq. (3.5) implies that

$$0 < b_{ii} \le \int_{\|v\|\le\sigma}\hat f(u^*+v)dv = M.$$

Therefore,

$$0 < \frac{b_{ii}}{M} \le 1, \quad i = 1,\ldots,m.$$

Therefore, from Eqs. (3.2) and (3.3) we have

$$\begin{aligned}&P(DT(x^*))P^T\\&= I - \tfrac{1}{M}P(-DF_\sigma(x^*))P^T = \begin{pmatrix} 1-\frac{b_{11}}{M} & & & \\ & 1-\frac{b_{22}}{M} & & 0 \\ & 0 & \ddots & \\ & & & 1-\frac{b_{mm}}{M}\end{pmatrix}\end{aligned} \tag{3.7}$$

with $0 \le 1 - \frac{b_{ii}}{M} < 1$, i=1,..., m.

Note that $DT(x^*)$ has the same eigenvalues as $P(DT(x^*))P^T$, we get

$$\|DT(x^*)\| = \rho(DT(x^*)) < 1,$$

which implies T is a contractive mapping on a small neighborhood of x^*. This can be seen from the following: According to the continuity of norm, there are $\delta > 0$ and $\alpha < 1$ such that for any $x \in B(x^*,\delta)$, $\|DT(x)\| < \alpha < 1$. Therefore, the Mean Value Theorem implies that for any x, $y \in B(x^*,\delta)$ there is $\xi \in B(x^*,\delta)$ such that

$$\|T(x) - T(y)\| = \|DT(\xi)(x-y)\| \le \|DT(\xi)\|\|(x-y)\| < \alpha\|(x-y)\|.$$

Therefore, T is a contract mapping on $B(x^*,\delta)$.

Therefore, the Contract Mapping Theorem implies that x^* is an asymptotically stable fixed point of T. This proves that every strongly strict local maximum of G_σ is an asymptotically stable fixed point of T. Similarly we can prove that every strongly strict local minimum of G_σ is a unstable fixed point of T. □

Noting that

$$G_\sigma(x) = \int_{\|y\| \le \sigma} (\sigma^2 - \|y\|^2) f(x+y)dy$$

is actually the weighted sum of values of density function f on $B(x, \sigma)$, we conclude that when σ is small, local maxima of G_σ must occur near local maxima of f and local minima of G_σ must occur near local minima of f. Therefore, in ideal situations, T has an asymptotically stable fixed point near a local maximum of f and a unstable fixed point near a local minimum of f. These can be seen from the next theorem.

Let

$$N = \int_{\|y\| \le \sigma} (\sigma^2 - \|y\|^2)dy,$$

then define

$$g_\sigma(x) = G_\sigma(x)/N \tag{3.8}$$

Note that for a given σ, N is a constant. Therefore, G_σ and g_σ have the same local maxima and minima.

Theorem 3.3 *If f is uniformly continuous, then $g_\sigma \to f$ uniformly as $\sigma \to 0$.*

Proof Applying the Mean Value Theorem, we have

$$g_\sigma(x) = G_\sigma(x)/N = \frac{\int_{\|y\| \le \sigma} (\sigma^2 - \|y\|^2) f(x+y)dy}{\int_{\|y\| \le \sigma} (\sigma^2 - \|y\|^2)dy}$$
$$= \frac{f(x+\xi)\int_{\|y\| \le \sigma} (\sigma^2 - \|y\|^2)dy}{\int_{\|y\| \le \sigma} (\sigma^2 - \|y\|^2)dy} = f(x+\xi),$$

where $\|\xi\| < \sigma$. Since f is uniformly continuous, for any given $\varepsilon > 0$ there is $\delta > 0$ such that when $\|h\| < \delta$ we have

$$|f(x+h) - f(x)| < \varepsilon \quad \text{for any } x.$$

Therefore, when $\sigma < \delta$, we have

$$|g_\sigma(x) - f(x)| = |f(x+\xi) - f(x)| < \varepsilon$$

for any x. This implies that $g_\sigma \to f$ uniformly as $\sigma \to 0$. □

4 Rate of convergence

This section gives a rough estimate to the rate of convergence of IMC when near a strongly strict local maximum x^* of G_σ. We know that the rate of convergence depends on $\rho(DT(x^*))$. Therefore, what we need to do is to compute $\rho(DT(x^*))$. From Eq. (3.7), we have that

$$\rho(DT(x^*)) = \max_i (1 - \frac{b_{ii}}{M}).$$

But from Eqs. (3.4), (3.5) and (3.6),

$$b_{ii} = M - \int_{\|v\|=\sigma} \hat{f}(u^* + v)v_i n_i \, dA = M - \int_{\|v\|=\sigma} \hat{f}(u^* + v)\sigma n_i^2 \, dA,$$

where we have used the fact that $v_i = \sigma n_i$. Therefore,

$$1 - \frac{b_{ii}}{M} = \frac{1}{M} \int_{\|v\|=\sigma} \hat{f}(u^* + v)\sigma n_i^2 \, dA,$$

and applying the Mean Value Theorem,

$$M\sum_i(1-\frac{b_{ii}}{M}) = \int_{\|v\|=\sigma} \hat{f}(u^*+v)\sigma\, dA = \sigma \hat{f}(u^*+\xi)\int_{\|v\|=\sigma} dA$$
$$= \sigma\hat{f}(u^*+\xi)(\sigma^{m-1}S_m) = \hat{f}(u^*+\xi)(\sigma^m S_m),$$

where $\|\xi\| = \sigma$ and S_m is the area of unit sphere in $\mathbb{R}^m$. But

$$M = \int_{\|v\|\le\sigma} \hat{f}(u^*+v)dv = \hat{f}(u^*+\eta)\int_{\|v\|\le\sigma} dv = \hat{f}(u^*+\eta)(\sigma^m V_m),$$

where $\|\eta\| < \sigma$ and V_m is the volume of unit ball in $\mathbb{R}^m$. Note that $V_m = S_m/m$, therefore we have

$$\sum_i(1-\frac{b_{ii}}{M}) = \frac{1}{M}\hat{f}(u^*+\xi)(\sigma^m S_m) = \frac{m\hat{f}(u^*+\xi)}{\hat{f}(u^*+\eta)}.$$

Assume that $\hat{f}$ is symmetric on u^* on $B(u^*,\sigma)$, then $\hat{f}$ is constant on the sphere $\|v\| = \sigma$. Therefore, $b_{ii} = b_{jj}$ for all i, j. Therefore,

$$\rho(DT(x^*)) = \frac{1}{m}\sum_i(1-\frac{b_{ii}}{M}) = \frac{\hat{f}(u^*+\xi)}{\hat{f}(u^*+\eta)}.$$

Note that $\hat{f}(x) = f(P^T x)$, we have

$$\rho(DT(x^*)) = \frac{f(x^*+\xi')}{f(x^*+\eta')},$$

where $\xi' = P^T\xi$, $\eta' = P^T\eta$ and therefore $\|\xi'\| = \sigma$, $\|\eta'\| < \sigma$. When σ is small, we can approximate $f(x^*+\eta')$ by $f(x^*)$. Note that we have assumed $\hat{f}$ is constant on the sphere $\|v\| = \sigma$, which implies that f is constant on the sphere $\|x\| = \sigma$. Therefore we get

$$\rho(DT(x^*)) \approx \frac{f(x^*+\sigma e_1)}{f(x^*)},$$

where $e_1 = (1, 0, \ldots, 0)$.

References

[1] Aggarwal, C. C., Procopiuc, C., Wolf, J. L., Yu, P. S. & Park, J. S. *Fast algorithms for projected clustering*, SIGMOD '99, Philadelphia, PA, 1999, pp. 61-72.

[2] Agrawal, R., Gehrke, J., Gunopilos, D. & Raghavan, P. *Automatic subspace clustering of high dimensional data for data mining applications*, SIGMOD '98, Seattle, WA, 1998, pp. 94-105.

[3] Cao, Y. & Wu, J. *Projective ART for clustering data sets in high dimensional spaces*, Neural Networks **15** (2002), 105-120.

[4] Cao, Y. & Wu, J. *Iterative mean clustering: theory, method and neural network implementation*, Neural Networks, to appear.

[5] Carpenter, G. A. & Grossberg, S. *A massively parallel architecture for a self-organizing neural pattern recognition machine*, Computer Vision, Graphics, and Image Processing **37** (1987), 54-115.

[6] Carpenter, G. A. & Grossberg, S. *ART2: Self-organization of stable category recognition codes for analog input patterns*, Applied Optics **26** (1987), 4919-4930.

[7] Carpenter, G. A., Grossberg, S. & Rosen, D. B. *Fuzzy ART: Fast stable learning and categorization of analog patterns by an adaptive resonance system*, Neural Networks **4** (1991), 759-771.

[8] Ester, M., Kriegel, H.-P., Sander, J. & Xu, X. *A Density-Based Algorithm for Discovering Clusters in Large Spatial Databases with Noise*, Proceedings of the 2nd International Conference on Knowledge Discovery and Data Mining, Portland, Oregon, 1996, pp. 226-231.

[9] Everitt, B., Landau, S. & Leese, M. *Cluster analysis*, 4th ed., Oxford University Press, New York, 2001.

[10] Jain, A. K. & Dubes, R. C. *Algorithms for clustering data*, Prentice-Hall, Upper Saddle River, NJ, 1988.

[11] Jain, A. K., Murty, M. N. & Flynn, P. J. *Data clustering: a review*, ACM Computing Surveys **31** (1999), 264-323.

[12] Kaufman, L. & Rousseeuw, P. *Finding groups in data - an introduction to cluster analysis*, John Wiley & Sons, New York, 1990.

[13] Kohonen, T. *Self-Organizing Maps*, 3rd ed., Springer, Berlin, 2001.

[14] Ng, R. T. & Han, J. *Efficient and effective clustering methods for spatial data mining*, Proceedings of the 20th VLDB conference, Santiago, Chile, 1994, pp. 144-155.

[15] Zhang, T., Ramakrishnan, R. & Livny, M. (1996). *BIRCH: an efficient data clustering method for very large databases*, SIGMOD '96, Montreal, Canada, 1996, pp. 103-114.

Fields Institute Communications
Volume **42**, 2004

Global Attractivity in a Model of Bobwhite Quail with Delay

Anping Chen
Department of Mathematics
Xiangnan University
Chenzhou, Hunan 423000, China
chenap@263.net

Lihong Huang
College of Mathematics and Econometrics
Hunan University
Changsha, Hunan 410082, China
lhhuang@hnu.net.cn

Dedicated to Professor Zhicheng Wang on the occasion of his seventieth birthday

Abstract. Consider the bobwhite quail single-species growth model

$$x_{n+1} = \alpha x_n + \sum_{i=1}^{m} \frac{\beta_i x_n}{1 + x_{n-\tau_i}^k}, \quad n = 0, 1, 2, \cdots,$$

where $0 < \alpha < 1 < \alpha + \beta$, $k > 0, \beta_i > 0, \beta = \sum_{i=1}^{m} \beta_i, m$ and $\tau_i (i = 1, 2, \cdots, m)$ are positive integers. We obtain new sufficient conditions for the global attractivity of the positive equilibrium of the model equation. Our result improves and extends some known results, and partly answers an open problem proposed by G. Ladas in 1994.

1 Introduction

In 1990, J. G. Miltin and J. Belair [6] raised the difference equation

$$x_{n+1} = \alpha x_n + \frac{\beta x_n}{1 + x_n^k}, \quad n = 0, 1, 2, \cdots \tag{1.1}$$

as a population model for the growth of the bobwhite quail population of northern Wisconsin, where α, β and k are positive constants. In 1994, G. Ladas [3] proposed the following open problems:

(i). Investigate the boundedness, persistence, permanence, oscillation, periodicity, and global attractivity of Eq.(1.1);

2000 *Mathematics Subject Classification.* Primary 39A11; Secondary 92B05.

Supported by the NNSF of China (10371034) and Foundation for University Key Teacher by the Ministry of Education of China and also by the Foundation of professor project of Chenzhou Teachers College.

(ii). Extend Eq.(1.1) by introducing delays into the model, and investigate the delayed equation.

In 1996, A. P. Chen [1], H. S. Tang and C. H. Ou [7] considered the global stability of (1.1). In 1999, by introducing delays τ (positive integer) into the model, A. P. Chen [2] investigated the stability and oscillation of the delayed equation

$$x_{n+1} = \alpha x_n + \frac{\beta x_n}{1 + x_{n-\tau}^k}, \quad n = 0, 1, 2, \cdots, \tag{1.2}$$

and obtained the following result:

Theorem A. *Assume that* $\frac{1}{\beta} k(\tau+1)(1-\alpha)(\alpha+\beta-1) < 1$, *and the condition*

(C1). $k\beta(\tau+1) < 4\alpha$

is satisfied. Then every positive solution $\{x_n\}_{n\geq -\tau}$ *of Eq.(1.2) tends to the positive equilibrium* $N = \left(\frac{\alpha+\beta-1}{1-\alpha}\right)^{\frac{1}{k}}$.

In 2000, L. S. Liao and A. P. Chen [5] discussed the delayed model

$$x_{n+1} = \alpha x_n + \frac{\beta x_n}{1 + \sum_{i=1}^{p} \delta_i x_{n-i}^k}, \quad n = 0, 1, 2, \cdots, \tag{1.3}$$

where α, δ_i, β and k are positive constants, and p is a positive integer. They gave the following global stability result:

Theorem B. *Assume that*

$$k^2(p+1)^2 < \frac{2(2\alpha+\beta)}{(1-\alpha)(\alpha+\beta-1)}$$

and

$$k^2(p+1)^2 \leq \frac{2(2\alpha+\beta)\sqrt{(2\alpha+\beta)^2 + 11\alpha(\alpha+\beta)}}{\beta^2}.$$

Then every positive solution $\{x_n\}_{n\geq -p}$ *of Eq.(1.3) tends to the positive equilibrium* $N^* = \left(\frac{\alpha+\beta-1}{(1-\alpha)\sum_{i=1}^{p}\delta_i}\right)^{\frac{1}{k}}$.

In this paper, we consider the following delayed model

$$x_{n+1} = \alpha x_n + \sum_{i=1}^{m} \frac{\beta_i x_n}{1 + x_{n-\tau_i}^k}, \quad n = 0, 1, 2, \cdots, \tag{1.4}$$

where α, k and β_i are positive constants, m and τ_i are positive integers, $i = 1, 2, \cdots, m$, and

(H0). $0 < \alpha < 1 < \alpha + \beta, \quad k > 0$, where $\beta = \sum_{i=1}^{m} \beta_i$.

It is easy to see that Eq.(1.4) also has a unique positive equilibrium $N = (\frac{\alpha+\beta-1}{1-\alpha})^{\frac{1}{k}}$.

By a solution of (1.4), we mean a sequence $\{x_n\}_{n\geq -\tau}$ which satisfies (1.4) for $n \geq 0$, where $\tau = \max_{1\leq i\leq m}\{\tau_i\}$. Clearly, for any given constants $a_{-\tau}, a_{-\tau+1}, \cdots, a_{-1}, a_0$, Eq.(1.4) has a unique solution satisfying the initial condition:

$$x_i = a_i, \quad i = -\tau, -\tau+1, \cdots, 0.$$

Moreover, it is easy to see that the solution of Eq.(1.4) is positive if the initial values $a_i \geq 0 (i = -\tau, -\tau + 1, \cdots, 0)$ and $\sum_{i=-\tau}^{0} a_i \neq 0$.

In this paper, Our main result is the following theorem.

Theorem C. *Assume that (H0) hold and $\frac{1}{\beta}k(\tau + 1)(1 - \alpha)(\alpha + \beta - 1) < 1$, and one of the following conditions is satisfied,*

(H1). $k\beta(\tau + 1) < 4\sqrt{\alpha(\alpha + \beta)}$;
(H2). $k\beta(\tau + 1) < 3\sqrt[3]{\alpha(\alpha + \beta)(2\alpha + \beta)}$;
(H3). $k\beta(\tau + 1) < \sqrt{2\alpha(\alpha + \beta) + (2\alpha + \beta)^2 + 6\sqrt[3]{\alpha^2(\alpha + \beta)^2(2\alpha + \beta)^2}}$.

Then every positive solution $\{x_n\}_{n\geq -\tau}$ of Eq. (1.4) tends to the positive equilibrium N.

Clearly, Eq. (1.2) is a special case of Eq. (1.4) with $m = 1$. It should be noted that (C1) implies (H1) if $m = 1$, but (H1) does not imply (C1), and that (H2) is also different from (C1) even for such a special case. Therefore, Theorem C improves and generalizes Theorem A.

On the other hand, if let $\delta_1 = \delta_2 = \cdots = \delta_{p-1} = 0, \delta_p = 1$ and $p = \tau$ in Eq. (1.3), and let $m = 1, \tau_1 = \tau$ and $\beta_1 = \beta$ in Eq. (1.4), then equations (1.2), (1.3) and (1.4) are the same. But Theorem C is also different from Theorem B even for this special case. For example, in Eq. (1.2), we take $k = 1, \tau = 16, \alpha = 0.95$ and $\beta = 0.25$, then

$$\frac{1}{\beta}k(\tau + 1)(1 - \alpha)(\alpha + \beta - 1) = 0.68 < 1$$

and

$$k\beta(\tau + 1) = 4.25 < 4.2708 \simeq 4\sqrt{\alpha(\alpha + \beta)}.$$

Hence the conditions of Theorem C hold. However, we have

$$k^2(p + 1)^2 = k^2(\tau + 1)^2 = 289 < \frac{2(2\alpha + \beta)}{(1 - \alpha)(\alpha + \beta - 1)} = 430$$

and

$$k^2(p + 1)^2 = 289 > 285.022217 \simeq \frac{2(2\alpha + \beta)\sqrt{(2\alpha + \beta)^2 + 11\alpha(\alpha + \beta)}}{\beta^2}.$$

This implies that the conditions of Theorem B are not satisfied.

On the other hand, if we take $k = 0.5, \tau = 10, \alpha = 0.8$ and $\beta = 0.85$ in Eq. (1.3), then

$$k^2(p + 1)^2 = k^2(\tau + 1)^2 = 30.35 < \frac{2(2\alpha + \beta)}{(1 - \alpha)(\alpha + \beta - 1)} = 37.6923$$

and

$$k^2(p + 1)^2 = 30.25 < 30.723689 \simeq \frac{2(2\alpha + \beta)\sqrt{(2\alpha + \beta)^2 + 11\alpha(\alpha + \beta)}}{\beta^2}.$$

Therefore, the conditions of Theorem B hold. However, we have

$$\frac{1}{\beta}k(\tau + 1)(1 - \alpha)(\alpha + \beta - 1) = 0.841176 < 1$$

and

$$k\beta(\tau + 1) = 4.6750 > 4.59565 \simeq \sqrt{\alpha(\alpha + \beta)},$$

which implies that the conditions of Theorem C do not hold.

2 Some lemmas

In this section, we give some preliminary lemmas which will be very useful in the proof of our main result.

Lemma 2.1 *Assume that* $\mu_1(t), \mu_2(t) \in C^1[a,b], \mu_1'(t) < 0, \mu_2'(t) < 0$ *and* $\mu_1(t) < \mu_2(t)$ *for* $t \in (a,b)$, *and that* $\mu_1(a) = \mu_2(a)$ *and* $\mu_1'(a) \neq 0$. *Then* $\frac{\mu_2'(a)}{\mu_1'(a)} \leq 1$.

Proof By Cauchy mean value theorem, for $\forall t \in [a,b]$, there exits a $\xi_t \in (a,b)$ such that

$$\frac{\mu_2(a) - \mu_2(t)}{\mu_1(a) - \mu_1(t)} = \frac{\mu_2'(\xi_t)}{\mu_1'(\xi_t)} < 1.$$

Let $t \to a+$. Then the proof is complete. □

Lemma 2.2 *The system of inequalities*

$$\lambda \geq (\tau+1)\ln\left\{\alpha + \frac{\beta}{1+N^k e^{k\mu}}\right\}, \tag{2.1}$$

$$\mu \leq (\tau+1)\ln\left\{\alpha + \frac{\beta}{1+N^k e^{k\lambda}}\right\} \tag{2.2}$$

has a unique solution $\lambda = \mu = 0$ *in the region* $D = \{(\lambda,\mu); -\infty < \lambda \leq 0 \leq \mu < +\infty\}$ *if and only if so does the following system of equations,*

$$\lambda = (\tau+1)\ln\left\{\alpha + \frac{\beta}{1+N^k e^{k\mu}}\right\},$$

$$\mu = (\tau+1)\ln\left\{\alpha + \frac{\beta}{1+N^k e^{k\lambda}}\right\}.$$

The proof is similar to that of the paper [4], and is omitted.

Lemma 2.3 *Assume that* $\frac{1}{\beta}k(\tau+1)(1-\alpha)(\alpha+\beta-1) < 1$, *and one of the conditions (H1), (H2) and (H3) is satisfied. Then the system of equations*

$$\lambda = (\tau+1)\ln\left\{\alpha + \frac{\beta}{1+N^k e^{k\mu}}\right\}, \tag{2.3}$$

$$\mu = (\tau+1)\ln\left\{\alpha + \frac{\beta}{1+N^k e^{k\lambda}}\right\} \tag{2.4}$$

has a unique solution $(0,0)$ *in the region* D .

Proof We denote curves C_1 and C_2 respectively as the followings.

$$C_1 : \mu_1 = \mu_1(\lambda), \quad \mu_1 = (\tau+1)\ln\left\{\alpha + \frac{\beta}{1+N^k e^{k\lambda}}\right\};$$

$$C_2 : \mu_2 = \mu_2(\lambda), \quad \lambda = (\tau+1)\ln\left\{\alpha + \frac{\beta}{1+N^k e^{k\mu_2}}\right\}.$$

Notice that $N = (\frac{\alpha+\beta-1}{1-\alpha})^{\frac{1}{k}}$, it is obvious that C_1 and C_2 pass the point $(0,0)$. Since $|\mu_1(\lambda)| \leq (\tau+1)\ln(\alpha+\beta)$ for $\lambda \in (-\infty, 0]$ and $\lim_{\lambda\to-\infty} \mu_1(\lambda) = (\tau+1)\ln(\alpha+\beta)$, the curve C_1 is bounded in the region D. On the other hand, the curve C_2 has an asymptotic line $\lambda = (\tau+1)ln\alpha$ in the region D, which implies that C_2 is unbounded and $\lim_{\lambda\to(\tau+1)ln\alpha+} \mu_2(\lambda) = +\infty$. By some simple calculations, we obtain

$$\frac{d\mu_1}{d\lambda} = -\frac{kN^k e^{k\lambda}(\tau+1)\beta}{(1+N^k e^{k\lambda})(\beta+\alpha(1+N^k e^{k\lambda}))} < 0, \tag{2.5}$$

$$\frac{d\mu_1}{d\lambda}|_{(0,0)} = -\frac{k(\tau+1)(1-\alpha)(\alpha+\beta-1)}{\beta}, \tag{2.6}$$

$$\frac{d\mu_2}{d\lambda} = -\frac{(1+N^k e^{k\mu_2})(\beta+\alpha(1+N^k e^{k\mu_2}))}{kN^k e^{k\mu_2}(\tau+1)\beta} < 0, \tag{2.7}$$

$$\frac{d\mu_2}{d\lambda}|_{(0,0)} = -\frac{\beta}{k(\tau+1)(1-\alpha)(\alpha+\beta-1)}. \tag{2.8}$$

In view of $\frac{1}{\beta}k(\tau+1)(1-\alpha)(\alpha+\beta-1) < 1$, we have

$$\frac{d\mu_2}{d\lambda}|_{(0,0)} < \frac{d\mu_1}{d\lambda}|_{(0,0)} < 0.$$

Therefore, there exits a constant $\varepsilon > 0$ such that $\frac{d\mu_2}{d\lambda} < \frac{d\mu_1}{d\lambda}$ for $\lambda \in (-\varepsilon, 0)$. This, together with $\mu_2(0) = \mu_1(0) = 0$, implies that

$$\mu_2(\lambda) > \mu_1(\lambda) \quad \text{for } \lambda \in (-\varepsilon, 0).$$

If the system of (2.3) and (2.4) has another solution except (0,0) in the region D, then it is certain that there exists $a < -\varepsilon$ such that $\mu_2(a) = \mu_1(a)$ and $\mu_2(\lambda) > \mu_1(\lambda)$ for $\lambda \in (a, 0)$. Again by (2.5) and (2.7), we have $\mu_2'(\lambda) < 0$ and $\mu_1'(\lambda) < 0$ for $\lambda \in (a, 0)$. From Lemma 2.1, it follows that $\frac{\mu_2'(a)}{\mu_1'(a)} \leq 1$. On the other hand, from (2.5) and (2.7), we obtain

$$\frac{\mu_2'(a)}{\mu_1'(a)} = \frac{(\alpha+\beta+\alpha N^k e^{kb})(1+N^k e^{kb})(\alpha+\beta+\alpha N^k e^{ka})(1+N^k e^{ka})}{\beta^2(\tau+1)^2 k^2 (N^k)^2 e^{ka} e^{kb}}. \tag{2.9}$$

If (H1) holds, then by using mean value inequality and (2.9), we obtain

$$\begin{aligned}\frac{\mu_2'(a)}{\mu_1'(a)} &\geq \frac{16\sqrt{(\alpha+\beta)\alpha N^k e^{kb}}\sqrt{N^k e^{kb}}\sqrt{(\alpha+\beta)\alpha N^k e^{ka}}\sqrt{N^k e^{ka}}}{\beta^2(\tau+1)^2 k^2 (N^k)^2 e^{ka} e^{kb}} \\ &= \frac{16\alpha(\alpha+\beta)}{\beta^2(\tau+1)^2 k^2} > 1.\end{aligned}$$

This is a contradiction.

If (H2) holds, then it follows from (2.9) that

$$\begin{aligned}\frac{\mu_2'(a)}{\mu_1'(a)} &= \frac{[(\alpha+\beta)+(2\alpha+\beta)N^k e^{kb}+\alpha(N^k)^2(e^{kb})^2]}{\beta(\tau+1)kN^k e^{kb}} \\ &\quad \cdot \frac{[(\alpha+\beta)+(2\alpha+\beta)N^k e^{ka}+\alpha(N^k)^2(e^{ka})^2]}{\beta(\tau+1)kN^k e^{ka}} \\ &\geq \frac{9\sqrt[3]{\alpha(\alpha+\beta)(2\alpha+\beta)(N^k)^3(e^{kb})^3}\sqrt[3]{\alpha(\alpha+\beta)(2\alpha+\beta)(N^k)^3(e^{ka})^3}}{\beta^2(\tau+1)^2 k^2 (N^k)^2 e^{ka} e^{kb}} \\ &= \frac{9\sqrt[3]{\alpha^2(\alpha+\beta)^2(2\alpha+\beta)^2}}{\beta^2(\tau+1)^2 k^2} > 1,\end{aligned}$$

which is a contradiction.

Finally, we consider the case where (H3) holds. Set

$$\Omega = (\alpha+\beta+\alpha N^k e^{kb})(1+N^k e^{kb})(\alpha+\beta+\alpha N^k e^{ka})(1+N^k e^{ka}).$$

Then, we have

$$\begin{aligned}\Omega &= \alpha(\alpha+\beta)(N^k)^2((e^{kb})^2+(e^{ka})^2)+(2\alpha+\beta)^2(N^k)^2e^{kb}e^{ka}\\ &\quad +(\alpha+\beta)^2+\alpha(2\alpha+\beta)(N^k)^3e^{ka}(e^{kb})^2+\alpha(2\alpha+\beta)(N^k)^3e^{kb}(e^{ka})^2\\ &\quad +(\alpha+\beta)(2\alpha+\beta)N^ke^{kb}+(\alpha+\beta)(2\alpha+\beta)N^ke^{ka}+\alpha^2(N^k)^4(e^{kb})^2(e^{ka})^2\\ &\geq 2\alpha(\alpha+\beta)(N^k)^2e^{ka}e^{kb}+(2\alpha+\beta)^2(N^k)^2e^{kb}e^{ka}\\ &\quad +3\sqrt[3]{(\alpha+\beta)^2\alpha^2(2\alpha+\beta)^2(N^k)^6(e^{ka})^3(e^{kb})^3}\\ &\quad +3\sqrt[3]{(\alpha+\beta)^2\alpha^2(2\alpha+\beta)^2(N^k)^6(e^{ka})^3(e^{kb})^3}\\ &= \{2\alpha(\alpha+\beta)+(2\alpha+\beta)^2+6\sqrt[3]{(\alpha+\beta)^2\alpha^2(2\alpha+\beta)^2}\}(N^k)^2e^{ka}e^{kb}.\end{aligned}$$

Therefore, it follows from (2.9) and (H3) that

$$\frac{\mu_2'(a)}{\mu_1'(a)} \geq \frac{2\alpha(\alpha+\beta)+(2\alpha+\beta)^2+6\sqrt[3]{(\alpha+\beta)^2\alpha^2(2\alpha+\beta)^2}}{(\beta k(\tau+1))^2} > 1,$$

which is a contradiction. □

Remark 1. In Lemma 2.3, the condition $\frac{1}{\beta}k(\tau+1)(1-\alpha)(\alpha+\beta-1)<1$ is almost necessary for the existence of a unique solution of the system of equations (2.3) and (2.4) in the region D. In fact, from the proof of Lemma 2.3, if $\frac{1}{\beta}k(\tau+1)(1-\alpha)(\alpha+\beta-1)>1$, then it follows from (2.6) and (2.8) that

$$\frac{d\mu_2}{d\lambda}|_{(0,0)} > \frac{d\mu_1}{d\lambda}|_{(0,0)}.$$

Therefore, there exists a constant $\varepsilon>0$ such that $\frac{d\mu_2}{d\lambda} > \frac{d\mu_1}{d\lambda}$ for $\lambda\in(-\varepsilon,0)$, which, together with $\mu_2(0)=\mu_1(0)=0$, implies that

$$\mu_2(\lambda) < \mu_1(\lambda) \quad \text{for } \lambda\in(-\varepsilon,0).$$

Again noting that $\lim_{\lambda\to(\tau+1)ln\alpha+}\mu_2(\lambda)=+\infty, |\mu_1(\lambda)|\leq(\tau+1)\ln(\alpha+\beta)$ for $\lambda\in(-\infty,0]$, and that $\lim_{\lambda\to-\infty}\mu_1=(\tau+1)\ln(\alpha+\beta)$, it is easy to see that there exists $a<-\varepsilon$ such that $\mu_2(a)=\mu_1(a)$, which means $(a,\mu_1(a))$ is another solution of the system of equations (2.3) and (2.4) except (0,0) in the region D.

3 Proof of main result

In this section, we give the proof of Theorem C.

Assume that $\{x_n\}_{n\geq-\tau}$ is a positive solution of Eq.(1.4). Let $\ln\frac{x_n}{N}=y_n$. Then it follows from (1.4) that

$$y_{n+1}-y_n=\ln\left\{\alpha+\sum_{i=1}^{m}\frac{\beta_i}{1+N^ke^{ky_{n-\tau_i}}}\right\}. \tag{3.1}$$

We distinguish three cases to prove that $\lim_{n\to+\infty}y_n=0$.

Case 1. $y_n\geq0$ eventually holds. For this case, there exists a positive integer n_0 such that $y_{n-\tau_i}\geq0$ for $n\geq n_0$ and $i=1,2,\cdots,m$. From (3.1), we have $\Delta y_n=y_{n+1}-y_n\leq\ln\{\alpha+\frac{\beta}{1+N^k}\}=\ln1=0$, which implies that y_n is a decreasing bounded sequence. Therefore, $\lim_{n\to+\infty} y_n$ exists, say $\lim_{n\to+\infty} y_n = y^*$. Let $n \to +\infty$ in Eq.(3.1). Then we obtain $y^* = 0$, which means $\lim_{n\to+\infty} y_n = 0$.

Case 2. $y_n \le 0$ eventually holds. For this case, the proof is similar to that of Case 1, and is omitted.

Case 3. y_n be a strictly oscillatory solution of (3.1). For this case, it is certain that for arbitrary given positive integer n there exist positive integers $n_1 > n$ and $n_2 > n$ such that $y_{n_1} \ge 0$ and $y_{n_2} \le 0$. Let $\liminf_{n\to+\infty} y_n = \lambda$ and $\limsup_{n\to+\infty} y_n = \mu$. Then

$$-\infty \le \lambda \le 0 \le \mu \le +\infty.$$

Next, we will show that $\lambda \ne -\infty$ and $\mu \ne +\infty$.

If $\lambda = -\infty$, then there exists a strictly increasing positive integer sequence $\{n_j\}$ with $\lim_{j\to+\infty} n_j = +\infty$ such that $y_{n_j} = \min\{y_n; n \le n_j\}$ satisfies $\lim_{j\to+\infty} y_{n_j} = -\infty$. From (3.1), we have

$$y_{n+1} - y_n \ge \ln\alpha, \quad i.e. \quad y_{n+1} \ge \ln\alpha + y_n.$$

It follows that

$$y_{n_j} \ge (\tau_i + 1)\ln\alpha + y_{n_j-\tau_i-1} \quad \text{for } i = 1, 2, \cdots, m.$$

Thus, we have

$$\lim_{j\to+\infty} y_{n_j-\tau_i-1} = -\infty \quad \text{for } i = 1, 2, \cdots, m.$$

Therefore, for $\forall P > 0$, there exits a positive integer Q such that

$$y_{n_j-\tau_i-1} < -P \quad \text{for } i = 1, 2, \cdots, m.$$

It follows that

$$\ln\left\{\alpha + \sum_{i=1}^{m} \frac{\beta_i}{1 + N^k e^{k y_{n_j-\tau_i-1}}}\right\} > \ln\left\{\alpha + \frac{\beta}{1 + N^k e^{-kP}}\right\}.$$

Note that $y_{n_j} = \min\{y_n; n \le n_j\}$, by (3.1), we have

$$\ln\left\{\alpha + \sum_{i=1}^{m} \frac{\beta_i}{1 + N^k e^{k y_{n_j-\tau_i-1}}}\right\} \le 0.$$

This implies that $\ln\{\alpha + \frac{\beta}{1+N^k e^{-kP}}\} < 0$. Let $P \to +\infty$. Then we obtain $\beta \le 1-\alpha$. This contradicts the fact that $\alpha + \beta > 1$.

If $\mu = +\infty$, then there exists a strictly increasing positive integer sequence $\{n_j\}$ with $\lim_{j\to+\infty} n_j = +\infty$ such that $y_{n_j+1} = \max\{y_n; n \le n_j+1\}$ satisfies $\lim_{j\to+\infty} y_{n_j+1} = +\infty$. From (3.1), we have

$$\ln\left\{\alpha + \sum_{i=1}^{m} \frac{\beta_i}{1 + N^k e^{k y_{n_j-\tau_i}}}\right\} = y_{n_j+1} - y_{n_j} \ge 0$$

and

$$\sum_{i=1}^{m} \frac{\beta_i}{1 + N^k e^{k y_{n_j-\tau_i}}} \ge 1 - \alpha. \tag{3.2}$$

On the other hand,

$$y_{n+1} - y_n \le \ln(\alpha + \beta), \quad i.e. \quad y_{n+1} \le \ln(\alpha + \beta) + y_n.$$

Therefore,

$$y_{n_j+1} \leq (\tau_i + 1)\ln(\alpha+\beta) + y_{n_j-\tau_i} \quad \text{for } i = 1, 2, \cdots, m.$$

It follows that

$$\lim_{j\to+\infty} y_{n_j-\tau_i} = +\infty \quad \text{for } i = 1, 2, \cdots, m.$$

Thus, for $\forall I > 0$, there exits a positive integer J such that

$$y_{n_j-\tau_i} > I \quad \text{for } j > J \text{ and } i = 1, 2, \cdots, m.$$

It follows that

$$\ln\left\{\alpha + \sum_{i=1}^{m} \frac{\beta_i}{1+N^k e^{k y_{n_j-\tau_i}}}\right\} < \ln\left\{\alpha + \frac{\beta}{1+N^k e^{kI}}\right\}. \tag{3.3}$$

By (3.2) and (3.3), we have $\ln\{\alpha + \frac{\beta}{1+N^k e^{kI}}\} > 0$. Let $I \to +\infty$. Then we have $\ln \alpha \geq 0$. This contradicts the fact that $\alpha < 1$.

Now, we prove that $\lambda = \mu = 0$. It is easy to see that for $\forall \varepsilon > 0$ there exists a positive integer n^* such that

$$\lambda_\varepsilon = \lambda - \varepsilon < y_{n-\tau_i} < \mu + \varepsilon = \mu_\varepsilon \quad \text{for } n > n^*. \tag{3.4}$$

From (3.1), we have

$$y_{n+1} - y_n \leq \ln\left\{\alpha + \sum_{i=1}^{m} \frac{\beta_i}{1+N^k e^{k\lambda_1}}\right\} \tag{3.5}$$

and

$$y_{n+1} - y_n \geq \ln\left\{\alpha + \sum_{i=1}^{m} \frac{\beta_i}{1+N^k e^{k\mu_1}}\right\}. \tag{3.6}$$

Since $\{y_n\}$ is a strictly oscillatory solution of (3.1), there exists a positive integer sequence $\{n_i\}$ with $\lim_{i\to+\infty} n_i = +\infty$ such that $y_{n_i} < 0$ and $y_{n_i+1} \geq 0$. Moreover, in the open interval $(n_i, n_i + 1)$, there exist both of the positive and negative terms of the sequence $\{y_n\}$. Let m_i and M_i be integers in $(n_i, n_i + 1)$ such that

$$y_{m_i+1} = \min\{y_j; n_i < j < n_{i+1}\} < 0, \Delta y_{m_i} \leq 0, \lim_{i\to+\infty} y_{m_i+1} = \lambda;$$

$$y_{M_i+1} = \max\{y_j; n_i < j < n_{i+1}\} > 0, \Delta y_{m_i} \geq 0, \lim_{i\to+\infty} y_{M_i+1} = \mu.$$

Again in view of (3.1), we have

$$\ln\left\{\alpha + \sum_{i=1}^{m} \frac{\beta_i}{1+N^k e^{k y_{m_i-\tau_i}}}\right\} \leq 0, \tag{3.7}$$

$$\ln\left\{\alpha + \sum_{i=1}^{m} \frac{\beta_i}{1+N^k e^{k y_{M_i-\tau_i}}}\right\} \geq 0. \tag{3.8}$$

It follows from (3.7) that

$$\sum_{i=1}^{m} \left\{\frac{\beta_i}{1+N^k e^{k y_{m_i-\tau_i}}} - \frac{\beta_i}{1+N^k}\right\} \leq 0.$$

Thus, there exits at least an integer $i_0 \in [1, m]$ such that $y_{m_i-\tau_{i_0}} > 0$. Similarly, there exits at least an integer $i_0^* \in [1, m]$ such that $y_{M_i-\tau_{i_0^*}} < 0$. Assume that m_i^*

and M_i^* are the largest integers such that $y_{m_i-\tau_{i_0}} > 0$ and $y_{M_i-\tau_{i_0^*}} < 0$ respectably. Then we have

$$\max\{n_i, m_i-\tau\} \le m_i^* < m_i - 1, y_{m_i^*} \ge 0, y_j < 0, j = m_i^*+1, \cdots, m_i+1;$$
$$\max\{n_i, M_i-\tau\} \le M_i^* < M_i - 1, y_{M_i^*} \le 0, y_j > 0, j = M_i^*+1, \cdots, M_i+1.$$

By summing (3.1) from $n = m_i^*$ to m_i and using (3.4), we obtain

$$\begin{aligned} y_{m_i+1} &= y_{m_i^*} + \sum_{n=m_i^*}^{m_i} \ln\left\{\alpha + \sum_{i=1}^{m} \frac{\beta_i}{1+N^k e^{k y_{n-\tau_i}}}\right\} \qquad (3.9)\\ &\ge y_{m_i^*} + (m_i+1-m_i^*)\ln\left\{\alpha + \frac{\beta}{1+N^k e^{k\mu_\varepsilon}}\right\}\\ &\ge (m_i+1-m_i^*)\ln\left\{\alpha + \frac{\beta}{1+N^k e^{k\mu_\varepsilon}}\right\}\\ &\ge (m_i+1-m_i^*+\tau)\ln\left\{\alpha + \frac{\beta}{1+N^k e^{k\mu_\varepsilon}}\right\}\\ &\ge (\tau+1)\ln\left\{\alpha + \frac{\beta}{1+N^k e^{k\mu_\varepsilon}}\right\}. \end{aligned}$$

Let $i \to +\infty$. Then we have

$$\lambda \ge (\tau+1)\ln\left\{\alpha + \frac{\beta}{1+N^k e^{k\mu_\varepsilon}}\right\}.$$

Again note that $\mu_\varepsilon = \mu - \varepsilon$. Let $\varepsilon \to 0$. It follows that

$$\lambda \ge (\tau+1)\ln\left\{\alpha + \frac{\beta}{1+N^k e^{k\mu}}\right\}. \qquad (3.10)$$

By a similar argument as the above, we can obtain

$$u \le (\tau+1)\ln\left\{\alpha + \frac{\beta}{1+N^k e^{k\lambda}}\right\}. \qquad (3.11)$$

In view of Lemmas 2.2 and 2.3, the system of the inequalities (3.9) and (3.10) has a unique solution $(0,0)$ in the region D. This complete the proof of Theorem C.

References

[1] A.P. Chen, *Oscillation and global attractivity of the systems of Bobwhite Quail Populations*, Annals of Mathematics (Hunan),**16(1)**,(1996),112-116.(in Chinese)

[2] A. P. Chen, *Oscillation and global stability in a bobwhite quail model with delay*, Journal of Biomathematis , **14(2)**,(1999), 157-164 (in Chinese).

[3] G. Ladas, *Open problems and conference on difference equations and application*, San Antouis Texas, Gordon and Breach Publishers, Inc, **may** 25-28, 1994.

[4] J.W. Li, *Global stability in a discrete model of Nicholson's blowflies*, Ann. of Diff. Eqs. , **12(2)**,(1996),173-182.

[5] L. S. Liao. A. P. Chen *Global Attractivity in a model of bobwhite quail model with delay*, Ann. of Diff. Eqs., **16(2)**,(2000),123-133 (in Chinese).

[6] J. G. Milton and J. Belair, *Chaos, noise, and extinction in models of population growth*, Theor. Pop. Biol. **37**, (1990), 273-290.

[7] H. S. Tang and C. H. Ou, *Some properties of a kind of difference equation*, Ann. Of Diff. Equs, **12(3)**, (1996),335-340.

[8] J. S. Yu, *Global stabiuty in a Discrete Delay Logistic Model*, Acta. Math. Appl. Sinic.A, **14(1)**, (1998), 80-85.

Fields Institute Communications
Volume **42**, 2004

Linear Time Discrete Periodic Diffusion Networks

Sui Sun Cheng
Department of Mathematics
Tsing Hua University
Hsinchu, Taiwan 30043
P.R. China
sscheng@math.nthu.edu.tw

Chen-Wei Chen
Department of Mathematics
Tsing Hua University
Hsinchu, Taiwan 30043
P.R. China

Tzong Yann Wu
Department of Mathematics
Tsing Hua University
Hsinchu, Taiwan 30043
P.R. China

Abstract. An elementary dynamic economic model is built to simulate the asymptotic behaviors of the wealth of n-nations which engage in local and uniform trading among neighbors. Detailed mathematical analyses show a variety of conclusions that can help us understand the dynamics involved in more realistic situations.

1 Introduction

The wealth of a nation in a certain time period depends on many factors. Indeed, many dynamical economic models have been built to illustrate such dependency (see e.g. [1-5]). In this paper, we build a dynamical model in which there are n nations and trading among them is carried out in 'local' and 'uniform' manners. Although our model is elementary, we will see that the subsequent theoretical conclusions yield insights that are helpful in understanding the realistic situations, and in building more realistic economic evolution models.

To be more precise, let there be n ($n \geq 2$) nations and let $u_i^{(t)}$ be the monetary value of the total assets and liabilities of a nation in the time period $t \in \{0, 1, 2, ...\}$. We remark that since liabilities may exceed assets, $u_i^{(t)}$ may take on negative values.

2000 *Mathematics Subject Classification.* Primary 39A11, 37N40; Secondary 91B02.
Key words and phrases. Economic networks, diffusion, trading propensity, wealth distribution, oscillatory instability.

We assume that the nations are arranged in a circular network so that u_1 has exactly two neighbors u_n and u_2, u_2 has exactly two neighbors u_1 and u_3, etc. Such a situation occurs when the nations are situated at the vertices of a regular polygon. During the time period t, if the wealth $u_1^{(t)}$ of the first nation is higher than $u_2^{(t)}$, money will tend to flow from the first nation to the second. The subsequent change of wealth of the second nation is $u_2^{(t+1)} - u_2^{(t)}$, and it is reasonable to postulate that it is proportional to the difference $u_1^{(t)} - u_2^{(t)}$, say, $\gamma(u_1^{(t)} - u_2^{(t)})$, where γ is a proportionality constant which reflects the collective exchange propensity. Similarly, money will flow from the third nation to the second if $u_3^{(t)} > u_2^{(t)}$. Thus, it is reasonable that the total effect is

$$u_2^{(t+1)} - u_2^{(t)} = \gamma\left(u_1^{(t)} - u_2^{(t)}\right) + \gamma\left(u_3^{(t)} - u_2^{(t)}\right) = \gamma\left(u_3^{(t)} - 2u_2^{(t)} + u_1^{(t)}\right).$$

By similar considerations, we may then obtain the system of equations

$$\begin{aligned} u_1^{(t+1)} - u_1^{(t)} &= \gamma\left(u_n^{(t)} - 2u_1^{(t)} + u_2^{(t)}\right), \\ u_2^{(t+1)} - u_2^{(t)} &= \gamma\left(u_1^{(t)} - 2u_2^{(t)} + u_3^{(t)}\right), \\ \cdots &= \cdots \\ u_j^{(t+1)} - u_j^{(t)} &= \gamma\left(u_{j-1}^{(t)} - 2u_j^{(t)} + u_{j+1}^{(t)}\right), \\ \cdots &= \cdots \\ u_n^{(t+1)} - u_n^{(t)} &= \gamma\left(u_{n-1}^{(t)} - 2u_n^{(t)} + u_1^{(t)}\right). \end{aligned}$$

Our system can also be written as

$$u^{(t+1)} = (I - \gamma A_n)u^{(t)} \tag{1.1}$$

where $u^{(t)}$ is the column vector $\left(u_1^{(t)}, \cdots, u_n^{(t)}\right)^T$ and A_n is the matrix given by

$$A_2 = \begin{pmatrix} 2 & -2 \\ -2 & 2 \end{pmatrix}$$

or

$$A_n = \begin{pmatrix} 2 & -1 & & & 0 & -1 \\ -1 & 2 & -1 & & & 0 \\ & \cdot & \cdot & \cdot & & \\ & & \cdot & \cdot & \cdot & \\ 0 & & & -1 & 2 & -1 \\ -1 & 0 & & & -1 & 2 \end{pmatrix}_{n\times n}, \quad n \geq 3.$$

Although our model is simple, we believe that a detailed mathematical analysis is prerequisite to the investigation of more complicated models such as

$$\begin{aligned} u_1^{(t+1)} - u_1^{(t)} &= \gamma\left(u_n^{(t)} - 2u_1^{(t)} + u_2^{(t)}\right) + \mu u_1^{(t)}\left(1 - u_1^{(t)}\right) \\ u_2^{(t+1)} - u_2^{(t)} &= \gamma\left(u_1^{(t)} - 2u_2^{(t)} + u_3^{(t)}\right) + \mu u_2^{(t)}\left(1 - u_2^{(t)}\right) \\ \cdots &= \cdots \\ u_n^{(t+1)} - u_n^{(t)} &= \gamma\left(u_{n-1}^{(t)} - 2u_n^{(t)} + u_1^{(t)}\right) + \mu u_n^{(t)}\left(1 - u_n^{(t)}\right) \end{aligned}$$

where the term $\mu u_i^{(t)}\left(1 - u_i^{(t)}\right)$ denotes the contribution from internal logistic growth when no trading is present.

What we intend to do is to find out the asymptotic behavior of each individual $\left\{u_i^{(t)}\right\}_{t=0}^{\infty}$. Such behaviors clearly depend on the number n, the parameter γ and the initial distribution $\left(u_1^{(0)}, ..., u_n^{(0)}\right)$ of wealth. After much computer experimentation, we find that each $\left\{u_i^{(t)}\right\}$ can either be a constant, convergent, periodic, 'asymptotically periodic', or divergent and oscillatory sequence. Furthermore, cooperative behaviors are also possible under proper initial distributions and parameter values. Such observations will be made clear and explained in our later discussions.

Since the matrix A_n is a circulant matrix [6] and since our model is a linear one, some standard techniques from linear algebra theory such as the power method can be applied to yield our observations. Therefore, some of the details will be skipped from our following presentations.

In the sequel, N denotes $\{0, 1, 2, \cdots\}$ and Z denotes $\{\cdots, -1, 0, 1, 2, \cdots\}$. The inner product of two n-vectors u and v will be denoted by $\langle u, v\rangle$, while the transpose of a vector x is denoted by x^T. Given a set V of vectors in R^n, the linear space for all linear combinations of the vectors in V will be denoted by $\langle V\rangle$. Let Ω be a set of vectors in R^n and let $u \in R^n$. The notation $\langle u, \Omega\rangle = 0$ means that $\langle u, v\rangle = 0$ for every $v \in \Omega$. The average of n numbers will appear many times in the sequel, therefore, for any $u = (u_1, ..., u_n)^T \in R^n$, we set

$$M(u) = \frac{1}{n}\sum_{j=1}^{n} u_j.$$

To avoid trivial cases, we will also assume that $n \geq 2$ and $\gamma \neq 0$ throughout the rest of our discussions. Indeed, when $\gamma = 0$, there will be no trading whatsoever between the nations so that $u^{(t+1)} = (I - \gamma A_n)\, u^{(t)} = u^{(t)} = ... = u^{(0)}$ at all times $t \in N$. We remark that γ is allowed to be negative. Although this is unlikely in normal situations, such a phenomena may be seen during trade wars.

2 Eigenvalues, eigenvectors and general solution

First of all, the eigenvalues of A_n and $\tilde{A}_n = I_n - \gamma A_n$ can be found by solving the difference equation

$$u_{j-1} - 2u_j + u_{j+1} + \lambda u_j = 0,$$

and are respectively

$$\lambda^{(k)} = 4\sin^2\frac{k\pi}{n},\ 1 \leq k \leq n,$$

and

$$1 - \gamma\lambda^{(k)} = 1 - 4\gamma\sin^2\frac{k\pi}{n},\ 1 \leq k \leq n.$$

Once they are found, we can also check directly that the corresponding orthonormal eigenvectors

$$v^{(k)} = \left(v_1^{(k)}, \cdots, v_m^{(k)}, \cdots, v_n^{(k)}\right)^T,\ 1 \leq k \leq n,$$

are given by

$$v_m^{(k)} = \frac{1}{\sqrt{n}}\left(\cos\frac{2mk\pi}{n} - \sin\frac{2mk\pi}{n}\right),\ 1 \leq m \leq n, 1 \leq k \leq n.$$

For instance, when $k = n$, the corresponding eigenvalue $\lambda^{(n)}$ is 0 and the corresponding eigenvector is $v^{(n)} = n^{-1/2}(1, 1, 1, ..., 1)^T$ and when n is even and

$k = n/2$, the corresponding eigenvalue $\lambda^{(k)}$ is 4 and the corresponding eigenvector is $v^{(n/2)} = n^{-1/2}(-1, 1, -1, 1, ..., -1, 1)^T$.

In view of the periodicity properties of the trigonometric functions, if we define

$$\lambda^{(0)} = 4\sin^2 \frac{0 \cdot \pi}{n} = 0,$$

and

$$v_m^{(0)} = \frac{1}{\sqrt{n}}\left(\cos\frac{2m \cdot 0 \cdot \pi}{n} + \sin\frac{2m \cdot 0 \cdot \pi}{n}\right) = \frac{1}{\sqrt{n}}, \ 1 \le m \le n,$$

then $\lambda^{(n)} = \lambda^{(0)}$ and $v^{(n)} = v^{(0)}$. The sets $\{\lambda^{(0)}, ..., \lambda^{(n)}\}$ and $\{v^{(0)}, ..., v^{(n)}\}$ have several interesting properties. For examples,

$$v_m^{(n-k)} = \frac{1}{\sqrt{n}}\left(\cos\frac{2mk\pi}{n} - \sin\frac{2mk\pi}{n}\right), \ 0 \le k \le n,$$

and

$$\begin{aligned} 0 = \lambda^{(0)} &= \lambda^{(n)} < ... < \lambda^{(k)} = \lambda^{(n-k)} < ... < \lambda^{(n/2-1)} = \lambda^{(n/2+1)} < \lambda^{(n/2)} \\ &= \lambda^{(n-n/2)} = 4 \end{aligned}$$

if n is even, and

$$\begin{aligned} 0 = \lambda^{(0)} &= \lambda^{(n)} < \lambda^{(1)} = \lambda^{(n-1)} < ... < \lambda^{(j)} = \lambda^{(n-j)} < ... < \lambda^{((n-1)/2)} \\ &= \lambda^{((n+1)/2)} < 4 \end{aligned}$$

if n is odd.

Therefore, if we let

$$V_k = \left\{v^{(k)}, v^{(n-k)}\right\}, \ k = 0, 1, 2, \cdots, [n/2],$$

then $\langle V_0 \rangle, \langle V_1 \rangle, \cdots, \langle V_{[n/2]} \rangle$ are the eigenspaces associated with A_n, furthermore, the dimension of $\langle V_0 \rangle$ is 1, the dimension of $\langle V_k \rangle$ is 2 for $k \in \{1, \cdots, [n/2] - 1\}$ and the dimension of $\langle V_{[n/2]} \rangle$ is 1 when n is even and is 2 when n is odd.

Since $\{v^{(1)}, \ldots, v^{(n)}\}$ is an orthonormal basis of R^n, for any $u^{(0)} = (u_1^{(0)}, \ldots, u_n^{(0)})^T$ in R^n, we can write

$$u^{(0)} = \sum_{k=0}^{n-1} \left\langle v^{(k)}, u^{(0)} \right\rangle v^{(k)}.$$

Thus given $u^{(0)}$, we obtain from (1.1) that

$$\begin{aligned} u^{(t)} &= \tilde{A}_n^t u^{(0)} = \sum_{k=0}^{n-1} \left\langle v^{(k)}, u^{(0)} \right\rangle \tilde{A}_n^t v^{(k)} = \sum_{k=0}^{n-1} (1 - \lambda^{(k)}\gamma)^t \left\langle v^{(k)}, u^{(0)} \right\rangle v^{(k)} \\ &= \sum_{k=0}^{[n/2]} \sum_{w \in V_k} (1 - \lambda^{(k)}\gamma)^t \left\langle w, u^{(0)} \right\rangle w = \sum_{k=0}^{[n/2]} (1 - \lambda^{(k)}\gamma)^t \sum_{w \in V_k} \left\langle w, u^{(0)} \right\rangle w, \end{aligned}$$

and

$$u_m^{(t)} = \sum_{k=0}^{[n/2]} (1 - \lambda^{(k)}\gamma)^t \left\langle \sum_{w \in V_k} w_m w, u^{(0)} \right\rangle, \ m = 1, ..., n,$$

for $t = 1, ...$, where $u^{(t)} = \left(u_1^{(t)}, ..., u_n^{(t)}\right)^T$. The sequence $\{u^{(t)}\}_{t=0}^{\infty}$ is said to be the solution of (1.1) determined by $u^{(0)}$.

Let us set

$$w^{(k,m)} = \sum_{w \in V_k} w_m w, \; m = 1, ..., n; k = 0, 1, ..., [n/2].$$

Since

$$\left\langle v^{(k)}, u^{(0)} \right\rangle v_m^{(k)} = \left\langle v_m^{(k)} v^{(k)}, u^{(0)} \right\rangle = \sum_{j=1}^{n} v_m^{(k)} v_j^{(k)} u_j^{(0)}$$

$$= \frac{1}{n} \sum_{j=1}^{n} \left(\cos \frac{2mk\pi}{n} + \sin \frac{2mk\pi}{n} \right) \left(\cos \frac{2jk\pi}{n} + \sin \frac{2jk\pi}{n} \right) u_j^{(0)}$$

$$= \frac{1}{n} \sum_{j=1}^{n} \left(\cos \frac{2(m-j)k\pi}{n} + \sin \frac{2(m+j)k\pi}{n} \right) u_j^{(0)},$$

thus

$$\left\langle w^{(k,m)}, u^{(0)} \right\rangle = \begin{cases} M\left(u^{(0)}\right) & k = 0 \\ \frac{1}{n} \sum_{j=1}^{n} (-1)^{m+j} u_j^{(0)} & k = n/2, n \text{ even} \\ \frac{2}{n} \sum_{j=1}^{n} \left(\cos \frac{2(m-j)k\pi}{n} \right) u_j^{(0)} & 0 < k < n/2 \end{cases} \tag{2.1}$$

for $m = 1, 2, ..., n$.

As a consequence, for $t = 0, 1, ..,$

$$u_m^{(t)} = M\left(u^{(0)}\right) + \sum_{k=1}^{[n/2]} (1 - \lambda^{(k)}\gamma)^t \left\langle w^{(k,m)}, u^{(0)} \right\rangle, \; m = 1, ..., n. \tag{2.2}$$

3 Location indexes

A solution of (1.1) depends on the dimension n, the parameter γ, and the initial vector $u^{(0)}$. Therefore the location of $u^{(0)}$ will be important. For this reason, we introduce two index functions. Given $u = (u_1, ..., u_n)^T \in R^n$, we define

(i) $\kappa(u) = [n/2]$ if and only if $\langle u, V_{[n/2]} \rangle \neq 0$,

(ii) $\kappa(u) = k \in \{1, 2, ..., [n/2] - 1\}$ if and only if

$$\langle u, V_{[n/2]} \rangle = \langle u, V_{[n/2]-1} \rangle = ... = \langle u, V_{k+1} \rangle = 0$$

and $\langle u, V_k \rangle \neq 0$, and

(iii) $\kappa(u) = 0$ if and only if $\langle u, V_{[n/2]} \rangle = \langle u, V_{[n/2]-1} \rangle = ... = \langle u, V_1 \rangle = 0$.

Similarly, given $u = (u_1, ..., u_n)^T \in R^n$, we define

(a) $\kappa_m(u) = [n/2]$ if and only if $\langle u, w^{([n/2],m)} \rangle \neq 0$,

(b) $\kappa_m(u) = k \in \{1, 2 \cdots, [n/2] - 1\}$ if and only if

$$\left\langle u, w^{([n/2],m)} \right\rangle = \left\langle u, w^{([n/2]-1,m)} \right\rangle = ... = \left\langle u, w^{(k+1,m)} \right\rangle = 0$$

and $\langle u, w^{(k,m)} \rangle \neq 0$, and

(c) $\kappa_m(u) = 0$ if and only if

$$\left\langle u, w^{([n/2],m)} \right\rangle = \left\langle u, w^{([n/2]-1,m)} \right\rangle = ... = \left\langle u, w^{(1,m)} \right\rangle = 0.$$

As an example, let $u = (3, 3, 3, 3, 3)^T$. Since $v^{(0)} = (1/\sqrt{5})[1, 1, 1, 1, 1]^T$, so $u \in \langle V_0 \rangle$, which implies $\langle u, v^{(1)} \rangle = \langle u, v^{(2)} \rangle = \langle u, v^{(3)} \rangle = \langle u, v^{(4)} \rangle = 0$. Thus by definition, $\kappa(u) = 0$ and for any $m \in \{1, 2, ..., 5\}$, $\kappa_m(u) = 0$.

As another example, if n is even and $u = (u_1, .., u_n)^T$ satisfies $\sum_{j=1}^{n/2} u_{2j-1}^{(0)} = \sum_{j=1}^{n/2} u_{2j}^{(0)}$ then $\langle u, v^{([n/2])}\rangle = \langle u, V_{[n/2]}\rangle = 0$, so that $\kappa(u) < n/2$.

An immediate consequence of (2.2) is

$$u_m^{(t)} = M\left(u^{(0)}\right) + \sum_{k=1}^{\kappa_m\left(u^{(0)}\right)} (1-\lambda^{(k)}\gamma)^t \left\langle w^{(k,m)}, u^{(0)}\right\rangle. \tag{3.1}$$

There are additional properties of $\kappa(u)$ and $\kappa_m(u)$.

Lemma 3.1 $\kappa(u) = \max\{\kappa_1(u), \kappa_2(u), \cdots, \kappa_n(u)\}$.

Proof If $\kappa_m(u) = k \in \{1, ..., [n/2]\}$, then $\langle w^{(k,m)}, u\rangle \neq 0$. Thus $\langle V_k, u\rangle \neq 0$. This shows that $\kappa(u) \geq k$. Conversely, if $\max\{\kappa_1(u), \kappa_2(u), \cdots, \kappa_n(u)\} < \kappa(u)$ then for any $m \in \{1, \cdots, n\}$, $\langle w^{(\kappa(u),m)}, u\rangle = 0$. Thus

$$\sum_{w\in V_{\kappa(u)}} \langle w, u\rangle\, w = 0.$$

Since $V_{\kappa(u)}$ is a linearly independent set, hence $\langle w, u\rangle = 0$ for $w \in V_{\kappa(u)}$, which is contrary to the definition of $\kappa(u)$. The proof is complete. □

Next, note that for $k \in \{1, ..., [n/2]\}\backslash\{n/2\}$, we have

$$\left\langle w^{(k,m)}, u\right\rangle = v_m^{(k)}\left\langle v^{(k)}, u\right\rangle + v_m^{(n-k)}\left\langle v^{(n-k)}, u\right\rangle$$

and

$$\left\langle w^{(k,j)}, u\right\rangle = v_j^{(k)}\left\langle v^{(k)}, u\right\rangle + v_j^{(n-k)}\left\langle v^{(n-k)}, u\right\rangle$$

by definition, thus $\left(\langle v^{(k)}, u\rangle, \langle v^{(n-k)}, u\rangle\right)$ is always a solution pair of the following system of equations:

$$\begin{aligned} v_m^{(k)}x + v_m^{(n-k)}y &= \langle w^{(k,m)}, u\rangle, \\ v_j^{(k)}x + v_j^{(n-k)}y &= \langle w^{(k,j)}, u\rangle. \end{aligned} \tag{3.2}$$

Note further that

$$\begin{aligned} \begin{vmatrix} v_m^{(k)} & v_m^{(n-k)} \\ v_j^{(k)} & v_j^{(n-k)} \end{vmatrix} &= v_m^{(k)}v_j^{(n-k)} - v_m^{(n-k)}v_j^{(k)} \\ &= \frac{2}{n}\left(\sin\frac{2mk\pi}{n}\cos\frac{2jk\pi}{n} - \cos\frac{2mk\pi}{n}\sin\frac{2jk\pi}{n}\right) \\ &= \frac{2}{n}\sin\frac{2(m-j)k\pi}{n}. \end{aligned} \tag{3.3}$$

Thus, if $2(m-j)k/n \notin Z$, then $\sin 2(m-j)k\pi/n \neq 0$. Hence the system (3.2) has a unique solution pair (x, y) so that

$$\begin{aligned} x &= \left\langle v^{(k)}, u\right\rangle = \frac{\langle w^{(k,m)}, u\rangle v_j^{(n-k)} - \langle w^{(k,j)}, u\rangle v_m^{(n-k)}}{2\sin 2(m-j)k\pi/n}, \\ y &= \left\langle v^{(n-k)}, u\right\rangle = \frac{\langle w^{(k,j)}, u\rangle v_m^{(k)} - \langle w^{(k,m)}, u\rangle v_j^{(k)}}{2\sin 2(m-j)k\pi/n}. \end{aligned}$$

If $2(m-j)k/n \in Z$, then $\sin 2(m-j)k\pi/n = 0$ so that (3.2) has infinitely many solution pairs.

Lemma 3.2 *Given $u \in R^n$, $m, j \in \{1, \cdots, n\}$ and $k \in \{1, \cdots, [n/2]\} \setminus \{n/2\}$. The system (3.2) either has exactly one solution pair or has infinitely many solution pairs. The system (3.2) has a unique solution pair if, and only if, $2(m-j)k/n \notin Z$ (and under this condition, $\langle w^{(k,m)}, u\rangle = \langle w^{(k,j)}, u\rangle = 0$ if, and only if, $\langle u, V_k\rangle = 0$). The system (3.2) has infinitely many solutions if, and only if, $2(m-j)k/n \in Z$ (and under this condition, $\langle w^{(k,m)}, u\rangle = 0$ if, and only if, $\langle w^{(k,j)}, u\rangle = 0$).*

For any $u \in R^n$, either $\kappa_j(u) = n/2$ for some $j \in \{1, ..., n\}$, or $\kappa_m(u) < n/2$ for all $m \in \{1, ..., n\}$. If $\kappa_j(u) = n/2$ for some $j \in \{1, ..., n\}$, we assert that $\kappa_i(u) = n/2$ for all $i \in \{1, ..., n\}$. Indeed, if $\kappa_m(u) < n/2$ for some $m \in \{1, ..., n\}$, then n is even and

$$0 = \left\langle w^{([n/2],m)}, u\right\rangle = \frac{(-1)^m}{\sqrt{n}}\left\langle v^{(n/2)}, u\right\rangle,$$

so that $\langle V_{n/2}, u\rangle = 0$ and $\kappa(u) \neq n/2$, which is a contradiction. If $\kappa_m(u) < n/2$ for all $m \in \{1, ..., n\}$, we assert that

$$\kappa_i(u) < \kappa_j(u) \Rightarrow 2(j-i)\kappa_j(u)/n \notin Z,$$

and

$$\kappa_s(u) > \max\{\kappa_i(u), \kappa_j(u)\} \Rightarrow 2(i-j)\kappa_s(u)/n \in Z.$$

Indeed, if $\kappa_i(u) < \kappa_j(u)$, then

$$\left\langle w^{(\kappa_j(u),j)}, u\right\rangle \neq 0, \ \left\langle w^{(\kappa_j(u),i)}, u\right\rangle = 0.$$

so that, by Lemma 3.2, $2(i-j)\kappa_j(u)/n \notin Z$. If $\kappa_s(u) > \max\{\kappa_i(u), \kappa_j(u)\}$, then

$$\left\langle w^{(\kappa_s(u),s)}, u\right\rangle \neq 0, \ \left\langle w^{(\kappa_s(u),i)}, u\right\rangle = 0, \ \left\langle w^{(\kappa_s(u),j)}, u\right\rangle = 0.$$

Therefore, if it were the case that $2(i-j)\kappa_s(u)/n \notin Z$, then the last two conditions, in view of Lemma 3.2, imply $\langle u, V_{\kappa_s(u)}\rangle = 0$. This is contrary to the definition of $\kappa_s(u)$. We have thus prove the first part of the next result.

Theorem 3.3 *For $u \in R^n$, either $\kappa_m(u) = n/2$ for (some and hence) all $m \in \{1, ..., n\}$, or else $\kappa_s(u) > \max\{\kappa_i(u), \kappa_j(u)\}$ implies $2(i-j)\kappa_s(u)/n \in Z$, $2(s-i)\kappa_s(u)/n \notin Z$ and $2(s-j)\kappa_s(u)/n \notin Z$. Conversely, if $a_1, a_2, ..., a_n \in \{0, \cdots, [n/2]\}$ such that either $a_1 = a_2 = \cdots = a_n$ or $a_s > \max\{a_i, a_j\}$ implies $2(i-j)a_s/n \in Z$, $2(s-i)a_s/n \notin Z$ and $2(s-j)a_s/n \notin Z$ for any $s, i, j \in \{1, ..., n\}$, then a vector $u \in R^n$ can be found such that $\kappa_m(u) = a_m$ for $m \in \{1, ..., n\}$.*

Proof We only need to prove the second part of our result. First of all, note that if $a_m = n/2$ for some m, then n is even and since $a_m > a_j$ implies, by our assumptions, the contradiction $2(m-j)a_m/n = (m-j) \notin Z$, thus $a_j = a_m = n/2$ for all $j \in \{1, ..., n\}$. But in this case, we may choose $u = v^{(n/2)}$ since $\langle w^{(n/2,m)}, v^{(n/2)}\rangle = v_m^{(n/2)}\langle v^{(n/2)}, v^{(n/2)}\rangle = (-1)^m \neq 0$ for $m \in \{1, ..., n\}$.

We therefore assume that $a_m < n/2$ for $m \in \{1, ..., n\}$ in the rest of our proof. Now note that since $u = \sum_{k=0}^{n-1}\langle v^{(k)}, u\rangle v^{(k)}$, it suffices to find the coefficient $\langle v^{(k)}, u\rangle$ for each $k = 0, 1, ..., n-1$, so that the resulting u satisfies $\kappa_m(u) = a_m$ for $m = 1, ..., n$. To achieve our goal, we first choose the coefficient $\langle v^{(0)}, u\rangle$ in an arbitrary manner. For the k-th coefficient $\langle v^{(k)}, u\rangle$ where $k \in \{1, ..., n-1\}$, we proceed by considering two cases: (A) $a_m \neq k$ for every $m \in \{1, ..., n\}$, (B) $a_s = k$

for $s \in \Omega$, where Ω is a nonempty set $\{m_1, m_2, ..., m_\beta\}$ of (mutually distinct) integers in $\{1, ...n\}$.

Suppose case (A) holds. Then when $a_m > k$ for every $m \in \{1, ..., n\}$, we choose $\langle u, v^{(k)} \rangle$ and $\langle u, v^{(n-k)} \rangle$ in arbitrary manners, and when $a_s < k$ for some $s \in \{1, ..., n\}$, we let $\langle u, v^{(k)} \rangle = \langle u, v^{(n-k)} \rangle = 0$. Note that in this case,

$$\left\langle w^{(k,s)}, u \right\rangle = v_s^{(k)} \left\langle v^{(k)}, u \right\rangle + v_s^{(n-k)} \left\langle v^{(n-k)}, u \right\rangle = 0$$

for any s such that $a_s < k$.

Suppose case (B) holds. If there exist $i, j \in \{1, ..., n\} \backslash \Omega$ such that $a_i, a_j < k$, then by our assumptions, $2(i-j)k/n \in Z$, $2(s-i)k/n \notin Z$ and $2(s-j)k/n \notin Z$. Thus, in view of (3.3), for each $s \in \Omega$, the system of equations

$$\begin{aligned} v_j^{(k)} x + v_j^{(n-k)} y &= 0, \\ v_s^{(k)} x + v_s^{(n-k)} y &= 0, \end{aligned}$$

has the unique solution pair $(0, 0)$, and the system of equations

$$\begin{aligned} v_i^{(k)} x + v_i^{(n-k)} y &= 0, \\ v_j^{(k)} x + v_j^{(n-k)} y &= 0, \end{aligned}$$

has infinitely many solution pairs. Choose a nontrivial pair (α, β) in R^2 such that

$$v_i^{(k)} \alpha + v_i^{(n-k)} \beta = 0,$$

and let $\langle v^{(k)}, u \rangle = \alpha$ and $\langle v^{(n-k)}, u \rangle = \beta$. Then $\langle w^{(k,j)}, u \rangle = 0$ for all j such that $a_j < k$, and $\langle w^{(k,s)}, u \rangle \neq 0$ for all $s \in \Omega$.

Finally, under (B), if $a_i > k$ for every $i \in \{1, ..., n\} \backslash \Omega$, pick a point (c, d) such that

$$v_s^{(k)} c + v_s^{(n-k)} d \neq 0, \ \ s \in \Omega,$$

and let $\langle v^{(k)}, u \rangle = c$ and $\langle v^{(n-k)}, d \rangle = d$. Then $\langle w^{(k,s)}, u \rangle \neq 0$ for $s \in \Omega$.

We assert that the vector $u = \sum_{k=0}^{n-1} \langle v^{(k)}, u \rangle v^{(k)}$ defined above will satisfy $\kappa_m(u) = a_m$ for every $m \in \{1, ..., n\}$. Indeed, by our previous constructions, if $k > a_m$ then $\langle w^{(k,m)}, u \rangle = 0$ and if $k = a_m \neq 0$, then $\langle w^{(k,m)}, u \rangle \neq 0$. The proof is complete. □

As an example, let $n = 15$ in (1.1). We assert that there cannot be a vector $u \in R^{15}$ such that $\kappa(u) = 6$, $\kappa_1(u) = \kappa_2(u) = 5$, $\kappa_6(u) = 3$, $\kappa_{11}(u) = 5$. Indeed, by Theorem 3.3, if $\kappa_i(u^{(0)}) < \kappa(u^{(0)})$ and $\kappa_j(u^{(0)}) < \kappa(u^{(0)})$, then

$$\frac{2(i-j)\kappa\left(u^{(0)}\right)}{n} = \frac{(i-j)4}{5} \in Z,$$

thus $i - j$ must be an integer multiple of 5. But for $i = 1$ and $j = 2$, $(j-i)4/5 = 4/5 \notin Z$. This is a contradiction.

Theorem 3.4 *Suppose $\kappa(u) \neq 0$. If $\kappa(u) = n/2$, then $\kappa_m(u) = \kappa(u)$ for all $m \in \{1, \cdots, n\}$, and if $\kappa(u) > \max\{\kappa_i(u), \kappa_j(u)\}$, then*

$$2(i-j)\kappa(u)/n \in Z. \tag{3.4}$$

Proof If $\kappa(u) = n/2$, then by Lemma 3.1, $\kappa(u) = \kappa_i(u)$ for some $i \in \{1, ..., n\}$. In view of Theorem 3.3, $\kappa_m(u) = \kappa_i(u)$ for any $m \in \{1, \cdots, n\}$. Next, if $\kappa(u) > \max\{\kappa_i(u), \kappa_j(u)\}$, then there is $\kappa_s(u) = \kappa(u) > \max\{\kappa_i(u), \kappa_j(u)\}$, so that Theorem 3.3 can be applied to yield $2(i-j)\kappa(u)/n \in Z$. The proof is complete. □

As an example, let $n = 15$, we assert that there is no $u \in R^{15}$ such that $\kappa(u) = 6$, $\kappa_1(u) = 3$ and $\kappa_4(u) = 2$. Indeed, the contrary case would imply

$$\frac{2(4-1)6}{15} = \frac{12}{5} \in Z,$$

which is not true.

Corollary 3.5 *If n is a prime, then there is at most one $m \in \{1, ..., n\}$ such that $\kappa_m(u) < \kappa(u)$.*

Proof We may suppose $\kappa(u) > 0$. If $n = 2$, then $\kappa(u) = 1 = n/2$ and hence our assertion is true by Theorem 2. Suppose n is odd. If there exist $i, j \in \{1, ..., n\}$ such that $\kappa_i(u) < \kappa(u)$ and $\kappa_j(u) < \kappa(u)$, then by Theorem 2, $2(i-j)\kappa(u)/n \in Z$. Since $0 < 2\kappa(u) < n$ and n is odd, we see that $i = j$. The proof is complete. □

Corollary 3.6 *Suppose n is a prime number which is greater than 2. Let $s \in \{1, ..., n\}$ and $a, b \in \{0, \cdots, [n/2]\}$ such that $a < b$. There exists $u \in R^n$ such that $\kappa_s(u) = a$ and $\kappa(u) = b$.*

Proof Let $a_m = b$ for $m \in \{1, \cdots, n\} \backslash \{s\}$ and $a_s = a$. If $a_\tau > \max\{a_i, a_j\}$, then $a_\tau = b$, $i = j$ and $a_i = a$. Since n is a prime number which is greater then 2, since $n > b > a$ and since $n > |\tau - i|$, we see that $2(i-j)a_\tau/n = 0 \in Z$, $2(\tau - i)a_\tau/n = 2(\tau - j)a_\tau/n = 2(\tau - i)b/n \notin Z$. The proof now follows from Theorem 3.3. □

As an example, when $n = 17$, there exists a vector $u \in R^{17}$ such that $\kappa(u) = 6$ and $\kappa_1(u) = 2$. Since n is odd, we see from Corollary 3.5 that $\kappa_m(u^{(0)}) = \kappa(u^{(0)}) = 6$ for $m = 2, 3, \cdots, 17$. To construct such a vector u, we follow the proof of Theorem 3.3 and let $a_1 = 2$ and $a_2 = a_3 = ... = a_{17} = 6$. Note that $a_m < 17/2$ for every $m \in \{1, ..., 17\}$. We first let $\left\langle v^{(k)}, u \right\rangle = 0$ for $k \neq 2, 6, 11, 15$. Next we determine $\left\langle v^{(6)}, u \right\rangle$ and $\left\langle v^{(11)}, u \right\rangle$: pick $\left\langle v^{(6)}, u \right\rangle = -v_1^{(11)} \neq 0$ and $\left\langle v^{(11)}, u \right\rangle = v_1^{(6)}$, in which case,

$$0 = v_1^{(6)} \left\langle v^{(6)}, u \right\rangle + v_1^{(11)} \left\langle v^{(11)}, u \right\rangle = \left\langle w^{(6,1)}, u \right\rangle,$$

so that $\kappa(u) = 6$ and $\kappa_1(u) = 2 < 3$. Next we determine $\left\langle v^{(2)}, u \right\rangle$ and $\left\langle v^{(15)}, u \right\rangle$: pick $\left\langle v^{(2)}, u \right\rangle = v_1^{(2)}$ and $\left\langle v^{(15)}, u \right\rangle = v_1^{(15)}$, in which case,

$$\left\langle w^{(2,1)}, u \right\rangle = v_1^{(2)} \left\langle v^{(2)}, u \right\rangle + v_1^{(15)} \left\langle v^{(15)}, u \right\rangle = 2/17.$$

Then it is easy to verify that

$$u = -v_1^{(11)} v^{(6)} + v_1^{(6)} v^{(11)} + v_1^{(2)} v^{(2)} + v_1^{(15)} v^{(15)}$$

is the required vector.

4 Individual asymptotic behaviors

We will need critical numbers to distinguish the dependence of a solution and its component sequences on the parameter γ. Given $u \in R^n$ such that $\kappa_m(u) > 0$. Let

$$\gamma_{m,n}(u) = \frac{2}{\lambda^{(\kappa_m(u))}} = \frac{1}{2\sin^2(\kappa_m(u)\pi/n)}.$$

If no confusion arises, we will write γ_m instead of $\gamma_{m,n}$. Some of these critical numbers are listed below:

n	$\gamma_{1,n}$	$\gamma_{2,n}$	$\gamma_{3,n}$	$\gamma_{4,n}$	$\gamma_{5,n}$	$\gamma_{6,n}$	$\gamma_{7,n}$
2	0.5						
3	0.66667						
4	1	0.5					
5	1.4472	0.55279					
6	2	0.66667	0.5				
7	2.656	0.81798	0.52605				
8	3.4142	1	0.58579	0.5			
9	4.2743	1.2101	0.66667	0.51555			
10	5.2361	1.4472	0.76393	0.55279	0.5		
11	6.2994	1.7106	0.87542	0.60428	0.51034		
12	7.4641	2	1	0.66667	0.5359	0.5	
13	8.7303	2.3152	1.1371	0.73822	0.57192	.50737	
14	10.098	2.656	1.2862	0.81798	0.61596	.52605	0.5
15	11.567	3.0223	1.4472	0.90536	0.66667	0.55279	0.50552

As an immediate consequence of (3.1), if $u^{(0)} \in R^n$ such that $\kappa_m\left(u^{(0)}\right) = 0$, then the solution $\left\{u^{(t)}\right\}_{t=0}^{\infty}$ of (1.1) determined by $u^{(0)}$ will satisfy

$$u_m^{(t)} = M\left(u^{(0)}\right), \ m = 1, 2, ..., n,$$

for $t \in N$, regardless of the magnitude of γ. Therefore, in the following discussions, we will assume that $\kappa_m\left(u^{(0)}\right) \neq 0$. Note that $\kappa_m\left(u^{(0)}\right) \neq 0$ implies $\left\langle w^{(\kappa_m(u^{(0)}),m)}, u^{(0)}\right\rangle \neq 0$.

We consider the following cases: (i) $\gamma < 0$, (ii) $0 < \gamma < \gamma_m\left(u^{(0)}\right)$, (iii) $\gamma = \gamma_m\left(u^{(0)}\right)$, and (iv) $\gamma > \gamma_m\left(u^{(0)}\right)$.

Theorem 4.1 *Let $m \in \{1, ..., n\}$. Let $u^{(0)} \in R^n$ such that $\kappa_m(u^{(0)}) \neq 0$. Let $\left\{u^{(t)}\right\}_{t=0}^{\infty}$ be the solution of (1.1) determined by $u^{(0)}$. If $\gamma < 0$, then $\left\{u_m^{(t)}\right\}$ will diverge to $+\infty$ or $-\infty$, and*

$$u_m^{(t)} \sim (1 - \lambda^{(\kappa_m(u^{(0)}))}\gamma)^t \left\langle w^{(\kappa_m(u^{(0)}),m)}, u^{(0)}\right\rangle$$

for all large t.

Proof Note that if $\gamma < 0$, then for $0 \leq k < \kappa_m(u^{(0)})$,

$$1 - \lambda^{(\kappa_m(u^{(0)}))}\gamma > 1 - \lambda^{(k)}\gamma > 0.$$

In view of (3.1), we see that

$$\begin{aligned}
&\lim_{t\to\infty} \frac{u_m^{(t)}}{(1-\lambda^{(\kappa_m(u^{(0)}))}\gamma)^t \left\langle w^{(\kappa_m(u^{(0)}),m)}, u^{(0)}\right\rangle} \\
&= \sum_{k=0}^{k_m(u^{(0)})} \lim_{t\to\infty} \frac{(1-\lambda^{(k)}\gamma)^t \langle w^{(k,m)}, u^{(0)}\rangle}{(1-\lambda^{(\kappa_n(u^{(0)}))}\gamma)^t \left\langle w^{(\kappa_m(u^{(0)}),m)}, u^{(0)}\right\rangle} \\
&= \sum_{k=0}^{k_m(u^{(0)})-1} \lim_{t\to\infty} \left(\frac{1-\lambda^{(k)}\gamma}{1-\lambda^{(\kappa_m(u^{(0)}))}\gamma}\right)^t \frac{\langle w^{(k,m)}, u^{(0)}\rangle}{\left\langle w^{(\kappa_m(u^{(0)}),m)}, u^{(0)}\right\rangle} + 1 \\
&= 1.
\end{aligned}$$

Equivalently, we get

$$u_m^{(t)} \sim (1-\lambda^{(\kappa_m(u^{(0)}))}\gamma)^t \left\langle w_{\kappa_m(u^{(0)}),m}, u^{(0)}\right\rangle$$

for all large t. The proof is complete. □

Theorem 4.2 *Let $m \in \{1, ..., n\}$ and $u^{(0)} \in R^n$ such that $\kappa_m(u^{(0)}) \neq 0$. Let $\{u^{(t)}\}_{t=0}^{\infty}$ be the solution of (1.1) determined by $u^{(0)}$. If $0 < \gamma < \gamma_m(u^{(0)})$, then $\lim_{t\to\infty} u_m^{(t)} = M(u^{(0)})$.*

Proof If $\kappa_m(u^{(0)}) \neq 0$, then since $0 < \gamma < 2/\lambda^{(\kappa_m(u^{(0)}))}$, for $1 \le k \le \kappa_m(u^{(0)})$, we have $0 < \lambda^{(k)}\gamma \le \lambda^{(\kappa_m(u^{(0)}))}\gamma < 2$ so that

$$\left|1-\lambda^{(k)}\gamma\right| < 1,\ 1 \le k \le \kappa_m(u^{(0)}).$$

In view of (3.1) again,

$$\lim_{t\to\infty} u_m^{(t)} = M\left(u^{(0)}\right) + \sum_{k=1}^{k_m(u^{(0)})} \lim_{t\to\infty} (1-\lambda^{(k)}\gamma)^t \left\langle w^{(k,m)}, u^{(0)}\right\rangle = M\left(u^{(0)}\right).$$

The proof is complete. □

Theorem 4.3 *Let $m \in \{1, ..., n\}$ and let $u^{(0)} \in R^n$ such that $\kappa_m(u^{(0)}) \neq 0$. Let $\{u^{(t)}\}_{t=0}^{\infty}$ be the solution of (1.1) determined by $u^{(0)}$. If $\gamma = \gamma_m(u^{(0)})$, then*

$$\lim_{t\to\infty} u_m^{(2t)} = M\left(u^{(0)}\right) + \left\langle w^{(\kappa_m(u^{(0)}),m)}, u^{(0)}\right\rangle,\ m = 1, ..., n,$$

and

$$\lim_{t\to\infty} u_m^{(2t+1)} = M\left(u^{(0)}\right) - \left\langle w^{(\kappa_m(u^{(0)}),m)}, u^{(0)}\right\rangle,\ m = 1, ..., n.$$

Proof If $\gamma = \gamma_m(u^{(0)}) = 2/\lambda^{(\kappa_m(u^{(0)}))}$, then

$$1-\lambda^{(\kappa_m(u^{(0)}))}\gamma = -1$$

and

$$1-\lambda^{(k)}\gamma = 1-2(\lambda^{(k)}/\lambda^{(\kappa_m(u^{(0)}))}).$$

Since $0 < \lambda^{(k)} < \lambda^{(\kappa_m(u^{(0)}))}$ for $k = 1, 2, \cdots, \kappa_m(u^{(0)})-1$, we see that

$$1 > 1-2(\lambda^{(k)}/\lambda^{(\kappa_m(u^{(0)}))}) > -1,$$

thus by (3.1),

$$\begin{aligned}\lim_{t\to\infty} u_m^{(2t)} &= M(u^{(0)}) + \sum_{k=1}^{k_m\left(u^{(0)}\right)} \lim_{t\to\infty} (1-\lambda^{(k)}\gamma)^{2t} \left\langle w^{(k,m)}, u^{(0)} \right\rangle \\ &= M(u^{(0)}) + \lim_{t\to\infty} (1-\lambda^{(\kappa_m(u^{(0)}))}\gamma)^{2t} \left\langle w^{(\kappa_m(u^{(0)}),m)}, u^{(0)} \right\rangle \\ &= M(u^{(0)}) + \left\langle w^{(\kappa_m(u^{(0)}),m)}, u^{(0)} \right\rangle\end{aligned}$$

and

$$\begin{aligned}\lim_{t\to\infty} u_m^{(2t+1)} &= M(u^{(0)}) + \sum_{k=1}^{k_m\left(u^{(0)}\right)} \lim_{t\to\infty} (1-\lambda^{(k)}\gamma)^{2t+1} \left\langle w^{(k,m)}, u^{(0)} \right\rangle \\ &= M(u^{(0)}) + \lim_{t\to\infty} (1-\lambda^{(\kappa_m(u^{(0)}))}\gamma)^{2t+1} \left\langle w^{(\kappa_m(u^{(0)}),m)}, u^{(0)} \right\rangle \\ &= M(u^{(0)}) - \left\langle w^{(\kappa_m(u^{(0)}),m)}, u^{(0)} \right\rangle .\end{aligned}$$

The proof is complete. □

We remark that since $\left\langle w^{(\kappa_m\left(u^{(0)}\right),m)}, u^{(0)} \right\rangle \neq 0$, $\lim_{t\to\infty} u_m^{(2t)} \neq \lim_{t\to\infty} u_m^{(2t+1)}$, so each $\left\{u_m^{(t)}\right\}_{t=0}^{\infty}$ will be 'asymptotically 2-periodic' and not 'asymptotically 1-periodic'. We remark further that true periodicity may occur. Indeed, such phenomenon occurs when $n = 2$ or 3, and will be shown in the next section.

Theorem 4.4 *Let $m \in \{1, ..., n\}$ and let $u^{(0)} \in R^n$ such that $\kappa_m(u^{(0)}) \neq 0$. Let $\{u^{(t)}\}_{t=0}^{\infty}$ be the solution of (1.1) determined by $u^{(0)}$. If $\gamma > \gamma_m(u^{(0)})$, then $\left\{u_m^{(t)}\right\}$ will diverge, and*

$$u_m^{(t)} \sim (1-\lambda^{(\kappa_m(u^{(0)}))}\gamma)^t \left\langle w^{(\kappa_m\left(u^{(0)}\right),m)}, u^{(0)} \right\rangle$$

for all large t.

Proof If $\gamma > \gamma_m\left(u^{(0)}\right) = 2/\lambda^{\left(\kappa_m\left(u^{(0)}\right)\right)}$, then $1-\lambda^{(\kappa_m(u^{(0)}))}\gamma < -1$. Thus for $0 \le k < \kappa_m(u^{(0)})$,

$$\left|1-\lambda^{(\kappa_m(u^{(0)}))}\gamma\right| > 1 \ge 1-\lambda^{(k)}\gamma > 1-\lambda^{(\kappa_m(u^{(0)}))}\gamma \ge -\left|1-\lambda^{(\kappa_m(u^{(0)}))}\gamma\right|,$$

so that

$$\left|\frac{1-\lambda^{(k)}\gamma}{1-\lambda^{(\kappa_m(u^{(0)}))}\gamma}\right| < 1.$$

In view of (3.1),

$$\begin{aligned}&\lim_{t\to\infty} \frac{u_m^{(t)}}{(1-\lambda^{(\kappa_m(u^{(0)}))}\gamma)^t \left\langle w^{(\kappa_m\left(u^{(0)}\right),m)}, u^{(0)} \right\rangle} \\ &= \sum_{k=0}^{\kappa_m\left(u^{(0)}\right)-1} \lim_{t\to\infty} \left(\frac{1-\lambda^{(k)}\gamma}{1-\lambda^{(\kappa_m(u^{(0)}))}\gamma}\right)^t \frac{\langle w^{(k,m)}, u^{(0)} \rangle}{\left\langle w^{(\kappa_m\left(u^{(0)}\right),m)}, u^{(0)} \right\rangle} + 1 \\ &= 1.\end{aligned}$$

Equivalently, we get

$$u_m^{(t)} \sim (1 - \lambda^{(\kappa_m(u^{(0)}))}\gamma)^t \left\langle w^{(\kappa_m(u^{(0)}),m)}, u^{(0)} \right\rangle$$

for all large t. The proof is complete. □

5 Cooperative asymptotic behaviors

We will need critical numbers to distinguish the dependence of a solution on the parameter γ. Given $u \in R^n$ such that $\kappa(u) > 0$, we set

$$\gamma_n^* = \frac{2}{\lambda^{([n/2])}} = \begin{cases} \frac{1}{1+\cos \pi/n} & n \text{ odd} \\ \frac{1}{2} & n \text{ even} \end{cases}.$$

If no confusion arises, we will also write γ^* instead of γ_n^*. Note that

$$\frac{1}{1 + \cos \pi/n} = \frac{1}{2\cos^2 \frac{\pi}{2n}}.$$

First of all, since (1.1) represents a closed system, the following is expected.

Theorem 5.1 *Let $\{u^{(t)}\}_{t=0}^{\infty}$ be a solution of (1.1). Then $M\left(u^{(t)}\right) = M\left(u^{(0)}\right)$ for all $t \in N$.*

Indeed, this follows from adding the equations in (1.1):

$$\sum_{j=1}^n u_j^{(t+1)} = \gamma \sum_{j=1}^n u_j^{(t)} + (1 - 2\gamma) \sum_{j=1}^n u_j^{(t)} + \gamma \sum_{j=1}^n u_j^{(t)} = \sum_{j=1}^n u_j^{(t)}.$$

Corollary 5.2 *Let $\{u^{(t)}\}_{t=0}^{\infty}$ be a solution of (1.1). Then $\lim_{t\to\infty} u^{(2t)}$ exists if, and only if $\lim_{t\to\infty} u^{(2t+1)}$ exists. Furthermore, if $\lim\limits_{t\to\infty} u^{(2t)}$ or $\lim\limits_{t\to\infty} u^{(2t+1)}$ exists, then $\lim\limits_{t\to\infty} (u^{(2t)} + u^{(2t+1)}) = 2M\left(u^{(0)}\right)(1,1,...,1)^T$.*

Proof From (1.1), we have

$$u^{(2t+1)} - u^{(2t)} = -\gamma A_n u^{(2t)},$$

and

$$u^{(2t+2)} - u^{(2t+1)} = -\gamma A_n u^{(2t+1)}.$$

Therefore, $\lim_{t\to\infty} u^{(2t)}$ exists if, and only if $\lim_{t\to\infty} u^{(2t+1)}$ exists. Furthermore, adding these two equations together, we obtain

$$u^{(2t+2)} - u^{(2t)} = -\gamma A_n \left(u^{(2t)} + u^{(2t+1)}\right).$$

If $\lim\limits_{t\to\infty} u^{(2t)}$ exists, then $\lim\limits_{t\to\infty} u^{(2t)} = \lim\limits_{t\to\infty} u^{(2t+2)}$ so that

$$0 = \lim_{t\to\infty} u^{(2t+2)} - \lim_{t\to\infty} u^{(2t)} = -\gamma A_n \lim_{t\to\infty} (u^{(2t)} + u^{(2t+1)}),$$

which implies

$$\lim_{t\to\infty} (u^{(2t)} + u^{(2t+1)}) \in \langle V_0 \rangle = \left\langle (1,1,...,1)^T \right\rangle.$$

In other words,

$$\lim_{t\to\infty} u_j^{(2t)} + \lim_{t\to\infty} u_j^{(2t+1)} = \beta, \; 1 \le j \le n,$$

for some constant β. But in view of Theorem 7,

$$\lim_{t\to\infty} 2\sum_{j=1}^n u_j^{(0)} = \lim_{t\to\infty} \sum_{j=1}^n \left(u_j^{(2t)} + u_j^{(2t+1)}\right) = n\beta,$$

thus $\beta = 2M\left(u^{(0)}\right)$. If $\lim_{t\to\infty} u^{(2t+1)}$ exists the proof is similar. The proof is complete. □

Next we note that if $\kappa\left(u^{(0)}\right) = 0$, then $\kappa_m\left(u^{(0)}\right) = 0$ for $m \in \{1, 2, ..., n\}$, thus the solution $\left\{u^{(t)}\right\}$ of (1.1) determined by $u^{(0)}$ will satisfy

$$u^{(t)} = M\left(u^{(0)}\right)(1, 1, ..., 1)^T$$

for all $t \in N$, regardless of the magnitude of γ. Therefore, we may restrict our attention to the case where $\kappa\left(u^{(0)}\right) \neq 0$ in the following discussions.

We consider the following cases: (i) $\gamma < 0$, (ii) $0 < \gamma < \gamma^*$, (iii) $\gamma = \gamma^*$ and (iv) $\gamma > \gamma^*$.

Theorem 5.3 *Suppose $\gamma < 0$. Let $\left\{u^{(t)}\right\}$ be the solution of (1.1) determined by $u^{(0)} \in R^n$ which satisfies $\kappa\left(u^{(0)}\right) \neq 0$. Then some component sequence $\left\{u_m^{(t)}\right\}$ is divergent to $+\infty$ or to $-\infty$, and thus $\left\{\left\|u^{(t)}\right\|_2\right\}$ is also divergent.*

Indeed, if $\kappa\left(u^{(0)}\right) \neq 0$, then in view of Lemma 3.1, $\kappa_m\left(u^{(0)}\right) \neq 0$ for some m. Thus by Theorem 4.1, $\left\{u_m^{(t)}\right\}$ is divergent.

We remark that if $\left\{u_m^{(t)}\right\}$ diverges to $+\infty$, there must be another component sequence $\left\{u_j^{(t)}\right\}$ which diverges to $-\infty$. This is a cooperative phenomenon which follows from Theorem 5.1.

Theorem 5.4 *Suppose $0 < \gamma < \gamma_n^*$. Then for any $u^{(0)} \in R^n$, the solution $\left\{u^{(t)}\right\}_{t=0}^{\infty}$ determined by $u^{(0)}$ will satisfy $\lim_{t\to\infty} u_m^{(t)} = M(u^{(0)})$ for $m \in \{1, \cdots, n\}$.*

Proof Let $m \in \{1, ..., n\}$. If $\kappa_m(u^{(0)}) = 0$, then we have already shown that $u_m^{(t)} = M\left(u^{(0)}\right)$ for $t \geq 0$. Suppose $\kappa_m(u^{(0)}) \neq 0$. If n is odd, then

$$0 < \gamma < \gamma_n^* = \frac{1}{2\cos^2(\frac{\pi}{2n})} = \frac{2}{\lambda^{((n-1)/2)}} \leq \frac{2}{\lambda^{(\kappa_m(u^{(0)}))}} = \gamma_m(u^{(0)}),$$

and if n is even, then

$$0 < \gamma < \gamma_n^* = \frac{1}{2} = \frac{2}{\lambda^{(n/2)}} \leq \frac{2}{\lambda^{(\kappa_m(u^{(0)}))}} = \gamma_m(u^{(0)}).$$

In both cases, Theorem 4.2 yields the desired conclusion. The proof is complete. □

Roughly speaking, the above result says that when the marginal propensity constant is a small positive number (or, when money are only allowed to flow slowly in normal situations), each country will ultimately be equal in wealth! However, Theorem 4.2 is concerned with the ultimate behavior of an individual nation and indicates that such cooperative phenomenon may not be true in general. Such examples will be given later.

Theorem 5.5 *Suppose $\gamma = \gamma^*$. Let $\left\{u^{(t)}\right\}_{t=0}^{\infty}$ be the solution of (1.1) determined by $u^{(0)} \in R^n$. Then*

$$\lim_{t\to\infty} u_m^{(2t)} = M(u^{(0)}) + \left\langle w^{([n/2],m)}, u^{(0)}\right\rangle, \quad m = 1, 2, ..., n,$$

and

$$\lim_{t\to\infty} u_m^{(2t+1)} = M(u^{(0)}) - \left\langle w^{([n/2],m)}, u^{(0)}\right\rangle, \quad m = 1, 2, ..., n.$$

Proof If $\gamma = \gamma^* = 2/\lambda^{([n/2])}$, then

$$1 - \lambda^{([n/2]))}\gamma = -1$$

and

$$1 - \lambda^{(k)}\gamma = 1 - 2(\lambda^{(k)}/\lambda^{([n/2])}).$$

Since $0 < \lambda^{(k)} < \lambda^{([n/2])}$ for $k = 1, 2, \cdots, [n/2] - 1$, we see that

$$1 > 1 - 2(\lambda^{(k)}/\lambda^{([n/2])}) > -1,$$

thus by (2.2),

$$\begin{aligned}
\lim_{t\to\infty} u_m^{(2t)} &= M(u^{(0)}) + \sum_{k=1}^{[n/2]} \lim_{t\to\infty} (1 - \lambda^{(k)}\gamma)^{2t} \left\langle w^{(k,m)}, u^{(0)} \right\rangle \\
&= M(u^{(0)}) + \lim_{t\to\infty} (1 - \lambda^{([n/2])}\gamma)^{2t} \left\langle w^{([n/2],m)}, u^{(0)} \right\rangle \\
&= M(u^{(0)}) + \left\langle w^{([n/2],m)}, u^{(0)} \right\rangle
\end{aligned}$$

and

$$\begin{aligned}
\lim_{t\to\infty} u_m^{(2t+1)} &= M(u^{(0)}) + \sum_{k=1}^{[n/2]} \lim_{t\to\infty} (1 - \lambda^{(k)}\gamma)^{2t+1} \left\langle w^{(k,m)}, u^{(0)} \right\rangle \\
&= M(u^{(0)}) + \lim_{t\to\infty} (1 - \lambda^{([n/2])}\gamma)^{2t+1} \left\langle w^{([n/2],m)}, u^{(0)} \right\rangle \\
&= M(u^{(0)}) - \left\langle w^{([n/2],m)}, u^{(0)} \right\rangle
\end{aligned}$$

for $m = 1, 2, ..., n$. The proof is complete. □

It is interesting to note from (2.1) that when n is even,

$$\left\langle w^{([n/2],m)}, u^{(0)} \right\rangle = \frac{1}{n} \sum_{i=1}^{n} (-1)^{m+i} u_i^{(0)}$$

so that

$$M\left(u^{(0)}\right) + \left\langle w^{([n/2],m)}, u^{(0)} \right\rangle = \begin{cases} \frac{2}{n} \sum_{i=1}^{n/2} u_{2i-1}^{(0)} & m = 1, 3, ..., n-1 \\ \frac{2}{n} \sum_{i=1}^{n/2} u_{2i}^{(0)} & m = 2, 4, ..., n \end{cases},$$

and

$$M\left(u^{(0)}\right) - \left\langle w^{([n/2],m)}, u^{(0)} \right\rangle = \begin{cases} \frac{2}{n} \sum_{i=1}^{n/2} u_{2i-1}^{(0)} & m = 2, 4, ..., n \\ \frac{2}{n} \sum_{i=1}^{n/2} u_{2i}^{(0)} & m = 1, 3, ..., n-1 \end{cases}.$$

Similarly, when n is odd

$$\left\langle w^{([n/2],m)}, u^{(0)} \right\rangle = \frac{2}{n} \sum_{i=1}^{n} (-1)^{i-m} u_i^{(0)} \cos\frac{(i-m)\pi}{n}, \ m = 1, 2, ..., n.$$

In both cases, the solution $\left\{u^{(t)}\right\}$ is 'asymptotically 2-periodic'. We remark that true periodicity may occur. Indeed, such phenomena occur when $n = 2$ or 3.

Theorem 5.6 *When $n = 2$ or 3, and $\gamma = \gamma_n^*$, every solution $\left\{u^{(t)}\right\}_{t=0}^{\infty}$ of (1.1) is 2-periodic.*

Indeed, suppose $n = 2$ and $\gamma = \gamma_2^* = 1/2$. Then in view of (1.1), we have

$$u^{(t+1)} = \left(I - \frac{1}{2}A_2\right)u^{(t)} = \begin{bmatrix} 0 & 1 \\ 1 & 0 \end{bmatrix} u^{(t)}$$

and

$$u^{(t+2)} = \left(I - \frac{1}{2}A_2\right)u^{(t+1)} = Iu^{(t)} = u^{(t)}.$$

Similarly, where $n = 3$ and $\gamma = \gamma_3^* = 2/3$, then

$$A_3 = \begin{bmatrix} 2 & -1 & -1 \\ -1 & 2 & -1 \\ -1 & -1 & 2 \end{bmatrix}$$

and

$$u^{(t+2)} = \left(I - \frac{2}{3} \cdot A_3\right)u^{(t+1)} = \left(I - \frac{2}{3} \cdot A_3\right)^2 u^{(t)} = Iu^{(t)} = u^{(t)}.$$

Theorem 5.7 *Suppose $\gamma > \gamma^*$. Let $\{u^{(t)}\}$ be the solution of (1.1) determined by $u^{(0)} \in R^n$ which satisfies $\kappa\left(u^{(0)}\right) \neq 0$. If $\kappa\left(u^{(0)}\right) = [n/2]$, then*

$$u_m^{(t)} \sim (1 - \lambda^{([n/2])}\gamma)^t \left\langle w^{([n/2],m)}, u^{(0)} \right\rangle, \quad m = 1, 2, ..., n, \tag{5.1}$$

for all large t. If $\kappa\left(u^{(0)}\right) < [n/2]$, then when $\gamma < 2/\lambda^{\left(\kappa\left(u^{(0)}\right)\right)}$, we have

$$\lim_{t\to\infty} u_m^{(t)} = M\left(u^{(0)}\right);$$

when $\gamma = 2/\lambda^{\left(\kappa\left(u^{(0)}\right)\right)}$, we have

$$\lim_{t\to\infty} u_m^{(2t)} = M\left(u^{(0)}\right) + \left\langle w^{\left(\kappa\left(u^{(0)}\right),m\right)}, u^{(0)} \right\rangle, \quad m = 1, 2, ..., n,$$

and

$$\lim_{t\to\infty} u_m^{(2t+1)} = M\left(u^{(0)}\right) - \left\langle w^{\left(\kappa\left(u^{(0)}\right),m\right)}, u^{(0)} \right\rangle, \quad m = 1, 2, ..., n,$$

and when $\gamma > 2/\lambda^{\left(\kappa\left(u^{(0)}\right)\right)}$, $u_m^{(t)}$ diverges for $m = 1, 2, .., n$ and

$$u_m^{(t)} \sim (1 - \lambda^{\left(\kappa\left(u^{(0)}\right)\right)}\gamma)^t \left\langle w^{\left(\kappa\left(u^{(0)}\right),m\right)}, u^{(0)} \right\rangle, \quad m = 1, 2, ..., n,$$

for all large t.

Proof If $\gamma > \gamma^* = 2/\lambda^{([n/2])}$, then $1 - \lambda^{([n/2])}\gamma < -1$. When $\kappa\left(u^{(0)}\right) = [n/2]$, for $0 \leq k < \kappa(u^{(0)}) = [n/2]$,

$$\left|1 - \lambda^{([n/2]))}\gamma\right| > 1 \geq 1 - \lambda^{(k)}\gamma > 1 - \lambda^{([n/2]))}\gamma \geq -\left|1 - \lambda^{([n/2])}\gamma\right|,$$

so that

$$\left|\frac{1 - \lambda^{(k)}\gamma}{1 - \lambda^{([n/2]))}\gamma}\right| < 1.$$

In view of (2.2),

$$\begin{aligned}
&\lim_{t\to\infty} \frac{u_m^{(t)}}{(1 - \lambda^{([n/2])}\gamma)^t \left\langle w^{([n/2],m)}, u^{(0)} \right\rangle} \\
&= \sum_{k=0}^{[n/2]-1} \lim_{t\to\infty} \left(\frac{1 - \lambda^{(k)}\gamma}{1 - \lambda^{([n/2]))}\gamma}\right)^t \frac{\left\langle w^{(k,m)}, u^{(0)} \right\rangle}{\left\langle w^{([n/2],m)}, u^{(0)} \right\rangle} + 1 \\
&= 1.
\end{aligned}$$

Equivalently, we get

$$u_m^{(t)} \sim (1-\lambda^{([n/2])}\gamma)^t \left\langle w^{([n/2],m)}, u^{(0)} \right\rangle, \ m = 1, 2, ..., n,$$

for all large t.

Next, suppose $\kappa\left(u^{(0)}\right) < [n/2]$. Then in view of (3.1),

$$u_m^{(t)} = M\left(u^{(0)}\right) + \sum_{k=1}^{\kappa\left(u^{(0)}\right)} \left(1-\lambda^{(k)}\gamma\right)^t \left\langle w^{(k,m)}, u^{(0)} \right\rangle, \ m = 1, ..., n.$$

If $\gamma < 2/\lambda^{\left(\kappa\left(u^{(0)}\right)\right)}$, then

$$\left|1-\lambda^{(k)}\gamma\right| < 1, \ k = 1, ..., \kappa\left(u^{(0)}\right),$$

hence,

$$\lim_{t\to\infty} u_m^{(t)} = M\left(u^{(0)}\right).$$

If $\gamma = 2/\lambda^{\left(\kappa\left(u^{(0)}\right)\right)}$, then

$$1-\lambda^{\left(\kappa\left(u^{(0)}\right)\right)} = -1$$

and

$$\left|1-\lambda^{(k)}\gamma\right| < 1, \ k = 1, 2, ..., \kappa\left(u^{(0)}\right) - 1,$$

thus

$$\lim_{t\to\infty} u_m^{(2t)} = M\left(u^{(0)}\right) + \left\langle w^{\left(\kappa\left(u^{(0)}\right),m\right)}, u^{(0)} \right\rangle, \ m = 1, 2, ..., n,$$

$$\lim_{t\to\infty} u_m^{(2t+1)} = M\left(u^{(0)}\right) - \left\langle w^{\left(\kappa\left(u^{(0)}\right),m\right)}, u^{(0)} \right\rangle, \ m = 1, 2, ..., n.$$

If $\gamma > 2/\lambda^{\left(\kappa\left(u^{(0)}\right)\right)}$, then

$$1-\lambda^{\left(\kappa\left(u^{(0)}\right)\right)} \le -1,$$

and

$$1 > 1-\lambda^{(k)}\gamma > 1-\lambda^{\left(\kappa\left(u^{(0)}\right)\right)}\gamma,$$

thus

$$u_m^{(t)} \sim (1-\lambda^{\left(\kappa\left(u^{(0)}\right)\right)}\gamma)^t \left\langle w^{\left(\kappa\left(u^{(0)}\right),m\right)}, u^{(0)} \right\rangle, \ m = 1, 2, ..., n,$$

for all large t. The proof is complete. □

We remark that when n is even and $\kappa(u^{(0)}) = n/2$, we have $\lambda^{([n/2])} = 4$ and

$$\left\langle w^{([n/2],m)}, u^{(0)} \right\rangle = \frac{1}{n}\sum_{j=1}^{n}(-1)^{m+j}u_j^{(0)},$$

so that (5.1) becomes

$$u_m^{(t)} \sim (1-4\gamma)^t \frac{1}{n}\sum_{j=1}^{n}(-1)^{m+j}u_j^{(0)}, \ m = 1, 2, ..., n,$$

for all large t. In particular,

$$\lim_{t\to\infty} u^{(2t)} = (\pm\infty, \mp\infty, \pm\infty, \mp\infty, ..., \pm\infty, \mp\infty)^T$$

and

$$\lim_{t\to\infty} u^{(2t+1)} = (\mp\infty, \pm\infty, ..., \mp\infty, \pm\infty)^T.$$

Similar results hold when n is odd.

6 Examples

In the example following Corollary 3.6, we have constructed a vector $u^{(0)} \in R^{17}$ such that $\kappa_1(u^{(0)}) = 2$ and $\kappa_m(u^{(0)}) = 6$ for each $m \in \{1, ..., 17\}$. Now note that

$$\gamma_{1,17}(u^{(0)}) = 1/(2\sin^2(2\pi/17)) \approx 3.8315,$$

and for $m = 2, 3, \cdots, 17$,

$$\gamma_{m,17}(u^{(0)}) = 1/(2\sin^2(6\pi/17)) \approx 0.4067.$$

Therefore, if we let $n = 17$ and $\gamma = 2$ in (1.1), then

$$\gamma_{m,17}(u^{(0)}) < \gamma < \gamma_{1,17}(u^{(0)})$$

for $m = 2, 3, \cdots, 17$. Let $\{u^{(t)}\}$ be the solution of (1.1) determined by $u^{(0)}$. Then $u_1^{(t)}$ will converge to $M\left(u^{(0)}\right) = \frac{1}{17}\sum_{j=1}^{17} u_j^{(0)}$ by Theorem 4.2 and $u_m^{(t)}$ will diverge for $m = 2, 3, \cdots, 17$ by Theorem 4.4. The following data support our theoretical conclusion:

$$\begin{aligned}
u^{(0)} &= (\mathbf{0.1176}, -0.0069, 0.1240, -0.1134, -0.1776, 0.0171, -0.1115, 0.0308,\\
&\quad 0.2150, 0.0044, 0.0741, 0.0471, -0.2172, -0.0537, -0.0284, -0.1023,\\
&\quad 0.1808)^T,\\
u^{(13)} &= 10^8(\mathbf{0.0000}, 3.1968, -3.8530, 1.4471, 2.1089, -3.9889, 2.6988, 0.7361,\\
&\quad -3.5860, 3.5860, -0.7361, -2.6988, 3.9889, -2.1089, -1.4471, 3.8530,\\
&\quad -3.1968)^T,\\
u^{(14)} &= 10^9(-\mathbf{0.0000}, -1.7297, 2.0847, -0.7830, -1.1410, 2.1582, -1.4602,\\
&\quad -0.3983, 1.9402, -1.9402, 0.3983, 1.4602, -2.1582, 1.1410, 0.7830,\\
&\quad -2.0847, 1.7297)^T,\\
u^{(15)} &= 10^{10}(\mathbf{0.0000}, 0.9358, -1.1279, 0.4236, 0.6173, -1.1677, 0.7900, 0.2155,\\
&\quad -1.0498, 1.0498, -0.2155, -0.7900, 1.1677, -0.6173, -0.4236, 1.1279,\\
&\quad -0.9358)^T.
\end{aligned}$$

As another example, observe first from Corollary 3.6, there is a vector $u^{(0)} \in R^5$ such that $\kappa(u^{(0)}) = 2$ and $\kappa_1(u^{(0)}) = 0$. By Corollary 3.5 $\kappa_m(u^{(0)}) = \kappa(u^{(0)}) = 2$ for $m = 2, 3, 4, 5$. By the proof of Theorem 3.3, let $\langle v^{(0)}, u^{(0)}\rangle = \langle w^{(0,1)}, u^{(0)}\rangle = 1$, $\langle v^{(1)}, u^{(0)}\rangle = \langle v^{(4)}, u^{(0)}\rangle = 0$, $\langle v^{(2)}, u^{(0)}\rangle = -v_1^{(3)} \neq 0$ and $\langle v^{(3)}, u^{(0)}\rangle = v_1^{(2)}$. Then

$$\left\langle w^{(2,1)}, u^{(0)}\right\rangle = v_1^{(2)}\left\langle v^{(2)}, u^{(0)}\right\rangle + v_1^{(3)}\left\langle v^{(3)}, u^{(0)}\right\rangle = 0,$$

and

$$u^{(0)} = v^{(0)} - v_1^{(3)} v^{(2)} + v_1^{(2)} v^{(3)}$$

satisfies $\kappa_1(u^{(0)}) = 0$ and $\kappa_m(u^{(0)}) = \kappa(u^{(0)}) = 2$ for $m = 2, 3, 4, 5$. If we let $\gamma = -1$, $n = 5$ in (1.1) and let $\{u^{(t)}\}$ be the solution of (1.1) determined by $u^{(0)}$, then $u_1^{(t)} = \frac{1}{n}\sum_{j=1}^{5} u_j^{(0)}$ for all t and $u_m^{(t)}$ will diverge by Theorem 4.1. The following data support our conclusion:

$$\begin{aligned} u^{(0)} &= (\mathbf{0.4472}, 0.2121, 0.8276, 0.0668, 0.6823)^T, \\ u^{(15)} &= 10^9(\mathbf{0.0000}, -2.1784, 3.5247, -3.5247, 2.1784)^T, \\ u^{(16)} &= 10^{10}(\mathbf{0.0000}, -1.0060, 1.6277, -1.6277, 1.0060)^T, \\ u^{(17)} &= 10^{10}(\mathbf{0.0000}, -4.6457, 7.5169, -7.5169, 4.6457)^T. \end{aligned}$$

As another example, let $n = 15$ in (1.1). We want to find a $u^{(0)} \in R^{15}$ such that $\kappa(u^{(0)}) = 6$, $\kappa_1(u^{(0)}) = 5$, $\kappa_6(u^{(0)}) = 3$, $\kappa_{11}(u^{(0)}) = 5$. By Theorem 3.3, if $\kappa_i(u^{(0)}) < \kappa(u^{(0)})$ and $\kappa_j(u^{(0)}) < \kappa(u^{(0)})$, then

$$\frac{2(i-j)\kappa\left(u^{(0)}\right)}{n} = \frac{2(i-j)6}{15} = \frac{4}{5}(i-j) \in Z,$$

so that $i-j$ must be an integer multiple of 5. Thus $\kappa_m(u^{(0)}) = \kappa(u^{(0)}) = 6$ for $m \neq 1, 6, 11$. Let $a_1 = 5$, $a_6 = 3$, $a_{11} = 5$, and $a_m = 6$ for $m = 2, 3, 4, 5, 7, 8, 9, 10, 12, 13, 14, 15$. Then it is easily verified that the assumptions in Theorem 3.3 are satisfied so that the desired vector $u^{(0)} \in R^{15}$ can be found. Furthermore, by the methods described in the proof of Theorem 3.3, such a vector is

$$\begin{aligned} u^{(0)} &= -v_1^{(9)} v^{(6)} + v_1^{(6)} v^{(9)} + \left(\frac{v_6^{(10)}}{2\sin 10\pi/3}\right) v^{(5)} + \left(\frac{-v_6^{(5)}}{2\sin 10\pi/3}\right) v^{(10)} \\ &\quad + v_6^{(3)} v^{(3)} + v_6^{(12)} v^{(12)}. \end{aligned}$$

Next recall that

$$\gamma_{1,15}(u^{(0)}) = 1/(2\sin^2(5\pi/15)) = 2/3 \approx 0.6667,$$

$$\gamma_{6,15}(u^{(0)}) = 1/(2\sin^2(3\pi/15)) \approx 1.4472,$$

$$\gamma_{11,15}(u^{(0)}) = 1/(2\sin^2(5\pi/15)) = 2/3 \approx 0.6667,$$

and

$$\gamma_{m,15}(u^{(0)}) = 1/(2\sin^2(6\pi/15)) \approx 0.5528$$

for $m \neq 1, 6, 11$. If we take $\gamma = 0.6$, then

$$\gamma_{m,15}(u^{(0)}) < \gamma < \gamma_{1,15}(u^{(0)}) = \gamma_{11,15}(u^{(0)}) < \gamma_{6,15}(u^{(0)}).$$

Let $\{u^{(t)}\}$ be the solution of (1.1) determined by $u^{(0)}$, then $u_1^{(t)}$, $u_6^{(t)}$ and $u_{11}^{(t)}$ will converge to $\frac{1}{15}\sum_{j=1}^{15} u_j^{(0)}$ and $u_m^{(t)}$ will diverge for $m \neq 1, 6, 11$. The corresponding numerical data are

$$\begin{aligned} u^{(0)} &= (\mathbf{0.0667}, 0.0295, 0.0189, -0.3013, 0.1862, \mathbf{0.1333}, -0.1038, \\ &\quad 0.0856, -0.2347, 0.0529, \mathbf{0.2000}, -0.0372, -0.0477, -0.1680, 0.1196)^T \\ u^{(90)} &= 10^5(\mathbf{0.0000}, -1.1436, 1.8504, -1.8504, 1.1436, -\mathbf{0.0000}, -1.1436, \\ &\quad 1.8504, -1.8504, 1.1436, \mathbf{0.0000}, -1.1436, 1.8504, -1.8504, 1.1436)^T \\ u^{(91)} &= 10^5(-\mathbf{0.0000}, 1.3390, -2.1665, 2.1665, -1.3390, \mathbf{0.0000}, 1.3390, \\ &\quad -2.1665, 2.1665, -1.3390, -\mathbf{0.0000}, 1.3390, -2.1665, 2.1665, -1.3390)^T \\ u^{(92)} &= 10^5(\mathbf{0.0000}, -1.5677, 2.5366, -2.5366, 1.5677, -\mathbf{0.0000}, -1.5677, \\ &\quad 2.5366, -2.5366, 1.5677, \mathbf{0.0000}, -1.5677, 2.5366, -2.5366, 1.5677)^T. \end{aligned}$$

If we pick $\gamma = 1$ then $\gamma_{m,15}(u^{(0)}) < \gamma < \gamma_{6,15}(u^{(0)})$. In such a case, $u_6^{(t)}$ will converge to $\frac{1}{15}\sum_{j=1}^{15} u_j^{(0)}$ and $u_m^{(t)}$ will diverge for $m \neq 6$. The supportive data are now

$$\begin{aligned}
u^{(0)} &= (0.0667, 0.0295, 0.0189, -0.3013, 0.1862, \mathbf{0.1333}, -0.1038,\\
&\quad 0.0856, -0.2347, 0.0529, 0.2000, -0.0372, -0.0477, -0.1680, 0.1196)^T\\
u^{(25)} &= 10^9(0.0022, 2.2034, -3.5688, 3.5711, -2.2079, \mathbf{0.0000}, 2.2079,\\
&\quad -3.5711, 3.5688, -2.2034, -0.0022, 2.2057, -3.5666, 3.5666, -2.2057)^T\\
u^{(26)} &= 10^9(-0.0045, -5.7700, 9.3433, -9.3478, 5.7790, \mathbf{-0.0000}, -5.7790,\\
&\quad 9.3478, -9.3433, 5.7700, 0.0045, -5.7745, 9.3389, -9.3389, 5.7745)^T\\
u^{(27)} &= 10^{10}(0.0009, 1.5109, -2.4461, 2.4470, -1.5127, \mathbf{0.0000}, 1.5127,\\
&\quad -2.4470, 2.4461, -1.5109, -0.0009, 1.5118, -2.4452, 2.4452, -1.5118)^T.
\end{aligned}$$

As our last example, let $n = 15$ in (1.1). We assert that there cannot be any solution $\{u^{(t)}\}$ of (1.1) such that $\left\{u_1^{(t)}\right\}$ and $\left\{u_5^{(t)}\right\}$ both converge to $M\left(u^{(0)}\right)$ while $\left\{u_2^{(t)}\right\}$ diverges. Indeed, if $\left\{u_2^{(t)}\right\}$ diverges, then $0 < \kappa_2\left(u^{(0)}\right)$ and in view of Theorem 4.2, $\gamma \leq 0$ or $\gamma \geq \gamma_{2,15} \approx 3.0223$. But the fact that $\left\{u_1^{(t)}\right\}$ and $\left\{u_5^{(t)}\right\}$ converge to $M\left(u^{(0)}\right)$ imply $0 < \gamma < \min\{\gamma_{1,15}, \gamma_{5,15}\} = 2/3$, which is a contradiction.

Other examples related to the cooperative behaviors of (1.1) can be constructed by similar considerations.

7 Concluding remarks

According to the above discussions, we see that although the total wealth of the n-nations in the model (1.1) is a constant, the wealth of an individual nation can change in different time periods. Indeed, let $\left\{u_i^{(t)}\right\}$ stands for the wealth of the i-th nation in the time period t, then it is possible that $u_i^{(t)} = u_i^{(0)}$ for all $t \in N$, or $u_i^{(t)}$ converges to the mean of the initial wealth distribution, or oscillates between excessive assets and liabilities.

The cooperative behaviors can be quite complicated but depend upon the exchange propensity and the initial wealth distributions. In particular, it is possible that every country will have the same wealth in the long run, or (as seen in the examples given in the last section) it is possible that one country has a stable economy while every other countries suffers from excessive oscillatory instability.

We have thus given a complete mathematical analysis of our simple model and also indicated the relations between our results and some of the phenomena observed in reality.

In view of these analyses and observations, we may conclude that our ad hoc model needs much modifications. For instance, we will need to make modifications so that the aggregate wealth across all nations cannot be invariant over time. Such modifications may be achieved by imposing contributions from internal logistic growth or technological breakthrough. Also we need modifications so as to terminate the economic activity of any nation that exhibits excessive oscillatory instability since such a nation in reality will be bankrupted already.

It is hoped that such realistic models can be dealt with in the near future. General results will also be expected by means of techniques such as the bifurcation analysis. Complete mathematical analysis for these models similar to those given in this paper, however, will be difficult, if not impossible.

References

[1] G. Gandolfo, Economic Dynamics: Methods and Models, North-Holland Publishing Co., 1971.

[2] F. Tohmé and S. London, A mathematical representation of economic evolution, Math. Comput. Modelling, 27(8)(1998), 29–40.

[3] J. Aschenwald, S. Fink and G. Tappeiner, Brave new modeling: cellular automata and artificial neural networks for mastering complexity in economics, Complexity, 7(1)(2001), 39–47.

[4] A. S. Elgazzar, A model for the evolution of economic systems in social networks, Phys. A, 303(3-4)(2002), 543–551.

[5] M. O. Jackson and A. Watts, The evolution of social and economic networks, J. Econom. Theory, 106(2)(2002), 265–295.

[6] P. J. Davis, Circulant Matrices, John Wiley & Sons, 1979,

Fields Institute Communications
Volume **42**, 2004

Convergence of Solutions for a Discrete-Time System

Binxiang Dai
College of Mathematics and Econometrics
Hunan University
Changsha, Hunan 410082, P. R. China
bxdai@hnu.net.cn

Xiangzhen Qian
College of Mathematics and Econometrics
Hunan University
Changsha, Hunan 410082, P. R. China
xzqian@hnu.net.cn

Lihong Huang
College of Mathematics and Econometrics
Hunan University
Changsha, Hunan 410082, P. R. China
lhhuang@hnu.net.cn

Dedicated to Professor Zhicheng Wang on the occasion of his seventieth birthday

Abstract. Consider a nonlinear discrete-time system, we obtain the sufficient condition that all solutions of this system converge to equilibrium.

1 Introduction

We consider the following nonlinear discrete-time system:

$$\begin{cases} x_n = \mu x_{n+1} - \alpha f(y_n - \beta y_{n-k}), \\ y_n = \mu y_{n-1} + \alpha f(x_n - \beta x_{n-k}), \end{cases} \quad n \geq 0. \tag{1.1}$$

where $\mu \in (0,1), \alpha > 0, \beta \geq 0$ are given constants and $f : R \to R$ is continuous, with $f(0) = 0, xf(x) \neq 0, |f(x)| < |x|$ for all $x \neq 0$.

2000 *Mathematics Subject Classification.* Primary 39A10, 39A12; Secondary 92B20.

This work partially supported by Science Foundation of Hunan University and Hunan Province Natural Science Foundation(02JJY2012).

System (1.1) can be regarded as the discrete analog of the following neural network of two neurons with dynamical threshold effects:

$$\begin{cases} \frac{dx(t)}{dt} = -x(t) + \alpha \cdot f(y(t) - \beta y(t-\tau)), \\ \frac{dy(t)}{dt} = -y(t) + \alpha \cdot f(x(t) - \beta x(t-\tau)). \end{cases} \tag{1.2}$$

System (1.2) has found interesting applications in, for example, temporal erolution of sublaffice magnetizations (see [1]). Recently, the dynamics of (1.2) and some related models have been discussed in [2-6]. In System (1.1), a particular case $f(\xi)$=tanh ξ is used in neural network theory.

In this paper, we derive sufficient conditions for the system (1.1) to have unique equilibrium such that the equilibrium is the globally attractor.

We let

$$\begin{cases} u_n = y_n - \beta y_{n-k}, \\ v_n = x_n - \beta x_{n-k}. \end{cases} \tag{1.3}$$

and obtain from (1.1) that

$$\begin{cases} u_n = \mu u_{n-1} + \alpha f(v_n) - \alpha\beta f(v_{n-k}), \\ v_n = \mu v_{n-1} + \alpha f(u_n) - \alpha\beta f(u_{n-k}), \end{cases} \quad n \geq 0. \tag{1.4}$$

We assume that μ, α, β are such that

$$|\alpha(1-\beta)| < 1 - \mu. \tag{1.5}$$

2 Main results and proofs

It is now elementary to note that (1.4) has the unique equilibrium (0,0) under (1.5). We first prove that the trivial solution of (1.4) is the global attractor when $k = 0$.

Theorem 2.1 *Suppose that condition (1.5) holds.If $k = 0$, then all solutions of system (1.4) satisfy*

$$\lim_{n\to\infty} u_n = 0, \ \lim_{n\to\infty} v_n = 0.$$

Proof When $k = 0$, system (1.4) becomes to

$$\begin{cases} u_n = \mu u_{n-1} + \alpha(1-\beta) f(v_n), \\ v_n = \mu v_{n-1} + \alpha(1-\beta) f(u_n), \end{cases} \quad n \geq 0. \tag{2.1}$$

Using the properties of function f, we have

$$\begin{pmatrix} |u_n| \\ |v_n| \end{pmatrix} \leq \mu \begin{pmatrix} |u_{n-1}| \\ |v_{n-1}| \end{pmatrix} + \begin{pmatrix} 0 & \alpha(1-\beta) \\ \alpha(1-\beta) & 0 \end{pmatrix} \begin{pmatrix} |u_n| \\ |v_n| \end{pmatrix},$$

i.e.

$$\begin{pmatrix} 1 & -\alpha(1-\beta) \\ -\alpha(1-\beta) & 1 \end{pmatrix} \begin{pmatrix} |u_n| \\ |v_n| \end{pmatrix} \leq \mu \begin{pmatrix} |u_{n-1}| \\ |v_{n-1}| \end{pmatrix}. \tag{2.2}$$

Let

$$A = \begin{pmatrix} 1 & -\alpha(1-\beta) \\ -\alpha(1-\beta) & 1 \end{pmatrix}.$$

Then

$$\det A = 1 - \alpha^2(1-\beta)^2 = (1 + \alpha(1-\beta))(1 - \alpha(1-\beta)) > 0.$$

Therefore A is reversible, and

$$A^{-1} = \frac{1}{1-\alpha^2(1-\beta)^2} \begin{pmatrix} 1 & \alpha(1-\beta) \\ \alpha(1-\beta) & 1 \end{pmatrix}.$$

From (2.2), we obtain

$$\begin{pmatrix} |u_n| \\ |v_n| \end{pmatrix} \leq \mu A^{-1} \begin{pmatrix} |u_{n-1}| \\ |v_{n-1}| \end{pmatrix}. \tag{2.3}$$

Since the eigenvalues of μA^{-1} satisfy the equation

$$(\lambda - \frac{\mu}{1-\alpha^2(1-\beta)^2})^2 - (\frac{\mu\alpha(1-\beta)}{1-\alpha^2(1-\beta)^2})^2 = 0,$$

we find that the eigenvalues of μA^{-1} are

$$\lambda_1 = \frac{\mu}{1-\alpha(1-\beta)}, \lambda_2 = \frac{\mu}{1+\alpha(1-\beta)}.$$

By the hypothesis (1.5), then $\lambda_1 < 1, \lambda_2 < 1$, hence $\|\mu A^{-1}\| < 1$. From (2.3) we have

$$\left\| \begin{pmatrix} |u_n| \\ |v_n| \end{pmatrix} \right\| \leq \|\mu A^{-1}\|^n \left\| \begin{pmatrix} |u_0| \\ |v_0| \end{pmatrix} \right\|$$

which yields to

$$\lim_{n\to\infty} u_n = 0, \ \lim_{n\to\infty} v_n = 0.$$

This completes the proof. □

Corollary 2.2 *Under the assumptions of Theorem 2.1, all solutions of (1.1) satisfy*

$$\lim_{n\to\infty} x_n = 0, \ \lim_{n\to\infty} y_n = 0.$$

Proof From (1.1) and (1.3), we have

$$\begin{cases} x_n = \mu x_{n-1} + \alpha f(u_n), \\ y_n = \mu y_{n-1} + \alpha f(v_n), \end{cases} \quad n \geq 0. \tag{2.4}$$

which leads to

$$\begin{cases} x_n = \mu^n x_0 - \sum\limits_{i=1}^{n} \alpha\mu^{n-i} f(u_i), \\ y_n = \mu^n y_0 + \sum\limits_{i=1}^{n} \alpha\mu^{n-i} f(v_i). \end{cases} \quad n \geq 0. \tag{2.5}$$

Let

$$\begin{cases} \bar{u}_n &= \sum\limits_{i=1}^{n} \alpha\mu^{n-i} f(u_i) = \alpha\mu^n \sum\limits_{i=1}^{n} (\frac{1}{\mu})^i f(u_i), \\ \bar{v}_n &= \sum\limits_{i=1}^{n} \alpha\mu^{n-i} f(v_i) = \alpha\mu^n \sum\limits_{i=1}^{n} (\frac{1}{\mu})^i f(v_i), \end{cases}$$

From (2.5), we have

$$\begin{cases} x_n = \mu^n x_0 + \bar{u}_n, \\ y_n = \mu^n y_0 + \bar{v}_n. \end{cases} \tag{2.6}$$

Since $0 < \mu < 1$, it is obvious that $\lim\limits_{n\to\infty} \mu^n = 0$.
Now, let us prove that $\lim\limits_{n\to\infty} \bar{u}_n = 0$.

If $\sum\limits_{i=1}^{\infty} (\frac{1}{\mu})^i f(u_i) < \infty$, then by the expression of $\bar{u}_n$, we have $\lim\limits_{n\to\infty} \bar{u}_n = 0$.

If $\sum_{i=1}^{\infty}(\frac{1}{\mu})^i f(u_i)=\infty$, then by Stoze Formular, we have

$$\begin{aligned}
\lim_{n\to\infty}\bar{u}_n &= \lim_{n\to\infty}\frac{\alpha\sum_{i=1}^{n}(\frac{1}{\mu})^i f(u_i)}{(\frac{1}{\mu})^n}\\
&= \alpha\lim_{n\to\infty}\frac{\sum_{i=1}^{n}(\frac{1}{\mu})^i f(u_i)-\sum_{i=1}^{n-1}(\frac{1}{\mu})^i f(u_i)}{(\frac{1}{\mu})^n-(\frac{1}{\mu})^{n-1}}\\
&= \alpha\lim_{n\to\infty}\frac{(\frac{1}{\mu})^n f(u_n)}{(\frac{1}{\mu})^n\cdot(1-\mu)}\\
&= \frac{\alpha}{1-\mu}\lim_{n\to\infty}f(u_n).
\end{aligned}$$

From Theorem 2.1 we know that $\lim_{n\to\infty}u_n=0$. By the continuity of f, we have $\lim_{n\to\infty}f(u_n)=0$, which implies that $\lim_{n\to\infty}\bar{u}_n=0$.

Thus, by the expression of x_n in (2.6), we shows that $\lim_{n\to\infty}x_n=0$.

Similarly, we can obtain that $\lim_{n\to\infty}y_n=0$. This completes the proof. □

In the following, we shall give the sufficient conditions which guarantee that the equilibrium point $(0,0)$ of system (1.1) is a global attractor of all solutions when $k\neq 0$.

Theorem 2.3 *Suppose that condition (1.5) holds. If*

$$\alpha(1+\beta)<1-\mu, \tag{2.7}$$

then all solutions of discrete system (1.4) satisfy

$$\lim_{n\to\infty}u_n=0,\qquad \lim_{n\to\infty}v_n=0.$$

Proof We have from (1.4), on using the properities of f,

$$\begin{pmatrix}|u_n|\\|v_n|\end{pmatrix}\le\mu\begin{pmatrix}|u_{n-1}|\\|v_{n-1}|\end{pmatrix}+\begin{pmatrix}0&\alpha\\\alpha&0\end{pmatrix}\begin{pmatrix}|u_n|\\|v_n|\end{pmatrix}+\begin{pmatrix}0&\alpha\beta\\\alpha\beta&0\end{pmatrix}\begin{pmatrix}|u_{n-k}|\\|v_{n-k}|\end{pmatrix},$$

which yields to

$$\begin{pmatrix}1&-\alpha\\-\alpha&1\end{pmatrix}\begin{pmatrix}|u_n|\\|v_n|\end{pmatrix}\le\mu\begin{pmatrix}|u_{n-1}|\\|v_{n-1}|\end{pmatrix}+\begin{pmatrix}0&\alpha\beta\\\alpha\beta&0\end{pmatrix}\begin{pmatrix}|u_{n-k}|\\|v_{n-k}|\end{pmatrix}. \tag{2.8}$$

Let

$$B=\begin{pmatrix}1&-\alpha\\-\alpha&1\end{pmatrix},C=\begin{pmatrix}0&\alpha\beta\\\alpha\beta&0\end{pmatrix}.$$

From conditions (1.5) and (2.7), we have $0<\alpha<1$, which implies that

$$\det B=1-\alpha^2>0.$$

Therefore B is reversible, and

$$B^{-1}=\frac{1}{1-\alpha^2}\begin{pmatrix}1&\alpha\\\alpha&1\end{pmatrix}.$$

We rewrite (2.8) to

$$\begin{pmatrix}|u_n|\\|v_n|\end{pmatrix}\le\mu B^{-1}\begin{pmatrix}|u_{n-1}|\\|v_{n-1}|\end{pmatrix}+B^{-1}C\begin{pmatrix}|u_{n-k}|\\|v_{n-k}|\end{pmatrix}. \tag{2.9}$$

The eigenvalues of μB^{-1} are given by the roots of

$$(\lambda - \frac{\mu}{1-\alpha^2})^2 - (\frac{\mu\alpha}{1-\alpha^2})^2 = 0.$$

We find that the eigenvalues of μB^{-1} are

$$\lambda_1 = \frac{\mu}{1-\alpha}, \lambda_2 = \frac{\mu}{1+\alpha}.$$

By the hypothesis(2.7), we know that $\mu < 1-\alpha-\alpha\beta \leq 1-\alpha$, hence $\lambda_1 < 1, \lambda_2 < 1$, it will follow that $\|\mu B^{-1}\| < 1$.Similarly, the eigenvalues of $B^{-1}C$ are

$$\mu_1 = -\frac{\alpha\beta}{1+\alpha}, \mu_2 = \frac{\alpha\beta}{1-\alpha}.$$

which are inside the unit circle, hence $\|B^{-1}C\| < 1$.

By induction, from (2.9), we have

$$\left\| \begin{pmatrix} |u_n| \\ |v_n| \end{pmatrix} \right\| \to 0 \text{ as } n \to \infty.$$

Therefore, all nontrivial solutions of (1.4) converge to the equilibrium and this convergence is independent of magnitude of the delay parameter k.This completes the proof. □

By an analogous argument of Corollary 2.2, we can obtain the following.

Corollary 2.4 *Assume that conditions (1.5) and (2.7) are satisfied. Then all solutions of system (1.1) satisfy*

$$\lim_{n\to\infty} x_n = 0, \qquad \lim_{n\to\infty} y_n = 0,$$

and this convergence holds for all $k > 0$.

Remark. Corlloary 2.4 indicates that, when hypotheses (1.5) and (2.7) hold, the unique equilibrium $(0,0)$ is the global attractor of system (1.1).

References

[1] R.Kuhn, J.L.Van Hemmen, Temporal Association: Models of Neural Networks (E.Domany, J.L.Van Hemmen and K.Schulten, Eds.), New York: Springer–Verlag, 1991.

[2] S.G.Ruan and J.Wei, Periodic solutions of planar systems with two delays, *Proc. Royal Soc. Edinburgh Ser. A.* **129** 1017-1032 (1999).

[3] K.Gopalsamy, I.K.C.Leung, Convergence under dynamical thresholds with delays, *IEEE Trans. Neural Networks*, **8** 341–348 (1997).

[4] K.Gopalsamy, I.K.C.Leung, Delay induced periodicity in a neural netlet of excitation and inhibition, *Phy. D*, **89** 395–426 (1996).

[5] B.Muller, J.Reinhardt, *Neural Networks*, Berlin: Springer, 1991.

[6] M.Z.Baptistini, On the existence and global bifurcation of periodic solutions to planar differential delay equations, *J. Diff. Eqns.*, **127** 391-425 (1996).

Fields Institute Communications
Volume **42**, 2004

Stabilities in Volterra Difference Equations on a Banach Space

Tetsuo Furumochi
Department of Mathematics
Shimane University
Matsue 690-8504, Japan
furumochi2math.shimane-u.ac.jp

Satoru Murakami
Department of Applied Mathematics
Okayama University of Science
1-1 Ridaicho, Okayama 700-0005, Japan
murakami@xmath.ous.ac.jp

Yutaka Nagabuchi
Department of Systems and Control Engineering
Anan National College of Technology
Anan Tokushima 774-0017, Japan
nagabuci@anan-nct.ac.jp

Abstract. We obtain a sufficient condition on the spectrum of the characteristic operator for the zero solution of Volterra difference equations on a Banach space to be exponentially asymptotically stable. Moreover, we apply some results obtained here to derive stability properties for some differential equations with piecewise continuous delays.

1 Introduction

In this paper we are concerned with the linear Volterra difference equation on a Banach space X

$$x(n+1) = \sum_{j=-\infty}^{n} Q(n-j)x(j), \qquad n \in \mathbf{Z}^+ := \{0, 1, 2, \cdots\}, \tag{1.1}$$

and the nonlinear Volterra difference equation

$$x(n+1) = \sum_{j=-\infty}^{n} Q(n-j)x(j) + f(x_n), \qquad n \in \mathbf{Z}^+, \tag{1.2}$$

2000 *Mathematics Subject Classification.* Primary: 39A10, 39A11; Secondary: 34K20, 35B35.
The second author was supported in part by the Grant-in-Aid for Scientific Research (C), No.13640197, of the Japanese Ministry of Education, Culture, Sports, Sciences and Technology.

where $Q(n)$, $n \in \mathbf{Z}^+$, are bounded linear operators on X such that $\sum_{n=0}^{\infty} \|Q(n)\| e^{\gamma n} < \infty$ for some positive constant γ, $x_n : \mathbf{Z}^- := \{\cdots, -2, -1, 0\} \to X$ is a function defined by $x_n(\theta) = x(n+\theta)$, $\theta \in \mathbf{Z}^-$, and f is a nonlinear term satisfying $f(0) = 0$, together with some conditions. The purpose of this paper is to study some stability properties of the zero solution for (1.1) and (1.2). When X is finite dimensional, there are several results on stabilities in Eq. (1.1); see e.g. [2, 3] and the references therein. In section 2, we will give a condition under which the zero solution of (1.1) is exponentially asymptotically stable, and extend some results in the case of finite dimensional X to the case of infinite dimensional X (Theorem 2.1). Also, in section 3 we will establish a representation formula in the phase space for solutions of perturbed equations (Proposition 3.1), and apply the formula to derive a result on the exponential asymptotic stability of the zero solution of the nonlinear equation (1.2) (Theorem 3.2). Finally, as an application, we will discuss some differential equations with piecewise continuous delays and investigate some stability properties for the equations.

2 Stabilities in linear Volterra difference equations

Let us denote by $\mathbf{Z}$, $\mathbf{Z}^+$ and $\mathbf{Z}^-$ the set of all integers, the set of all nonnegative integers and the set of all nonpositive integers, respectively. For any Banach space X (with norm $|\cdot|$) over the field $\mathbf{C}$ of all the complex numbers, we denote by $L(X)$ the space of all bounded linear operators on X, and define the norm of any Q belonging to $L(X)$ by

$$\|Q\| = \sup\{|Qx| : x \in X,\ |x| = 1\}.$$

We now consider the linear Volterra difference equation on X

$$x(n+1) = \sum_{j=-\infty}^{n} Q(n-j)x(j), \qquad n \in \mathbf{Z}^+, \tag{2.1}$$

where $Q(n) \in L(X)$. Throughout the paper, we assume the following condition on (2.1):

$$\exists \gamma > 0,\ l \geq 1 : \quad \sum_{n=0}^{\infty} \|Q(n)\| e^{\gamma n} \leq l < \infty.$$

Let us define the Banach space $\mathbf{B}^\gamma$ by

$$\mathbf{B}^\gamma = \{\phi : \mathbf{Z}^- \to X \mid \sup_{\theta \in \mathbf{Z}^-} |\phi(\theta)| e^{\gamma\theta} < \infty\},$$

which is equipped with the norm

$$\|\phi\| = \sup_{\theta \in \mathbf{Z}^-} |\phi(\theta)| e^{\gamma\theta}, \qquad \phi \in \mathbf{B}^\gamma.$$

For any $\phi \in \mathbf{B}^\gamma$, it follows that

$$\begin{aligned} \sum_{j=-\infty}^{-1} \|Q(n-j)\phi(j)\| &\leq \left(\sum_{j=-\infty}^{-1} \|Q(n-j)\| e^{\gamma(n-j)} \right) \|\phi\| e^{-\gamma n} \\ &\leq l\|\phi\| e^{-\gamma n}, \quad n \in \mathbf{Z}^+. \end{aligned}$$

Hence, if we set

$$p(n) := \sum_{j=-\infty}^{-1} Q(n-j)\phi(j), \qquad n \in \mathbf{Z}^+,$$

then $p(n)$ is well-defined, and there exists a unique solution x of (2.1) on $\mathbf{Z}^+$ such that $x(\theta) \equiv \phi(\theta)$, $\theta \in \mathbf{Z}^-$. We denote such a solution x by $x(\cdot;\phi)$ or $x(\cdot;\phi,0)$.

The zero solution of (2.1) is said to be (globally) *exponentially asymptotically stable*, if there exist positive constants K and α with the property that

$$|x(n;\phi)| \leq Ke^{-\alpha n}\|\phi\|, \qquad (\forall\, n \in \mathbf{Z}^+,\ \forall\, \phi \in \mathbf{B}^\gamma).$$

In the following, we will establish a result on the exponential asymptotic stability of the zero solution of (2.1) in connection with the invertibility of the operator $zI - \tilde{Q}(z)$, which is called *the characteristic operator* of (2.1). Here, I is the identity operator on X, and $\tilde{Q}(z)$ is the Z-transform of $Q(n)$, $n \in \mathbf{Z}^+$, that is,

$$\tilde{Q}(z) = \sum_{n=0}^{\infty} Q(n)z^{-n}.$$

Notice that $\tilde{Q}(z)$ exists in the domain $|z| > e^{-\gamma}$ of the complex plane, and it is analytic in the domain. For integro-differential equations of Volterra type, several results on stability properties have been established in connection with the invertibility of the corresponding characteristic operator expressed by the Laplace transform instead of the Z-transform; see e.g. [7, 10, 6, 13] and the references therein. Corresponding to the results for integro-differential equations, we will derive the following result for difference equations.

Theorem 2.1 *Suppose that the following condition holds true:*

$$\left(zI - \tilde{Q}(z)\right)^{-1} \in L(X) \qquad (\forall\, |z| \geq 1). \tag{2.2}$$

Then the zero solution of (2.1) is exponentially asymptotically stable.

Conversely, in case where $Q(j)$, $j = 0, 1, 2, \cdots$ are compact operators, if the zero solution of (2.1) is exponentially asymptotically stable, then the condition (2.2) holds true.

Before establishing the theorem, we will provide some preparatory results.

For (2.1) we define $R(n)$, $n \in \mathbf{Z}^+$, in $L(X)$ by the relation

$$R(n+1) = \sum_{j=0}^{n} Q(n-j)R(j), \qquad n = 0, 1, 2, \cdots, \tag{2.3}$$

and $R(0) = I$. $R(n)$ is called *the fundamental solution* of (2.1). By the virtue of the inequality

$$\begin{aligned} \|R(n+1)\| &= \|Q(n)R(0) + Q(n-1)R(1) + \cdots + Q(0)R(n)\| \\ &\leq l \max_{0 \leq s \leq n} \|R(s)\|, \end{aligned}$$

we get the following estimate on $\|R(n)\|$:

$$\|R(n)\| \leq l^n, \qquad n \in \mathbf{Z}^+.$$

Therefore, the Z-transform of $R(n)$, $n \in \mathbf{Z}^+$, exists in the domain $|z| > l$ of the complex plane, and it is analytic in the domain.

Utilizing the fundamental solution of (2.1), one can write the solution x of the perturbed equation

$$x(n+1) = \sum_{j=0}^{n} Q(n-j)x(j) + q(n), \qquad n \in \mathbf{Z}^+, \tag{2.4}$$

with the perturbed term $q : \mathbf{Z}^+ \to X$. Indeed, it is straightforward to check the following relation

$$x(n) = R(n)x(0) + \sum_{k=0}^{n-1} R(n-k-1)q(k), \quad n \in \mathbf{Z}^+.$$

Here and hereafter, we employ the convention $\sum_{0}^{-1} = 0$. In particular, it follows that

$$x(n,\phi) = R(n)\phi(0) + \sum_{k=0}^{n-1} R(n-k-1)\Big(\sum_{j=-\infty}^{-1} Q(k-j)\phi(j)\Big), \quad n \in \mathbf{Z}^+. \tag{2.5}$$

We say that $R(n)$ decays exponentially, if there exist positive constants M and ϵ such that

$$\|R(n)\| \le Me^{-\epsilon n}, \qquad n \in \mathbf{Z}^+.$$

Proposition 2.2 *If the condition (2.2) holds true, then the fundamental solution $R(n)$ of (2.1) decays exponentially.*

Proof Let us consider the Z-transform of the both sides of (2.3) to get

$$z\tilde{R}(z) - zI = \tilde{Q}(z)\tilde{R}(z) \qquad (|z| > l).$$

The operators $zI - \tilde{Q}(z)$, $|z| \ge 1$, are invertible by virtue of the condition (2.2), and hence we get the following relation

$$z(zI - \tilde{Q}(z))^{-1} = \tilde{R}(z) \qquad (|z| > l). \tag{2.6}$$

Notice that $zI - \tilde{Q}(z)$ is invertible for each $|z| = 1$, and consequently so is it for any point belonging to some open neighborhood of z with $|z| = 1$. Since the set $|z| = 1$ is compact in the complex plane, there is a positive constant $\delta_1 \in (0, 1-e^{-\gamma})$ such that the operator $zI - \tilde{Q}(z)$ is invertible in $L(X)$ for any z such that $|z| \ge 1-\delta_1$. Then the function $G(z) := z(zI - \tilde{Q}(z))^{-1}$ is analytic on the domain $|z| > 1-\delta_1$, and it can be expanded into the Laurent series

$$G(z) = \sum_{m\in\mathbf{Z}} b(m)z^m \qquad (\forall\ |z| > 1-\delta_1),$$

where

$$b(m) = \frac{1}{2\pi i}\int_{|z|=L} \frac{G(z)}{z^{m+1}}dz \qquad (L > 1-\delta_1).$$

Let $|z| = L > l$. Then $|z|e^{\gamma} > (1-\delta_1)e^{\gamma} > 1$, and

$$\begin{aligned}\|\tilde{Q}(z)\| = \left\|\sum_{n=0}^{\infty} Q(n)z^{-n}\right\| &\le \sum_{n=0}^{\infty} \|Q(n)\||z|^{-n} \\ &\le \sum_{n=0}^{\infty} \|Q(n)\|e^{\gamma n} = l < |z|,\end{aligned}$$

which yields that

$$\|G(z)\| = \left\|\Big(I - \frac{\tilde{Q}(z)}{z}\Big)^{-1}\right\| = \left\|\sum_{n=0}^{\infty}\Big(\frac{\tilde{Q}(z)}{z}\Big)^n\right\| \le \sum_{n=0}^{\infty}\Big(\frac{l}{L}\Big)^n = \frac{L}{L-l}.$$

Therefore, for $n = 1, 2, \cdots$ we have

$$\|b(n)\| = \left\| \frac{1}{2\pi i} \int_{|z|=L} \frac{G(z)}{z^{n-1}} dz \right\| \leq \left[\sup_{|z|=L} \|G(z)\| \right] L^{-n}$$
$$\leq \frac{L}{L^n(L-l)} \to 0 \quad (L \to \infty).$$

Then $b(n) = 0 \ (n = 1, 2, \cdots)$, and hence

$$G(z) = \sum_{n=0}^{\infty} b(-n) z^{-n} \qquad (\forall\ |z| > 1 - \delta_1).$$

In particular, the series $\sum_{n=0}^{\infty} b(-n)\left(1 - \frac{\delta_1}{2}\right)^{-n}$ is convergent. Consequently, it follows that $\sup_{n \in \mathbf{Z}^+} \|b(-n)\left(1 - \frac{\delta_1}{2}\right)^{-n}\| =: M < \infty$, and

$$\|b(-n)\| \leq M\left(1 - \frac{\delta_1}{2}\right)^n, \qquad n \in \mathbf{Z}^+. \tag{2.7}$$

Let us take an L with $L > l$. Since

$$\sum_{n=0}^{\infty} R(n) z^{-n} = \tilde{R}(z) = G(z) = \sum_{n=0}^{\infty} b(-n) z^{-n} \qquad (\forall\ |z| = L)$$

by (2.6), it follows that

$$R(n) = \frac{1}{2\pi i} \int_{|z|=L} \tilde{R}(z) z^{n-1} dz = \frac{1}{2\pi i} \int_{|z|=L} G(z) z^{n-1} dz = b(-n), \quad n \in \mathbf{Z}^+.$$

Put $\epsilon = -\log(1 - \frac{\delta_1}{2})$. Then $\epsilon > 0$ and

$$\|R(n)\| = \|b(-n)\| \leq M\left(1 - \frac{\delta_1}{2}\right)^n = Me^{-\epsilon n}, \qquad n \in \mathbf{Z}^+,$$

by (2.7), which shows that $R(n)$ decays exponentially. □

Proof of Theorem 2.1 Assume that the condition (2.2) holds. Since $R(n)$ decays exponentially by virtue of Proposition 2.2, there exist positive constants M and ϵ with $\epsilon \in (0, \gamma)$ such that

$$\|R(n)\| \leq Me^{-\epsilon n}, \qquad n \in \mathbf{Z}^+.$$

Then it follows from (2.5) that

$$\begin{aligned} |x(n;\phi)| &\leq \|R(n)\||\phi(0)| + \sum_{k=0}^{n-1} \|R(n-k-1)\| \left(\sum_{j=-\infty}^{-1} \|Q(k-j)\||\phi(j)| \right) \\ &\leq Me^{-\epsilon n}\|\phi\| + \sum_{k=0}^{n-1} Me^{-\epsilon(n-k-1)} \left(l\|\phi\|e^{-\gamma k} \right) \\ &\leq Me^{-\epsilon n}\|\phi\| \left(1 + \frac{le^{\epsilon}}{1 - e^{\epsilon - \gamma}} \right), \quad n \in \mathbf{Z}^+, \end{aligned}$$

which shows the exponential asymptotic stability of the zero solution of (2.1).

Conversely, assume that the zero solution of (2.1) is exponentially asymptotically stable and that $Q(j)$, $j = 0, 1, 2, \cdots$, are compact. For any $\bar{x} \in X$, $R(\cdot)\bar{x}$ is a solution of (2.1) through $(0, \phi)$ with $\phi(0) = \bar{x}$ and $\phi(\theta) \equiv 0$ for $\theta = -1, -2, \cdots$. Hence it follows that

$$|R(n)\bar{x}| \leq Ke^{-\alpha n}\|\phi\| = Ke^{-\alpha n}|\bar{x}|, \quad n \in \mathbf{Z}^+,$$

and hence

$$\|R(n)\| \leq Ke^{-\alpha n}, \quad n \in \mathbf{Z}^+.$$

Thus $R(\cdot)$ decays exponentially. Without loss of generality, one can assume that $0 < \alpha < \gamma$. Now suppose that the condition (2.2) does not hold true, that is, the operator $z_0 I - \tilde{Q}(z_0)$ is not invertible for some z_0 with $|z_0| \geq 1$. Observe that $\sum_{j=0}^{\infty} Q(j)z_0^{-j}$ is a compact operator, because it is the limit of compact operators $\sum_{j=0}^{n} Q(j)z_0^{-j}$ in the operator norm as $n \to \infty$. Then the Riesz-Schauder theorem on compact operators (e.g., [14, Theorem 4.25]) implies that z_0 belongs to the point spectrum of $\tilde{Q}(z_0)$, and hence there exists a nonzero element $\bar{x} \in X$ such that $(z_0 I - \tilde{Q}(z_0))\bar{x} = 0$, or $z_0\bar{x} = \sum_{j=0}^{\infty} Q(j)z_0^{-j}\bar{x}$.

Let us consider a function $x : \mathbf{Z} \to X$ defined by $x(n) = z_0^n\bar{x}$, $n \in \mathbf{Z}$. For any $n \in \mathbf{Z}^+$, we get

$$\begin{aligned} x(n+1) = z_0^{n+1}\bar{x} = z_0^n\left(\sum_{j=0}^{\infty} Q(j)z_0^{-j}\bar{x}\right) &= \sum_{j=0}^{\infty} Q(j)x(n-j) \\ &= \sum_{j=0}^{n} Q(n-j)x(j) + q(n), \end{aligned}$$

where $q(n) := \sum_{j=-\infty}^{-1} Q(n-j)z_0^j\bar{x}$. It follows from (2.5) that

$$\begin{aligned} x(n) &= R(n)\bar{x} + \sum_{k=0}^{n-1} R(n-k-1)q(k) \\ &= R(n)\bar{x} + \sum_{k=0}^{n-1} R(n-k-1)\left(\sum_{j=-\infty}^{-1} Q(k-j)z_0^j\bar{x}\right), \qquad n \in \mathbf{Z}^+. \end{aligned}$$

Since $|x(n)| \geq |\bar{x}|$ ($\forall\, n \geq 0$) and since $|z_0|^j \leq 1$ ($\forall\, j \leq 0$), it follows that

$$0 < |\bar{x}| \leq |x(n)| \leq \left\{\|R(n)\| + \sum_{k=0}^{n-1} \|R(n-k-1)\|\left(\sum_{j=-\infty}^{-1} \|Q(k-j)\|\right)\right\}|\bar{x}|,$$

or

$$\begin{aligned} 1 &\leq \left\{\|R(n)\| + \sum_{k=0}^{n-1} \|R(n-k-1)\|\left(\sum_{j=-\infty}^{-1} \|Q(k-j)\|\right)\right\} \\ &\leq Ke^{-\alpha n}\left\{1 + e^{\alpha}\sum_{k=0}^{n-1} e^{(\alpha-\gamma)k}\left(\sum_{j=-\infty}^{-1} \|Q(k-j)\|e^{\gamma(k-j)}\right)\right\} \\ &\leq Ke^{-\alpha n}\left(1 + \frac{le^{\alpha}}{1-e^{\alpha-\gamma}}\right), \qquad n \in \mathbf{Z}^+. \end{aligned}$$

Letting $n \to \infty$ in the above inequality, we get $1 \leq 0$, a contradiction. Thus the condition (2.2) must hold true. This completes the proof of the theorem. $\square$

For any $\phi \in \mathbf{B}^\gamma$ and $m \in \mathbf{Z}^+$, we consider a mapping $V(m)\phi : \mathbf{Z}^- \to X$ defined by

$$V(m)\phi = x(\cdot + m; \phi).$$

$V(m)$ is called *the solution operator* of (2.1). It follows from (2.5) that

$$[V(m)\phi](\theta) = \begin{cases} \phi(m+\theta) & (\theta \leq -m), \\ R(m+\theta)\phi(0) \\ \quad + \sum_{r=0}^{m+\theta-1} R(m+\theta-r-1)\Big(\sum_{j=-\infty}^{-1} Q(r-j)\phi(j)\Big) & (-m \leq \theta \leq 0). \end{cases}$$

By this formula we see that $V(m)$ is a bounded linear operator on $\mathbf{B}^\gamma$. Furthermore, one can see that $V(m)$ satisfies the following semigroup property:

$$V(0) = I, \qquad V(m)V(n) = V(m+n) \qquad (\forall\, m,\ n \geq 0).$$

Corollary 2.3 *Assume that the condition (2.2) holds true. Then $V(n)$ decays exponentially, that is,*

$$\exists K_0 > 0,\ \beta > 0 : \ \|V(n)\| \leq K_0 e^{-\beta n}, \qquad n \in \mathbf{Z}^+.$$

Proof By virtue of Theorem 2.1, the zero solution of (2.1) is exponentially asymptotically stable. Then there exist some positive constants K and α with $K \geq 1$ and $\alpha < \gamma$ such that for any $\phi \in \mathbf{B}^\gamma$, $x(n) := x(n;\phi)$ satisfies

$$|x(n)| \leq K e^{-\alpha n}\|\phi\|, \qquad n \in \mathbf{Z}^+.$$

Therefore it follows that

$$\begin{aligned} \|V(n)\phi\| &= \max\Big(\sup_{-n\leq\theta\leq 0} |x(n+\theta)|e^{\gamma\theta},\ \sup_{\theta\leq -n} |x(n+\theta)|e^{\gamma\theta}\Big) \\ &\leq \max\Big(\sup_{-n\leq\theta\leq 0} K e^{-\alpha(n+\theta)} e^{\gamma\theta}\|\phi\|,\ \sup_{\theta\leq -n} |\phi(n+\theta)|e^{\gamma\theta}\Big) \\ &\leq \|\phi\| \max\Big(K e^{-\alpha n},\ e^{-\gamma n}\Big) \\ &= K e^{-\alpha n}\|\phi\|, \end{aligned}$$

and hence

$$\|V(n)\| \leq K e^{-\alpha n}, \quad n \in \mathbf{Z}^+.$$

Thus $V(n)$ decays exponentially. $\square$

3 Stabilities in perturbed nonlinear difference equations

For a mapping $p : \mathbf{Z}^+ \to X$, we consider the perturbed equation

$$x(n+1) = \sum_{j=-\infty}^{n} Q(n-j)x(j) + p(n), \qquad n \in \mathbf{Z}^+, \tag{3.1}$$

and denote by $x(\cdot;\phi,p)$ the solution x of (3.1) such that $x(\theta)\equiv\phi(\theta)$, $\theta\in\mathbf{Z}^-$. Let us define an operator $\mathbf{E}:X\to\mathbf{B}^\gamma$ by

$$(\mathbf{E}\bar{x})(\theta)=\begin{cases}\bar{x} & (\theta=0),\\ 0 & (\theta=-1,-2,-3,\cdots)\end{cases}$$

for any $\bar{x}\in X$. Clearly, $\mathbf{E}$ is a bounded linear operator mapping X into $\mathbf{B}^\gamma$, and it satisfies

$$\|\mathbf{E}\bar{x}\|=|\bar{x}|,\qquad \bar{x}\in X.$$

Recall that for any mapping $x:\mathbf{Z}\to X$ and $n\in\mathbf{Z}$, x_n is a mapping defined by $x_n(\theta)\equiv x(n+\theta)$, $\theta\in\mathbf{Z}^-$. The following result yields a representation formula for $x(\cdot;\phi,p)$ in the phase space $\mathbf{B}^\gamma$, cf. [11].

Proposition 3.1

$$x_n(\phi,p)=V(n)\phi+\sum_{j=0}^{n-1}V(n-j-1)(\mathbf{E}p(j)),\qquad n\in\mathbf{Z}^+.$$

Proof It is sufficient to certify the following relation:

$$\Big[x_n(\phi,p)-V(n)\phi\Big](\theta)=\sum_{j=0}^{n-1}\Big[V(n-j-1)(\mathbf{E}p(j))\Big](\theta),\qquad \theta\in\mathbf{Z}^-. \tag{3.2}$$

Let us recall that

$$x(n;\phi,p)=\begin{cases}\phi(n) & (n\le 0),\\ R(n)\phi(0)+\sum_{s=0}^{n-1}R(n-s-1)p(s) & \\ \quad +\sum_{s=0}^{n-1}R(n-s-1)\Big(\sum_{j=-\infty}^{-1}B(s-j)\phi(j)\Big) & (n\ge 0).\end{cases}$$

The case of $\theta\le -n$: Since $n+\theta\le 0$, we get

$$\begin{aligned}\Big[x_n(\phi,p)-V(n)\phi\Big](\theta) &= x(n+\theta;\phi,p)-[V(n)\phi](\theta)\\ &= \phi(n+\theta)-\phi(n+\theta)=0.\end{aligned}$$

Also, for $j=0,1,\cdots,n-1$, we get $n-j-1+\theta\le n-1+\theta\le -1$, and hence

$$\sum_{j=0}^{n-1}\Big[V(n-j-1)(\mathbf{E}p(j))\Big](\theta)=\sum_{j=0}^{n-1}\Big[\mathbf{E}p(j)\Big](n-j-1+\theta)=0.$$

Thus the relation (3.2) holds true.

The case of $-n+1 \le \theta \le 0$: Since $1 \le n+\theta$, we get

$$\begin{aligned}
&[x_n(\phi,p) - V(n)\phi](\theta) \\
&= x(n+\theta;\phi,p) - x(n+\theta;\phi,0) \\
&= R(n+\theta)\phi(0) + \sum_{s=0}^{n+\theta-1} R(n+\theta-s-1)\Big(\sum_{j=-\infty}^{-1} B(s-j)\phi(j) + p(s)\Big) \\
&\quad -\Big[R(n+\theta)\phi(0) + \sum_{r=0}^{n+\theta-1} R(n+\theta-r-1)\Big(\sum_{j=-\infty}^{-1} B(r-j)\phi(j)\Big)\Big] \\
&= \sum_{s=0}^{n+\theta-1} R(n+\theta-s-1)p(s).
\end{aligned}$$

On the one hand, since $0 \le n+\theta-1 \le n-1$, it follows that

$$\begin{aligned}
\sum_{j=0}^{n-1}\Big[V(n-j-1)(\mathbf{E}p(j))\Big](\theta) &= \sum_{j=0}^{n-1} x(n-j-1+\theta;\mathbf{E}p(j),0) \\
&= \sum_{j=0}^{n+\theta-1} x(n-j-1+\theta;\mathbf{E}p(j),0) \\
&\quad + \sum_{j=n+\theta}^{n-1} x(n-j-1+\theta;\mathbf{E}p(j),0).
\end{aligned}$$

If $0 \le j \le n+\theta-1$, then

$$\begin{aligned}
&x(n-j-1+\theta;\mathbf{E}p(j),0) \\
&= R(n-j-1+\theta)[\mathbf{E}p(j)](0) \\
&\quad + \sum_{r=0}^{n-j+\theta-2} R(n-j+\theta-r-2)\Big(\sum_{s=-\infty}^{-1} B(r-s)(\mathbf{E}p(j))(s)\Big) \\
&= R(n-j-1+\theta)p(j).
\end{aligned}$$

If $n+\theta \le j \le n-1$, then

$$x(n-j-1+\theta;\mathbf{E}p(j),0) = [\mathbf{E}p(j)](n-j-1+\theta) = 0$$

because of $n-j-1+\theta \le -1$. Thus we get

$$\sum_{j=0}^{n-1}\Big[V(n-j-1)\mathbf{E}p(j)\Big](\theta) = \sum_{j=0}^{n+\theta-1} R(n-j-1+\theta)p(j),$$

which shows the relation (3.2). This completes the proof. □

Now we consider the following nonlinear equation

$$x(n+1) = \sum_{j=-\infty}^{n} Q(n-j)x(j) + f(x_n), \qquad n \in \mathbf{Z}^+, \tag{3.3}$$

where $f : \mathbf{B}^\gamma \to X$ is assumed to be a continuous function such that $f(0) = 0$ and $f(\phi) = o(\|\phi\|)$, that is,

$$\lim_{\|\phi\|\to 0} \frac{|f(\phi)|}{\|\phi\|} = 0.$$

Theorem 3.2 *Let $f : \mathbf{B}^\gamma \to X$ be the one cited above, and assume that the condition (2.2) is satisfied. Then the zero solution of (3.3) is locally exponentially asymptotically stable; that is, there exist some positive constants C, δ and λ with the property that*

$$|x(n;\phi,f)| \le Ce^{-\lambda n}\|\phi\|, \qquad (\forall\ n \in \mathbf{Z}^+,\ \|\phi\| \le \delta).$$

Proof Let M and ϵ with $M \ge 1$ and $\epsilon < \gamma$ be the ones ensured by the fact that $R(n)$ decays exponentially. Let us take a positive constant μ so that $\mu \le (1-e^{-\epsilon})/(4M)$. Since $f(\phi) = o(\|\phi\|)$, one can choose a positive constant η satisfying

$$|f(\phi)| \le \mu\|\phi\| \qquad (\forall\ \|\phi\| \le \eta).$$

Define a positive constant δ by the relation

$$2M\left(1 + \frac{le^\epsilon}{1-e^{\epsilon-\gamma}}\right)\delta = \eta,$$

and let ϕ be any element in $\mathbf{B}^\gamma$ satisfying $\|\phi\| \le \delta\ (< \eta)$. We first claim that

$$|x(n;\phi,f)| \le \eta, \qquad (n \in \mathbf{Z}^+). \tag{3.4}$$

Suppose the claim is false. Then there exists some $n \in \mathbf{Z}^+$ such that $\sup_{0\le k\le n-1} |x(k)| \le \eta$ and $|x(n)| > \eta$, where $x(\cdot) := x(\cdot;\phi,f)$. For any k with $1 \le k \le n-1$, we get

$$\begin{aligned} \|x_k\| &= \max\left(\sup_{\theta\le -k} |x(k+\theta)|e^{\gamma\theta},\ \sup_{-k\le\theta\le 0} |x(k+\theta)|e^{\gamma\theta}\right) \\ &\le \max(\|\phi\|,\ \eta) = \eta \end{aligned}$$

which implies $|f(x_k)| \le \mu\|x_k\| \le \mu\eta$, and hence

$$\begin{aligned} |x(n)| &= \left| R(n)\phi(0) + \sum_{k=0}^{n-1} R(n-k-1)\left(\sum_{j=-\infty}^{-1} Q(k-j)\phi(j) + f(x_k)\right)\right| \\ &\le Me^{-\epsilon n}|\phi(0)| + \sum_{k=0}^{n-1} Me^{-\epsilon(n-k-1)}\left(l\|\phi\|e^{-\gamma k} + \mu\eta\right) \\ &\le Me^{-\epsilon n}\|\phi\|\left(1 + \frac{le^\epsilon}{1-e^{\epsilon-\gamma}}\right) + \sum_{k=0}^{n-1} M\mu\eta e^{-\epsilon(n-1-k)} \\ &\le M\delta\left(1 + \frac{le^\epsilon}{1-e^{\epsilon-\gamma}}\right) + \frac{M\mu\eta}{1-e^{-\epsilon}} \\ &\le \frac{\eta}{2} + \frac{\eta}{4} < \eta. \end{aligned}$$

This is a contradiction to $|x(n)| > \eta$. Thus the claim (3.4) must hold true.

Now, by virtue of Corollary 2.3, $\|V(n)\|$ decays exponentially. Let K_0 and β be the constants satisfying $\|V(n)\| \le K_0 e^{-\beta n}$, $n \in \mathbf{Z}^+$. Replacing μ by a smaller number if necessary, we may assume that

$$0 < \beta - \log(1 + K_0\mu e^\beta) =: \lambda.$$

It follows from Proposition 3.1 that

$$x_n = V(n)\phi + \sum_{j=0}^{n-1} V(n-j-1)(\mathbf{E}f(x_j)), \qquad n \in \mathbf{Z}^+,$$

and hence

$$\|x_n\| \le K_0 e^{-\beta n}\|\phi\| + \sum_{j=0}^{n-1} K_0 e^{-\beta(n-j-1)}\mu\|x_j\|$$

because of $\|x_j\| \le \eta$, $j = 0, 1, \cdots, n-1$, by (3.4). Put $g(n) = \|x_n\|e^{\beta n}$. Then

$$g(n) \le K_0\|\phi\| + K_0\mu e^{\beta}\sum_{j=0}^{n-1} g(j),$$

and consequently

$$g(n) \le K_0\|\phi\|\left(1 + K_0\mu e^{\beta}\right)^n$$

by applying the discrete Gronwall's inequality. Thus we obtain

$$|x(n;\phi,f)| \le \|x_n\| \le K_0\|\phi\|\left(\frac{1+K_0\mu e^{\beta}}{e^{\beta}}\right)^n \le K_0\|\phi\|e^{-\lambda n}, \qquad n \in \mathbf{Z}^+,$$

which proves the exponential asymptotic stability of the zero solution of (3.3). □

4 Volterra difference equations induced by differential equations with piecewise continuous delays

One of the ways deriving Volterra difference equations on a Banach space is based on some abstract differential equations with piecewise continuous delays. In fact, in [15] some differential equations on a Banach space is derived from some differential equations with piecewise continuous delays on a Banach space. In what follows, following the idea in [15], we explain how Volterra difference equations on a Banach space are canonically induced by some abstract differential equations with piecewise continuous delays. Moreover, we analyze the spectrum of the characteristic operator for the induced Volterra difference equations to obtain a condition under which some results established in the previous section are applicable.

In what follows, we employ the notation $[\cdot]$ to denote the Gaussian symbol, and consider a differential equation

$$\dot{u}(t) = Au(t) + \sum_{k=0}^{\infty} B(k)u([t-k]), \quad t \ge 0 \tag{4.1}$$

on a Banach space X, which contains piecewise continuous delays $t-[t-k]$, $k = 0, 1, 2, \cdots$. Here and hereafter, we assume that A is the infinitesimal generator of a strongly continuous semigroup $T(t)$, $t \ge 0$, of bounded linear operators on X, and $B(k)$, $k = 0, 1, 2, \cdots$, are bounded linear operators on X such that

$$\sum_{k=0}^{\infty} \|B(k)\|e^{\gamma k} < \infty \tag{4.2}$$

for some $\gamma > 0$. A function $u : \Big(\mathbf{Z}^- \cup [0,\infty)\Big) \to X$ such that $u_0 \in \mathbf{B}^\gamma$ (that is, $\sup_{\theta\in\mathbf{Z}^-} |u(\theta)|e^{\gamma\theta} < \infty$) is called a (mild) solution of (4.1) on $[0,\infty)$, if u is continuous on $[0,\infty)$, and it satisfies the following relation

$$u(t) = T(t-\sigma)u(\sigma) + \int_\sigma^t T(t-s)\Big(\sum_{k=0}^{\infty} B(k)u([s-k])\Big)ds, \qquad t \geq \sigma \geq 0.$$

In case of $n \leq t < n+1$ for some nonnegative integer n, the above relation yields that

$$\begin{aligned} u(t) &= T(t-n)u(n) + \int_n^t T(t-s)\Big(\sum_{k=0}^{\infty} B(k)u([s-k])\Big)ds \\ &= T(t-n)u(n) + \sum_{k=0}^{\infty}\Big(\int_n^t T(t-s)B(k)u(n-k)ds\Big). \end{aligned}$$

Letting $t \to n+1$ in this equation, we get Volterra difference equation

$$u(n+1) = \sum_{k=0}^{\infty} Q(k)u(n-k), \quad n \in \mathbf{Z}^+, \tag{4.3}$$

where $Q(k)$, $k \in \mathbf{Z}^+$, are bounded linear operators on X defined by

$$Q(0)x = T(1)x + \int_0^1 T(\tau)B(0)xd\tau, \qquad Q(k)x = \int_0^1 T(\tau)B(k)xd\tau, \ k = 1,2,\cdots \tag{4.4}$$

for $x \in X$.

Conversely, if u satisfies (4.3) with $\sup_{\theta\in\mathbf{Z}^-} |u(\theta)|e^{\gamma\theta} < \infty$, then the function u extended to non-integers t by the relation

$$u(t) = T(t-n)u(n) + \sum_{k=0}^{\infty}\Big(\int_n^t T(t-s)B(k)u(n-k)ds\Big), \quad n < t < n+1, \ n \in \mathbf{Z}^+,$$

is a (mild) solution of (4.1). Thus, some abstract differential equations such as (4.1) lead to Volterra difference equations on X. Sometimes, we call (4.3) *the induced Volterra difference equation* of (4.1).

A strongly continuous semigroup $T(t)$ on X is said to be *compact* whenever $T(t)$ is a compact operator on X for $t > 0$. It is known [12] that if the semigroup $T(t)$ is compact, then $T(t)$ is continuous in $t > 0$ with respect to the operator norm. The following result is an immediate consequence of [4, p.525, Theorem C.7].

Proposition 4.1 *If $T(t)$ is a compact semigroup on X and $B \in L(X)$, then the operator $Q \in L(X)$ defined by the relation*

$$Qx = \int_0^1 T(\tau)Bxd\tau, \quad x \in X,$$

is compact.

By virtue of this proposition, $Q(k)$, $k \in \mathbf{Z}^+$, defined by the relation (4.4) are compact operators on X whenever $T(t)$ is a compact semigroup on X. Moreover, it follows from (4.2) that $Q(k),\ k \in \mathbf{Z}^+$ satisfy the condition

$$\sum_{k=0}^{\infty} \|Q(k)\| e^{\gamma k} < \infty.$$

In what follows, we restrict our consideration to the case where $B(k),\ k \in \mathbf{Z}^+$ are scalar, that is, $B(k) \equiv b(k)I,\ k \in \mathbf{Z}^+$, for some $b(k) \in \mathbf{C}$, and investigate the spectrum of the characteristic operator $zI - \tilde{Q}(z) := zI - \sum_{k=0}^{\infty} Q(k)z^{-k}$ of (4.3).

Theorem 4.2 *Let $T(t)$ be a compact semigroup on X, and assume that $B(k) \equiv b(k)I,\ k \in \mathbf{Z}^+$, where $b(k)$ is a scalar function satisfying $\sum_{k=0}^{\infty} |b(k)|e^{\gamma k} < \infty$ for some positive constant γ. Then the spectrum of the characteristic operator $zI - \tilde{Q}(z)$ with $|z| \geq 1$ of (4.3) is given by*

$$\sigma(zI - \tilde{Q}(z)) = \left(\{z\} \cup \{z - e^{\nu} - \tilde{b}(z) \int_0^1 e^{\nu\tau} d\tau \mid \nu \in \sigma(A)\}\right). \tag{4.5}$$

Proof We first claim that

$$\lim_{n\to\infty} \left\| \frac{1}{n} \sum_{k=1}^{n} \{T(1/n)\}^k - \int_0^1 T(\tau)d\tau \right\| = 0. \tag{4.6}$$

Indeed, for any $x \in X$ it follows that

$$\begin{aligned}
&\left| \left(\int_0^1 T(\tau)d\tau \right) x - \left(\frac{1}{n} \sum_{k=1}^{n} \{T(1/n)\}^k \right) x \right| \\
&= \left| \left(\int_0^1 T(\tau) x d\tau \right) - \left(\frac{1}{n} \sum_{k=1}^{n} T(k/n)x \right) \right| \\
&= \left| \sum_{k=1}^{n} \left(\int_{\frac{k-1}{n}}^{\frac{k}{n}} T(\tau)x d\tau - \int_{\frac{k-1}{n}}^{\frac{k}{n}} T(k/n)x d\tau \right) \right| \\
&\leq \sum_{k=1}^{n} \int_{\frac{k-1}{n}}^{\frac{k}{n}} \|T(\tau) - T(k/n)\| d\tau |x|.
\end{aligned}$$

Hence we get

$$\left\| \frac{1}{n} \sum_{k=1}^{n} \{T(1/n)\}^k - \int_0^1 T(\tau)d\tau \right\| \leq \sum_{k=1}^{n} \int_{\frac{k-1}{n}}^{\frac{k}{n}} \|T(\tau) - T(k/n)\| \, d\tau.$$

For any $\epsilon > 0$, take a positive constant η_0, $\eta_0 < 1/2$, such that $2C\eta_0 < \epsilon/3$, where $C := \sup\{\|T(t)\| : 0 \leq t \leq 1\}$. Recall that $T(t)$ is continuous in $t \in [\eta_0, 1]$ with respect to the operator norm. Then there exists a natural number n_0 satisfying $n_0 > 1/\eta_0$ and $\sup\{\|T(\tau) - T(s)\| : \eta_0 \leq s \leq \tau \leq 1,\ |\tau - s| \leq 1/n_0\} < \epsilon/3$. Let $n \geq n_0$. Putting $[n\eta_0] + 1 = k_0$, we get $k_0 \geq 2$ and $(k_0 - 1)/n \leq \eta_0 < 1/2$; hence

$k_0 < n$. Then

$$\begin{aligned}
&\left\|\frac{1}{n}\sum_{k=1}^{n}\{T(1/n)\}^k - \int_0^1 T(\tau)d\tau\right\| \\
\le\ & \sum_{k=1}^{k_0}\int_{\frac{k-1}{n}}^{\frac{k}{n}} \|T(\tau)-T(k/n)\|d\tau + \sum_{k=k_0+1}^{n}\int_{\frac{k-1}{n}}^{\frac{k}{n}} \|T(\tau)-T(k/n)\|d\tau \\
\le\ & \frac{2Ck_0}{n} + \sup\{\|T(\tau)-T(s)\| :\ \eta_0 \le s \le \tau \le 1,\ |\tau - s| \le 1/n\} \\
<\ & 2C(\eta_0 + \frac{1}{n}) + \frac{\epsilon}{3} < \epsilon.
\end{aligned}$$

This proves the claim (4.6).

Now, set $S = \{T(t) :\ 0 \le t \le 1\}$. Since S commutes, $\mathbb{A} := \Gamma(\Gamma(S))$ is a commutative Banach algebra containing S, see [14, Theorem 11.22]. Here, for any subset Ω of $L(X)$, $\Gamma(\Omega)$ denotes the centralizer of Ω; that is,

$$\Gamma(\Omega) = \{v \in L(X) : vw = wv \text{ for every } w \in \Omega\}.$$

Let Δ be the maximal ideal space of $\mathbb{A}$. Let us denote by $\hat{a}$ the Gelfand transform of $a \in \mathbb{A}$. It is known [14, pp.268–270] that $\hat{a}$ is a function from Δ (which is equipped with the Gelfand topology) into $\mathbf{C}$ with the properties that the range of $\hat{a}$ is the spectrum $\sigma(a)$ of a and that

$$\|\hat{a}\|_\infty \le \|a\|, \quad a \in \mathbb{A},$$

where $\|\hat{a}\|_\infty$ is the maximum of $|\hat{a}(\xi)|$ on Δ. Moreover, the Gelfand transform is a homomorphism mapping $\mathbb{A}$ into a subspace of $C(\Delta;\mathbf{C})$, the space of all the complex valued continuous functions on Δ. Let $|z| \ge 1$, and put

$$a = zI - \tilde{Q}(z) = zI - T(1) - \left(\int_0^1 T(\tau)d\tau\right)\tilde{b}(z)$$

and

$$a_n = zI - W^n - \left(\frac{1}{n}\sum_{k=1}^{n} W^k\right)\tilde{b}(z)$$

for each $n = 1, 2, \cdots$, where $W := T(1/n)$. Then $\{a, a_1, a_2, \cdots\} \subset \mathbb{A}$, and by (4.6) we get

$$\|\widehat{(a_n)} - \hat{a}\|_\infty \le \|a_n - a\| = \|\frac{1}{n}\sum_{k=1}^{n}\{T(1/n)\}^k - \int_0^1 T(\tau)d\tau\||\tilde{b}(z)| \to 0$$

as $n \to \infty$. Thus

$$\lim_{n\to\infty} \widehat{(a_n)}(\xi) = \hat{a}(\xi), \qquad \xi \in \Delta. \tag{4.7}$$

Observe that $\widehat{(a_n)}(\xi) = z - \left(\hat{W}(\xi)\right)^n - \frac{1}{n}\sum_{k=1}^{n}\left(\hat{W}(\xi)\right)^k \tilde{b}(z)$. Since the operator $T(1/n)$ is compact, the Riesz-Schauder theorem implies that $\sigma(T(1/n)) = P_\sigma(T(1/n)) \cup \{0\}$. Also, it follows from [12, Theorems 2.2.3–2.2.4] that

$$\exp\Big((1/n)\sigma(A)\Big) \subset \sigma(T(1/n)), \qquad P_\sigma(T(1/n)) \cup \{0\} = \exp\Big((1/n)P_\sigma(A)\Big) \cup \{0\}.$$

Therefore we get $\sigma(W) = \sigma(T(1/n)) = \exp\Big((1/n)P_\sigma(A)\Big) \cup \{0\}$. From these observations, we see that the range of $\widehat{(a_n)}$ is identical with the set

$$\{z\} \cup \left(\{z - e^\nu - \frac{1}{n}\sum_{k=1}^{n} e^{(k/n)\nu}\tilde{b}(z) \mid \nu \in \sigma(A)\}\right).$$

Note that $\lim_{n\to\infty}(1/n)\sum_{k=1}^{n} e^{(k/n)\nu} = \int_0^1 e^{\nu\tau}d\tau$. Therefore, combining this fact with (4.7) we conclude that the set in the right hand side of (4.5) is identical with the range of $\hat{a}$ which is equal to $\sigma(a) = \sigma(zI - \tilde{Q}(z))$. This completes the proof. □

Combining Theorem 4.2 with Theorem 2.1, we get the following result:

Corollary 4.3 *Let $T(t)$ be a compact semigroup on X, and assume that $B(k) \equiv b(k)I$, $k \in \mathbf{Z}^+$, where $b(k)$ is a scalar function satisfying $\sum_{k=0}^{\infty}|b(k)|e^{\gamma k} < \infty$ for some positive constant γ. Then the following three statements are equivalent:*

(i) The condition (2.2) for (4.3) holds true:

(ii) The zero solution of (4.3) is exponentially asymptotically stable:

(iii) $z \neq e^\nu + \tilde{b}(z)\int_0^1 e^{\nu\tau}d\tau, \qquad (\forall\, |z| \geq 1,\ \nu \in \sigma(A)).$

Example 4.4 Consider the following scalar partial differential equation with piecewise continuous delays

$$\frac{\partial v}{\partial t}(t,\xi) = \frac{\partial^2 v}{\partial \xi^2}(t,\xi) + \delta v([t],\xi)\left(1 - (1/K)\sum_{j=0}^{\infty} c(j)v([t-j],\xi)\right), \quad t>0,\ 0<\xi<\pi, \tag{4.8}$$

which may be thought as a semi-discretization of integro-differential equation

$$\frac{\partial v}{\partial t}(t,\xi) = \frac{\partial^2 v}{\partial \xi^2}(t,\xi) + \delta v(t,\xi)\left(1 - (1/K)\int_0^\infty c(s)v(t-s,\xi)\right), \quad t>0,\ 0<\xi<\pi,$$

which is a model for the dynamics of growth of a one-species population taking diffusion effects into consideration. In what follows, we impose the Neumann boundary condition

$$\frac{\partial v}{\partial \xi}(t,0) = \frac{\partial v}{\partial \xi}(t,\pi) = 0, \quad t>0 \tag{4.9}$$

on (4.8), and investigate stability properties of the positive constant solution of (4.8)-(4.9) as an application of the results obtained in the previous sections. In the above, we assume that δ and K are positive constants, and $c(j)$, $j \in \mathbf{Z}^+$, is a nonnegative sequence such that $\sum_{j=0}^{\infty} c(j) = 1$ and $\sum_{j=0}^{\infty} c(j)e^{\gamma j} < \infty$ for some positive constant γ. Obviously, $v(t,\xi) \equiv K$ is a solution of (4.8)-(4.9). In order to investigate the stability of the equilibrium $v \equiv K$, we set $u = v - K$. Then (4.8)-(4.9) is transformed to the equation

$$\frac{\partial u}{\partial t}(t,\xi) = \frac{\partial^2 u}{\partial \xi^2}(t,\xi) - \frac{\delta}{K}\Big(u([t],\xi) + K\Big)\sum_{j=0}^{\infty} c(j)u([t-j],\xi), \quad t>0,\ 0<\xi<\pi \tag{4.10}$$

and

$$\frac{\partial u}{\partial \xi}(t,0) = \frac{\partial u}{\partial \xi}(t,\pi) = 0, \quad t>0. \tag{4.11}$$

Take $X = C([0,\pi];\mathbf{R})$ and set $A\chi = \chi''$ for $\xi \in \mathbf{D}(A) := \{\chi \in C^2([0,\pi];\mathbf{R}) : \chi'(0) = \chi'(\pi) = 0\}$. A generates a compact strongly continuous semigroup $T(t)$, $t \geq 0$, on $X = C([0,\pi];\mathbf{R})$. In fact, $T(t)$ is given by the formula

$$T(t)\phi \sim \sum_{n=0}^{\infty} e^{-n^2 t} a_n \cos n\xi, \quad t \geq 0,$$

for any $\phi \in X$ with the Fourier expansion $\sum_{n=0}^{\infty} a_n \cos n\xi$, and moreover we have $\sigma(A) = \{-n^2 : n = 0, 1, 2, \cdots\}$. Then (4.10)–(4.11) can be set up as a differential equation with piecewise continuous delays on the Banach space $X = C([0,\pi];\mathbf{R})$

$$\dot{u}(t) = Au(t) + \sum_{k=0}^{\infty} B(k)u([t-k]) + F(u_t),$$

where $B(n) = -\delta c(n)I$ and $F(\phi) = -(\delta/K)\phi([t]-t)\sum_{j=0}^{\infty} c(j)\phi([t-j]-t)$. This equation is equivalent to the nonlinear difference equation

$$u(n+1) = \sum_{k=-\infty}^{n} Q(n-k)u(k) + f(u_k),$$

where $Q(n)$, $n \in \mathbf{Z}^+$ are the ones given by (4.4), and

$$f(\phi) = \int_0^1 T(\tau)M(\phi(0), -(\delta/K)\sum_{k=0}^{\infty} c(k)\phi(-k))d\tau, \qquad \phi \in \mathbf{B}^\gamma,$$

with $M(\chi,\eta) := \chi(\xi)\eta(\xi)$, $\xi \in [0,\pi]$, for χ, $\eta \in X$. One can easily see that $f : \mathbf{B}^\gamma \to X$ is continuous, $f(0) = 0$ and $f(\phi) = o(\|\phi\|)$. Hence we get the following result on the stability of the constant solution of (4.8)–(4.9) combining Theorem 3.2 and Corollary 4.3.

Theorem 4.5 *Suppose that the following condition holds true:*

$$z \neq 1 - \delta\tilde{c}(z) \text{ and } z \neq e^{-n^2} + \frac{\delta(e^{-n^2}-1)}{n^2}\tilde{c}(z), \qquad (n = 1, 2, \cdots, |z| \geq 1). \quad (4.12)$$

Then the constant solution $v(t,\xi) \equiv K$ of (4.8)–(4.9) is exponentially asymptotically stable.

One can easily check that the condition (4.12) is satisfied in the case that $0 < \delta < 2$ and $c(0) = 1$ with $c(n) = 0$, $n = 1, 2, \cdots$, or in the case that $0 < \delta < 6$ and $c(n) = 2^{-n-1}$, $n \in \mathbf{Z}^+$.

References

[1] K.L. Cooke and J. Wiener, *Retarded differential equations with piecewise constant delays*, J. Math. Anal. Appl., **99** (1984), 265–297.

[2] S. Elaydi, *An Introduction to Difference Equations*, Springer-Verlag, New-York, 1996.

[3] S. Elaydi and S. Murakami, *Asymptotic stability versus exponential stability in linear Volterra difference equations of convolution type*, J. Difference Equations and Appl., **2** (1996), 401–410.

[4] K.J. Engel and R. Nagel,*One-Parameter Semigroups for Linear Evolution Equations*, Graduate Texts in Mathematics, **194**, Springer-Verlag, Berlin, 2000.

[5] K. Gopalsamy, *Stability and Oscillations in Delay Differential Equations of Population Dynamics*, Mathematics and its Applications, Kluwer Academic Publishers, 1992.

[6] G. Gripenberg, S.O. Londen and O.J. Staffans,*Volterra Integral and Functional Equations*, Cambridge Univ. Press, Cambridge, 1990.

[7] S.I. Grossman and R.K. Miller, *Nonlinear Volterra integrodifferential equations*, J. Differential Equations, **13** (1973), 551–566.

[8] I. Györi and G. Ladas, *Oscillation Theory of Delay Differential Equations with Applications*, Oxford Math. Monographs, Clarendon Press, Oxford, 1991.
[9] S. Mohamad and K. Gopalsamy, *Extreme stability and almost periodicity in a discrete logistic equation*, Tôhoku Math. J., **52** (2000), 107–125; Correction, Tôhoku Math. J., **53** (2001), 629–631.
[10] S. Murakami, *Exponential asymptotic stability for scalar linear Volterra equations*, Differential and Integral Equations, **4** (1991), 519–525.
[11] S. Murakami, *Representation of solutions of linear functional difference equations in phase space*, Nonlinear Anal. T. M. A., **30** (1997), 1153–1164.
[12] A. Pazy, *Semigroups of Linear Operators and Applications to Partial Differential Equations*, Applied Math. Sci. **44**, Springer-Verlag, Berlin-New York, 1983.
[13] J. Prüss, *Evolutionary Integral Equations and Applications*, Birkhäuser, Basel, 1993.
[14] W. Rudin, *Functional Analysis*, McGraw-Hill, New Delhi, 1988.
[15] J. Wiener, *Generalized Solutions of Functional Differential Equations*, World Scientific, Singapore, 1993.

Fields Institute Communications
Volume **42**, 2004

Periodic Oscillation for a Class of Delay Difference Systems

Shangjiang Guo
College of Mathematics and Econometrics
Hunan University
Changsha, Hunan 410082, P. R. China
shangjguo@etang.com

Lihong Huang
College of Mathematics and Econometrics
Hunan University
Changsha, Hunan 410082, P. R. China
lhhuang@hnu.net.cn

Dedicated to Professor Zhicheng Wang on the occasion of his seventieth birthday

Abstract. By means of the coincidence theory, we give some sufficient conditions for the existence of periodic solutions of a class of delay difference systems, which can be regarded as the discrete analog of the artificial neural network model of three neurons. Our results can be applied to neural networks with a broad range of activation functions assuming neither differentiability nor strict monotonicity. Moreover, these conclusions are presented in terms of system parameters and can be easily verified.

1 Introduction

In this paper, we consider the following delay difference system

$$\begin{cases} \Delta x_n = -a_1 x_n + f_1(y_{n-k_1}) + f_2(z_{n-k_2}) + I_n, \\ \Delta y_n = -a_2 y_n + f_3(x_{n-k_3}) + f_4(z_{n-k_4}) + J_n, \\ \Delta z_n = -a_3 z_n + f_5(x_{n-k_5}) + f_6(y_{n-k_6}) + K_n, \end{cases} \quad n \in \mathbb{N} \tag{1.1}$$

where Δ is a forward difference operator, i.e., $\Delta x_n = x_{n+1} - x_n$, $k_i, i = 1, 2 \cdots, 6$ are positive integers, $0 \le a_i < 1, i = 1, 2, 3$ are constants, $f_i \in C(\mathbb{R}, \mathbb{R}), i = 1, 2, \cdots, 6$, $\{I_n\}_{n=0}^{\infty}$, $\{J_n\}_{n=0}^{\infty}$ and $\{K_n\}_{n=0}^{\infty}$ are periodic sequences with common period $\omega > 0$. System (1.1) can be regarded as the discrete analog of the following artificial neural

---

2000 *Mathematics Subject Classification.* Primary 39A11; Secondary 92B20.

The work was supported in part by National Natural Science Foundation of P. R. China (No. 10071016) and by the Foundation for University Excellent Teacher by the Ministry of Education.

network of three neurons

$$\begin{cases} \dot{x}(t) = -a_1x(t) + f_1(y(t-\tau_1)) + f_2(z(t-\tau_2)) + I(t), \\ \dot{y}(t) = -a_2y(t) + f_3(x(t-\tau_3)) + f_4(z(t-\tau_4)) + J(t), \\ \dot{z}(t) = -a_3z(t) + f_5(x(t-\tau_5)) + f_6(y(t-\tau_6)) + K(t), \end{cases} \tag{1.2}$$

where $\dot{x}$, $\dot{y}$ and $\dot{z}$ are replaced by the forward difference Δx_n and Δy_n and Δz_n, respectively. Such a network has been found in a variety of neural structures, such as neocortex [6], cerebellum [3] and hippocampus [1].

To simplify the presentation, let $\mathbb{Z}$ denote the set of all integers. for any $a, b \in \mathbb{Z}$, define $N(a) = \{a, a+1, \cdots\}$, $N(a,b) = \{a, a+1, \cdots, b\}$ when $a \leq b$, $\mathbb{N} = N(0)$. By a solution of (1.1), we mean a sequence $\{(x_n, y_n, z_n)^T\}_{n=-k}^{\infty}$, where $k = \max_{i=1,2,\cdots,6}\{k_i\}$, and satisfies (1.1) for $n \in \mathbb{N}$. Clearly, for any three k-dimensional vectors $X = (X_0, \cdots, X_{k-1}), Y = (Y_0, \cdots, Y_{k-1})$ and $Z = (Z_0, \cdots, Z_{k-1})$, system (1.1) has a unique solution $\{(x_n, y_n, z_n)^T\}_{n=-k}^{\infty}$ satisfying the initial conditions

$$x_i = X_{i+k},\ y_i = Y_{i+k},\ z_i = Z_{i+k} \text{ for } i \in N(-k, -1).$$

Our purpose is to discuss the existence of periodic solutions of system (1.1). We note that such an existence problem is highly nontrivial and, to the best of our knowledge, while some progress has been made for differential equation [2, 4] and little has been done for nonlinear delay difference systems. Our approach to the existence problem here is based on the coincidence degree. Moreover, our results can be applied to neural networks with a broad range of activation functions assuming neither differentiability nor strict monotonicity. Moreover, These conclusions are presented in terms of system parameters and can be easily verified.

2 Global existence of periodic solutions

We shall first summarize a few concepts and results from [5] that will be basic for this section.

Let X be a Banach space, $L : DomL \subset X \to X$ a linear mapping, and $N : X \to X$ a continuous mapping. The mapping L is said to be a Fredholm mapping of index zero if $\dim KerL = codimImL < +\infty$ and ImL is closed in X. If L is a Fredholm mapping of index zero, then there exist continuous projectors $P : X \bigcap DomL \to KerL$ and $Q : X \to X/ImL$ such that $ImP = KerL$, $ImL = KerQ = Im(I-Q)$. It follows that $L|_{DomL \bigcap KerP} : (I-P)X \to ImL$ is invertible. We denote the inverse of that mapping by K_P. Let Ω be an open bounded subset of X. The mapping N will be called L-compact on $\overline{\Omega}$ if $QN(\overline{\Omega})$ is bounded and $K_P(I-Q)N : \overline{\Omega} \to X$ is compact. Since ImQ is isomorphic to $KerL$, there exists an isomorphism $J : ImQ \to KerL$.

In the proof of our existence theorem, we will use the continuation theorem of Gaines and Mawhin [5].

Lemma 2.1 *Let X be a Banach space and L a Fredholm mapping of index zero. Assume that $N : \overline{\Omega} \to X$ is L-compact on $\overline{\Omega}$ with Ω open bounded in X. Furthermore,*

1. *for each $\lambda \in (0,1)$, $x \in \partial\Omega \bigcap DomL$, $Lx \neq \lambda Nx$;*
2. *$QNx \neq 0$ for each $x \in \partial\Omega \bigcap KerL$;*
3. *$deg(JQN, \Omega \bigcap KerL, 0) \neq 0$, where deg denotes the Brouwer degree.*

Then $Lx = Nx$ has at least one solution in $\overline{\Omega}$.

Throughout the remaining part of this paper, we let X be a set of all 3-dimensional real vector ω-periodic sequences $u = \{(x_n\ y_n, z_n)^T\}$ equipped with the usual supremum norm $\|\cdot\|$, i.e.,

$$\|u\| = \|x| + \|y\| + \|z\|, \text{ for } u \in X$$

where

$$\begin{aligned} \|x\| &= \max_{n\in N(0,\omega-1)} \{|x_n|\}, \\ \|y\| &= \max_{n\in N(0,\omega-1)} \{|y_n|\} \end{aligned}$$

and

$$\|z\| = \max_{n\in N(0,\omega-1)} \{|z_n|\}.$$

It is easy to see that X is a Banach space.

We consider two subspaces X^0 and X^c of X defined by

$$X^0 = \left\{ u = \{(x_n, y_n, z_n)^T\}; \sum_{n=1}^{\omega} x_n = 0, \sum_{n=1}^{\omega} y_n = 0, \sum_{n=1}^{\omega} z_n = 0 \right\}$$

and

$$X^c = \left\{ u = \{(x_n, y_n, z_n)^T\}; x_n = \beta_1, y_n = \beta_2, z_n = \beta_3 \right\},$$

respectively. Denote by $L : X \to X$ the difference operator given by $Lu = \{(Lu)_n\}$ with

$$(Lu)_n^{(1)} = \Delta x_n, (Lu)_n^{(2)} = \Delta y_n, (Lu)_n^{(3)} = \Delta z_n, \tag{2.1}$$

where $(Lu)_n^{(i)}$ denotes the i-th component of vector $(Lu)_n$, $i = 1, 2, 3$. Let the linear operator $S : X \to X^c$ be defined by $Su = \{(Su)_n\}$ with

$$(Su)_n^{(1)} = \frac{1}{\omega}\sum_{n=1}^{\omega} x_n, (Su)_n^{(2)} = \frac{1}{\omega}\sum_{n=1}^{\omega} y_n, (Su)_n^{(3)} = \frac{1}{\omega}\sum_{n=1}^{\omega} z_n. \tag{2.2}$$

Thus, we have

Lemma 2.2 1. *Both X^0 and X^c are closed linear subspaces of X and $X = X^0 \oplus X^c, dim X^c = 3$;*
2. *L is a bounded linear operator with $KerL = X^c$ and $ImL = X^0$.*

Proof It is easy to check that both X^0 and X^c are closed. For any $u \in X$, let $u^c = Su$ and $u^0 = u - Su$. Obviously, $u^c \in X^c$ and $u^0 \in X^0$. Hence, $X = X^c + X^0$. On the other hand, if $u = \{(x_n, y_n, z_n)^T\} \in X^c \bigcap X^0$, then $u \in X^c$ implies that there exists a constant vector $\beta \in \mathbb{R}^3$ such that $(x_n, y_n, z_n)^T = \beta$ for $n \in \mathbb{Z}$ and $u \in X^0$ gives $\beta = Su = 0$, and hence, $u = 0$. This completes the proof of the first conclusion.

It is trivial to see that $KerL = X^c$, so we only need to prove that $ImL = X^0$. Since $SLu = 0$ for any $u \in X$, we have $ImL \subseteq X^0$. For any $v = \{v_n\} \in X^0$, where $v_n = (X_n, Y_n, Z_n)^T$, define $u = \{u_n\} = \{(x_n, y_n, z_n)^T\}$ to be

$$u_n = \begin{cases} \sum\limits_{i=1}^{n-1} v_n, & \text{if } n \geq 1, \\ 0, & \text{if } n = 0, \\ -\sum\limits_{i=n}^{-1} v_n, & \text{if } n \leq -1. \end{cases}$$

Then, it is easy to see that $u_{n+\omega} - u_n = \omega S v$ for $n \geq -\omega$ and $u_{n+\omega} - u_n = -\omega S v$ for $n \leq -(\omega+1)$. This, combined with with $v \in X^0$, implies that $u_{n+\omega} = u_n$ for all $\mathbb{Z}$, i.e., $u \in X$. Moreover, $Lu = v$. Therefore, $X^0 \subseteq ImL$, and hence, the second conclusion is proved. □

Our main result on the global existence of a positive periodic solution of (1.1) is stated in the following theorem.

Theorem 2.3 *Assume that there exist constants* $0 \leq \alpha < \frac{1}{2}\min\{a_1, a_2, a_3\}$ *and* $\beta > 0$ *such that for any* $x \in \mathbb{R}$,

$$|f_i(x)| \leq \alpha|x| + \beta, \quad i = 1, 2, \cdots, 6.$$

Then system (1.1) has at least one ω*-periodic solution.*

Proof Let operator $L : DomL \subset X \to X$ be defined as (2.1). It follows from Lemma 2.2 that L is a Fredholm operator of index 0. Set

$$(Nu)_n = \begin{pmatrix} -a_1 x_n + f_1(y_{n-k_1}) + f_2(z_{n-k_2}) + I_n \\ -a_2 y_n + f_3(x_{n-k_3}) + f_4(z_{n-k_4}) + J_n \\ -a_3 z_n + f_5(x_{n-k_5}) + f_6(y_{n-k_6}) + K_n \end{pmatrix}$$

and

$$Pu = Qu = Su, u \in X,$$

where operator K is defined as (2.2). Obviously, $ImP = KerL$, $ImL = KerQ = Im(I - Q)$.

Furthermore, the generalized inverse (of L) $K_P : ImL \to KerP \bigcap DomL$ reads

$$(K_P u)_n = \begin{pmatrix} \sum\limits_{i=0}^{n-1} x_i - \frac{1}{\omega} \sum\limits_{n=1}^{\omega} \sum\limits_{s=0}^{n-1} x_s \\ \sum\limits_{i=0}^{n-1} y_i - \frac{1}{\omega} \sum\limits_{n=1}^{\omega} \sum\limits_{s=0}^{n-1} y_s \\ \sum\limits_{i=0}^{n-1} z_i - \frac{1}{\omega} \sum\limits_{n=1}^{\omega} \sum\limits_{s=0}^{n-1} z_s \end{pmatrix}.$$

Thus,

$$\begin{aligned} QNu &= \frac{1}{\omega} \begin{pmatrix} -a_1 \sum\limits_{n=1}^{\omega} x_n + \sum\limits_{n=1}^{\omega} f_1(y_{n-k_1}) + \sum\limits_{n=1}^{\omega} f_2(z_{n-k_2}) + \sum\limits_{n=1}^{\omega} I_n \\ -a_2 \sum\limits_{n=1}^{\omega} y_n + \sum\limits_{n=1}^{\omega} f_3(x_{n-k_3}) + \sum\limits_{n=1}^{\omega} f_4(z_{n-k_4}) + \sum\limits_{n=1}^{\omega} J_n \\ -a_3 \sum\limits_{n=1}^{\omega} z_n + \sum\limits_{n=1}^{\omega} f_5(x_{n-k_5}) + \sum\limits_{n=1}^{\omega} f_6(y_{n-k_6}) + \sum\limits_{n=1}^{\omega} K_n \end{pmatrix} \\ &= \frac{1}{\omega} \sum_{n=1}^{\omega} \begin{pmatrix} -a_1 x_n + f_1(y_n) + f_2(z_n) + I_n \\ -a_2 y_n + f_3(x_n) + f_4(z_n) + J_n \\ -a_3 z_n + f_5(x_n) + f_6(y_n) + K_n \end{pmatrix} \end{aligned}$$

and

$$
\begin{aligned}
&(K_P(I-Q)Nu)_n \\
&= (K_PNu)_n - (K_PQNu)_n \\
&= \sum_{i=0}^{n-1}\begin{pmatrix} -a_1x_i + f_1(y_{i-k_1}) + f_2(z_{i-k_2}) + I_i \\ -a_2y_i + f_3(x_{i-k_3}) + f_4(z_{i-k_4}) + J_i \\ -a_3z_i + f_5(x_{i-k_5}) + f_6(y_{i-k_6}) + K_i \end{pmatrix} \\
&\quad -\frac{1}{\omega}\sum_{n=1}^{\omega}\sum_{s=0}^{n-1}\begin{pmatrix} -a_1x_s + f_1(y_{s-k_1}) + f_2(z_{s-k_2}) + I_s \\ -a_2y_s + f_3(x_{s-k_3}) + f_4(z_{s-k_4}) + J_s \\ -a_3z_s + f_5(x_{s-k_5}) + f_6(y_{s-k_6}) + K_s \end{pmatrix} \\
&\quad -\left(\frac{n}{\omega} - \frac{\omega+1}{2\omega}\right)\sum_{n=1}^{\omega}\begin{pmatrix} -a_1x_n + f_1(y_{n-k_1}) + f_2(z_{n-k_2}) + I_n \\ -a_2y_n + f_3(x_{n-k_3}) + f_4(z_{n-k_4}) + J_n \\ -a_3z_n + f_5(x_{n-k_5}) + f_6(y_{n-k_6}) + K_n \end{pmatrix}.
\end{aligned}
$$

Obviously, both K_P and $(I-Q)N$ are bounded operators. Define $\varphi : KerP\bigcap DomL \to \mathbb{R}^{3\omega}$,

$$
\varphi(u) = \varphi(\{(x_n, y_n, z_n)^T\}) = \begin{pmatrix} x_1 & \cdots & x_\omega \\ y_1 & \cdots & y_\omega \\ z_1 & \cdots & z_\omega \end{pmatrix}
$$

for all $u \in KerP\bigcap DomL$. Clearly, $\varphi : KerP\bigcap DomL \cong \mathbb{R}^{3\omega}$ is a linear isomorphism. Hence $K_p : ImL \to KerP\bigcap DomL$ is compact. Thus, $\overline{K_P(I-Q)N(\overline{\Omega})}$ is compact for any open bounded subset $\Omega \subset X$. Moreover, $QN(\overline{\Omega})$ is clearly bounded. Thus, N is L-compact on $\overline{\Omega}$ with any open bounded set $\Omega \subset X$.

Now, we reach the point where we search for an appropriate open bounded subset Ω for the application of continuation theorem (Lemma 2.1).

Consider the operator equation $Lu = \lambda Nu, \lambda \in (0,1)$, that is

$$
\begin{cases} \Delta x_n = -\lambda a_1 x_n + \lambda f_1(y_{n-k_1}) + \lambda f_2(z_{n-k_2}) + \lambda I_n, \\ \Delta y_n = -\lambda a_2 y_n + \lambda f_3(x_{n-k_3}) + \lambda f_4(z_{n-k_4}) + \lambda J_n, \\ \Delta z_n = -\lambda a_3 z_n + \lambda f_5(x_{n-k_5}) + \lambda f_6(y_{n-k_6}) + \lambda K_n. \end{cases} \tag{2.3}
$$

Let $u = \{(x_n, y_n, z_n)^T\}$ be a solution of Eq.(2.3) for a certain $\lambda \in (0,1)$. It follows from the first equation of (2.3) that

$$
\begin{aligned}
\|x\| &= \max_{n\in N(0,\omega-1)} |x_{n+1}| \\
&\le \max_{n\in N(0,\omega-1)} \{(1-\lambda a_1)|x_n| + \lambda|f_1(y_{n-k_1})| + \lambda|f_2(z_{n-k_2})| + \lambda|I_n|\} \\
&\le \max_{n\in N(0,\omega-1)} \{(1-\lambda a_1)|x_n| + \lambda\alpha|y_{n-k_1}| + \lambda\alpha|z_{n-k_2}| + 2\beta + \lambda|I_n|\} \\
&\le (1-\lambda a_1)\|x\| + \lambda\alpha\|y\| + \lambda\alpha\|z\| + 2\lambda\beta + \lambda\|I\|,
\end{aligned}
$$

that is,

$$
a_1\|x\| \le \alpha\|y\| + \alpha\|z\| + (2\beta + \|I\|). \tag{2.4}
$$

Similarly, we can obtain

$$
a_2\|y\| \le \alpha\|x\| + \alpha\|z\| + (2\beta + \|J\|) \tag{2.5}
$$

and

$$
a_3\|z\| \le \alpha\|x\| + \alpha\|y\| + (2\beta + \|K\|), \tag{2.6}
$$

where $M_1^* = \alpha(2\beta + \|K\|) + a_3(2\beta + \|J\|)$ and $M_2^* = \alpha(2\beta + \|K\|) + a_3(2\beta + \|I\|)$. Substituting (2.6) into (2.5) and (2.4), we have

$$(a_2a_3 - \alpha^2)\|y\| \leq (a_3\alpha + \alpha^2)\|x\| + M_1^* \tag{2.7}$$

and

$$(a_1a_3 - \alpha^2)\|x\| \leq (a_3\alpha + \alpha^2)\|y\| + M_2^*. \tag{2.8}$$

Substituting (2.7) into (2.8) gives

$$(a_1a_2a_3^2 - (a_1+a_2)a_3\alpha^2 - 2a_3\alpha^3 - a_3^2\alpha^2)\|x\| \leq M_3^* := (a_3\alpha+\alpha^2)M_1^* + (a_2a_3 - \alpha^2)M_2^*.$$

It follows from Condition (ii) that $a_1a_2a_3^2 - (a_1 + a_2)a_3\alpha^2 - 2a_3\alpha^3 - a_3^2\alpha^2 > 0$. Hence,

$$\|x\| \leq d_1 := \frac{M_3^*}{a_1a_2a_3^2 - (a_1 + a_2)a_3\alpha^2 - 2a_3\alpha^3 - a_3^2\alpha^2}.$$

Similarly, we have

$$\|y\| \leq d_2 \quad \text{and} \quad \|z\| \leq d_3.$$

Clearly, $d_i (i = 1, 2, 3)$ are independent of λ. Denote $M^* = d_1 + d_2 + d_3 + C$, where $C > 0$ is taken sufficiently large such that

$$\min\{(a_1 - 2\alpha), (a_2 - 2\alpha), (a_3 - 2\alpha)\}M^* > 6\beta + \|I\| + \|J\| + \|K\|.$$

We take $\Omega = \{u = \{(x_n, y_n, z_n)^T\} \in X; \|u\| < M^*\}$. This satisfies condition (1) of Lemma 2.1 . If $(x, y, z)^T \in \partial\Omega \bigcap KerL = \partial\Omega \bigcap \mathbb{R}^3$, then $(x, y, z)^T$ is a constant vector in $\mathbb{R}^3$ with $|x| + |y| + |z| = M^*$, and we have

$$QN\begin{pmatrix} x \\ y \\ z \end{pmatrix} = \begin{pmatrix} -a_1x + f_1(y) + f_2(z) + \frac{1}{\omega}\sum\limits_{n=1}^{\omega} I_n \\ -a_2y + f_3(x) + f_4(z) + \frac{1}{\omega}\sum\limits_{n=1}^{\omega} J_n \\ -a_3z + f_5(z) + f_6(y) + \frac{1}{\omega}\sum\limits_{n=1}^{\omega} K_n \end{pmatrix}.$$

Therefore,

$$\begin{aligned}
\|QN(x,y,z)^T\| &= |-a_1x+f_1(y)+f_2(z)+\frac{1}{\omega}\sum_{n=1}^{\omega}I_n| \\
&\quad +|-a_2y+f_3(x)+f_4(z)+\frac{1}{\omega}\sum_{n=1}^{\omega}J_n| \\
&\quad +|-a_3z+f_5(z)+f_6(y)+\frac{1}{\omega}\sum_{n=1}^{\omega}K_n| \\
&\geq a_1|x|-|f_1(y)|-|f_2(z)|-\frac{1}{\omega}\sum_{n=1}^{\omega}|I_n| \\
&\quad +a_2|y|-|f_3(x)|-|f_4(z)|-\frac{1}{\omega}\sum_{n=1}^{\omega}|J_n| \\
&\quad +a_3|z|-|f_5(z)|-|f_6(y)|-\frac{1}{\omega}\sum_{n=1}^{\omega}|K_n| \\
&\geq (a_1-2\alpha)|x|+(a_2-2\alpha)|y|+(a_3-2\alpha)|z| \\
&\quad -6\beta-(\|I\|+\|J\|+\|K\|) \\
&\geq \min\{(a_1-2\alpha),(a_2-2\alpha),(a_3-2\alpha)\}M^* \\
&\quad -6\beta-(\|I\|+\|J\|+\|K\|)>0.
\end{aligned}$$

Consequently,

$$QN(x,y,z)^T \neq (0,0,0)^T \text{ for } (x,y,z)^T \in \partial\Omega \bigcap KerL.$$

This satisfies condition (2) of Lemma 2.1.

Define $\psi : \Omega\bigcap KerL \times [0,1] \to X/ImL = X^c$ by

$$\begin{aligned}
\psi(u,\mu) &= \mu diag(-a_1,-a_2,-a_3)u+(1-\mu)QNu \\
&= \mu\begin{pmatrix}-a_1x\\-a_2y\\-a_3z\end{pmatrix}+(1-\mu)\begin{pmatrix}-a_1x+f_1(y)+f_2(z)+\frac{1}{\omega}\sum\limits_{n=1}^{\omega}I_n\\-a_2y+f_3(x)+f_4(z)+\frac{1}{\omega}\sum\limits_{n=1}^{\omega}J_n\\-a_3z+f_5(z)+f_6(y)+\frac{1}{\omega}\sum\limits_{n=1}^{\omega}K_n\end{pmatrix}
\end{aligned}$$

for all $u=(x,y,z)^T \in \Omega\bigcap KerL$ and $\mu \in [0,1]$.

When $u \in \partial\Omega \bigcap KerL$ and $\mu \in [0,1]$, $u = (x,y,z)^T$ is a constant vector in $\mathbb{R}^3$ with $|x|+|y|+|z| = M^*$. Thus

$$\begin{aligned}
\|\psi(u,\mu)\| &= |-\mu a_1 x + (1-\mu)(-a_1 x + f_1(y) + f_2(z) + \frac{1}{\omega}\sum_{n=1}^{\omega} I_n)| \\
&\quad +|-\mu a_2 y + (1-\mu)(-a_2 y + f_3(x) + f_4(z) + \frac{1}{\omega}\sum_{n=1}^{\omega} J_n)| \\
&\quad +|-\mu a_3 z + (1-\mu)(-a_3 z + f_5(z) + f_6(y) + \frac{1}{\omega}\sum_{n=1}^{\omega} K_n)| \\
&\geq a_1|x| - |f_1(y)| - |f_2(z)| - \frac{1}{\omega}\sum_{n=1}^{\omega} |I_n| \\
&\quad +a_2|y| - |f_3(x)| - |f_4(z)| - \frac{1}{\omega}\sum_{n=1}^{\omega} |J_n| \\
&\quad +a_3|z| - |f_5(z)| - |f_6(y)| - \frac{1}{\omega}\sum_{n=1}^{\omega} |K_n| \\
&\geq (a_1 - 2\alpha)|x| + (a_2 - 2\alpha)|y| + (a_3 - 2\alpha)|z| \\
&\quad -6\beta - (\|I\| + \|J\| + \|K\|) \\
&\geq \min\{(a_1 - 2\alpha), (a_2 - 2\alpha), (a_3 - 2\alpha)\}M^* \\
&\quad -6\beta - (\|I\| + \|J\| + \|K\|) > 0.
\end{aligned}$$

Therefore,

$$\psi(u,\mu) \neq 0, \quad \text{for any } u \in \partial\Omega \bigcap KerL.$$

Using the property of topological degree and taking $J : ImQ \to KerL, (x,y,z)^T \to (x,y,z)^T$, we have

$$\begin{aligned}
deg(JQN, \Omega\bigcap KerL, 0) &= deg(\psi(\cdot,0), \Omega\bigcap KerL, 0) \\
&= deg(\psi(\cdot,1), \Omega\bigcap KerL, 0) \\
&= deg(diag(-a_1,-a_2,-a_3), \Omega\bigcap KerL, 0) = 1.
\end{aligned}$$

This completes the proof of condition (3) of Lemma 2.1. Therefore, system (1.1) has at least one ω-periodic solution. The proof is complete. □

If we further assume that $f_i, i = 1,2,\cdots,6$ are globally Lipschitz, then we shall obtain the global exponential stability of the ω-periodic solution. Namely,

Theorem 2.4 *Assume that there exists a constant* $0 \leq \alpha < \frac{1}{2}\min\{a_1,a_2,a_3\}$ *such that for any* $x, y \in \mathbb{R}$,

$$|f_j(x) - f_j(y)| \leq \alpha|x-y|, \quad i = 1,2,\cdots,6.$$

Then system (1.1) has exactly one ω-periodic solution and all other solutions converge exponentially to it as $n \to \infty$. Moreover, the global exponential convergent rate λ can be estimated by the following inequality.

$$2\alpha\lambda^{k+1} + (1-a^0)\lambda < 1, \tag{2.9}$$

where $a^0 = \min\{a_1,a_2,a_3\}$.

Proof In view of $0 \leq \alpha < \frac{1}{2}\min\{a_1, a_2, a_3\}$, there must exist a sufficiently small constant $1 < \lambda < \min\{(1-a_i)^{-1}; i = 1, 2, 3\}$ such that

$$\max_{1\leq i\leq 3}\left\{\frac{2\alpha}{1-(1-a_i)\lambda}\lambda^{k+1}\right\} < 1. \tag{2.10}$$

It follows from (2.9) that $|f_j(x)| \leq \alpha|x| + |f_j(0)|$, $j = 1, 2, \cdots, 6$. Hence, all the conditions in Theorem 2.1 hold. Thus, system (1.1) has at least one ω-periodic solution. Let $u^* = \{(x_n^*, y_n^*, z_n^*)^T\}_{n=-k}^{\infty}$ be a ω-periodic solution of system (1.1) and $u = \{(x_n, y_n, z_n)^T\}_{n=-k}^{\infty}$ an arbitrary solution of system (1.1). Then it follows that for $n \in \mathbb{N}$,

$$\begin{cases} \Delta(x_n - x_n^*) = -a_1(x_n - x_n^*) + f_1(y_{n-k_1}) - f_1(y^*_{n-k_1}) + f_2(z_{n-k_2}) - f_2(z^*_{n-k_2}), \\ \Delta(y_n - y_n^*) = -a_2(y_n - y_n^*) + f_3(x_{n-k_3}) - f_3(x^*_{n-k_3}) + f_4(z_{n-k_4}) - f_4(z^*_{n-k_4}), \\ \Delta(z_n - z_n^*) = -a_3(z_n - z_n^*) + f_5(x_{n-k_5}) - f_5(x^*_{n-k_5}) + f_6(y_{n-k_6}) - f_6(y^*_{n-k_6}). \end{cases}$$

Set

$$X_n = \begin{cases} \mid x_n - x_n^* \mid \lambda^n, & \text{if } n \in \mathbb{N}, \\ \mid x_n - x_n^* \mid, & \text{if } n \in N(-k, 0), \end{cases} \tag{2.11}$$

$$Y_n = \begin{cases} \mid y_n - y_n^* \mid \lambda^n, & \text{if } n \in \mathbb{N}, \\ \mid y_n - y_n^* \mid, & \text{if } n \in N(-k, 0) \end{cases} \tag{2.12}$$

and

$$Z_n = \begin{cases} \mid z_n - z_n^* \mid \lambda^n, & \text{if } n \in \mathbb{N}, \\ \mid z_n - z_n^* \mid, & \text{if } n \in N(-k, 0). \end{cases} \tag{2.13}$$

Then we have

$$\begin{cases} X_n \leq [(1-a_1)\lambda]^n x_0 + \sum\limits_{s=0}^{n-1}[(1-a_1)\lambda]^{n-1-s}\lambda^{k+1}\alpha(Y_{s-k_1} + Z_{s-k_2}), \\ Y_n \leq [(1-a_2)\lambda]^n Y_0 + \sum\limits_{s=0}^{n-1}[(1-a_2)\lambda]^{n-1-s}\lambda^{k+1}\alpha(X_{s-k_3} + Z_{s-k_4}), \\ Z_n \leq [(1-a_3)\lambda]^n Z_0 + \sum\limits_{s=0}^{n-1}[(1-a_3)\lambda]^{n-1-s}\lambda^{k+1}\alpha(X_{s-k_5} + Y_{s-k_6}). \end{cases} \tag{2.14}$$

for $n \in \mathbb{N}$.

Set $M = \max_{n\in N(-k,0)}\{|x_n - x_n^*|, |y_n - y_n^*|, |z_n - z_n^*|\}$. We shall claim that for any sufficiently small constant $\varepsilon > 0$,

$$X_n < M - \varepsilon, Y_n < M + \varepsilon, Z_n < M + \varepsilon, \quad n \in \mathbb{N}. \tag{2.15}$$

By a way of contrary. Without loss of generality, we suppose that there exists some $n_1 \in \mathbb{N}$ such that

$$X_{n_1} = M + \varepsilon, \quad Y_n \leq M + \varepsilon, \quad Z_n \leq M + \varepsilon, \quad n \in N(0, n_1). \tag{2.16}$$

It follows from (2.14) that

$$\begin{aligned} M + \varepsilon &= X_{n_1} \\ &\leq [(1-a_1)\lambda]^{n_1} X_0 + \sum_{s=0}^{n_1-1}[(1-a_1)\lambda]^{n-1-s}\lambda^{k+1}\alpha(y_{s-k_1} + z_{s-k_2}) \\ &\leq [(1-a_1)\lambda]^{n_1}(M+\varepsilon) + 2\alpha\lambda^{k+1}\frac{1-[(1-a_1)\lambda]^{n_1}}{1-(1-a_1)\lambda}(M+\varepsilon) \\ &< [(1-a_1)\lambda]^{n_1}(M+\varepsilon) + \{1-[(1-a_1)\lambda]^{n_1}\}(M+\varepsilon) \\ &= M + \varepsilon, \end{aligned}$$

which is a contradiction. Therefore, (2.15) holds. Let $\varepsilon \to 0$. Then we have

$$X_n \leq M, Y_n \leq M, Z_n \leq M \quad \text{for} \quad n \in \mathbb{N}.$$

Thus, it follows from (2.11)-(2.13) that

$$|x_n - x_n^*| \leq M\lambda^{-n}, |y_n - y_n^*| \leq M\lambda^{-n}, |z_n - z_n^*| \leq M\lambda^{-n} \quad \text{for} \quad n \in \mathbb{N},$$

which implies that the ω-periodic solution $\{(x_n^*, y_n^*, z_n^*)^T\}_{n=-k}^{\infty}$ is globally exponentially stable. □

References

[1] Anderson, P., Gross, O. and Lomo, T., *et al.*, Participation of inhibitory interneurons in the control of hippocampal cortical output, in *"The Interneuron"* (M. Brazier ed.), Los Angeles: University of California Press, 1969.

[2] Cao, J. & Wang, L. Periodic oscillatory solution of bidirectional associative memory networks with delays. *Physical Review E*, **61** 1825-1828 (2000).

[3] Eccles, J. C., Ito, M., and Szenfagothai, J., *The Cerebellum as Neuronal Machine*, New York: Springer, 1967.

[4] Fang, H. & Li, J. Global exponential stability and periodic solutions of cellular neural networks with delay. *Physical Review E*, **61** 4212-4217 (2000).

[5] Gaines, D. R. E. and Mawhin, J. L. *Coincidence Degree and Nonlinear Differential Equations*, Springer-Verlag, Berlin, 1977.

[6] Szenfagothai, J. The "module-concept" in cerebral cortex architecture, *Brain Research*, **95** (1967), 475-496.

Fields Institute Communications
Volume **42**, 2004

Applications of Critical Point Theory to Difference Equations

Zhiming Guo
Department of Applied Mathematics
Hunan University
Changsha, 410082, P. R. China
gzm100@21cn.com

Jianshe Yu
Department of Applied Mathematics
Hunan University
Changsha, 410082, P. R. China
jsyu@hnu.net.cn

Dedicated to Professor Zhicheng Wang on the occasion of his seventieth birthday

Abstract. In this paper, we give some applications of critical point theory to the existence of periodic and subharmonic solutions for some kinds of difference equations. Some examples are given to illustrate the obtained results. In addition, several open problems are posed, which are expected to be solved by critical point theory.

1 Introduction

The theory of nonlinear difference equations has been widely used to study discrete models appearing in many fields such as Computer science, Economics, Neural network, Ecology and Cybernetics etc. In the last decade, there has been much literature on qualitative properties of difference equations, those studies have been deeply inspired by the development of the study of differential equations. For recent literature on difference equations, we refer to [1, 12, 13, 20, 30, 31] for results on stability and attractivity, and to [1, 13, 29] for study on oscillation. For the general background of difference equations, one can refer to books [1, 9, 13, 20, 26]. However, the results on the existence and multiplicity of periodic solutions, subharmonic solutions, homoclinic and heteroclitic solutions of difference equations are relatively rare, see for example [10], just because there is scarce techniques studying the existence of periodic solutions of discrete systems.

2000 *Mathematics Subject Classification.* Primary 39A11; Secondary 39A12.
This project was supported by TRATOYT of P. R. China and by Trans-Century Training Programme Foundation for the Talents by the State Education Commission.

As is known to us, the development of the study of periodic solution of differential equations is relatively rapid. The existence and multiplicity of periodic solutions of differential equations have been studied by many authors using various approches, e.g. Kaplan-Yorke method, critical point theory (which includes minimax theory, geometrical index theory and Morse theory), coincidence degree theory and so on, see for example [8, 17, 19, 21, 23, 25, 27]. Among these approches, critical point theory seems to be a very powerful tool to deal with such problems, see for example [21, 23, 25, 27]. Therefore, it is natural for us to ask whether the critical point theory and coincidence degree theory can be applied to tackle the existence of periodic solutions of difference equations. We think that it is necessary and possible to develop a new approach to the existence and multiplicity of periodic solutions to difference equations by using critical point theory. Indeed, we have successfully obtained some meaningful results by this approach, and this paper is essentially a survey of the results in our recent papers [14, 15, 16].

2 Some results obtained by critical point theory

For our convenience, denote by $\mathbf{N}, \mathbf{Z}, \mathbf{R}$ the set of all natural numbers, integers and real numbers repectively. For $a, b \in \mathbf{Z}$, define $\mathbf{Z}(a) = \{a, a+1, \cdots, \}$, $\mathbf{Z}(a,b) = \{a, a+1, \cdots, b\}$ when $a \leq b$.

2.1 The second-order superquadratic nonlinear difference equations. Consider the nonlinear second order difference equation

$$\Delta^2 x_{n-1} + f(n, x_n) = 0, n \in \mathbf{Z} \tag{2.1}$$

where $f \in C(\mathbf{R} \times \mathbf{R}, \mathbf{R})$ and there exists a positive integer m such that $\forall (t,z) \in \mathbf{R} \times \mathbf{R}$, $f(t+m, z) = f(t,z), \Delta x_n = x_{n+1} - x_n, \Delta^2 x_n = \Delta(\Delta x_n)$.

We are interested in the existence and multiplicity of m-periodic solutions for equation (2.1). Let p be a given positive integer. We also need to study the existence of pm-periodic solutions for (2.1). As usual, such a periodic solution will be called a subharmonic solution.

We may think of (2.1) as being a discrete analogue of the following second order differential equation

$$x'' + f(t, x) = 0, \quad t \in \mathbf{R}, \tag{2.2}$$

where $f \in C(\mathbf{R} \times \mathbf{R}, \mathbf{R})$, $f(t+T, z) = f(t,z)$ for any $(t,z) \in \mathbf{R} \times \mathbf{R}$.

Although many excellent researches have been worked out on the existence of periodic and subharmonic solutions for (2.2) (see for example [21, 23, 25, 27]), it seems that no similar results were obtained in the literature for (2.1). In our paper [14], we show that the critical point theory is an effective approach to tackle the existence of periodic solutions of difference equations. For the general background of difference equations, one can refer to books [1, 9, 13, 20, 26].

Now let us state the main results of [14].

Theorem 2.1 *Suppose that $f(t,z)$ satisfies the following conditions:*

($\mathbf{f}_1$) $f(t,z) \in C(\mathbf{R} \times \mathbf{R}, \mathbf{R})$ and there exists a positive integer m such that $\forall (t,z) \in \mathbf{R} \times \mathbf{R}$, $f(t+m, z) = f(t,z)$;

($\mathbf{f}_2$) For any $z \in \mathbf{R}$, $\int_0^z f(t,s)ds \geq 0$ and $f(t,z) = o(z)$ (as $z \to 0$);

($\mathbf{f}_3$) There exist some constants $R > 0, \beta > 2$ such that for any $|z| \geq R$,

$$zf(t,z) \geq \beta \int_0^z f(t,s)ds > 0.$$

Then for any given positive integer $p > 0$, (2.1) possesses at least three periodic solutions with period pm.

Corollary Suppose that $f(t, z)$ satisfies ($\mathbf{f}_1$)-($\mathbf{f}_3$), then for any given positive integer $p > 0$, there exists at least two nontrivial periodic solutions for (2.1) with period pm.

Remark 2.2 By ($\mathbf{f}_2$), it is easy to see that

$$\lim_{z\to 0} \frac{f(t,z)}{z} = 0.$$

This means that $f(t, z)$ grows more rapidly than linearly at zero. By integrating the inequality $zf(t,z) \geq \beta \int_0^z f(t,s)ds > 0$, there exist constants $a_1 > 0, a_2 > 0$ such that for any $z \in \mathbf{R}$, $\int_0^z f(t,s)ds \geq a_1|z|^\beta - a_2$. So we have

($\mathbf{f}_3$)$'$ There exist constants $a_1 > 0, a_2 > 0$ such that

$$\int_0^z f(t,s)ds \geq a_1|z|^\beta - a_2, \forall z \in \mathbf{R}.$$

By ($\mathbf{f}_3$) and ($\mathbf{f}_3$)$'$, we have,

$$zf(t,z) \geq \beta \int_0^z f(t,s)ds \geq \beta a_1|z|^\beta - \beta a_2, \forall |z| \geq R.$$

Then

$$\lim_{|z|\to +\infty} \frac{f(t,z)}{z} = +\infty.$$

So ($\mathbf{f}_3$) implies that $f(t, z)$ grows superlinearly at infinity. From this point, the assumptions of Theorem 2.1 correspond the case that $f(t, z)$ grows more rapidly than linearly both at zero and at infinity.

In the case that $f(t, z)$ is independent of t, we consider autonomous second-order difference equation

$$\Delta^2 x_{n-1} + f(x_n) = 0, \tag{2.3}$$

where $f \in C(\mathbf{R}, \mathbf{R})$.

In this instance, if there exist constants $R_1 > 0$ and $\beta > 2$ such that for any $|z| \geq R_1$, $zf(z) \geq \beta \int_0^z f(s)ds > 0$, then $f(R_1) > 0$ and $f(-R_1) < 0$. Obviously, there is some $x_0 \in (-R_1, R_1)$ such that $f(x_0) = 0$. Let $x_n = x_0, \forall n = 1, 2, \cdots$, then for any positive integer m, $\{x_n\}$ is a periodic solution of (2.3) with period m. Consequently, we are interested in the existence of nontrivial periodic solution for (2.3). In this case, we have

Theorem 2.3 *Soppose that $f(z)$ satisfies:*

($\mathbf{F}_1$) $f \in C(\mathbf{R}, \mathbf{R})$;

($\mathbf{F}_2$) $\forall z \in \mathbf{R}$, $\int_0^z f(s)ds \geq 0$ and $f(z) = o(z)$ (as $z \to 0$);

($\mathbf{F}_3$) There exist constants $R_1 > 0, \beta > 2$ such that,

$$zf(z) \geq \beta \int_0^z f(s)ds > 0, \forall |z| \geq R_1.$$

Then for any given positive integer m (2.3) possesses at least two nontrivial periodic solutions with period m.

In order to apply critical point theory, we introduce some appropriate variational frameworks. First, we give some notations.

Let S be the vector space of sequence $x = \{x_n\}_{n\in \mathbf{Z}}$, i. e.

$$S = \{\{x_n\}|x_n \in \mathbf{R}, n \in \mathbf{Z}\}.$$

We can rewrite x as $x = (\cdots, x_{-n}, x_{-n+1}, \cdots, x_{-1}, x_0, x_1, x_2, \cdots, x_n, \cdots)$.

For any given positive integer p, m, E_{pm} is defined as follows:

$$E_{pm} = \{x = \{x_n\} \in S | x_{n+pm} = x_n, n \in \mathbf{Z}\}.$$

Then E_{pm} is a subspace of S and is isomorphic to $\mathbf{R}^{pm}$. E_{pm} can be equiped with the inner product

$$< x, y >_{E_{pm}} = \sum_{j=1}^{pm} x_j y_j, \forall x, y \in E_{pm}$$

by which the norm $\|\cdot\|_{E_{pm}}$ can be induced by

$$\|x\|_{E_{pm}} = \left(\sum_{j=1}^{pm} x_j^2\right)^{\frac{1}{2}}, \forall x \in E_{pm}. \tag{2.4}$$

It is easy to see that $(E_{pm}, < \cdot, \cdot >)$ is a finite dimensional Hilbert space and is linear homoemorphic to $\mathbf{R}^{pm}$.

Consider the functional I defined on E_{pm}

$$I(x) = \sum_{n=1}^{pm} [\frac{1}{2}(\Delta x_n)^2 - F(n, x_n)], \forall z \in E_{pm}, \tag{2.5}$$

where $F(t, z) = \int_0^z f(t, s)ds$.

In view of $x_{n+pm} = x_n, n \in \mathbf{Z}, \forall x \in E_{pm}$, (2.5) can be rewritten as

$$I(x) = \sum_{n=1}^{pm} [(x_n^2 - x_n x_{n+1}) - F(n, x_n)]. \tag{2.6}$$

By the continuity of $f(t, z)$, $I \in C^1(E_{pm}, \mathbf{R})$ and $I'(x) = 0$ if and only if $\frac{\partial I(x)}{\partial x_n} = 0, n \in \mathbf{Z}(1, pm)$.

Set $x_0 = x_{pm}$, then

$$\frac{\partial I(x)}{\partial x_n} = 2x_n - x_{n+1} - x_{n-1} - f(n, x_n), n \in \mathbf{Z}(1, pm).$$

or

$$\frac{\partial I(x)}{\partial x_n} = -(\Delta^2 x_{n-1} + f(n, x_n)), n \in \mathbf{Z}(1, pm).$$

Consequently, $x \in E_{pm}$ is a critical point of I, i. e. $I'(x) = 0$ if and only if

$$\Delta^2 x_{n-1} + f(n, x_n) = 0, n \in \mathbf{Z}(1, pm).$$

Since $\{x_n\}$ is pm-periodic in n, and $f(t, z)$ is m periodic in t, $x \in E_{pm}$ is a critical point of I if and only if $\forall n \in \mathbf{Z}, \Delta^2 x_{n-1} + f(n, x_n) = 0$ and therefore $x = \{x_n\}$ is a pm-periodic solution of (2.1). Thus the problem of finding the pm-periodic solution for (2.1) is reduced to the one of seeking the critical point of functional I(see (2.6)) on E_{pm}.

By Linking Theorem (see [27]) and elaborate analytical techniques, we can obtained the critical point of functional I, for detail, one can refer to [14].

Remark 2.4 As an application of Theorem 2.1 and Theorem 2.3, we give a example to illustrate our results.

Suppose

$$f(t,z) = (az|z|^{\mu} + bz|z|^{\nu})(\varphi(t) + M),$$

where $a \geq 0, b \geq 0$ and a, b do not equal to zero simultaneously, $\mu > 0, \nu > 0, M > 0$, $\varphi(t)$ is a continuously m-periodic function and $|\varphi(t)| < M$.

Under these conditions, $f(t,z)$ satisfies all the assumptions of Theorem 2.1; if $\varphi(t) \equiv 0$, then $f(z)$ satisfies the assumptions of Theorem 2.3. Consequently, The conclusion of Theorem 2.1 (or Theorem 2.3) holds.

2.2 The second-order subquadratic nonlinear difference equations. Consider the nonlinear second order difference equation

$$\Delta^2 x_{n-1} + f(n, x_n) = 0, \quad n \in \mathbf{Z}, \tag{2.7}$$

where $f \in C(\mathbf{R} \times \mathbf{R}^k, \mathbf{R}^k)$, $f(t+m, z) = f(t,z)$ for any $(t,z) \in \mathbf{R} \times \mathbf{R}^k$, m is a positive integer, $\Delta x_n = x_{n+1} - x_n, \Delta^2 x_n = \Delta(\Delta x_n)$.

Let p be a given positive integer. We still study the existence of pm-periodic solutions for (2.7).

Since $\Delta^2 x_n = x_{n+1} - 2x_n + x_{n-1}$, (2.7) can be rewritten as

$$x_{n+1} - 2x_n + x_{n-1} + f(n, x_n) = 0, \quad n \in \mathbf{Z}. \tag{2.8}$$

Throughout this subsection, we suppose that there exists a continuously differentiable function $F(t,z) \in C^1(\mathbf{R} \times \mathbf{R}^k)$ such that $\nabla_z F(t,z) = f(t,z)$ for any $(t,z) \in \mathbf{R} \times \mathbf{R}^k$, where $\nabla_z F(t,z)$ denotes the gradient of $F(t,z)$ in z.

Theorem 2.5 *Suppose that $F(t,z)$ satisfies*

($\mathbf{F}_1$) There exists positive integer m such that $F(t+m,z) = F(t,z), \forall (t,z) \in \mathbf{R} \times \mathbf{R}^k$;

($\mathbf{F}_2$) There is a constant $M_0 > 0$ such that $\forall (t,z) \in \mathbf{R} \times \mathbf{R}^k$, $|f(t,z)| \leq M_0$.

($\mathbf{F}_3$) $F(t,z) \to +\infty$ uniformly for $t \in \mathbf{R}$ as $|z| \to \infty$.

Then for any given positive integer $p > 0$, equation (2.7) possesses at least one pm-periodic solution.

Theorem 2.6 *Suppose $F(t,z)$ satisfies ($\mathbf{F}_1$) and*

($\mathbf{F}_4$) There exist constants $R_1 > 0$ and $\alpha, 1 < \alpha < 2$ such that for any $(t,z) \in \mathbf{R} \times \mathbf{R}^k, |z| \geq R_1$,

$$0 < z \cdot f(t,z) \leq \alpha F(t,z) \tag{2.9}$$

($\mathbf{F}_5$) There exist constants $a_1 > 0, a_2 \geq 0$ and $\gamma, 1 < \gamma \leq \alpha$ such that

$$F(t,z) \geq a_1|z|^{\gamma} - a_2, \forall (t,z) \in \mathbf{R} \times \mathbf{R}^k. \tag{2.10}$$

Then for any positive integer $p > 0$, equation (2.7) possesses at least one pm-periodic solution.

Remark 2.7 By integrating inequality (2.9), we have

($\mathbf{F}_4'$) There are constants $a_3 > 0, a_4 > 0$ such that

$$F(t,z) \leq a_3|z|^{\alpha} + a_4. \tag{2.11}$$

By ($\mathbf{F}_4$) and ($\mathbf{F}_4'$), one gets $\lim\limits_{|z| \to +\infty} \frac{F(t,z)}{z^2} = 0$, which means that $F(t,z)$ is subquadratic at infinity. Corresponding to second-order Hamiltonian systems (2.2), $F(t,z)$ will be also called the potential of equation (2.7). In this case, we call equation (2.7) is subquadratic at infinity.

Remark 2.8 The results of Theorem 2.5 and Theorem 2.6 ensure that the equation (2.7) possesses at least one pm-periodic solution. However, in some cases, we are interested in the existence of nontrivial periodic solution. For example, when $f(t,z)$ is independent of t, (2.7) is the following autonomous difference system

$$\Delta^2 x_{n-1} = f(x_n). \tag{2.12}$$

If $F \in C^1(\mathbf{R}^k, \mathbf{R}), \nabla F(z) = f(z)$ and the corresponding assumption $(\mathbf{F}_3)$ of Theorem 2.5 or $(\mathbf{F}_5)$ of Theorem 2.6 is satisfied, then it is clear that $F(z)$ attains its minimum at some point $z_0 \in \mathbf{R}^k$. This implies $f(z_0) = 0$ and $x_n = z_0, \forall n \in \mathbf{Z}(1)$ is a trivial perioidc solution of (2.12). For this reason, we need to give some results of the existence of nontrivial priodic solutions for (2.7) and (2.12).

For simplicity, we only give results for equation (2.7). By similar arguments, the associated results for (2.12) can be obtained. Moreover, in the case of autonomous difference systems, for any given positive integer m, we obtain m-periodic solutions for (2.12).

Theorem 2.9 *Suppose $F(t,z)$ satisfies $(\mathbf{F}_1)$ and*

$(\mathbf{F}_6)$ $F(t,0) = 0, \forall t \in \mathbf{R}$;

$(\mathbf{F}_7)$ There exists a constant $\alpha, 1 < \alpha < 2$ such that for any $(t,z) \in \mathbf{R} \times \mathbf{R}^k$, $|z| \neq 0$,

$$0 < z \cdot f(t,z) \le \alpha F(t,z) \tag{2.13}$$

$(\mathbf{F}_8)$ There exist constants $a_5 > 0$ and $\gamma, 1 < \gamma \le \alpha$ such that

$$F(t,z) \ge a_5 |z|^\gamma, \forall (t,z) \in \mathbf{R} \times \mathbf{R}^k. \tag{2.14}$$

Then for any positive integer $p > 0$, equation (2.7) possesses at least one nontrivial pm-periodic solution.

Theorem 2.10 *Suppose that $F(t,z)$ satisfies $(\mathbf{F}_1)$-$(\mathbf{F}_3)$ and*

$(\mathbf{F}_9)$ For any $t \in \mathbf{R}, F(t,0) = 0$ and $f(t,z) = 0$ if and only if $z = 0$.

$(\mathbf{F}_{10})$ There exists constants $a_6 > 0$ and $\theta, 0 < \theta < 2$ such that $F(t,z) \ge a_6 |z|^\theta, \forall (t,z) \in \mathbf{R} \times \mathbf{R}^k$.

Then for any given positive integer $p > 0$, equation (2.7) possesses at least one nontrivial pm-periodic solution.

Now, let's give an example to illustrate the results of Theorem 2.9.

Example 2.11 Let us consider the following difference equations

$$\Delta^2 x_{n-1} = [a(\beta_1 + 2) x_n |x_n|^{\beta_1} + b(\beta_2 + 2) x_n |x_n|^{\beta_2}](\varphi(n) + G), \tag{2.15}$$

where $a > 0, b \ge 0, -1 < \beta_1 \le \beta_2 < 0, G > 0$, $\varphi(t)$ is a bounded continuous m-periodic function and $|\varphi(t)| < G$.

In this case,

$$f(t,z) = [a(\beta_1 + 2) z|z|^{\beta_1} + b(\beta_2 + 2) z|z|^{\beta_2}](\varphi(t) + G),$$

and

$$F(t,z) = (a|z|^{\beta_1 + 2} + b|z|^{\beta_2 + 2})(\varphi(t) + G).$$

Let $\alpha = \beta_2 + 2, \gamma = \beta_1 + 2$. It is easy to check that all the assumptions of Theorem 2.9 have been satisfied. Hence, for any given positive integer p, the equation (2.15) possesses at least one nontrivial pm-periodic solution.

To apply the critical point theory to prove above theorems, we still need to establish the variational structrue.

Let S be the set of sequence $x = (\cdots, x_{-n}, \cdots, x_{-1}, x_0, x_1, x_2, \cdots, x_n, \cdots) = \{x_n\}$, where $x_n = (x_{n1}, x_{n2}, \cdots, x_{nk})^T$.

For any $x, y \in S, a, b \in \mathbf{R}$, $ax + by$ is defined by

$$ax + by \triangleq \{ax_n + by_n\}_{n=-\infty}^{+\infty}.$$

Then S is a vector space.

For any given positive integer p, m, E_{pm} is defined as a subspace of S by

$$E_{pm} = \{x = \{x_n\} \in S | x_{n+pm} = x_n, n \in \mathbf{Z}(1)\}.$$

E_{pm} can be equiped with norm $\|\cdot\|_{E_{pm}}$ and inner product $< \cdot, \cdot >_{E_{pm}}$ as follows

$$\|x\|_{E_{pm}} \triangleq \left(\sum_{n=1}^{pm} |x_n|^2\right)^{\frac{1}{2}}, \forall x = \{x_n\} \in E_{pm}, \tag{2.16}$$

and

$$< x, y >_{E_{pm}} \triangleq \sum_{n=1}^{pm} x_n \cdot y_n, \forall x = \{x_n\} \in E_{pm}, y = \{y_n\} \in E_{pm}, \tag{2.17}$$

where $|\cdot|$ denotes the usual norm in $\mathbf{R}^k$, and $\cdot$ denotes the usual scalar product in $\mathbf{R}^k$.

Define map

$$\begin{array}{rccc} \varphi : & E_{pm} & \rightarrow & \mathbf{R}^{pmk} \\ & x = \{x_n\} & \longmapsto & (x_1^T, x_2^T, \cdots, x_{pm}^T)^T, \end{array} \tag{2.18}$$

where x_l^T denote the transpose of vector x_l.

It is easy to see that the map φ defined in (2.18) is a linear homeomorphism and $(E_{pm}, < \cdot, \cdot >_{E_{pm}})$ is a Hilbert space, which can be identified with $\mathbf{R}^{pmk}$.

Consider the functional J defined in E_{pm}

$$J(x) = \sum_{n=1}^{pm} [-\frac{1}{2}|\Delta x_n|^2 + F(n, x_n)]. \tag{2.19}$$

In view of $x_{n+pm} = x_n, \forall n \in \mathbf{Z}(1), \forall x \in E_{pm}$, (2.19) can be rewriten as

$$J(x) = \sum_{n=1}^{pm} [(x_n x_{n+1} - |x_n|^2) + F(n, x_n)]. \tag{2.20}$$

Since E_{pm} is linear homeomorphic to $\mathbf{R}^{pmk}$, by the continuity of $f(t, z)$, $J(x)$ can be viewed as a continuously differentiable functional defined on a finite dimensional Hilbert space. That is $J \in C^1(E_{pm}, \mathbf{R})$. Furthermore, $J'(x) = 0$ if and only if $\frac{\partial J(x)}{\partial x_{n,l}} = 0$, $n \in \mathbf{Z}(1, pm)$, $l \in \mathbf{Z}(1, k)$. Let's denote $x_0 = x_{pm}$, then

$$\frac{\partial J(x)}{\partial x_{n,l}} = x_{n+1,l} + x_{n-1,l} - 2x_{n,l} - f_l(n, x_n), l \in \mathbf{Z}(1, k), n \in \mathbf{Z}(1, pm).$$

or

$$\frac{\partial J(x)}{\partial x_{n,l}} = (\Delta^2 x_{n-1,l} + f_l(n, x_n)), l \in \mathbf{Z}(1, k), n \in \mathbf{Z}(1, pm).$$

Therefore, $x \in E_{pm}$ is a critical point of J, i.e. $J'(x) = 0$ if and only if $\forall n \in \mathbf{Z}(1, pm), l \in \mathbf{Z}(1, k)$,

$$\Delta^2 x_{n-1,l} + f_l(n, x_n) = 0.$$

That is,

$$\Delta^2 x_{n-1} + f(n, x_n) = 0, \forall n \in \mathbf{Z}(1, pm).$$

On the other hand, $\{x_n\} \in E_{pm}$ is pm-periodic in n, and $f(t, z)$ is m-periodic in t. So, $x \in E_{pm}$ is a critical point of J if and only if $\Delta^2 x_{n-1} + f(t, x_n) = 0, \forall n \in \mathbf{Z}$, and x is a pm-periodic solution of (2.7). Thus we reduce the problem of finding pm-periodic solution of (2.7) to the one of seeking the critical point of functional J in E_{pm}.

As for how to find the critical point of J by Saddle Point Theorem (see [27]), please see [15].

2.3 The existence of periodic and subharmonic solutions for superquadratic discrete Hamiltonian systems. Consider the discrete Hamiltonian systems

$$\begin{cases} \Delta x_1(n) = -H_{x_2}(n, x_1(n+1), x_2(n)) \\ \Delta x_2(n) = H_{x_1}(n, x_1(n+1), x_2(n)) \end{cases} \tag{2.21}$$

where $x_1, x_2 \in \mathbf{R}^k$, $H \in C^1(\mathbf{R} \times \mathbf{R}^k \times \mathbf{R}^k, \mathbf{R})$. $\Delta x_i(n) = x_i(n+1) - x_i(n)$, $i = 1, 2$.

Suppose that $H(t, x_1, x_2)$ is m-periodic in the first variable t, m is a positive integer. For any positive integer p, We want to obtain the existence and multiplicity results of pm-periodic solutions of (2.21).

For the existence of multiple periodic solutions and subharmonic solutions of Hamiltonian system

$$\begin{cases} \frac{dx}{dt} = -H_y(t, x(t), y(t)) \\ \frac{dy}{dt} = H_x(t, x(t), y(t)) \end{cases} \tag{2.22}$$

there have been many studies in the literatrue using various approaches. Especially, by critical point theory, many authors such as P. H. Rabinowitz, J. Mahwin, K. C. Chang and Y. M. Long obtained some significant results for (2.22). For details, one can refer to [5, 6, 22, 23, 27] and references therein.

Discrete Hamiltonian system (2.21), as the discretization of Hamiltonian system (2.22), also has been investigated by many scholars. Many researches have been made on the disconjugacy, boundary value problems, oscillations and asymptotic behavior of (2.21)(see for example [2, 4, 7, 11, 18]). However, there are few researches done on the existence of periodic solutions for discrete Hamiltonian systems. From this point, let's introduce some results obtained in our another paper [16]. We expect that the approaches applied in this paper can be used to study the similar problems.

Let us denote by $z = (x_1^T, x_2^T)^T, H(t, z) = H(t, x_1, x_2)$. We assume that

$(\mathbf{H}_1)$ $H(t, z) \in C(\mathbf{R} \times \mathbf{R}^{2k}, \mathbf{R})$ and there exists a positive integer m such that $\forall (t, z) \in \mathbf{R} \times \mathbf{R}^{2k}$, $H(t + m, z) = H(t, z)$;

$(\mathbf{H}_2)$ For any $(t, z) \in \mathbf{R} \times \mathbf{R}^{2k}$, $H(t, z) \geq 0$ and $H(t, z) = o(|z|^2)$ (as $z \to 0$);

$(\mathbf{H}_3)$ There exist some constants $R > 0, \beta > 2$ such that for any $|z| \geq R$,

$$(z, H_z'(t, z)) \geq \beta H(t, z) > 0.$$

$(\mathbf{H}_4)$ $H(t, z)$ is even for the second variable z, namely, $H(t, -z) = H(t, z)$, $\forall (t, z) \in \mathbf{R} \times \mathbf{R}^{2k}$.

Remark 2.12 By $(\mathbf{H}_1)$ and $(\mathbf{H}_3)$, it is easy to see that
$(\mathbf{H}_3)'$ There exist constants $a_1 > 0, a_2 > 0$ such that

$$H(t, z) \geq a_1|z|^\beta - a_2, \forall (t, z) \in \mathbf{R} \times \mathbf{R}^{2k}.$$

So the assumptions $(\mathbf{H}_1)$-$(\mathbf{H}_3)$ imply that $H(t, z)$ grows superquadratically both at zero and at infinity.

Theorem 2.13 *Under the assumptions $(\mathbf{H}_1)$-$(\mathbf{H}_4)$, for any given positive integer p, there exist at least $(pm - 1)k$ geometrically distinct nontrivial periodic solutions of (2.21) with period pm.*

Without assumption $(\mathbf{H}_4)$, we have

Theorem 2.14 *Under the assumptions $(\mathbf{H}_1)$-$(\mathbf{H}_3)$, for any given positive integer p, there exist at least 2 nontrivial periodic solutions of (2.21) with period pm.*

Let S be the set of sequence $x = \{x(n)\}_{n \in \mathbf{Z}}$, i. e.

$$S = \{\{x(n)\} | x(n) = \begin{pmatrix} x_1(n) \\ x_2(n) \end{pmatrix} \in \mathbf{R}^{2k}, x_j(n) \in \mathbf{R}^k, j = 1, 2, n \in \mathbf{Z}\}.$$

We can rewrite x as

$$x = (\cdots, x^T(-n), x^T(-n+1), \cdots, x^T(-1), x^T(0), x^T(1), x^T(2), \cdots, x^T(n), \cdots)^T.$$

where $\cdot^T$ is the transpose of a vector or a matrix.

For any $x, y \in S$, $a, b \in \mathbf{R}$, $ax + by$ is defined by

$$\begin{aligned} ax + by \quad & \triangleq \{ax(n) + by(n)\} \\ & = (\cdots, ax^T(-n) + by^T(-n), \cdots, ax^T(-1) + by^T(-1), \\ & ax^T(0) + by^T(0), ax^T(1) + by^T(1), \cdots, ax^T(n) + by^T(n), \cdots)^T. \end{aligned}$$

Then S is a vector space.

For any given positive integer p, m, E_{pm} is defined as a subspace of S by

$$E_{pm} = \{x = \{x(n)\} \in S | x(n + pm) = x(n), n \in \mathbf{Z}\}.$$

E_{pm} can be equiped with the inner product $< \cdot, \cdot >_{E_{pm}}$ and norm $\|\cdot\|_{E_{pm}}$ as follows:

$$< x, y >_{E_{pm}} = \sum_{j=1}^{pm} (x(j), y(j)), \forall x, y \in E_{pm} \tag{2.23}$$

and

$$\|x\|_{E_{pm}} = \left(\sum_{j=1}^{pn} |x(j)|^2 \right)^{\frac{1}{2}}, \forall x \in E_{pm}. \tag{2.24}$$

where $(\cdot, \cdot)$ and $|\cdot|$ denote the usual inner product and the usual norm in $\mathbf{R}^{2k}$.

Define map

$$\begin{aligned} \varphi : \quad E_{pm} & \rightarrow \mathbf{R}^{2pmk} \\ x = \{x(n)\} & \rightarrow (x^T(1), x^T(2), \cdots, x^T(pm))^T \end{aligned} \tag{2.25}$$

It is easy to see that the map φ defined in (2.25) is a linear homeomorphism and $(E_{pm}, < \cdot, \cdot >)$ is a Hilbert space, which can be identified with $\mathbf{R}^{2pmk}$.

Consider functional F defined on E_{pm},

$$F(x) = -\frac{1}{2}\sum_{n=1}^{pm}(J\Delta Lx(n), x(n)) - \sum_{n=1}^{pm} H(n-1, x_1(n), x_2(n-1)), \forall x \in E_{pm}, \quad (2.26)$$

where $Lx(n) = \begin{pmatrix} x_1(n) \\ x_2(n-1) \end{pmatrix}$, $J = \begin{pmatrix} 0 & -I \\ I & 0 \end{pmatrix}$ is the standard $2k \times 2k$ sympletic matrix, I is the $k \times k$ identity matrix.

Let us denote by

$$F_1(x) = -\frac{1}{2}\sum_{n=1}^{pm}(J\Delta Lx(n), x(n)).$$

Then

$$\begin{aligned} F_1(x) &= -\frac{1}{2}\sum_{n=1}^{pm}\left(\begin{pmatrix} 0 & -I \\ I & 0 \end{pmatrix}\begin{pmatrix} x_1(n+1) - x_1(n) \\ x_2(n) - x_2(n-1) \end{pmatrix}, \begin{pmatrix} x_1(n) \\ x_2(n) \end{pmatrix}\right) \\ &= -\frac{1}{2}\sum_{n=1}^{pm}\left(\begin{pmatrix} x_2(n-1) - x_2(n) \\ x_1(n+1) - x_1(n) \end{pmatrix}, \begin{pmatrix} x_1(n) \\ x_2(n) \end{pmatrix}\right) \\ &= -\frac{1}{2}\sum_{n=1}^{pm}[(x_2(n-1), x_1(n)) - (x_2(n), x_1(n)) \\ &\quad + (x_1(n+1), x_2(n)) - (x_1(n), x_2(n))] \end{aligned}$$

The pm periodicity of $x = \{x(n)\}$ in n implies that

$$\sum_{n=1}^{pm}(x_2(n-1), x_1(n)) = \sum_{n=1}^{pm}(x_2(n), x_1(n+1)).$$

Thus we have

$$F_1(x) = \sum_{n=1}^{pm}[(x_1(n), x_2(n)) - (x_1(n+1), x_2(n))] = -\sum_{n=1}^{pm}(\Delta x_1(n), x_2(n)).$$

Or equivalently,

$$F_1(x) = \sum_{n=1}^{pm}[(x_1(n), x_2(n)) - (x_1(n), x_2(n-1))] = \sum_{n=1}^{pm}(x_1(n), \Delta x_2(n-1)).$$

Therefore,

$$F(x) = -\sum_{n=1}^{pm}(\Delta x_1(n), x_2(n)) - \sum_{n=1}^{pm} H(n-1, x_1(n), x_2(n-1)). \quad (2.27)$$

It is easy to verify that for any given $n \in Z(1, pm)$,

$$F'_{x_1(n)} = x_2(n) - x_2(n-1) - H'_{x_1}(n-1, x_1(n), x_2(n-1))$$

and

$$F'_{x_2(n)} = x_1(n) - x_1(n+1) - H'_{x_2}(n, x_1(n+1), x_2(n)).$$

Consequently, $F'_{x(n)} = 0$ if and only if

$$\begin{cases} \Delta x_2(n-1) &= H'_{x_1}(n-1, x_1(n), x_2(n-1)) \\ -\Delta x_1(n) &= H'_{x_2}(n, x_1(n+1), x_2(n)) \end{cases}$$

That is

$$\begin{cases} \Delta x_1(n) &= -H'_{x_2}(n, x_1(n+1), x_2(n)) \\ \Delta x_2(n) &= H'_{x_1}(n, x_1(n+1), x_2(n)) \end{cases}$$

Hence, $F'(x) = 0$ if and only if for any $n \in Z(1, pm)$,

$$\begin{cases} \Delta x_1(n) &= -H'_{x_2}(n, x_1(n+1), x_2(n)) \\ \Delta x_2(n) &= H'_{x_1}(n, x_1(n+1), x_2(n)) \end{cases}$$

By the periodicity of H and x, we see that x is a critical point of F if and only if $\{x(n)\}$ is a pm-periodic solution of (2.21), where $x(n+pm) = x(n), \forall n \in \mathbf{Z}$. From this point, the problem of finding pm-periodic solution of (2.21) has been reduced to the one of seeking the critical point of functional F defind on E_{pm}.

In order to prove Theorem 2.13 and Theorem 2.14, we need to exploit Z_2 geometrical index theory (i.e. genus theory) and Linking Theorem. To save the page, we omit the proofs here. For detail, one can refer to ([16])

3 Some problems seem to be sloved by critical point theory

In Section 2, we have studied second-order difference equations and discrete Hamiltonian systems by critical poiont theory. Of course, many kind of difference equations can be theoritically studied by critical point theory. The key problem lies in how to construct appropriate variational frameworks. In fact, the main principle of critical point theory is by studying the changes of the topological structure of two level sets of corresponding functional to investigate the existence and multiplicity of critical points. In view of this, in order to apply critical point theory, some new variational structures for certain kind of difference equations need to be established.

3.1 Delay difference equation with discrete variables. Consider the following difference equations:

$$x_{n+1} = f(n, x_n, x_{n-1}, \cdots, x_{n-l+1}), n \in Z \tag{3.1}$$

where $f \in C(R \times R^{lk}, R^k)$, $f(t+m, x, \cdots, x) = f(t, x, \cdots, x)$ for any $(t, x) \in R \times R^k$.

It is easy to see that equation (3.1) covers many kind of difference equations.

Define $y_n = (y_{n,1}, y_{n,2}, \cdots, y_{n,l})^T \in \mathbf{R}^{kl}$, where $y_{n,1} = x_n, y_{n,2} = x_{n-1}, \cdots, y_{n,l} = x_{n-l+1}$. Then we have

$$\begin{cases} y_{n+1,1} &= f(n, y_{n,1}, y_{n,2}, \cdots, y_{n,l}) \\ y_{n+1,2} &= y_{n,1} \\ y_{n+1,3} &= y_{n,2} \\ \cdots & \cdots\cdots \\ y_{n+1,l-1} &= y_{n,l-2} \\ y_{n+1,l} &= y_{n,l-1} \end{cases} \tag{3.2}$$

Let $g_1(n, y_n) = f(n, y_{n,1}, y_{n,2}, \cdots, y_{n,l})$, $g_2(n, y_n) = y_{n,1}$, $\cdots$, $g_l(n, y_n) = y_{n,l-1}$, and $g(n, y_n) = (g_1(n, y_n), g_2(n, y_n), \cdots, g_l(n, y_n))^T$. Then (3.2) becomes

$$y_{n+1} = g(n, y_n), \forall n \in \mathbf{Z}. \tag{3.3}$$

where $y_n = (y_{n,1}, y_{n,2}, \cdots, y_{n,l})^T \in \mathbf{R}^{kl}, g(n, y_n) \in C(\mathbf{R} \times \mathbf{R}^{kl}, \mathbf{R}^{kl})$.

By above argument, we only need to study the equation

$$x_{n+1} = f(n, x_n), \forall n \in \mathbf{Z}. \tag{3.4}$$

where $x_n \in \mathbf{R}^k, f(n, x_n) \in C(\mathbf{R} \times \mathbf{R}^k, \mathbf{R}^k)$.

For equation (3.4), if there exists a positive integre m such that $f(t+m,z) = f(t,z), \forall (t,z) \in \mathbf{R}\times\mathbf{R}^k$, then we are interested in the existence and multiplicity of m-periodic solutions and subharmonic solutions (For given positive integer p, the pm-periodic solution of (3.4), as usual, is also called a subharmonic solution of (3.4).)

We suppose that a sequnce $x = \{x_n\}_{n=-\infty}^{+\infty}$ is a m-priodic solution of (3.4), that is $x_{n+m} = x_n, \forall n \in \mathbf{Z}$. Let $y_1 = x_1, y_2 = x_2, \cdots, y_m = x_m, y_{m+1} = x_{m+1} = x_1$ and $y = (y_1, y_2, \cdots, y_m)^T$. Then

$$Py = F(y) \tag{3.5}$$

Or

$$y = P^{-1}F(y) \tag{3.6}$$

where

$$P = \begin{pmatrix} 0 & I & 0 & 0 & \cdots & 0 & 0 \\ 0 & 0 & I & 0 & \cdots & 0 & 0 \\ 0 & 0 & 0 & I & \cdots & 0 & 0 \\ \cdots & \cdots & \cdots & \cdots & \cdots & \cdots & \cdots \\ 0 & 0 & 0 & 0 & \cdots & 0 & I \\ I & 0 & 0 & 0 & \cdots & 0 & 0 \end{pmatrix}_{km\times km}$$

$F(y) = (F_1(y), F_2(y), \cdots, F_m(y))^T \in \mathbf{R}^{km}\times\mathbf{R}^{km}$, $F_1(y) = f(1,y_1)$, $F_2(y) = f(2,y_2), \cdots, F_m(y) = f(m,y_m)$.

Denote by E_m the set of sequence $\{y_n\}$ with period m, i. e.

$$E_m = \{y = \{y_n\}_{n=-\infty}^{+\infty} | y_{n+m} = y_n, \forall n \in \mathbf{Z}\}.$$

Then (3.6) can be seen as a operator equation defined on E_m. So one can obtain some existence results by fixed point theory or topological degree theory. However, if one can rewrite (3.6) as

$$Ly + h(y) = 0, \tag{3.7}$$

where L is a linear self-adjoint operator, and there exists a continuously differentiable function $H(y)$ such that $H'(y) = h(y)$, then critical point theory can be applied here. The rest study is almost similar to that we have done.

Unfortunately, in general, such a kind of assumptions are often can not be imposed on equation (3.6). Otherwise some contradictions will be lead to. Therefore, how to impose some reasonable assumptions on (3.6) so as to apply critical point theory is a very interesting and hard problem.

3.2 Delay difference equation with continuous variables. Let's consider another example.

$$x(t+1) = f(t, x(t), x(t-1), x(t-2), \cdots, x(t-l+1)) \tag{3.8}$$

where $f \in C(R\times R^{lk}, R^k)$, $f(t+m, x, \cdots, x) = f(t, x, \cdots, x)$ for any $(t,x) \in R\times R^k$.

Similar to the above discussion, we define $y(t) = (y_1(t), y_2(t), \cdots, y_l(t))^T \in \mathbf{R}^{kl}$, where $y_1(t) = x(t), y_2(t) = x(t-1), \cdots, y_l(t) = x(t-l+1)$. Then we have

$$\begin{cases} y_1(t+1) &= f(t, y_1(t), y_2(t), \cdots, y_l(t)) \\ y_2(t+1) &= y_1(t) \\ y_3(t+1) &= y_2(t) \\ \cdots & \cdots\cdots \\ y_{l-1}(t+1) &= y_{l-2}(t) \\ y_l(t+1) &= y_{l-1}(t) \end{cases} \tag{3.9}$$

Let $g_1(t, y(t)) = f(t, y_1(t), y_2(t), \cdots, y_l(t))$, $g_2(t, y(t)) = y_1(t)$, $\cdots$, $g_l(t, y(t)) = y_{l-1}(t)$, and $g(t, y(t)) = (g_1(t, y(t)), g_2(t, y(t)), \cdots, g_l(t, y(t)))^T$. Then (3.9) becomes

$$y(t+1) = g(t, y(t)), \forall t \in \mathbf{R}. \tag{3.10}$$

where $y(t) = (y_1(t), y_2(t), \cdots, y_l(t))^T \in \mathbf{R}^{kl}, g(t, y(t)) \in C(\mathbf{R} \times \mathbf{R}^{kl}, \mathbf{R}^{kl})$.

By similar argument, we only need to study the equation

$$x(t+1) = f(t, x(t)), \forall t \in \mathbf{R}. \tag{3.11}$$

where $x(t) \in \mathbf{R}^k, f(t, x(t)) \in C(\mathbf{R} \times \mathbf{R}^k, \mathbf{R}^k)$.

Now we want to study the existence and multiplicity of m-periodic solutions and subharmonic solutions for (3.11).

Denote by C_m the space of continuous periodic function defined on the whole line $\mathbf{R}$ with period m. P denote the linear operator from C_m to C_m, which is defined by

$$Px(t) = x(t+1), \forall x(t) \in C_m.$$

Clearly, P^{-1} exists and therefore (3.11) can be seen as the following operator equations defined on C_m:

$$x(t) = P^{-1} f(t, x(t)). \tag{3.12}$$

Now the problem is whether or not (3.12) can be reformulated as

$$Lx + h(t, x) = 0, \tag{3.13}$$

where L is a linear self-adjoint operator on C_m, and there exists a continuously differentiable function $H(t, x)$ such that $H'_x(t, x) = h(t, x)$. If it is true, the variational framework for (3.12) can be easily established.

3.3 Other problems. Besides the above two classes difference equations (3.1 and (3.8), there are still many kinds of difference equations needed to be studied. For example,

$$x(t+1) = f(t, x([t]), x([t-1]), \cdots, x([t-l+1])).$$

In addition, critical point theory can be applied to investigate the homoclinic and heteroclitic orbits for second-order differential equation and Hamiltonian systems (e. g. see [3, 28]). As for difference equations with variational structure, we think that the critical point theory is also valid. However, to our knowledge, the existence and multiplicity results of homoclinic and heteroclitic orbits for difference equations have not been seen in the literature obtained by critical point theory, or even by other approaches.

Consequently, we think that there have been many problems in this field needed to be solved and critical point theory will be play an important role in the development of qualitative theory of difference equations.

References

[1] R. P. Agarwal, Difference Equations and Inequalities: Theory, Methods and Applications, Marcel Dekker, New York, 1992.

[2] C. D. Ahlbrandt, Equivalence of discrete Euler equations and discrete Hamiltonian systems, *J. Math. Anal. Appl.*, 180(1993), 498-517.

[3] Gianni Arioli and Andrzej Szulkin, Homoclinic solutions of Hamiltonian systems with symmetry, *J. Differential Equations*, 158, 1999, 291-313.

[4] M. Bohner, Linear Hamiltonian difference systems: disconjugacy and Jacobi-type conditions, *J. Math. Anal. Appl.*, 199(1996), 804-826.

[5] K. C. Chang, Critical Point Theory and its Applications, Science and Technical Press, Shanghai, China, 1980.
[6] K. C. Chang, Infinite Dimensional Morse Theory and Multiple Solution Problems, Birkhäuser, Boston, 1993.
[7] S. Chen, Disconjugacy, disfocality, and oscillation of secind order difference equations, *J. Diff. Eqns.*, 107(1994), 383-394.
[8] A. Capietto, J. Mawhin and F. Zanolin, A continuation approach to superlinear periodic boundary value problems, *J. Differential Equations*, 88(1990), 347-395.
[9] S. N. Elaydi, An Introduction to Difference Equations, Springer-Verlag, New York, 1999.
[10] S. N. Elaydi and S. Zhang, Stability and periodicity of difference equations with finite delay, *Funkcialaj Ekvac.*, 37(1994), 401-413.
[11] L. H. Erbe and P. Yan, Disconjugacy for linear Hamiltonian difference systems, *J. Math. Anal. Appl.*, 167(1992), 355-367.
[12] L. H. Erbe, H. Xia and J. S. Yu, Global stability of a linear nonautonomous delay difference equations, *J. Diff. Equations Appl.*, 1(1995), 151-161.
[13] I. Gröri and G. Ladas, Oscillation Theory of Delay Differential Equations with Applications, Oxford University Press, Oxford, 1991.
[14] Zhiming Guo and Jianshe Yu, The Existence of Periodic and Subharmonic Solutions for Second-order Superlinear Difference Equations, *Science in China*, Series A, 46:4(2003), 506-515.
[15] Zhiming Guo and Jianshe Yu, The Existence of Periodic and Subharmonic Solutions to Subquadratic Second-order Difference Equations, *J. London Math. Soc.*, 68(2003), 419-430.
[16] Zhiming Guo and Jianshe Yu, The Existence of Periodic and Subharmonic Solutions for Superquadratic Discrete Hamiltonian Systems, *Nonlinear Analysis T. M. A.*, 55(2003), 969-983.
[17] J. K. Hale and J. Mawhin, Coincidence degree and periodic solutions of neutral equations, *J. Differential Equations*, 15(1974), 295-307.
[18] P. Hartman, Difference equations: disconjugacy, principal solutions, Green's functions, complete monotonicity, *Trans. Amer. Math. Soc.*, 246(1978), 1-30.
[19] J. L. Kaplan and J. A. Yorke, Ordinary differential equations which yield periodic solution of delay equations, *J. Math. Anal. Appl.*, 48(1974), 317-324.
[20] V. L. Kocic and G. Ladas, Global Behavior of Nonlinear Difference Equations of Higher Order with Applications, Kluwer Academic Publishers, Boston, 1993.
[21] J. Q. Liu and Z. Q. Wang, Remarks on subharmonics with minimal periods of Hamilonian systems, *Nonlinear. Anal. T. M. A.*, 1993, 7, 803-821.
[22] Y. M. Long, The Index Theory of Hamiltonian System and its Applications, Science Press, Beijing, 1993.
[23] J. Mawhin and M. Willem, Critical Point Theory and Hamiltonian Systems, Spinger-Verlag, New York Inc., 1989.
[24] H. Matsunaga, T. Hara and S. Sakata, Global attractivity for a nonlinear difference equation with variable delay, *Computer Math. Applic.*, 41(2001), 543-551.
[25] R. Michalek and G. Tarantello, Subharmonic solutions with prescribed minimal period for nonautonomous Hamiltonian systems, *J. Differential Equations*, 72(1988), 28-55.
[26] E. C. Pielou, An Introduction to Mathematical Ecology, Willey Interscience, New York, 1969.
[27] P. H. Rabinowitz, Minimax Methods in Critical Point Theory with Applications to Differential Equations, CBMS. AMS 65 (1986).
[28] P. H. Rabinowitz, Homoclinic orbits for a class of Hamiltonian systems, *Proc. of Royal Soc. of Edinburgh*, 114A, 1990, 33-38
[29] X. H. Tang and J. S. Yu, Ocillation of nonlinear delay difference equations, *J. Math. Anal. Appl.*, 249(2000), 476-490.
[30] J. S. Yu, Asymptotic stability for a linear difference equation with variable delay, *Computers Math. Applic.*, 36(10-12)(1998), 203-210.
[31] Z. Zhou and Q. Zhang, Uniform stability of nonlinear difference systems, *J. Math. Anal. Appl.*, 225(1998), 486-500.

Fields Institute Communications
Volume **42**, 2004

Lie Symmetries for Lattice Equations

D. Levi
Dipartimento di Fisica "E. Amaldi"
Universitá degli Studi Roma Tre and Sezione INFN, Roma Tre
Via della Vasca Navale 84, 00146 Roma, Italy
levi@fis.uniroma3.it

Abstract. Lie symmetries has been introduced by Sophus Lie to study differential equations. It has been one of the most efficient way for obtaining exact analytic solution of differential equations. Here we show how one can extend this technique to the case of differential difference and difference equations.

1 Introduction

Sophus Lie introduced the notion of Lie groups as a unifying tool to study differential equations [49, 50]. Lie groups have been used to solve differential equations, to classify them, and to establish properties of their solutions [53, 28, 54, 7, 22, 60].

In particular, they provided one of the most efficient methods for obtaining exact analytic solutions for partial differential equations, i.e. symmetry reduction. This method consists of a sequence of algorithmic steps the first of which is finding the Lie group $\mathcal{G}$ of local point transformations:

$$\begin{aligned} \tilde{\mathbf{x}} &= F_{\varsigma}(\mathbf{x}, u) = \mathbf{x} + g\xi(\mathbf{x}, u) + \ldots \\ \tilde{u} &= H_{\varsigma}(\mathbf{x}, u) = u + g\phi(\mathbf{x}, u) + \ldots \end{aligned} \tag{1.1}$$

where g is the set of group parameters. Given a partial differential equation of order k $E_k(\mathbf{x}, u, u_{x_i}, u_{x_{i,j}}, \ldots\ u_{x_{i_1,\ldots,i_k}}) = 0$, $\mathcal{G}$ is obtained requiring that the transformation (1.1) leaves the set of solutions invariant, i.e. defining the infinitesimal generator of the Lie point symmetry

$$\hat{X} = \xi_i \partial_{x_i} + \phi \partial_u, \tag{1.2}$$

we have

$$pr\hat{X} E_k|_{E_k=0} = 0. \tag{1.3}$$

where by $pr\hat{X}$ we mean the prolongation of $\hat{X}$ to all derivatives contained in $E_k = 0$. Then we look for solutions which are invariant under a subgroup $\mathcal{G}_0$ of $\mathcal{G}$. The finite group transformations, obtained by integrating the infinitesimal generators,

2000 *Mathematics Subject Classification*. Primary 70G65, 34K, 39K; Secondary 05A40.

transform solutions of E_k into solutions of the same equation. Other classes of exact solutions can be obtained by considering *conditional symmetries* [42].

We can consider the *evolutionary formalism* [53], an equivalent representation of the infinitesimal generator of the symmetry (1.2),

$$\hat{X}_e = Q(\mathbf{x}, \mathbf{u}, \mathbf{u_{x_1}}, \ldots, \mathbf{u_{x_k}})\partial_{\mathbf{u}}, \tag{1.4}$$
$$Q = \phi - \xi_i u_{x_i}, \quad pr\hat{X}_e E_k|_{E_k=0} = 0.$$

Let us notice that the existence of the evolutionary symmetry (1.4) implies

$$u_g = Q(\mathbf{x}, \mathbf{u}, \mathbf{u_{x_1}}, \ldots, \mathbf{u_{x_k}}), \tag{1.5}$$

i.e. that symmetries are equivalent to commuting flows. As Q is linear in the first derivatives of u eq. (1.5) is integrable on the characteristics being a quasilinear partial differential equation of first order and its solution gives the corresponding group transformations (1.1).

The notion of point symmetries can be extended [53] to the case of *generalized symmetries* by requiring that $Q = Q(\mathbf{x}, \mathbf{u}, \mathbf{u_{x_{i_1}}}, \ldots, \mathbf{u_{x_{i_1},\ldots,x_{i_s}}})$. In this case eq. (1.5) is a partial differential equation of order s and thus it cannot be solved explicitly to get the group transformation. However, in the case of variational symmetries, by Noether theorem [53], they still can provide conservation laws and are usually associated to *exactly integrable equations*.

The application of Lie group theory to discrete equations, like difference equations, differential-difference equations, or q-difference equations are much more recent [52, 43, 44, 45, 41, 26, 23, 30, 25, 46, 48, 36, 47, 37, 20, 21, 55, 56, 12, 5, 15, 16, 6, 13, 14, 18, 17, 40, 10, 34, 27].

By a difference equation we mean a functional relation, linear or non-linear, between functions calculated at different points of a lattice [29, 4, 59, 2, 19]. These systems appear in many applications. First of all they can be written down as discretizations of a differential equation when one is trying to solve it with a computer. In such a case one reduces the differential equation to a recurrence relation:

$$\frac{du}{dx} = f(x, u) \qquad \Rightarrow \qquad v(n+1) = g(n, v(n)).$$

On the other hand we can consider dynamical systems defined on a lattice, i.e. systems where the real independent fields depend on a set of independent variables which vary partly on the integers and partly on the reals. For example we can consider

$$\frac{d^2u(n,t)}{dt^2} = F(t, u(n,t), u(n-1,t), .., u(n-a,t), u(n+1,t), .., u(n+b,t)).$$

These kind of equations can appear in many different setting. Among them they are associated to the evolution of many body problems, to the study of crystals, to biological and economical systems, etc..

As an example of possible applications we consider the problem of the transmission of energy in one dimensional molecular system, problem which is of particular relevance for understanding the functioning of physical systems of biological interest[8]. This is a particularly hot topic as some relevant biological processes require the transport of energy with low dispersion along essentially one dimensional chains, such as the spines in an α helix [11]. A mechanism for the nondispersive transport of vibrational energy along hydrogenon bonded chains was proposed by Davydov and its continuos limit for small lattice spacing gave rise to a Nonlinear

Schrödinger equation (NLS) which has soliton solutions [58]. If such soliton like solutions are valid also at biological temperatures is an open problem. In the case of diatomic nonlinear lattices we can describe such systems by the equations

$$\begin{aligned} &M_1\ddot{x}_n - k_1(y_n - x_n) + k_2(x_n - y_{n-1}) - \epsilon\beta_1(y_n - x_n)^2 \\ &+\epsilon\beta_2(x_n - y_{n-1})^2 = 0 \\ &M_2\ddot{y}_n + k_1(y_n - x_n) - k_2(x_{n+1} - y_n) + \epsilon\beta_1(y_n - x_n)^2 \\ &-\epsilon\beta_2(x_{n+1} - y_n)^2 = 0 \end{aligned} \tag{1.6}$$

where M_1 and M_2 are the different values of the two atomic masses, ϵ is a small parameter while k_1, k_2, β_1 and β_2 are four constants of order 1. When $k_2 \ll k_1$ these equation represent a molecular chain with intramolecular interaction stronger than the intermolecular one. This is the case, for example, of an hydrogen-bonded polypeptide chain.

In ref.[8] this problem has been considered from two different points of view:

1. One has performed a molecular dynamics numerical simulation.
2. One has performed a reductive perturbative analysis of the dynamics of the diatomic chain.

The numerical results (see Figs.1-3 of [8]) show clearly the relevance of the nonlinear terms in this non-dispersive energy transport. Moreover one can show that localized initial excitations of generic form can evolve into solitonlike solutions that travel along the chain. The numerical calculations show that such solutions are robust with respect to thermal disorder, i.e. energy can be transported with no appreciable dispersion even at room temperature. In the second case the development led to a NLS equation and to a solitonlike solution that transfers energy without dispersion. However, as involve a continuum limit, are not sufficient to prove the result.

To conclude I would like to comment on the richness of the world of the differential difference equations with respect to that of partial differential equations by presenting a result recently obtained by MacKay and Aubry[51]. In a theorem they showed that dynamical chains are much richer than their continuum counterparts. Almost any Hamiltonian network of weakly coupled oscillators has a 'breather' solution while the existence of breathers for a nonlinear wave equation is rare. The results of this Theorem imply that the discrete world can be richer of interesting solutions and thus worthwhile studying by itself and symmetries are a simple and efficient way to do so.

2 Intrinsic Lie point symmetries for discrete equations

The first steps in the construction of Lie symmetries for difference equations were taken by Shiguro Maeda in 1980 [52] and later extended by many authors [43, 44, 45, 41, 26, 23, 30, 25, 46, 48, 36, 47, 37, 20, 21, 55, 56, 12, 5, 15, 16, 6, 13, 14, 18, 17, 40, 10, 34, 27].

For simplicity in the following I will consider just the case of a scalar equation in two independent variables but, equivalent results can be obtained in the case of N independent and M dependent variables.

A discrete equation in R^2 is a functional relation for a field $u(P)$ at different points P_i in R^2, i.e. $E = E(x, t, u(P_1), \dots, u(P_L)) = 0$. A differential difference equation is obtained by considering the points P_i uniformly spaces in one direction, say t, with spacing h_t, in such a way that we are allowed to consider the continuous limit when h_t goes to zero.

The points P_i in R^2 can be labelled by two discrete indexes which characterize the points with respect to two independent directions, $P_{n,m}$. For example, in cartesian coordinates we have:

$$P_{n,m} = (x_{n,m}, t_{n,m}) \tag{2.1}$$

and the function $u(P)$ reads

$$u_{P_{n,m}} = u(x_{n,m}, t_{n,m}) = u_{n,m}. \tag{2.2}$$

A difference scheme will be *a set of relations* among the values of $\{x, t, u(x, t)\}$ at a finite number, say L, of points in R^2 $\{P_1, \ldots, P_L\}$ around a reference point, say P_1. Some of these relations will define where the points are in R^2 and others how $u(P)$ transforms in R^2. In our case, as we have one only dependent variable and two independent variables we expect to have five equations, four which define the two independent variables in the two independent directions in R^2, and one the dependent variable in terms of the lattice points:

$$\begin{aligned} &E_a(\{x_{n+j,m+i}, t_{n+j,m+i}, u_{n+j,m+i}\}) = 0 \\ &1 \le a \le 5; \qquad -i_1 \le i \le i_2, \quad -j_1 \le j \le j_2 \quad (i_1, i_2, j_1, j_2) \varepsilon Z \\ &i_1 + i_2 = N, \qquad j_1 + j_2 = M. \end{aligned} \tag{2.3}$$

System (2.3) must be such that, starting from L points we are able to calculate $\{x, t, u\}$ in all points.

If a continuous limit of (2.3) exists, than one of the equations will go over to a partial differential equation and the others will be identically satisfied (generically $0 = 0$). We can also do partial continuous limits when only one of the independent variables become continuous while the other is still discrete. In this case only part of the lattice equations are identically satisfied and we obtain a differential difference equation for the dependent variable and an equation for the lattice variable.

Let us present now, to clarify the ideas, some examples of difference scheme.

Let as consider at first the case of the discrete heat equation on a uniform orthogonal lattice:

$$\frac{u_{n+1,m} - u_{n,m}}{t_{n+1,m} - t_{n,m}} = \frac{u_{n,m+2} - 2u_{n,m+1} + u_{n,m}}{(x_{n,m+1} - x_{n,m})^2} \tag{2.4}$$

$$\begin{aligned} &x_{n,m+1} - x_{n,m} = h_x; \qquad t_{n,m+1} - t_{n,m} = 0 \\ &x_{n+1,m} - x_{n,m} = 0; \qquad t_{n+1,m} - t_{n,m} = h_t, \end{aligned} \tag{2.5}$$

where h_x, h_t are two a priory fixed constants which define the spacing between two neighboring points in the two directions of the orthogonal lattice. The lattice equations (2.5) could be substituted by different ones which will provide different Lie point symmetries for the difference scheme while keeping the continuous limit. For example we can consider instead of eq. (2.5)

$$\begin{aligned} &x_{n,m+2} - 2x_{n,m+1} + x_{n,m} = 0; \qquad t_{n,m+1} - t_{n,m} = 0 \\ &x_{n+1,m} - x_{n,m} = 0; \qquad t_{n+2,m} - 2t_{n+1,m} + t_{n,m} = 0, \end{aligned} \tag{2.6}$$

or

$$\begin{aligned} &x_{n,m+2} - 2x_{n,m+1} + x_{n,m} = 0; \qquad t_{n,m+1} - t_{n,m} = 0 \\ &x_{n+1,m} - x_{n,m} = 0; \qquad t_{n+1,m} - t_{n,m} = c(x_{n,m+1} - x_{n,m})^2, \end{aligned} \tag{2.7}$$

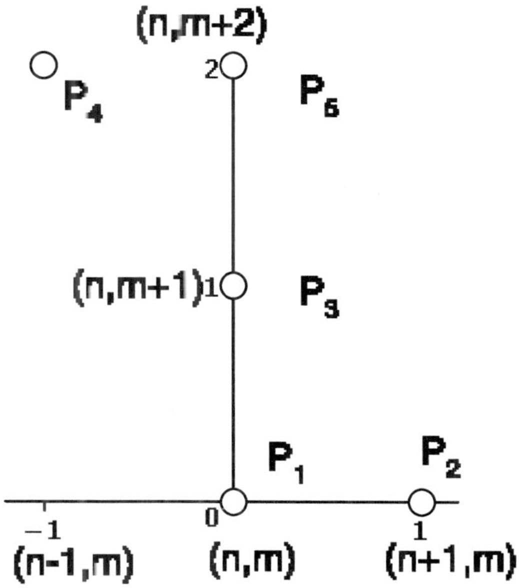

Figure 1 The set of points which are connected by the DTL (2.9).

or

$$x_{n,m+2} - 2x_{n,m+1} + x_{n,m} = 0; \qquad t_{n,m+1} - t_{n,m} = 0 \tag{2.8}$$
$$x_{n+1,m} - (1+c)x_{n,m} = 0; \qquad t_{n+1,m} - t_{n,m} = h.$$

Eq. (2.6) corresponds to a lattice scheme when the lattice spacing can be varied in the Lie point transformation, while the lattice (2.7) involves only four points as the heat equation itself and eq. (2.8) corresponds to an exponential lattice, i.e. a lattice that is neither equally spaced nor orthogonal.

A second example is provided by the Discrete-time Toda Lattice equation (DTL) [31]:

$$e^{u_{P_1}-u_{P_3}} - e^{u_{P_3}-u_{P_5}} = \alpha^2\{e^{u_{P_4}-u_{P_3}} - e^{u_{P_3}-u_{P_{12}}}\} \tag{2.9}$$

where α is a constant parameter. The DTL is a relation between 5 points in the two dimensional plane (see fig.1):

$$P_2 = (x_{n+1,m}, t_{n+1,m}); \quad P_3 = (x_{n,m+1}, t_{n,m+1}); \tag{2.10}$$
$$P_4 = (x_{n-1,m+2}, t_{n-1,m+2}); \quad P_5 = (x_{n,m+2}, t_{n,m+2}) \quad P_1 = (x_{n,m}, t_{n,m}).$$

i.e.

$$e^{u_{n,m}-u_{n,m+1}} - e^{u_{n,m+1}-u_{n,m+2}} = \alpha^2(e^{u_{n-1,m+2}-u_{n,m+1}} - e^{u_{n,m+1}-u_{n+1,m}}). \tag{2.11}$$

On an orthogonal uniform unchangeable lattice such that $t_{n,m} = t = mh_t$ and $x_{n,m} = nh_x$, setting $v_n(t) = v_{n,m}$ and $\alpha = h_t^2$, when h_t goes to zero, we get;

$$\ddot{v}_n = e^{v_{n-1}-v_n} - e^{v_n-v_{n+1}}, \tag{2.12}$$

the usual Toda Lattice [59].

2.1 Symmetries of a difference scheme. As we are interested in Lie point symmetries, we look for transformations of the form:

$$\begin{aligned} \tilde{x} &= F_g(x,t,u) = x + g\ \xi(x,t,u) + \dots \\ \tilde{t} &= G_g(x,t,u) = t + g\ \tau(x,t,u) + \dots \\ \tilde{u} &= H_g(x,t,u) = u + g\ \phi(x,t,u) + \dots \end{aligned} \tag{2.13}$$

where g, as before, is the group parameter. The transformation (2.13) is such that if $\{x,t,u\}$ satisfy the difference scheme $E_a = 0$, $\{\tilde{x},\tilde{t},\tilde{u}\}$ will be a solution of the same scheme. Such a transformation acts on the whole space of the independent and dependent variables $\{x,t,u\}$, at least in some neighborhood of P_1 including all points up to P_L. This means that the same set of functions F_g, G_g and H_g will determine the transformation in all points of the scheme. In the point P_1 we define the infinitesimal generator as:

$$\hat{X}_{P_1} = \xi(x,t,u)\partial_x + \tau(x,t,u)\partial_t + \phi(x,t,u)\partial_u \tag{2.14}$$

and than we prolong it to all other $L-1$ points of the scheme. Since the transformation is given by the same set of functions $\{F_g, G_g, H_g\}$ at all points, the *prolongation* of $\hat{X}_{P_j}$ is obtained simply by evaluating $\hat{X}_{P_1}$ at the corresponding points involved in the scheme. So

$$pr\hat{X} = \sum_{i=1}^{L} \hat{X}_{P_i}. \tag{2.15}$$

Consequently the invariance condition for the difference scheme is:

$$pr\hat{X}E_a|_{E_a=0} = 0. \tag{2.16}$$

Eq. (2.16) is a set of functional equations whose solution is obtained by turning them into differential equations by successive derivation with respect to the independent variables $\{x,t,u\}$ at the different points of the lattice [3].

The solution of (2.16) provide us with the function $\xi(x,t,u)$, $\tau(x,t,u)$ and $\phi(x,t,u)$, the infinitesimal coefficients of the local Lie point symmetry group. The transformation is obtained by integrating the vector field, i.e. by solving the following system of differential equations:

$$\begin{aligned} \frac{d\tilde{x}}{dg} &= \xi(\tilde{x},\tilde{t},\tilde{u}), \qquad & \tilde{x}|_{g=0} &= x, \\ \frac{d\tilde{t}}{dg} &= \tau(\tilde{x},\tilde{t},\tilde{u}), \qquad & \tilde{t}|_{g=0} &= t, \\ \frac{d\tilde{u}}{dg} &= \phi(\tilde{x},\tilde{t},\tilde{u}), \qquad & \tilde{u}|_{g=0} &= u. \end{aligned} \tag{2.17}$$

In general we expect the infinitesimal coefficients ξ and τ to be determined by the lattice equations. So according to the form of the lattice, different symmetries can appear.

In fact, in the case of eq. (2.5), by applying the infinitesimal generator (2.14) to the lattice equations we get:

$$\begin{aligned} \xi(x_{n,m+1}, t_{n,m+1}, u_{n,m+1}) &= \xi(x_{n,m}, t_{n,m}, u_{n,m}); \\ \xi(x_{n+1,m}, t_{n+1,m}, u_{n+1,m}) &= \xi(x_{n,m}, t_{n,m}, u_{n,m}). \end{aligned}$$

As $u_{n,m+1}$, $u_{n+1,m}$ and $u_{n,m}$ are independent functions we get $\xi = \xi(x,t)$. $t_{n,m+1} = t_{n,m}$ but $x_{n,m+1} \neq x_{n,m}$ and consequently $\xi = \xi(t)$. As $x_{n+1,m} = x_{n,m}$ but $t_{n+1,m} \neq$

$t_{n,m}$ we get that the only possible value for ξ in this case is ξ=costant. In a similar fashion we derive that also τ must be a constant and that $\phi=u+s(x,t)$, where $s(x,t)$ is a solution of the heat equation, the linear superposition formula. Summarizing we get that the infinitesimal generators of the symmetries for the heat equation are given by

$$\hat{P}_0 = \partial_t; \quad \hat{P}_1 = \partial_x; \quad \hat{W} = u\partial_u; \quad \hat{S} = s(x,t)\partial_u. \tag{2.18}$$

If we associate to the heat equation (2.4) the lattice (2.6) we get that we have the generators (2.18) and an extra generator, $\hat{D} = x\partial_x + 2t\partial_t$. See [39] for other examples.

3 Extensions

As one can see in the examples considered so far in the literature [43, 44, 45, 39], in the case of discrete equations we will, in general, get less symmetries than in the corresponding continuous system. In the case of discrete equations we rarely find symmetries which involve both dependent and independent variables. The extension to a variable lattice is not always sufficient to obtain them [17].

Research for more general symmetries requires the extension of the intrinsic Lie point ansatz. We have various possibilities.

Let us consider the fields u_{P_j} in different lattice points P_j as different independent fields related by the difference equation. Under such an hypothesis the difference equation is just an algebraic relation between the fields. The discrete variables are just indices of the dependent fields and the only independent variables are the continuous ones. In this case the infinitesimal generator, for one dependent field u and one independent variable t, reads:

$$\hat{X} = \tau(t, \{u_{P_i}\})\partial_t + \sum_{i=1}^{L} \phi_{P_i}(t, \{u_{P_i}\})\partial_{u_{P_i}} \tag{3.1}$$

This ansatz is equivalent to extending the dependence of ξ, τ and ϕ from the point P_1 to all points P_j of the lattice and thus no prolongation is necessary. As, by Taylor expansion, $u_{n,m+1}$ can be expressed in term of $u_{n,m}$ and all of its derivatives, the presence of other points in the definition of the infinitesimal generator of the symmetries is equivalent to consider generalized symmetries. This explains the difficulties in obtaining in this case nontrivial results [44].

On the other hand, as was shown in [20, 21], by allowing symmetry generators depending on different lattice points, linear difference equations admit a symmetry group which is isomorph to the one of the continuous equation. As a classical example, let us consider the discrete heat equation:

$$\Delta_t u = \Delta_{xx} u \tag{3.2}$$

where the discrete derivative Δ is defined as the incremental ratio,

$$\Delta_z^+ = \frac{T_z - 1}{h_z}, \qquad T_z f(z) = f(z + h_z), \tag{3.3}$$

with h_z the lattice spacing in the z direction. In [20, 21] it was shown that one can construct the following symmetry group:

$$\begin{aligned}
\hat{P}_0 &= \Delta_t^+ u\partial_u, \\
\hat{P}_1 &= \Delta_x^+ u\partial_u, \\
\hat{W} &= u\partial_u, \\
\hat{B} &= (2tT_t^{-1}\Delta_x^+ u + xT_x^{-1}u)\partial_u, \\
\hat{D} &= (2tT_t^{-1}\Delta_t^+ u + xT_x^{-1}\Delta_x^+ u + \frac{1}{2}u)\partial_u, \\
\hat{K} &= \{t^2T_t^{-1}\Delta_t^+ u - h_t tT_t^{-2}\Delta_t^+ u + txT_t^{-1}T_x^{-1}\Delta_x^+ u + \\
&+\frac{1}{4}x^2T_x^{-2}u - \frac{1}{4}h_x xT_x^{-2}u + \frac{1}{2}tT_t^{-1}u\}\partial_u.
\end{aligned} \tag{3.4}$$

In a sequent work [32] one was able to prove that the previous result can be extended to any discrete derivative, not only to that given by (3.3), for any linear partial differential equation by using the following prescriptions:

- we write down the vector fields in evolutionary form
- we substitute

$$\begin{aligned}
u_{,t} &\to \Delta_t u \\
u_{,x} &\to \Delta_x u \\
x &\to x\beta_x \\
t &\to t\beta_t
\end{aligned} \tag{3.5}$$

where β_x, β_t are functions of the same shift operators commuting with Δ_x and Δ_t and such that the following commutation relation is satisfied:

$$[\Delta_z, z\beta_z] = 1. \tag{3.6}$$

Equation (3.6) define completely the function β in terms of Δ. If by Δ we consider the continuous derivative than $\beta = 1$; for Δ given by eq. (3.3) than $\beta_z = T_z^{-1}$, completely in agreement with the results presented in (3.4). If we choose $\Delta_z^s = \frac{T_z - T_z^{-1}}{2h_z}$ than $\beta_z = 2(T_z + T_z^{-1})^{-1}$. From the correspondence (3.5) it follows that for any analytic solution of the linear partial differential equation we find a solution of the discrete counterpart. In the case of Δ^+, or $\Delta_z^- = \frac{1-T_z^{-1}}{h_z}$, the continuous case will give all solutions of the discrete one. However this will not be the case for the more general Δ^s, when by the correspondence we obtain just a subset of the possible solutions. The result contained in (3.5, 3.6) is a generalization of the *umbral calculus* [57, 38].

As pointed out before, as the symmetries depend on more points on the lattice, we are in a situation similar to the case of generalized symmetries and we are no more able to get the corresponding group transformations. However we can still use the symmetries to carry out a symmetry reduction and get explicit solutions. Let us see this on the example of the heat equation (3.1). In the case of a symmetry generator given by $\hat{P}_1 - a\hat{P}_0$ Eq. (1.5) reads:

$$\frac{du_{n,m}}{dg} = -(\frac{1}{h_x} - a\frac{1}{h_t})u_{n,m} + \frac{1}{h_x}u_{n,m+1} - \frac{a}{h_t}u_{n+1,m}. \tag{3.7}$$

Eq. (3.7) is a linear differential difference equation and to get the group transformations we need the explicit solution of its initial problem. The symmetry reduction is obtained by setting $\frac{du_{n,m}}{dg} = 0$. In this case we get:

$$u_{n,m} = c_0 + c_1(1 + a^2 h_t)^n (1 + a h_x)^m \tag{3.8}$$

which is the discrete representation of the continuous solution.

What happens in the case of nonlinear difference equations? In general we are no more able to obtain a symmetry group isomorphic to the continuous one, the application of the *umbral* correspondence to a nonlinear equation is questionable and research on it is in progress [35, 9].

In all cases, both for linear and nonlinear difference equations,

- we can still compute intrinsic Lie point symmetries
- generalized symmetries, i.e. symmetries depending on a finite number of points of the lattice, are usually associated to integrable equations and their form is usually very complicate and very difficult to predict apriori without any further information on the structure of the system at study. As an example of this let us show few examples of symmetries of the discrete Burgers equation [26]

$$\Delta_t u = \frac{1 + h_x u}{1 + h_t(\Delta_x u + u T_x u)} \Delta_x(\Delta_x u + u T_x u) \tag{3.9}$$

i.e.

$$u_{,\lambda_1} = [1 + h_t(\Delta_x u + u T_x u)]\Delta_t u \tag{3.10}$$

a time translation and

$$\begin{aligned} u_{,\lambda_2} = \quad & [1 + h_x u]\Delta_x\{2tT_t^{-1} \frac{u}{1 + h_t(\Delta_x u + u T_x u)} + \\ & +(x + \frac{h_x}{2})T_x^{-1} \frac{1}{1 + h_x u}\} \end{aligned} \tag{3.11}$$

a boost. In both cases they depend nonlinearly on the filed u at various points of the lattice. In the case of the time translation (3.10) polynomially, but sometimes, as one can see in eq. (3.11), also rationally.

As a further exemplification of the complicate structure of symmetries for nonlinear discrete equations, let us present the simplest symmetries one obtains for the Toda Lattice equation (2.12) from its integrability [25]:

$$\hat{Y}_{\mu_{-1}} = \partial_{v_n} \tag{3.12}$$

$$\hat{Y}_{\mu_0} = t\partial_{v_n} \tag{3.13}$$

$$\hat{Y}_{\mu_1} = (t\dot{v}_n - 2n)\partial_{v_n} \tag{3.14}$$

$$\hat{X}_{\epsilon_0} = \dot{v}_n \partial_{v_n} \tag{3.15}$$

$$\hat{X}_{\epsilon_1} = \{\dot{v}_n^2 + e^{v_{n-1} - v_n} + e^{v_n - v_{n+1}} - 2\}\partial_{v_n} \tag{3.16}$$

$$\begin{aligned} \hat{X}_{\epsilon_2} = & \{\dot{v}_n^3 - 2\dot{v}_n + e^{v_{n-1} - v_n}(\dot{v}_{n-1} + 2\dot{v}_n) + \\ & + e^{v_n - v_{n+1}}(\dot{v}_{n+1} + 2\dot{v}_n)\}\partial_{v_n} \end{aligned} \tag{3.17}$$

$$\begin{aligned} \hat{Y}_{\nu} = & [t\{\dot{v}_n^2 + e^{v_{n-1} - v_n} + e^{v_n - v_{n+1}} - 2\} - (2n-1)\dot{v}_n + w_n]\partial_{v_n} \\ & w_{n-1} - w_n = 2\dot{v}_n \end{aligned} \tag{3.18}$$

where (3.12 - 3.15) are intrinsic symmetries, (3.15 - 3.17) are associated to isospectral deformation of the discrete Schrödinger spectral problem associated to the Toda lattice while (3.12 - 3.14, 3.18) are associated to non isospectral deformations. Let us notice that the coefficient of t in eq. (3.18) is not equal to $\dot{v}_n$, as would be in the case of the Korteweg de Vries equation, but corresponds to a higher flow in the Toda Lattice hierarchy. The infinitesimal generators $\hat{X}$, for the isospectral deformations, and $\hat{Y}$, for the non isospectral ones, satisfy the following commutation relations:

$$[\hat{X}_{\epsilon_i}, \hat{X}_{\epsilon_j}] = 0; \qquad [\hat{X}_{\epsilon_1}, \hat{Y}_{nu}] = 4\hat{X}_{\epsilon_1} - 2\hat{X}_{\epsilon_3}. \tag{3.19}$$

In the case of the discrete Nonlinear Schrödinger equation [1, 33, 24]

$$i\dot{Q}_n + \frac{1}{h_x^2}[2Q_n - (1 - \epsilon|Q_n|^2)(Q_{n+1} + Q_{n-1})] = 0, \tag{3.20}$$

where $\epsilon = \pm 1$, the symmetries read

$$\hat{X}_1 = Q_n \partial_{Q_n} - Q_n^* \partial_{Q_n^*} \tag{3.21}$$

$$\hat{X}_2 = (1 - \epsilon|Q_n|^2)[Q_{n+1}\partial_{Q_n} - Q_{n-1}^* \partial_{Q_n^*}] \tag{3.22}$$

$$\hat{X}_3 = (1 - \epsilon|Q_n|^2)Q_{n-1}\partial_{Q_n} - (1 - \epsilon|Q_n|^2)Q_{n+1}^* \partial_{Q_n^*} \tag{3.23}$$

$$\hat{Y} = [-\frac{2t}{h_x^2}(1 - \epsilon|Q_n|^2)(Q_{n+1} - Q_{n+1}) + i(2n+1)Q_n]\partial_{Q_n} + \tag{3.24}$$

$$-[\frac{2t}{h_x^2}(1 - \epsilon|Q_n|^2)(Q_{n+1}^* - Q_{n+1}^*) + i(2n+1)Q_n^*]\partial_{Q_n^*}.$$

$\hat{X}_1$ and $\hat{X} = \hat{X}_2 + \hat{X}_3$ are intrinsic Lie point symmetries, $\{\hat{X}_1,\ \hat{X}_2$ and $\hat{X}_3\}$ are isospectral symmetries while $\hat{Y}$ is a nonisospectral symmetry and they satisfy similar commutation relations as (3.19). We can use $\frac{\hat{Y} - i\hat{X}_1}{2}$ to do a symmetry reduction for the Nonlinear Schrödinger equation (3.20). In the continuous limit this would correspond to a dilation symmetry which would give rise to elliptic functions or Painlevé solutions. This reduces to solving the Nonlinear Schrödinger equation (3.20) together with

$$-\frac{t}{h_x}(1 - \epsilon|Q_n|^2)(Q_{n+1} - Q_{n+1}) + inQ_n = 0.$$

Defining $Q_n = \rho_n e^{i\theta_n}$, under the assumption that $\rho_n^2 \neq \epsilon$ we get the following nonlinear reduced equation:

$$\sqrt{\rho_n^2 \rho_{n+1}^2 - h_x^4 c_0^2} + \sqrt{\rho_n^2 \rho_{n-1}^2 - h_x^4 c_0^2} = \frac{nh_x^2}{t} \frac{\rho_n^2}{1 - \epsilon \rho_n^2}. \tag{3.25}$$

Up to now we have shown that extra symmetries can be found in the case of linear and integrable difference equations and that we can use them to get solutions. In the following we want to suggest an answer in the generic case.

In [47] one considers the class of differential difference equations:

$$\dot{u}_n = F_n(u_{n+1}, u_n, u_{n-1}) \tag{3.26}$$

defined on a uniform unchangeable orthogonal lattice and look for symmetries depending linearly on t, as it is the case of the symmetries $\hat{Y}$ (3.18, 3.24) for the Toda Lattice and discrete Nonlinear Schrödinger equations. To exclude the case of integrable equations we assume symmetries of the form:

$$u_{n,\lambda} = t\dot{u}_n + H_n(u_{n+1}, u_n, u_{n-1}). \tag{3.27}$$

Under the assumption that the differential difference equation (3.26) is really nonlinear

$$(\frac{\partial F_n}{\partial u_{n+1}}, \frac{\partial F_n}{\partial u_{n-1}}) \neq (0, 0) \quad for \quad any \quad n \tag{3.28}$$

we can, by a simple transformation reduce in all generality the function H_n to the form

$$H_n = H_n(u_{n+1}, u_n) \tag{3.29}$$

with $\frac{\partial H_n}{\partial u_{n+1}} \neq 0$, as we are not interested in intrinsic Lie point symmetries.

We can state the following theoremm, proved in [47];

Theorem *If a nonlinear equation (3.26) has a symmetry of the form (3.28, 3.29) than it is equivalent, up to a Lie point transformation*

$$\tilde{t} = \omega t, \quad \tilde{u}_n = \phi_n(u_n); \quad \omega \neq 0, \quad \phi_n' \neq 0 \quad for \quad any \quad n \tag{3.30}$$

to an equation of the form

$$\begin{aligned} \dot{u}_n &= A_n + B_n \\ A_n &= a_{n+1}e^{u_{n+1}} - a_n e^{u_n} - 1 \\ B_n &= a_n e^{-u_n} - a_{n-1}e^{-u_{n-1}} - 1 \\ a_n^2 &= n^2 + \alpha n + \beta \quad for \quad any \quad n, \end{aligned} \tag{3.31}$$

where α and β are arbitrary constants. Than the symmetry is given by

$$u_{n,\lambda} = t\dot{u}_n + A_n \tag{3.32}$$

and the equation is linearizable.

As a consequence of the previous Theorem an equation of the form (3.26) can have dilation symmetries of the form (3.27) only if it is equivalent to a linear equation.

4 Conclusions

We have shown how one can construct in a coherent way Lie point symmetries for discrete equations.

If we want to extend the symmetries to the case when they depend on more point of the lattice than this can be done only in the case of linear, linearizable or integrable equations. This statement is very plausible but no complete proof of this statement has been given. In the case of linear, linearizable or integrable equations we can find generalized symmetries from which we are not able to construct group transformations but we can use them to get explicit solutions via symmetry reduction.

For a generic equation we can construct intrinsic Lie point and conditional symmetries [44] and we can use them, as for partial differential equations, to construct group transformations, to do symmetry reduction and obtain explicit solutions, to classify the discrete equations according to their symmetries, etc..

References

[1] M. J. Ablowitz and J. F. Ladik, *Nonlinear differential-difference equations*, J. Math. Phys. **16** (1975), 598–603.

[2] M. J. Ablowitz and H. Segur *Solutions and the Inverse Scattering Transform*, SIAM, Providence, R.I., 1981.

[3] J. Aczel, *Lectures on Functional Equations and their Applications*, Acad. Press, New York 1966, see chapter IV and references therein included.

[4] R. P. Agarwal, *Difference equations and inequalities: theory, methods and applications*, Dekker, New York 1992.

[5] W. F. Ames, R. L. Anderson, V. A. Dorodnitsyn, E. V. Ferapontov, R. K. Gazizov, N. H. Ibragimov and S. R. Svirshchevskii, *CRC Hand-book of Lie Group Analysis of Differential Equations*, ed. by N. H. Ibragimov, *Volume I: Symmetries, Exact Solutions and Conservation Laws*, CRC Press, Boca Raton, 1994.

[6] M. Bakirova, V. A. Dorodnitsyn, and R. Kozlov, *Invariant difference schemes for heat transfer equations with a source*, J. Phys. A: Math.Gen. **30** (1997), 8139–8155, see also V. A. Dorodnitsyn, R. Kozlov, *The complete set of symmetry preserving discrete versions of a heat transfer equation with a source*, Preprint NTNU, NUMERICS NO. 4/1997, Trondheim, Norway, 1997.

[7] G.W. Bluman and S. Kumei, *Symmetries and Differential Equations*, Springer, Berlin, 1989.

[8] A. Campa, A. Giansanti, A. Tenenbaum, D. Levi, and O. Ragnisco, *Quasi-solitons on a diatomic chain at room temperature*, Phys. Rev. **B 48** (1993), 10168–10182.

[9] B. Carl and C. Schiebold, *Ein direkter Ansatz zur Untersuchung von Solitongleichunghen*, Jahresber. Deutsch. Math.-Verein. **120** (2000), 102–148.

[10] P. A. Clarkson and F. W. Nijhoff (editors), *Symmetries and Integrability of Difference equations* CUP, Cambridge, UK, 1999.

[11] A. S. Davydov, *Solutions in molecular systems*, Phys. Scr. **20** (1979), 387–394; *Biology and Quantum Mechanics*, Pergamon, New York 1982.

[12] V. A. Dorodnitsyn, *Transformation groups in a space of difference variables*, VINITI Acad. Sci. USSR, Itogi Nauki i Techniki, **34** (1989), 149–190, (in Russian), translation in J. Sov. Math. **55** (1991), 1490–1517.

[13] V. A. Dorodnitsyn, *Finite-difference models entirely inheriting symmetry of original differential equations*, Modern Group Analysis: Advanced Analytical and Computational Methods in Mathematical Physics, Kluwer, Dordrecht, 1993, pp. 191–201.

[14] V. A. Dorodnitsyn, *Finite-difference analog of the Noether theorem*, Dokl. Akad. Nauk **328** (1993), 678–682, (in Russian) translation in Phys. Dokl. **38** (1993), 66–68 . V. A. Dorodnitsyn, *Noether-type theorems for difference equation*, IHES/M/98/27, Bures-sur-Yvette, France, 1998.

[15] V. A. Dorodnitsyn, *Finite-difference models entirely inheriting continuous symmetry of original differential equations.*, Int. J. Mod. Phys. C **5** (1994), 723–734.

[16] V. A. Dorodnitsyn, *Continuous symmetries of finite-difference evolution equations and grids*, Symmetries and Integrability of Difference Equations, eds. D. Levi, L. Vinet, and P. Winternitz, CRM/AMS, Providence, R.I.,1996, pp. 103–112, see also V.Dorodnitsyn, *Invariant discrete model for the Korteweg-de Vries equation*, Preprint CRM-2187, Montreal, Canada, 1994.

[17] V. A. Dorodnitsyn, R. Kozlov and P. Winternitz P, *Lie group classification of second order ordinary difference equations*, J. Math. Phys. **41** (2000), 480–504.

[18] V. A. Dorodnitsyn and P. Winternitz, *Lie point symmetry preserving discretizations for variable coefficient Korteweg - de Vries equations*, Nonlinear Dynam., **22** (2000), 49–59.

[19] S. N. Elaydi, *An Introduction to Difference Equations*, Springer, New York, 1999.

[20] R. Floreanini, J. Negro, L. M. Nieto and L. Vinet, *Symmetries of the heat equation on a lattice*, Lett. Math. Phys. **36** (1996), 351–355.

[21] R. Floreanini and L. Vinet, *Lie symmetries of finite-difference equations*, J. Math. Phys. **36** (1995), 7024–7042.

[22] G. Gaeta, *Nonlinear Symmetries and Nonlinear Equations*, Kluwer, Dordrecht, 1994.

[23] D. Gomez-Ullate, S. Lafortune, and P. Winternitz, *Symmetries of discrete dynamical systems involving two species*, J. Math. Phys. **40** (1999), 2782–2804.

[24] R. Hernandez Heredero and D. Levi, *The Discrete Nonlinear Schrödinger equation and its Lie Symmetries*, J. Non. Math. Phys., in press.

[25] R. Hernandez Heredero, D. Levi, M. A. Rodriguez and P. Winternitz P, *Lie algebra contractions and symmetries of the Toda hierarchy*, J. Phys. A: Math. Gen. **33** (2000), 5025–5040.

[26] R. Hernandez Heredero, D. Levi, and P. Winternitz, *Symmetries of the discrete Burgers equation*, J. Phys. A: Math. Gen. **32** (1999), 2685–2695.

[27] J. Hietarinta, F. W. Nijhoff and J. Satsuma (editors) *Symmetries and Integrability of Difference equations*, J. Phys. A: Math. Gen. **34** (2001), 10337–10744.

[28] N.H. Ibragimov, *Transformation Groups Applied to Mathematical Physics*, Reidel, Boston, 1985.

[29] W. G. Kelley and A. C. Peterson, *Difference equations: an introduction with applications*, Acad. Press, New York 1991.

[30] S. Lafortune, L. Martina and P. Winternitz, *Point symmetries of generalized Toda field theories*, J. Phys. A: Math. Gen. **33** (2000), 2419–2435.

[31] D. Levi and L. Martina, *Integrable Hierarchies of Nonlinear Difference Difference Equations and Symmetries*, J. Phys. A: Math. Gen. **34** (2001), 10357–10368.

[32] D. Levi, J. Negro and M. dell'Olmo, *Discrete derivatives and symmetries of difference equations*, J. Phys. A: Math. Gen. **34** (2001), 2023–2030.

[33] D. Levi and G. R. W. Quispel, *Discrete Painlevé equations for nonisospectral soliton equations*, Symmetries and Integrability of difference equations, CRM/AMS, Providence, R.I., 2000, pp. 363–366.

[34] D. Levi and O. Ragnisco (editors), *Symmetries and Integrability of Difference equations*, CRM/AMS, Providence, R.I., 2000.

[35] D. Levi, O. Ragnisco and P. Tempesta, work in progress.

[36] D. Levi, M. A. Rodriguez, *Symmetry group of partial differential equations and of differential-difference equations: the Toda lattice vs the Korteweg-de Vries equations*, J. Phys. A: Math. Gen., **25** (1992), L975–L979.

[37] D. Levi, M. A. Rodriguez, *Lie symmetries for integrable equations on the lattice*, J. Phys. A: Math. Gen. **32** (1999), 8303–8316.

[38] D. Levi, P. Tempesta and P. Winternitz, *Umbral Calculus, Difference Equations and the Discrete Schroedinger Equation*, nlin.SI/0305047, submitted to Comm. Math. Phys. for the publication.

[39] D. Levi, S. Tremblay and P. Winternitz, *Lie point symmetries of difference equations and lattices*, J. Phys. A: Math. Gen. **33** (2000), 8507–8523; *Lie symmetries of multidimensional difference equations*, J. Phys. A: Math. Gen. **34** (2001), 9507–9524.

[40] D. Levi, L. Vinet, and P. Winternitz (editors), *Symmetries and Integrability of Difference Equations*, CRM/AMS, Providence, R.I., 1996.

[41] D. Levi, L. Vinet, and P. Winternitz, *Lie group formalism for difference equations*, J. Phys. A: Math. Gen. **30** (1997), 633–649.

[42] D. Levi and P. Winternitz, *Non-classical symmetry reduction: example of the Boussinesq equation*, J. Phys. A: Math. Gen. **22**(1989), 2915–2924.

[43] D. Levi and P. Winternitz, *Continuous symmetries of discrete equations*, Phys. Lett. **A 152** (1991), 335–338.

[44] D. Levi and P. Winternitz, *Symmetries and conditional symmetries of differential-difference equations*, J. Math. Phys. **34** (1993), 3713–3730.

[45] D. Levi and P. Winternitz, *Symmetries of discrete dynamical systems*, J. Math. Phys. **37** (1996), 5551–5576.

[46] D. Levi and R. Yamilov, *Conditions for the existence of higher symmetries of evolutionary equations on the lattice*, J. Math. Phys. **38** (1997), 6648–6674.

[47] D. Levi, R. Yamilov, *Dilatation symmetries and equations on the lattice*, J. Phys. A: Math. Gen. **32** (1999), 8317–8323.

[48] D. Levi and R. Yamilov, *Non-point integrable symmetries for equations on the lattice*, J. Phys. A: Math. Gen. **33** (2000), 4809–4823.

[49] S. Lie, *Klassifikation und Integration von Gewohnlichen Differentialgleichungen zwischen x, y die eine Gruppe von Transformationen gestatten*, Math. Ann. **32** (1888), 213.

[50] S. Lie, *Theorie der Transformationgruppen*, B.G. Teubner, Leipzig, 1888, 1890, 1893.

[51] R. S. MacKay and S. Aubry, *Proof of the existence of breathers for time reversible or Hamiltonian networks of weakly coupled oscillators*, Nonlinearity **7** (1994), 1623–1643.

[52] S. Maeda, *Canonical structure and symmetries for discrete systems*, Math. Japan **25** (1980), 405–420.

[53] P.J. Olver, *Applications of Lie Groups to Differential Equations*, Springer, New York, 1993.

[54] L.V. Ovsiannikov, *Group Analysis of Differential Equations*, Academic Press, New York, 1982.

[55] G. R. W. Quispel, H. W. Capel, and R. Sahadevan, *Continuous symmetries of difference equations; the Kac-van Moerbeke equation and Painlevé reduction*, Phys. Lett. **A 170** (1992), 379–383.

[56] G. R. W. Quispel and R. Sahadevan, *Lie symmetries and integration of difference equations*, Phys. Lett. **A 184** (1993), 64–70.

[57] S. Roman and G. C. Rota, *The umbral calculus*, Advances in Math. **27** (1978), 95–188.
[58] A. C. Scott, *Davydov's soliton*, Phys. Rep. **217** (1992), 1–67.
[59] M. Toda, *Theory of Nonlinear Lattice*, Springer, New York 1988.
[60] P. Winternitz, *Group theory and exact solutions of nonlinear partial differential equations*, Integrable Systems, Quantum Groups and Quantum Field Theories, Kluwer, Dordrecht, 1993, pp. 429–495.

Fields Institute Communications
Volume **42**, 2004

Global Attractivity for a Logistic Equation with Piecewise Constant Arguments

Xuecheng Li
Department of Mathematics
Xuchang College
Xuchang, Henan 461000, P. R. China
xuechenli@sina.com

Zhicheng Wang
College of Mathematics and Econometrics
Hunan University
Changsha, Hunan 410082, P. R. China
zcwang@hnu.net.cn

Abstract. In this paper, we study the global attractivity for a logistic equation with piecewise constant arguments

$$x'(t) = r(t)y(t)\left\{1 - \frac{y([t])}{K}\right\}, \quad t \geq 0. \qquad (*)$$

We obtain a new sufficient condition for the positive steady state $y(t) = K$ of Eq. $(*)$ to be globally attractive. Our result generalizes and improves the corresponding results gave in [1-5].

1 Introduction and main results

We consider a logistic equation with piecewise constant argument

$$y'(t) = r(t)y(t)\left\{1 - \frac{y([t])}{K}\right\}, \quad t \geq 0. \qquad (1.1)$$

with the initial condition

$$y(0) = y_0 > 0,$$

where $[\cdot]$ denotes the greatest integer function, $r : [0, \infty) \to [0, \infty)$ is a continuous function and K is a positive constant. For any nonnegative integer $t, y'(t)$ means the right-hand derivative of $y(t)$. Equation (1.1) has been considered as a semi-discretization of the delay logistic equation

$$y'(t) = r(t)y(t)\left\{1 - \frac{y(t-1)}{K}\right\}, \quad t \geq 0$$

2000 *Mathematics Subject Classification*. Primary 34K20, 34K25; Secondary 34K60.

The second author was supported in part by NSFC Grant #10271044.

(see [1,2,3]).

Recently, Matsunaga, Hara and Sakata [4] gave the following results:

Theorem 1.1 ([4]) *Assume that*

$$\int_k^{k+1} r(s)ds \le 2 \quad for \quad k = 0, 1, \cdots \tag{1.2}$$

and

$$\int_0^{\infty} r(s)ds = \infty. \tag{1.3}$$

Then the positive steady state $y(t) = K$ of (1.1) is globally attractive.

Theorem 1.2 ([4]) *Let $r(t) \equiv r$. If $r > 2$, then the positive steady state $y(t) = K$ of (1.1) is not globally attractive.*

From the above theorems, one may naturally propose the following open problem:

Problem 1.3 Is it true that, if

$$\int_k^{k+1} r(s)ds > 2 \tag{1.4}$$

then the positive steady $y(t) = K$ of (1.1) is not globally attractive?

Solving the above problem is the main goal of this paper. We now state our main results.

Theorem 1.4 *Assume that*

$$\limsup_{k\to\infty} \int_k^{k+1} r(s)ds \le 2. \tag{1.5}$$

and

$$\int_0^{\infty} r(s)ds = \infty. \tag{1.6}$$

Then the positive steady state $y(t) = K$ of (1.1) is globally attractive.

Theorem 1.5 *Assume that*

$$\liminf_{k\to\infty} \int_k^{k+1} r(s)ds > 2. \tag{1.7}$$

Then the positive steady state $y(t) = K$ of (1.1) is not globally attractive.

The following example shows that the answer to Problem 1.3 is not true, and that our results generalize and improve the corresponding results in [1-5].

Example 1.6 Consider the logistic equation with piecewise constant argument

$$y'(t) = \left(2 + \frac{1}{t^2}\right) y(t) \left\{1 - \frac{y([t])}{K}\right\}, \ t \ge 1.$$

where $r(t) = 2 + \frac{1}{t^2}$. We have

$$\int_k^{k+1} \left(2 + \frac{1}{t^2}\right) dt = 2 + \frac{1}{k(k+1)} > 2,$$

$$\limsup_{k\to\infty} \int_k^{k+1} \left(2 + \frac{1}{t^2}\right) dt = 2$$

and

$$\int_1^{\infty} r(s)ds = \infty.$$

By Theorem 1.4, the positive steady state $y(t) = K$ of (1.1) is globally attractive.

2 Proof of theorems

First of all, by using the method of steps, we can see that every solution $y(t)$ of (1.1) is positive for all $t \geq 0$. Thus the change of variables $x(t) = \log(y(t)/K)$ reduces (1.1) to

$$x'(t) = -r(t)\{e^{x([t])} - 1\}, \quad t \geq 0. \tag{2.1}$$

Let k be a nonnegative integer. On any interval $[k, k+1)$, (2.1) is expressed as

$$x'(t) = -r(t)\{e^{x(k)} - 1\}. \tag{2.2}$$

It is clear that $x(t)$ is monotone on $[k, k+1]$ and that $x(k) = 0$ implies $x(t) \equiv 0$ on $[k, \infty)$. Also, integrating (2.2) from k to t, we have

$$x(t) - x(k) = -\{e^{x(k)} - 1\} \int_k^t r(s)ds. \tag{2.3}$$

By continuity solutions, we therefore obtain as $t \to k+1$ in (2.3),

$$x(k+1) = x(k) - \{e^{x(k)} - 1\} \int_k^{k+1} r(s)ds. \tag{2.4}$$

To prove Theorem 1.4, we need the following lemma on oscillatory solutions of (2.1).

Lemma 2.1 *Let $x(t)$ be an oscillatory solution of (2.1), $\{\xi_n : \xi_n < \xi_{n+1}, n = 1, 2, \cdots\}$ be the zeros of $x(t)$. Assume that (1.5) holds and there exists an increasing sequence of positive integers $\{t_n\}$ such that*

$$t_n \leq \xi_n < t_{n+1} \quad \textit{and} \quad x(t_n)x(t_n + 1) < 0 \quad \textit{for} \quad n = 1, 2, \cdots.$$

Without loss of generality, we may assume that

$$(-1)^n x(t_n) > 0 \quad \textit{for} \quad n = 1, 2, \cdots.$$

Then there exists a positive integer M_0, such that

$$-\frac{1}{2} < x(t_{2m+3}) < 0 < x(t_{2m+2}) < \frac{1}{3} \quad \textit{for} \quad m = M_0, M_0 + 1, \cdots.$$

Proof In view of the definition of $\{t_n\}$ and the monotonicity of $x(t)$ on $[k, k+1]$, we notice that if $t_n + 1 < t_{n+1}$ then $|x(t)|$ is nonincreasing on $[t_n + 1, t_{n+1}]$, which implies

$$\begin{array}{ll} x(t_n + 1) \leq x(t_{n+1}) < 0 & \text{if } n \text{ is even} \\ 0 < x(t_{n+1}) \leq x(t_n + 1) & \text{if } n \text{ is odd} \end{array} \tag{2.5}$$

Note that, for $m = 0, 1, \cdots$

$$x(t_{2m+1}) < 0 < x(t_{2m+1} + 1)$$

Let $A_m = -x(t_{2m+1})$; then $A_m > 0$. It follows from (1.5), there exists a positive integer M_0, such that

$$\int_k^{k+1} r(s)ds < 2 + \varepsilon_0 \quad \text{for} \quad k > M_0 \tag{2.6}$$

where $\varepsilon_0 = 0.01$. By (1.1), we obtain

$$\begin{aligned} x(t_{2m+1}+1) &= x(t_{2m+1}) - \left(e^{x(t_{2m+1})} - 1\right) \int_{t_{2m+1}}^{t_{2m+1}+1} r(s)ds \\ &\leq -A_m + (2+\varepsilon_0)(1-e^{-A_m}). \end{aligned} \tag{2.7}$$

Define $\phi(x) = -x + (2+\varepsilon_0)(1-e^{-x})$ and observe that $\phi'(x) = -1 + (2+\varepsilon_0)e^{-x}$. So it is easily seen that

$$\begin{aligned} \phi(A_m) \leq \phi(\log(2+\varepsilon_0)) &= -l\log(2+\varepsilon_0) + (2+\varepsilon_0)\left(1 - \frac{1}{2+\varepsilon_0}\right) \\ &= 1 + \varepsilon_0 - \log(2+\varepsilon_0) < 1 + \varepsilon_0 - \log 2 < \frac{1}{3}. \end{aligned}$$

Hence, by (2.5) and (2.7), we obtain

$$0 < x(t_{2m+2}) \leq x(t_{2m+1}+1) < \frac{1}{3} \quad \text{for} \quad m > M_0$$

Next, let $B_m = x(t_{2m+2})$; then $0 < B_m < \frac{1}{3}$, for $m > M_0$ and

$$\begin{aligned} x(t_{2m+2}+1) &= x(t_{2m+2}) - \{e^{x(t_{2m+2})} - 1\} \int_{t_{2m+2}}^{t_{2m+2}+1} r(s)ds \\ &\geq B_m - (2+\varepsilon_0)(e^{B_m} - 1) \end{aligned} \tag{2.8}$$

Define $\psi(x) = x - (2+\varepsilon_0)(e^x - 1)$. Since $\psi'(x) = 1 - (2+\varepsilon_0)e^x < 0$ for $x > 0$, it follows that

$$\psi(B_m) \geq \psi(\frac{1}{3}) = \frac{1}{3} - (2+\varepsilon_0)(e^{\frac{1}{3}} - 1) > -\frac{1}{2}.$$

Thus, by (2.5) and (2.8), we conclude that

$$0 > x(t_{2m+3}) \geq x(t_{2m+2}+1) > -\frac{1}{2} \quad \text{for} \quad m > M_0.$$

Thus proof is complete. □

Proof of Theorem 1.4 By virtue of the change of variables, it suffices to show that if (1.5) and (1.6) are satisfied, then every solution $x(t)$ of (2.1) tends to 0 as $t \to \infty$. For the case when $x(t)$ is nonoscillatory, (1.6) implies that $x(t)$ tend to 0 as $t \to \infty$ in the similar way to [3]. Therefore, we shall consider the case when $x(t)$ is oscillatory.

Let $\{\xi_n : \xi_n < \xi_{n+1}, n = 1, 2, \cdots\}$ be the zeros of $x(t)$. We only have to consider the case when ξ_n is not integer for $n = 1, 2, \cdots$. Then there exists a sequence of positive integers $\{t_n\}$ such that

$$t_n < \xi_n < t_{n+1} \quad \text{and} \quad x(t_n)x(t_n+1) < 0 \quad \text{for} \quad n = 1, 2, \cdots.$$

without loss of generality, we may assume that

$$(-1)^n x(t_n) > 0 \quad \text{for} \quad n = 1, 2, \cdots.$$

For $m = 0, 1, \cdots$, put $A_m \equiv -x(t_{2m+1})$ and $B_m \equiv x(t_{2m+2})$; then A_m and B_m are positive. From Lemma 2.1, we may also assume that

$$0 < A_m < \frac{1}{2}, \quad \text{for} \quad m > M_0.$$

We here claim that A_m tends 0 as $m \to \infty$.

For every $\varepsilon > 0, (\varepsilon < 0.01)$. It follows from (1.5), there exists a positive integer $M(\varepsilon) > M_0$, such that

$$\int_k^{k+1} r(s)ds < 2+\varepsilon \quad \text{for} \quad k > M \tag{2.9}$$

Using the fact $e^{-x} > 1 - x - \frac{1}{2}x^2 - \frac{1}{6}x^3$ for $x > 0$, we have from (2.7) and (2.9) that

$$\begin{aligned} B_m &\leq x(t_{2m+1}+1) \leq -A_m + (2+\varepsilon)(1-e^{-A_m}) \\ &< -A_m + (2+\varepsilon)\left(A_m - \frac{A_m^2}{2} + \frac{A_m^3}{6}\right) \\ &= (1+\varepsilon)A_m - \left(1+\frac{\varepsilon}{2}\right)A_m^2 + \left(\frac{1}{3}+\frac{\varepsilon}{6}\right)A_m^3 \equiv g^*(A_m). \end{aligned} \tag{2.10}$$

It is easy to see that $g^*(x)$ is a increasing function on $[0,\infty)$, and so

$$B_m < g^*(A_m) < g^*\left(\frac{1}{2}\right) < \frac{1}{3}. \tag{2.11}$$

Next, for $0 < x < \frac{1}{3}$, we get

$$\begin{aligned} e^x &= 1 + x + \frac{x^2}{2} + \frac{e^{\theta x}}{6}x^3 \quad \text{for some} \quad \theta \in (0,1) \\ &< 1 + x + \frac{x^2}{2} + \frac{x^3}{3} \end{aligned}$$

because of $e^x < 2$. Using the above fact, we have from (2.8) that

$$\begin{aligned} A_{m+1} &\leq -x(t_{2m+2}+1) \leq -B_m + (2+\varepsilon)(e^{B_m}-1) \\ &< -B_m + (2+\varepsilon)\left(B_m + \frac{B_m^2}{2} + \frac{B_m^3}{3}\right) \\ &= (1+\varepsilon)B_m + \left(1+\frac{\varepsilon}{2}\right)B_m^2 + \left(\frac{2}{3}+\frac{\varepsilon}{3}\right)B_m^3 \equiv h^*(B_m). \end{aligned} \tag{2.12}$$

Thus, (2.10) and (2.12) became

$$A_{m+1} < h^*(g^*(A_m)), \quad \text{for} \quad m > M_0 \tag{2.13}$$

Consequently, there exists $\alpha \in [0, \frac{1}{2}]$ such that $\limsup_{m\to\infty} A_m = \alpha$, by (2.13), we have

$$\alpha \leq h^*(g^*(\alpha)). \tag{2.14}$$

This implies

$$\begin{aligned} \alpha \leq\ & (1+\varepsilon)\left\{(1+\varepsilon)\alpha - \left(1+\frac{\varepsilon}{2}\right)\alpha^2 + \left[\frac{1}{3}+\frac{\varepsilon}{6}\right]\alpha^3\right\} \\ &+ \left(1+\frac{\varepsilon}{2}\right)\left\{(1+\varepsilon)\alpha - \left(1+\frac{\varepsilon}{2}\right)\alpha^2 + \left[\frac{1}{3}+\frac{\varepsilon}{6}\right]\alpha^3\right\}^2 \\ &+ \left(\frac{2}{3}+\frac{\varepsilon}{3}\right)\left\{(1+\varepsilon)\alpha - \left(1+\frac{\varepsilon}{2}\right)\alpha^2 + \left[\frac{1}{3}+\frac{\varepsilon}{6}\right]\alpha^3\right\}^3 \end{aligned}$$

Let $\varepsilon \to 0$, we obtain that

$$\alpha \leq (\alpha - \alpha^2 + \frac{1}{3}\alpha^3) + (\alpha - \alpha^2 + \frac{1}{3}\alpha^3)^2 + \frac{2}{3}(\alpha - \alpha^2 + \frac{1}{3}\alpha^3)^3 \tag{2.15}$$

Furthermore, for $0 < x \le \frac{1}{2}$, we get

$$\begin{aligned} & x - \{x - x^2 + \frac{1}{3}x^3) + (x - x^2 + \frac{1}{3}x^3)^2 + \frac{2}{3}(x - x^2 + \frac{1}{3}x^3)^3\} \\ & > x - (x - x^2 + \frac{1}{3}x^3) - (x - x^2 + \frac{1}{3}x^3)^2 - \frac{2}{3}x^3 \\ & = x^2(x - \frac{5}{3}x^2 + \frac{2}{3}x^3 - \frac{1}{9}x^4) \\ & = \frac{x^3}{3}(3 - 5x) + \frac{x^5}{9}(6 - x) > 0 \quad \text{for} \quad 0 < x \le \frac{1}{2}. \end{aligned} \tag{2.16}$$

It follows from (2.16), if $\alpha > 0$, we obtain

$$\alpha > \left(\alpha - \alpha^2 + \frac{1}{3}\alpha^3\right) + \left(\alpha - \alpha^2 + \frac{1}{3}\alpha^3\right)^2 + \frac{2}{3}\left(\alpha - \alpha^2 + \frac{1}{3}\alpha^3\right)^3. \tag{2.17}$$

which contradicts (2.15). Hence, we arrive at

$$A_m \to 0 \quad \text{as} \quad m \to \infty \tag{2.18}$$

Now we turn to find the behavior of $x(t)$. recall that $|x(t)|$ is nonincreasing on $[t_n + 1, t_{n+1}]$ if $t_n + 1 < t_{n+1}$ and that

$$|x(t_n)| = \max_{t_n \le t \le \xi_n} |x(t)| \quad \text{and} \quad |x(t_n + 1)| = \max_{\xi_n \le t \le t_n + 1} |x(t)|.$$

It follows from (2.10) and noticing that $g^*(x) < (1 + \varepsilon)x$, we have $0 < B_m < g^*(A_m) < (1 + \varepsilon)A_m$ and hence

$$|x(t)| \le (1 + \varepsilon)A_m \quad \text{for all} \quad t \in [t_{2m+1}, t_{2m+3}].$$

Finally, by (2.18), we therefore conclude that $x(t)$ tend to 0 as $t \to \infty$. The proof of Theorem 1.4 is complete. □

Proof of Theorem 1.5 By virtue of the change of variables, it suffices to show that the zero solution of (2.1) is not globally attractive.

It follows from (1.7), there exists a positive integer M and a real number $r > 2$, such that

$$\int_k^{k+1} r(s)ds \ge r > 2 \quad \text{for} \quad k \ge M. \tag{2.19}$$

If $x(M) > 0$. It follows from (2.4) and (2.19) that

$$\begin{aligned} x(M + 1) &= x(M) - (e^{x(M)} - 1)\int_M^{M+1} r(s)ds \\ &\le x(M) - r(e^{x(M) - 1}) \end{aligned}$$

Define $w(x) = x - r(e^x - 1)$ and observe that $w'(x) = 1 - re^x$, $w''(x) = -re^x < 0$. So it is easy to see that $w(x) > 0$ for $x > 0$. That is

$$x(M + 1) < 0 \quad \text{for} \quad x(M) > 0. \tag{2.20}$$

If $x(M) < 0$, we obtain that

$$\begin{aligned} x(M + 1) &= x(M) - (e^{x(M)} - 1)\int_M^{M+1} r(s)ds \\ &\ge x(M) - r(e^{x(M)} - 1) \\ &> 0, \quad \text{for} \quad -\log r < x(M) < 0. \end{aligned} \tag{2.21}$$

Without loss of generality, we may assume $-\log\ r < x(2M) < 0$. Hence, for $m = M, M+1, \cdots$, let $C_m = -x(2m)$, then $C_{m+1} > 0$ as long as $0 < C_m < \log\ r$.

To end the proof, we shall show that there exist $\lambda > 1$ and a sufficiently small $\varepsilon_0 \in (0, \log\ r)$ such that

$$C_{m+1} > \lambda C_m \quad \text{as long as} \quad 0 < C_m \leq \varepsilon_0.$$

Assume that $0 < C_m < \log\ r$ and put $D_m = x(2m+1)$; then $D_m > 0$. Using the inequality $e^{-x} < 1 - x + \frac{1}{2}x^2$ for $x > 0$, we have

$$\begin{aligned}
D_m &= x(2m+1) = x(2m) - \{e^{x(2m)} - 1\} \int_{2m}^{2m+1} r(s)ds \\
&= -C_m + (1 - e^{-C_m}) \int_{2m}^{2m+1} r(s)ds \\
& \quad -C_m + r(1 - e^{-C_m}) \\
&> -C_m + r\left(C_m - \frac{C_m^2}{2}\right) \\
&= (r-1)C_m - \frac{r}{2}C_m^2 \equiv g_*(C_m).
\end{aligned} \tag{2.22}$$

Also, using the inequality $e^x > 1 + x + \frac{1}{2}x^2$ for $x > 0$, we have

$$\begin{aligned}
C_{m+1} &= -x(2m+2) = -x(2m+1) + [e^{x(2m+1)} - 1] \int_{2m+1}^{2m+2} r(s)ds \\
&= -D_m + (e^{D_m} - 1) \int_{2m+1}^{2m+2} r(s)ds \\
& \quad -D_m + r(e^{D_m} - 1) \\
&> -D_m + r\left(D_m + \frac{D_m^2}{2}\right) \\
&= (r-1)D_m + \frac{r}{2}D_m^2 \equiv h_*(D_m).
\end{aligned} \tag{2.23}$$

Since $h_*(x)$ is an increasing function on $[0, \infty)$, (2.22) and (2.23) become

$$C_{m+1} > h_*(g_*(C_m)) \quad \text{for} \quad m > M. \tag{2.24}$$

Now it follows that

$$\begin{aligned}
h_*(g_*(x) &= (r-1)\left\{(r-1)x - \frac{r}{2}x^2\right\} + \frac{r}{2}\left\{(r-1) - \frac{r}{2}x^2\right\}^2 \\
&= (r-1)^2 x - \frac{r(r-1)}{2}x^2 + \frac{r}{2}\left\{(r-1)^2x^2 - r(r-1)x^3 + \frac{r}{4}x^4\right\} \\
&= (r-1)^2 x + \frac{r(r-1)(r-2)}{2}x^2 - \frac{r^2(r-1)}{2}x^3 + \frac{r^3}{8}x^4 \\
&= (r-1)^2 x + \frac{r(r-1)}{2}x^2\{(r-2) - rx\} + \frac{r^3}{8}x^4 \\
&> (r-1)^2 x \quad \text{for} \quad 0 < x \leq \frac{r-2}{r}.
\end{aligned} \tag{2.25}$$

Thus, by (2.24) and (2.25), we obtain

$$C_{m+1} > (r-1)^2 C_m \quad \text{as long as} \quad 0 < C_m < \frac{r-2}{r},$$

which implies that if $r > 2$ then the zero solution of (2.1) is not globally attractive. The proof of Theorem 1.5 is complete. □

References

[1] Carvalho, L. A. V. and Cooke, K. L. (1988). A nonlinear equation with piecewise constant argument. *Differential Integral Equations*, **1** 359–367.

[2] Gopalsamy, K., Kulenovic, M. R. S. and Ladas, G. (1991). On a logistic equation with piecewise constant arguments. *Differential Integral Equations*, **4**, 215–223.

[3] So J. W.-H. and Yu, J. S. (1995). Global stability in a logistic equation with piecewise constant arguments. *Hokkaido Math. J.*, **24**, 269–286.

[4] Matsunaga, N., Hara, T. and Sakata, S. (2001). Global attractivity for a logistic equation with piecewise constant argument, *Nonliear Differ. Equ. Appl.*, **8**, 45–52.

[5] Furumochi, T. (1992). Stability and oscillation for linear and nonlinear scalar equations with piecewise constant argument. *Appl. Anal.*, **44**, 113–125.

Fields Institute Communications
Volume **42**, 2004

Bounded Oscillation for a Class of Second Order Neutral Delay Difference Equations

Xiaoyan Lin
Department of Mathematics
Huaihua College
Huaihua, Hunan 418008, P. R. China
xiacyanlin98@hotmail.com

Abstract. In this paper, we investigate bounded oscillation for the second order neutral difference equation

$$\Delta^2(x_n - cx_{n-m}) = p_n x_{n-k}.$$

The sufficient conditions obtained in this paper improve the results in related literature.

1 Introduction

Recently, there have been many investigations on delay difference equations. In particular, an extensive literature now exists on the oscillation theory for delay difference equations. We refer to [4-7].

Consider the following second order neutral delay difference equation

$$\Delta^2(x_n - cx_{n-m}) = p_n x_{n-k}, \qquad n \geq n_0, \tag{1.1}$$

where c, p_n are real numbers with $p_n \geq 0, p_n \not\equiv 0$, $n \geq n_0$, m, k, n_0 are nonnegative integers and $m \geq 1$, Δ denotes the forward operator $\Delta x_n = x_{n+1} - x_n$.

As is customary, a solution $\{x_n\}$ of (1.1) is said to be eventually positive if $x_n > 0$ for all large n, and eventually negative if $x_n < 0$ for all large n. It is said to be oscillatory if it is neither eventually positive nor eventually negative. In [1,2], the authors have shown that Eq. (1.1) always has an unbounded and nonoscillatory solution. Therefore, we only need to find conditions for all bounded solutions of Eq. (1.1) to be oscillatory. On the other hands, papers [1, 2] also studied the bounded oscillation of (1.1) and obtained some sufficient conditions in the cases when $0 < c < 1$ and $c > 1$. The following results were established in [2] which improve the corresponding result in [1].

2000 *Mathematics Subject Classification.* Primary 39A11; Secondary 92D25.

Theorem 1.1 *Let $0 \le c < 1$, $k \ge 1$. Assume that*

$$\limsup_{n\to\infty} \sum_{i=0}^{k-1} (i+1) p_{n+i} > 1 - c. \tag{1.2}$$

Then every bounded solution of Eq. (1.1) oscillates.

Theorem 1.2 *Let $c < 0, k > m$. Assume that*

$$\limsup_{n\to\infty} \left(\frac{p_n}{p_{n-m}} \right) = \alpha \in (0, \infty), \tag{1.3}$$

and

$$\limsup_{n\to\infty} \sum_{i=1}^{k-m} i p_{n+i} > 1 - c\alpha. \tag{1.4}$$

Then every bounded solution of Eq. (1.1) oscillates.

In this paper, we use some new techniques to obtain the better sufficient conditions which guarantee that every bounded solution of Eq. (1.1) oscillates. Our results improve the above Theorem 1.1 and Theorem 1.2.

2 Main results

Theorem 2.1 *Let $0 \le c < 1$, $k \ge 1$. Assume that*

$$\limsup_{n\to\infty} \sum_{i=0}^{k-1} \sum_{j=n}^{n+k-i} \frac{p_{j+i}}{1 - c + p_{j+k+1}} \left(\prod_{s=j-k+i+1}^{j+1} \frac{1 - c + p_{s+k}}{1 - c - p_{s+k-1}} \right) > 1. \tag{2.1}$$

Then every bounded solution of Eq. (1.1) oscillates.

Proof For the sake of contradiction, assume that (1.1) has an eventually positive bounded solution $\{x_n\}$ such that $x_{n-m} > 0$, $x_{n-k} > 0$, $n \ge n_1 \ge n_0$. Set $y_n = x_n - c x_{n-m}$, then $\Delta^2 y_n = p_n x_{n-k}$. It is not difficult to prove that there exists $n_2 \ge n_1$ such that

$$\Delta^2 y_n \ge 0, \ \Delta y_n < 0, \ y_n > 0, \qquad n \ge n_2.$$

Let $h_n = \left[\frac{n-n_2}{m}\right]$, where $[\cdot]$ denotes the greatest integer function. Then we have for $n > n_2 + m$.

$$\begin{aligned} x_n &= y_n + c x_{n-m} \\ &= y_n + c y_{n-m} + c^2 y_{n-2m} + \cdots + c^{h_n - 1} y_{n-(h_n-1)m} + c^{h_n} x_{n-m h_n}. \end{aligned}$$

By using the decreasing nature of $\{y_n\}$, we have

$$x_n \ge (1 + c + c^2 + \cdots + c^{h_n - 1}) y_n = \frac{1 - c^{h_n}}{1 - c} y_n.$$

From (2.1) we know there exists some $\varepsilon (0 < \varepsilon < 1)$ such that

$$\begin{aligned} \limsup_{n\to\infty} \sum_{i=0}^{k-1} \sum_{j=n}^{n+k-i} & \frac{(1-\varepsilon) p_{j+i}}{1 - c + (1-\varepsilon) p_{j+i+1}} \\ & \times \left(\prod_{s=j-k+i+1}^{j+1} \frac{1 - c + (1-\varepsilon) p_{s+k}}{1 - c - (1-\varepsilon) p_{s+k+1}} \right) > 1. \end{aligned} \tag{2.2}$$

For this ε, there must exists an integer n_3 such that $c^{h_n} < \varepsilon$ for $n \geq n_3$. Thus $x_n \geq (1-\varepsilon)y_n/(1-c)$ for $n \geq n_3$. Substituting this into Eq. (1.1) we have

$$\Delta^2 y_n \geq \frac{1-\varepsilon}{1-c} p_n y_{n-k}, \qquad n \geq n_3 + k. \tag{2.3}$$

Summing (2.3) from n to $n+k+1$, we have

$$\Delta y_n + \frac{1-\varepsilon}{1-c} \sum_{i=0}^{k+1} p_{n+i} y_{n-k+i} \leq 0. \tag{2.4}$$

From (2.4) we have

$$y_{n+1} - y_n + \frac{1-\varepsilon}{1-c} p_{n+i} y_{n-k+i} \leq 0, \qquad 0 \leq i \leq k+1.$$

Then

$$y_{n+1} - y_n + \frac{1-\varepsilon}{1-c} p_{n+i} y_n \leq 0, \qquad 0 \leq i \leq k.$$

Hence

$$0 < y_{n+1} \leq \left(1 - \frac{1-\varepsilon}{1-c} p_{n+i}\right) y_n,$$

which implies

$$1 - \frac{1-\varepsilon}{1-c} p_{n+i} > 0, \qquad 0 \leq i \leq k. \tag{2.5}$$

On the other hand, it follows from (2.4) that

$$\left(1 + \frac{1-\varepsilon}{1-c} p_{n+k+1}\right) y_{n+1} - \left(1 - \frac{1-\varepsilon}{1-c} p_{n+k}\right) y_n + \frac{1-\varepsilon}{1-c} \sum_{i=0}^{k-1} p_{n+i} y_{n-k+i} \leq 0.$$

Thus there exists an integer $N \geq n_3 + k$ such that

$$y_{n+1} - \frac{1-c-(1-\varepsilon)p_{n+k}}{1-c+(1-\varepsilon)p_{n+k+1}} y_n + \sum_{i=0}^{k-1} \frac{(1-\varepsilon)p_{n+i}}{1-c+(1-\varepsilon)p_{n+k+1}} y_{n-k+i} \leq 0, \ n \geq N.$$

Multiplication both sides of the last inequality by $\prod_{i=N}^{n+1} \frac{1-c+(1-\varepsilon)p_{i+k}}{1-c-(1-\varepsilon)p_{i+k-1}}$, we have

$$\Delta\left(y_n \prod_{i=N}^{n} \frac{1-c+(1-\varepsilon)p_{i+k}}{1-c-(1-\varepsilon)p_{i+k-1}}\right) + \sum_{i=0}^{k-1}\left(\prod_{j=N}^{n+1} \frac{1-c+(1-\varepsilon)p_{j+k}}{1-c-(1-\varepsilon)p_{j+k-1}}\right)$$
$$\times \frac{(1-\varepsilon)p_{n+i}}{(1-c)+(1-\varepsilon)p_{n+k-1}} y_{n-k+i} \leq 0.$$

Set $z_n = y_n \prod_{i=N}^{n} \frac{1-c+(1-\varepsilon)p_{i+k}}{1-c-(1-\varepsilon)p_{i+k-1}}$, then we obtain

$$\Delta z_n + \sum_{i=0}^{k-1}\left(\prod_{j=n-k+i+1}^{n+1} \frac{1-c+(1-\varepsilon)p_{j+k}}{1-c-(1-\varepsilon)p_{j+k-1}}\right)$$
$$\times \left(\frac{(1-\varepsilon)p_{n+i}}{(1-c)+(1-\varepsilon)p_{n+k+1}} z_{n-k+i}\right) \leq 0, \qquad n \geq N \geq n_3 + k. \tag{2.6}$$

It is easy to see that (2.6) has an eventually positive solution and $\Delta z_n < 0$. Summing (2.6) from M to ∞, we have

$$\begin{aligned} z_M &\geq \sum_{n=M}^{\infty} \frac{1}{1-c+(1-\varepsilon)p_{n+k+1}} \\ &\quad \times \sum_{i=0}^{k-1} \left(\prod_{j=n-k+i+1}^{n+1} \frac{1-c+(1-\varepsilon)p_{j+k}}{1-c-(1-\varepsilon)p_{j+k-1}} \right) (1-\varepsilon)p_{n+i} z_{n-k+i} \\ &\geq \sum_{i=0}^{k-1} \sum_{n=M}^{M+k-i} \frac{1}{1-c+(1-\varepsilon)p_{n+k+1}} \\ &\quad \times \left(\prod_{j=n-k+i+1}^{n+1} \frac{1-c+(1-\varepsilon)p_{j+k}}{1-c-(1-\varepsilon)p_{j+k-1}} \right) (1-\varepsilon)p_{n+i} z_{n-k+i} \\ &\geq z_M \sum_{i=0}^{k-1} \sum_{n=M}^{M+k-i} \frac{(1-\varepsilon)p_{n+i}}{1-c+(1-\varepsilon)p_{n+k+1}} \left(\prod_{j=n-k+i+1}^{n+1} \frac{1-c+(1-\varepsilon)p_{j+k}}{1-c-(1-\varepsilon)p_{j+k-1}} \right) \end{aligned}$$

Thus, we have

$$\sum_{i=0}^{k-1} \sum_{n=M}^{M+k-i} \frac{(1-\varepsilon)p_{n+i}}{1-c+(1-\varepsilon)p_{n+k+1}} \left(\prod_{j=n-k+i+1}^{n+1} \frac{1-c+(1-\varepsilon)p_{j+k}}{1-c-(1-\varepsilon)p_{j+k-1}} \right) \leq 1.$$

Taking the limit superior as $M \to \infty$ we obtain

$$\limsup_{M\to\infty} \sum_{i=0}^{k-1} \sum_{n=M}^{M+k-i} \frac{(1-\varepsilon)p_{n+i}}{1-c+(1-\varepsilon)p_{n+k+1}} \left(\prod_{j=n-k+i+1}^{n+1} \frac{1-c+(1-\varepsilon)p_{j+k}}{1-c-(1-\varepsilon)p_{j+k-1}} \right) \leq 1.$$

that is

$$\limsup_{n\to\infty} \sum_{i=0}^{k-1} \sum_{j=n}^{n+k-i} \frac{(1-\varepsilon)p_{j+i}}{1-c+(1-\varepsilon)p_{j+k+1}} \left(\prod_{s=j-k+i+1}^{j+1} \frac{1-c+(1-\varepsilon)p_{s+k}}{1-c-(1-\varepsilon)p_{s+k-1}} \right) \leq 1,$$

This contradicts (2.2). The proof is complete. □

Remark 2.2 Theorem 2.1 improves Theorem 1.1. In fact, when $k = 1$, it is easy to verify that (2.1) implies (1.2). When $k \geq 2$, in view of Theorem A, if $\limsup_{n\to\infty} p_n \geq 2(1-c)/3$, then every bounded solution of Eq. (1.1) oscillates. Therefore, we only consider the case when $p_n \leq 2(1-c)/3$. In fact, we have

$$\begin{aligned} &\sum_{i=0}^{k-1} \sum_{j=n}^{n+k-i} \frac{p_{j+i}}{1-c+p_{i+k+1}} \left(\prod_{s=j-k+i+1}^{j+1} \frac{1-c+p_{s+k}}{1-c-p_{s+k-1}} \right) \\ &= \frac{1}{1-c} \sum_{i=0}^{k-1} \sum_{j=n}^{n+k-i} p_{j+i} \cdot \frac{1}{1-\frac{1}{1-c}p_{j+k}} \left(\prod_{s=j-k+i+1}^{j} \frac{1-c+p_{s+k}}{1-c-p_{s+k-1}} \right) \\ &> \frac{1}{1-c} \sum_{i=0}^{k-1} \sum_{j=n}^{n+k-i} p_{j+i} = \frac{1}{1-c} \sum_{i=0}^{k-1} (i+1) p_{n+i}. \end{aligned}$$

This shows that (2.1) also implies (1.2). On the other hand, the following example shows that condition (2.1) does not imply condition (1.2).

Example 2.3 Consider the following equation:

$$\Delta^2\left(x_n - \frac{1}{4}x_{n-3}\right) = p_n x_{n-2},$$

where $k = 2, m = 3, c = \frac{1}{4}, 1 - c = \frac{3}{4}$, and $p_{3n} = d = 1/3,\ p_{3n+1} = p_{3n+2} = 0$. It is not difficult to calculate

$$\limsup_{n\to\infty}\sum_{i=0}^{k-1}(i+1)p_{n+i} = \limsup_{n\to\infty}(p_n + 2p_{n+1}) = \frac{2}{3} < 1 - c,$$

Thus condition (1.2) doesn't hold. But

$$\begin{aligned}
&\limsup_{n\to\infty}\sum_{i=0}^{k-1}\sum_{j=n}^{n+k-i}\frac{p_{j+i}}{1-c+p_{j+k+1}}\left(\prod_{s=j-k+i+1}^{j+1}\frac{1-c+p_{s+k}}{1-c-p_{s+k-1}}\right)\\
&= \limsup_{n\to\infty}\sum_{i=0}^{1}\sum_{j=n}^{n+2-i}\frac{p_{j+i}}{1-c+p_{j+3}}\left(\prod_{s=j+i-1}^{j+1}\frac{1-c+p_{s+2}}{1-c-p_{s+1}}\right)\\
&= \limsup_{n\to\infty}\left[\sum_{j=n}^{n+2}\frac{p_j}{1-c+p_{j+3}}\left(\prod_{s=j-1}^{j+1}\frac{1-c+p_{s+2}}{1-c-p_{s+1}}\right)\right.\\
&\quad\left.+\sum_{j=n}^{n+1}\frac{p_{j+1}}{1-c+p_{j+3}}\left(\prod_{s=j}^{j+1}\frac{1-c+p_{s+2}}{1-c-p_{s+1}}\right)\right]\\
&= \frac{d}{1-c+d}\left[\frac{1-c}{1-c-d}\cdot\frac{1-c}{1-c}\cdot\frac{1-c+d}{1-c}\right] + \frac{d}{1-c}\cdot\frac{1-c}{1-c-d}\cdot\frac{1-c}{1-c}\\
&= \frac{2d}{1-c-d} = \frac{8}{5} > 1.
\end{aligned}$$

Thus condition (2.1) is satisfied.

In view of Theorem 2.1, every bounded solution of Eq. (1.1) oscillates.

Theorem 2.4 *Let $c < 0, k > m$. Assume that*

$$\limsup_{n\to\infty}\frac{p_n}{p_{n-m}} = \alpha \in (0,\infty), \tag{2.7}$$

and

$$\limsup_{n\to\infty}\sum_{i=0}^{k-m-1}\sum_{j=n}^{n+k-m-i}\frac{p_{j+i}}{1-c\alpha+p_{j+k-m+1}} \times\left(\prod_{s=j-k+m+i+1}^{j+1}\frac{1-c\alpha+p_{s+k-m}}{1-c\alpha-p_{s+k-m-1}}\right) > 1. \tag{2.8}$$

Then every bounded solution of Eq. (1.1) oscillates.

Proof Assume that $\{x_n\}$ is an eventually positive bounded solution. Set $y_n = x_n - cx_{n-m}$, then there exists $n_1 \geq n_0$ such that

$$\Delta^2 y_n \geq 0,\ \Delta y_n < 0,\ y_n > 0, \quad n \geq n_1.$$

In view of (2.8) there must exist a constant $\mu > 1$ such that

$$\limsup_{n\to\infty} \sum_{i=0}^{k-m-1} \sum_{j=n}^{n+k-m-i} \frac{p_{j+i}}{(1-\mu c\alpha)+p_{j+k-m+1}} \left(\prod_{s=j-k+m+i+1}^{j+1} \frac{1-\mu c\alpha + p_{s+k-m}}{1-\mu c\alpha - p_{s+k-m}} \right) > 1. \tag{2.9}$$

For this μ, it follows from (11) that there must exist an integer $n_2 \geq n_1$ such that

$$-c\frac{p_n}{p_{n-m}} \leq -\mu c\alpha, \qquad n \geq n_2. \tag{2.10}$$

So we have

$$\Delta^2 y_n - c\frac{p_n}{p_{n-m}}\Delta^2 y_{n-m} = p_n x_{n-k} - c\frac{p_n}{p_{n-m}} \cdot p_{n-m}x_{n-m-k},$$

that is

$$\Delta^2 y_n - c\frac{p_n}{p_{n-m}}\Delta^2 y_{n-m} = p_n y_{n-k}, \qquad n \geq n_2. \tag{2.11}$$

Substituting (2.10) into (2.11) we obtain

$$\Delta^2(y_n - \mu c\alpha y_{n-m}) \geq p_n y_{n-k}.$$

Set $z_n = y_n - \mu c\alpha y_{n-m}$, then

$$\Delta^2 z_n \geq p_n y_{n-k}, \qquad n \geq n_2, \tag{2.12}$$

where $\Delta^2 z_n \geq 0$, $\Delta z_n < 0$, $z_n > 0$.

On the other hand, we have

$$z_n = y_n - \mu c\alpha y_{n-m} \leq y_{n-m} - \mu c\alpha y_{n-m} = (1-\mu c\alpha)y_{n-m}.$$

Thus

$$y_n \geq \frac{1}{1-\mu c\alpha} z_{n+m}, \qquad n \geq n_2 + m. \tag{2.13}$$

Substituting (2.13) into (2.12) we have

$$\Delta^2 z_n \geq \frac{1}{1-\mu c\alpha} p_n z_{n-(k-m)}, \tag{2.14}$$

where

$$\Delta^2 z_n \geq 0,\ \Delta z_n < 0,\ z_n > 0, \qquad n \geq N.$$

Summing (2.14) from n to $n+(k-m)+1$ we have

$$\Delta z_n + \frac{1}{1-\mu c\alpha} \sum_{i=0}^{k-m+1} p_{n+i} z_{n-(k-m)+i} \leq 0, \qquad n \geq N. \tag{2.15}$$

Set $w_n = z_n \prod_{i=N}^{n} \frac{1-\mu c\alpha + p_{i+k-m}}{1-\mu c\alpha - p_{i+k-m-1}}$. Similar to the proof of Theorem 2.1, from (2.15) we have

$$\Delta w_n + \frac{1}{1-\mu c\alpha + p_{n+k-m+1}} \times \sum_{i=0}^{k-m-1} \left(\prod_{j=n-(k-m)+i+1}^{n+1} \frac{1-\mu c\alpha + p_{j+(k-m)}}{1-\mu c\alpha - p_{j+(k-m)-1}} \right) p_{n+i} z_{n+(k-m)+i} \leq 0. \tag{2.16}$$

This shows that inequality (20) has an eventually positive solution $\{w_n\}$ and $\Delta w_n < 0$. Summing (2.16) from N to ∞, we obtain

$$\begin{aligned}
w_N &\geq \sum_{n=N}^{\infty} \frac{1}{1-\mu c\alpha + p_{n+k-m+1}} \sum_{i=0}^{k-m-1} \left(\prod_{j=n-(k-m)+i+1}^{n+1} \frac{1-\mu c\alpha + p_{j+(k-m)}}{1-\mu c\alpha - p_{j+(k-m)-1}} \right) \\
&\quad \times p_{n+i} w_{n-k+m+i} \\
&\geq \sum_{i=0}^{k-m-1} \sum_{n=N}^{N+k-m-i} \frac{1}{1-\mu c\alpha + p_{n+k-m+1}} \left(\prod_{j=n-k+m+i+1}^{n+1} \frac{1-\mu c\alpha + p_{j+k-m}}{1-\mu c\alpha - p_{j+k-m-1}} \right) \\
&\quad \times p_{n+i} w_{n-k+m+i} \\
&\geq w_N \sum_{i=0}^{k-m-1} \sum_{n=N}^{N+(k-m)-i} \frac{p_{n+i}}{1-\mu c\alpha + p_{n+k-m+1}} \\
&\quad \times \left(\prod_{j=n-k+m+i+1}^{n+1} \frac{1-\mu c\alpha + p_{j+k-m}}{1-\mu c\alpha - p_{j+k-m-1}} \right).
\end{aligned}$$

It follows that

$$\sum_{i=0}^{k-m-1} \sum_{n=N}^{N+k-m-i} \frac{p_{n+i}}{1-\mu c\alpha + p_{n+k-m+1}} \left(\prod_{j=n-k+m+i+1}^{n+1} \frac{1-\mu c\alpha + p_{j+k-m}}{1-\mu c\alpha - p_{j+k-m-1}} \right) \leq 1.$$

Taking the limit superior as $N \to \infty$, we obtain

$$\begin{aligned}
\limsup_{N\to\infty} &\sum_{i=0}^{k-m-1} \sum_{n=N}^{N+k-m-i} \frac{p_{n+i}}{1-\mu c\alpha + p_{n+k-m+1}} \\
&\times \left(\prod_{j=n-(k-m)+i+1}^{n+1} \frac{1-\mu c\alpha + p_{j+k-m}}{1-\mu c\alpha - p_{j+k-m-1}} \right) \leq 1,
\end{aligned}$$

that is

$$\begin{aligned}
\limsup_{n\to\infty} &\sum_{i=0}^{k-m-1} \sum_{j=n}^{n+(k-m)-i} \frac{p_{j+i}}{1-\mu c\alpha + p_{j+k-m+1}} \\
&\times \left(\prod_{s=j-k+m+i+1}^{j+1} \frac{1-\mu c\alpha + p_{j+k-m}}{1-\mu c\alpha - p_{s+k-m-1}} \right) \leq 1,
\end{aligned}$$

which contradicts (2.7). The proof is complete. □

Remark 2.5 Similarly, we can prove that Theorem 2.4 improves Theorem 1.2.

Example 2.6 Consider the equation

$$\Delta^2 \left(x_n + \frac{1}{3} x_{n-1} \right) = p_n x_{n-2},$$

where $c = -\frac{1}{3}$, $m = 1, k = 2, k - m = 1$ and $p_{2n} = \frac{1}{3}, p_{2n+1} = 1$.

It is easy to see that

$$\limsup_{n\to\infty}\frac{p_n}{p_{n-m}}=\limsup_{n\to\infty}\frac{p_n}{p_{n-1}}=3=\alpha,$$
$$1-c\alpha=2,$$
$$\limsup_{n\to\infty}\sum_{i=1}^{k-m} ip_{n+i}=\limsup_{n\to\infty}p_{n+1}=1<1-c\alpha.$$

Thus, condition (1.4) doesn't hold. But

$$\begin{aligned}
&\limsup_{n\to\infty}\sum_{i=0}^{k-m-1}\sum_{j=n}^{n+k-m+i}\frac{p_{j+i}}{1-c\alpha+p_{j+k-m+1}}\\
&\qquad\times\left(\prod_{s=j-k+m+i+1}^{j+1}\frac{1-c\alpha+p_{s+k-m}}{1-c\alpha-p_{s+k-m-1}}\right)\\
&=\limsup_{n\to\infty}\sum_{j=n}^{n+1}\frac{p_j}{2+p_{j+2}}\left(\prod_{s=j}^{j+1}\frac{2+p_{s+1}}{2-p_s}\right)\\
&=\limsup_{n\to\infty}\left[\frac{p_n}{2+p_{n+2}}\cdot\frac{2+p_{n+1}}{2-p_n}\cdot\frac{2+p_{n+2}}{2-p_{n+1}}+\frac{p_{n+1}}{2+p_{n+3}}\cdot\frac{2+p_{n+2}}{2-p_{n+1}}\cdot\frac{2+p_{n+3}}{2-p_{n+2}}\right]\\
&=2>1.
\end{aligned}$$

Thus condition (2.8) holds. In view of Theorem 2.4 every bounded solution of Eq. (1.1) oscillates.

References

[1] B.S. Lalli and B.G. Zhang (1992). On existence of positive solutions and bounded oscillations for neutral difference equations. *J. Math. Anal. Appl.*, **166**, 272–287.

[2] J.H. Shen (1994). On second order neutral delay difference equations with variable coefficients. *Journal of Mathematical Study*, **27**, 60–70.

[3] J.H. Shen and I.P. Stavroulakis (2001). Oscillation criteria for delay difference equations. *Electron. J. Diff. Eqns.*, **10**, 1–15.

[4] X.H. Tang, J.S. Yu and D.H. Peng (2000). Oscillation and nonoscillation of neutral difference equations with positive and negative coefficients. *Computers Math. Applic.*, **139**, 169–181.

[5] X.H. Tang and J.S. Yu (1999). Oscillation of delay difference equations. *Computers Math. Applic.*, **37**, 11–20.

[6] X.H. Tang and J.S. Yu (2000). Oscillations of delay difference equations in a critical state. *Applied Math. Letters*, **13**, 9–15.

[7] L.H. Erbe and B.G. Zhang (1988). Oscillation of discrete analogues of delay equation, in *International Conference on Theory and Applications of Differential Equations*, pp. 21–25, Ohio.

[8] G. Ladas, Ch.G. Philos and Y.G. Sficas (1989). Sharp condition for the oscillation of delay difference equations. *J. Appl. Math. Simulation*, **2**, 101–112.

Fields Institute Communications
Volume **42**, 2004

Asymptotic Behavior in Nonlinear Discrete-Time Neural Networks with Delayed Feedback

Kaiyu Liu
College of Mathematics and Econometrics
Hunan University
Changsha 410082, P.R.China
liukyhnu@yahoo.com.cn

Hongqiang Zhang
Information and Compution Science Department
ChangSha Comm University
Changsha 410076, P.R. China

Abstract. We consider a delay difference system

$$\begin{cases} x_{n+1} = \beta x_n + g(y_{n-k}), \\ y_{n+1} = \beta y_n - g(x_{n-k}) \end{cases} \qquad (*)$$

arising as a discrete-time network of two neurons with McCulloch-Pitts nonlinearity, where $\beta \in (0,1)$. k is a positive integer, g is a signal transmission function with a threshold σ. Some interesting results are obtained for the asymptotic behavior of the system $(*)$. Our theorems improve the corresponding theorems in [5] by removing the restriction of the initial conditions.

1 Introduction

We consider the following nonlinear discrete-time system

$$\begin{cases} x_{n+1} = \beta x_n + g(y_{n-k}), \\ y_{n+1} = \beta y_n - g(x_{n-k}), \end{cases} \quad n \in N \tag{1.1}$$

where $\beta \in (0,1), k \in N$ and $g : R \to R$ is given by

$$g(x) = \begin{cases} -\rho & \text{if } x > \sigma, \\ \rho & \text{if } x \le \sigma, \end{cases} \tag{1.2}$$

for two parameters $\rho > 0$ and $\sigma \in R$.

2000 *Mathematics Subject Classification.* Primary 39A11, 92B20.
The first author was supported in part by the NSF of Hunan Province, P. R. China.

System (1.1) can be regarded as the discrete analog of the following artificial neural network of two neurons with delayed feedback

$$\begin{cases} \dot{x} = -\mu x(t) + g(y(t-\tau)), \\ \dot{y} = -\mu y(t) - g(x(t-\tau)), \end{cases} \tag{1.3}$$

where $\frac{dx}{dt}$ and $\frac{dy}{dt}$ are replaced by the forward difference $x_{n+1} - x_n$ and $y_{n+1} - y_n$ respectively. System (1.3) has been recently investigated (see [1,2,3] and references herein). In the discrete version (1.1), $\beta \in (0,1)$ is the internal decay rate, g is the McCulloch-Pitts signal function with the threshold σ and the synaptic weight $\rho > 0$, and k is the signal transmission delay. Such a network is said to be frustrated to describe the excitatory feedback from neuron y to x and the inhibitory feedback from neuron x to y.

Let Z denote the set of all integers. For any $a, b \in Z$ with $a \leq b$, define $N(a) = \{a, a+1, \cdots\}$ and $N(a,b) = \{a, a+1, \cdots, b\}$. Also, let

$$N = N(0).$$

By a solution of (1.1), we mean a sequence (x_n, y_n) of points in R^2 that is defined for all $n \in N(-k)$ and satisfies (1.1) for $n \in N$. Let X denote the set of mappings from $N(-k, 0)$ to R^2. Clearly, for any $\Phi = (\varphi, \psi) \in X$, (1.1) has a unique solution (x_n^Φ, y_n^Φ) satisfying the initial conditions

$$x_i^\Phi = \varphi_i, \quad y_i^\Phi = \psi_i \quad \text{for} \quad i \in N(-k, 0).$$

Recently, Zhou and Wu [5] have studied the dynamics of (1.1)-(1.2) in a special case when $\Phi \in X_\sigma$, where

$$X_\sigma = \{\Phi \in X; \Phi = (\varphi, \psi), \varphi \in R_\sigma^\pm \text{ and } \psi \in R_\sigma^\pm\}$$

with

$$R_\sigma^+ = \{\varphi; \varphi : N(-k,0) \to R \text{ and } \varphi_i - \sigma > 0 \text{ for } i \in N(-k,0)\}$$

and

$$R_\sigma^- = \{\varphi; \varphi : N(-k,0) \to R \text{ and } \varphi_i - \sigma \leq 0 \text{ for } i \in N(-k,0)\}$$

Our goal is to investigate the dynamics of (1.1)-(1.2) with $\sigma \leq -\frac{\rho}{1-\beta}$, for every $\Phi \in X$. This is done by using several lemmas stated in Section 2. These lemmas are concerned with the non-oscillatory property of $x_n - \sigma$ and $y_n - \sigma$ for each solution (x_n, y_n) of (1.1)-(1.2). In Section 3, we show that if $\sigma < -\frac{\rho}{1-\beta}$ then every solution of (1.1) converges to the unique equilibrium of (1.1). We also show that if $\sigma = -\frac{\rho}{1-\beta}$ then every solution converges to one of the two equilibria and we give necessary and sufficient conditions, in terms of k, σ and Φ, which guarantee (x_n^Φ, y_n^Φ) converging to each of two equilibria respectively. Our theorems improve the corresponding theorems in [5] by removing the restriction that $\Phi \in X_\sigma$.

2 Several lemmas

In this section, we will establish some lemmas, which play an important role in our main results in the next section.

Let (x_n, y_n) be a solution of (1.1) with a given initial value $\Phi \in X$ and set

$$z_n = (x_n - \sigma)(y_n - \sigma) \tag{2.1}$$

Lemma 2.1 *Suppose* $\sigma \leq -\frac{\rho}{1-\beta}$ *and*

$$x_n \not\equiv \sigma, \qquad y_n \not\equiv \sigma \quad \textit{if} \quad \sigma = -\frac{\rho}{1-\beta}. \tag{2.2}$$

Then either $z_n > 0$ *or* $z_n < 0$.

Proof Assume, for the sake of contradiction, that z_n is oscillatory. Then by (2.1), either $x_n - \sigma$ or $y_n - \sigma$ oscillates. In the following we will only consider the case which $y_n - \sigma$ is oscillatory. The case which $x_n - \sigma$ oscillates is similar. Let $\{n_j\}_{j=1}^{\infty}$ be a sequence of positive integers, $0 < n_1 < n_2 < \cdots < n_j < n_{j+1} \to \infty$ as $j \to \infty$, such that

$$y_{n_j} \leq \sigma, \quad y_{n_j+1} \geq \sigma \quad \text{for} \quad j = 1, 2, \cdots . \tag{2.3}$$

By (1.1), it is easy to show that

$$y_{n_j+1}\left(\frac{1}{\beta}\right)^{n_j+1} = y_{n_1}\left(\frac{1}{\beta}\right)^{n_1} - \sum_{l=n_1}^{n_j}\left(\frac{1}{\beta}\right)^{l+1} g(x_{l-k}) \tag{2.4}$$

and

$$y_{n_j}\left(\frac{1}{\beta}\right)^{n_j} = y_{n_1+1}\left(\frac{1}{\beta}\right)^{n_1+1} - \sum_{l=n_1+1}^{n_j-1}\left(\frac{1}{\beta}\right)^{l+1} g(x_{l-k}). \tag{2.5}$$

From (2.3) and (2.4), we have

$$\begin{aligned} \sigma\left[\left(\frac{1}{\beta}\right)^{n_j+1} - \left(\frac{1}{\beta}\right)^{n_1}\right] &\leq -\sum_{l=n_1}^{n_j}\left(\frac{1}{\beta}\right)^{l+1} g(x_{l-k}) \\ &\leq \frac{\rho}{1-\beta}\left[\left(\frac{1}{\beta}\right)^{n_j+1} - \left(\frac{1}{\beta}\right)^{n_1}\right]. \end{aligned} \tag{2.6}$$

By (2.5), we obtain

$$\begin{aligned} \sigma\left[\left(\frac{1}{\beta}\right)^{n_j} - \left(\frac{1}{\beta}\right)^{n_1+1}\right] &\geq -\sum_{l=n_1+1}^{n_j-1}\left(\frac{1}{\beta}\right)^{l+1} g(x_{l-k}) \\ &\geq -\frac{\rho}{1-\beta}\left[\left(\frac{1}{\beta}\right)^{n_j} - \left(\frac{1}{\beta}\right)^{n_1+1}\right]. \end{aligned} \tag{2.7}$$

Noting $0 < \beta < 1$, from (2.6) and (2.7) it follows that $|\sigma| \leq \frac{\rho}{1-\beta}$. Since $\sigma \leq -\frac{\rho}{1-\beta}$, we must have $\sigma = -\frac{\rho}{1-\beta}$.

By (2.7), we see that

$$g(x_{l-k}) = \rho \quad \text{for} \quad l \in N(n_1 + 1, n_j).$$

Letting $j \to \infty$ yields

$$g(x_{l-k}) = \rho \quad \text{for} \quad l \in N(n_1).$$

Substituting this into (1.1), we get

$$y_{n+1} = \beta y_n - \rho \quad \text{for} \quad n \in N(n_1 + 1).$$

By solving this equation, we obtain

$$y_n = \left(y_{n_1+1} + \frac{\rho}{1-\beta}\right)\beta^{n-n_1-1} - \frac{\rho}{1-\beta} \quad \text{for} \quad n \in N(n_1 + 2).$$

Clearly, y_n is monotone for $n \in N(n_1+2)$. Since $y_n - \sigma$ is oscillatory, it follows that

$$y_n = -\frac{\rho}{1-\beta} = \sigma \quad \text{for} \quad n \in N(n_1+2).$$

It contradicts (2.2).

This completes the proof. □

Lemma 2.2 *Let the assumption of Lemma 2.1 hold. Then*

$$\lim_{n\to\infty} x_n = x^* \quad \textit{and} \quad \lim_{n\to\infty} y_n = y^*$$

exist. Moreover, $z_n < 0$ *implies that* $\sigma = -\frac{\rho}{1-\beta}$ *and* $(x^*, y^*) = \left(-\frac{\rho}{1-\beta}, -\frac{\rho}{1-\beta}\right)$, $z_n > 0$ *implies that* $\sigma \le -\frac{\rho}{1-\beta}$ *and* $(x^*, y^*) = \left(-\frac{\rho}{1-\beta}, \frac{\rho}{1-\beta}\right)$.

Proof In view of Lemma 2.1, we have the following two possible cases

Case1. $z_n < 0$. In this case, we have

$$x_n < \sigma, \qquad y_n > \sigma \tag{2.8}$$

for all large n, namely, say for $n > n_0$. Then for $n > n_0 + k$, by (1.1), it follows that

$$\begin{cases} x_{n+1} = \beta x_n - \rho, \\ y_{n+1} = \beta x_n - \rho. \end{cases}$$

This implies that

$$\lim_{n\to\infty} x_n = -\frac{\rho}{1-\beta}, \qquad \lim_{n\to\infty} y_n = -\frac{\rho}{1-\beta}.$$

This together with (2.8), yields

$$-\frac{\rho}{1-\beta} = \lim_{n\to\infty} x_n \le \sigma, \qquad -\frac{\rho}{1-\beta} = \lim_{n\to\infty} y_n \ge \sigma,$$

that is, $\sigma = -\frac{\rho}{1-\beta}$.

Case 2. $z_n > 0$. For this case, we see that

$$x_n > \sigma, \qquad y_n > \sigma. \tag{2.9}$$

As in the proof of Case 1, we can easily show that (2.9) implies that

$$\lim_{n\to\infty} (x_n, y_n) = \left(-\frac{\rho}{1-\beta}, \frac{\rho}{1-\beta}\right), \qquad \sigma \le -\frac{1}{1-\beta}.$$

This completes the proof of the Lemma. □

Lemma 2.3 *Let* $\sigma = -\frac{\rho}{1-\beta}$ *and let*

$$n^- = \min\left\{n \in N;\ y_n > -\frac{\rho}{1-\beta}\right\}.$$

Then n^- *exists and we have*

(i) $z_n < 0$ *implies that* $x_{n^-+k} < -\frac{\rho}{1-\beta}$ *and* $y_k > -\frac{\rho}{1-\beta}$;

(ii) $z_n > 0$ *implies that* $x_{n^-+k} > -\frac{\rho}{1-\beta}$;

(iii) $x_{n^-} \equiv -\frac{\rho}{1-\beta}$ *implies that* $x_{n^-+k} = -\frac{\rho}{1-\beta}$ *and* $y_k > -\frac{\rho}{1-\beta}$.

Proof First, we note that if

$$n^- = \min\left\{n \in N; y_n > -\frac{\rho}{1-\beta}\right\}$$

exists, then $n^- \in N$ and $y_{n^-} > -\frac{\rho}{1-\beta}$. By (1.1) and (1.2), we have

$$y_{n+1} \geq \beta y_n - \rho \quad \text{for} \quad n \in N.$$

By a simple induction, we obtain

$$y_n > -\frac{\rho}{1-\beta} \quad \text{for} \quad n \in N(n^-). \tag{2.10}$$

Substituting this into (1.1), we get

$$x_{n+1} = \beta x_n - \rho \quad \text{for} \quad n \in N(n^- + k).$$

Solving this equation, yields

$$x_n = \left(x_{n^-+k} + \frac{\rho}{1-\beta}\right)\beta^{n-(n^-+k)} - \frac{\rho}{1-\beta} \quad \text{for} \quad n \in N(n^- + k + 1). \tag{2.11}$$

In view of Lemma 2.1, we have the following three possible cases:

Case (i) $z_n < 0$. In this case, by the proof of Lemma 2.2, we have

$$x_n < -\frac{\rho}{1-\beta}, \qquad y_n > -\frac{\rho}{1-\beta}, \tag{2.12}$$

this implies that n^- exists. From (2.11) and (2.12), we see that $x_{n^-+k} < -\frac{\rho}{1-\beta}$ and

$$x_n < -\frac{\rho}{1-\beta} \quad \text{for} \quad n \in N(n^- + k). \tag{2.13}$$

On the other hand, by (1.1), we have for $n \in N(0, n^- + k - 1)$,

$$x_{n^-+k}\left(\frac{1}{\beta}\right)^{n^-+k} = x_n\left(\frac{1}{\beta}\right)^n + \sum_{l=n}^{n^-+k-1}\left(\frac{1}{\beta}\right)^{l+1} g(y_{l-k}).$$

This implies that, for $n \in N(0, n^- + k - 1)$,

$$x_n = \beta^n\left[x_{n^-+k}\left(\frac{1}{\beta}\right)^{n^-+k} - \sum_{l=n}^{n^-+k-1}\left(\frac{1}{\beta}\right)^{l+1} g(y_{l-k})\right] \leq -\frac{\rho}{1-\beta}. \tag{2.14}$$

From (2.13) and (2.14), we have

$$x_n < -\frac{\rho}{1-\beta} \quad \text{for} \quad n \in N.$$

Hence

$$y_{n+1} = \beta y_n - \rho \quad \text{for} \quad n \in N(k)$$

and therefore,

$$y_n = \left(y_k + \frac{\rho}{1-\beta}\right)\beta^{n-k} - \frac{\rho}{1-\beta} \quad \text{for} \quad n \in N(k+1).$$

This together with (2.12), yields $y_k > -\frac{\rho}{1-\beta}$.

Case (ii) $z_n > 0$. For this case, we have

$$x_n > -\frac{\rho}{1-\beta}, \qquad y_n > -\frac{\rho}{1-\beta}.$$

Clearly, this implies that n^- exists and $x_{n^-+k} > -\frac{\rho}{1-\beta}$ (by (2.11)).

Case (iii) $x_n \equiv -\frac{\rho}{1-\beta}$ (eventually). By (1.1) and (1.2), we have $y_{n-k} > -\frac{\rho}{1-\beta}$. This implies that n^- exists and $x_{n^-+k} = -\frac{\rho}{1-\beta}$. Now we claim that $k \geq n^-$. If not, $n^- > k > 0$. By the definition of n^-, we see that

$$y_n \leq -\frac{\rho}{1-\beta} \quad \text{for} \quad n \in N(0, n^- - 1). \tag{2.15}$$

Letting $n = n^- - k - 1$ in (2.14) and noting (2.15), we find

$$x_{n^- -k-1} = \beta^{n^- -k-1} \left[-\frac{\rho}{1-\beta} \left(\frac{1}{\beta}\right)^{n^- +k} - \sum_{l=n^- -k-1}^{n^- +k-1} \left(\frac{1}{\beta}\right)^{l+1} g(y_{l-k}) \right] \leq -\frac{\rho}{1-\beta},$$

which yields

$$y_{n^-} = \beta y_{n^- -1} - \rho.$$

Noting $y_{n^-} > -\frac{\rho}{1-\beta}$, it follows that

$$y_{n^- -1} > \frac{1}{\beta}\left(-\frac{\rho}{1-\beta} + \rho\right) = -\frac{\rho}{1-\beta}.$$

This contradicts (2.15), so we have $k \geq n^+$. Therefore, it follows from (2.10) that $y_k > -\frac{\rho}{1-\beta}$.

This completes the proof. □

3 Main results

In this section, we will investigate the large time behavior of solutions of (1.1)-(1.2) with $\sigma \leq -\frac{\rho}{1-\beta}$, for every initial value $\Phi \in X$.

In view of Lemmas 2.1 and 2.2, we can immediately get the following theorem.

Theorem 3.1 *Let $\sigma < -\frac{\rho}{1-\beta}$ and let $\Phi \in X$. Then $(x_n^\Phi, y_n^\Phi) \to \left(-\frac{\rho}{1-\beta}, \frac{\rho}{1-\beta}\right)$ as $n \to \infty$.*

Theorem 3.1 gives asymptotic behavior of all solutions of (1.1)-(1.2) for the case where $\sigma < -\frac{\rho}{1-\beta}$. In the following we will consider the case $\sigma = -\frac{\rho}{1-\beta}$. First we give the following theorem.

Theorem 3.2 *Let $\sigma = -\frac{\rho}{1-\beta}$ and let $\Phi \in X$. Then $(x_n^\Phi, y_n^\Phi) \to \left(-\frac{\rho}{1-\beta}, -\frac{\rho}{1-\beta}\right)$ or $\left(-\frac{\rho}{1-\beta}, \frac{\rho}{1-\beta}\right)$ as $n \to \infty$.*

Proof By Lemma 2.1, we have the following three possible cases:

(i) $z_n < 0$;

(ii) $z_n > 0$;

(iii) $x_n \equiv -\frac{\rho}{1-\beta}$. In case (i) and (ii), by Lemma 2.2, we see that

$$\lim_{n\to\infty} (x_n, y_n) = (x^*, y^*) = \begin{cases} \left(-\frac{\rho}{1-\beta}, -\frac{\rho}{1-\beta}\right) & \text{if } z_n < 0, \\ \left(-\frac{\rho}{1-\beta}, \frac{\rho}{1-\beta}\right) & \text{if } z_n > 0. \end{cases}$$

In case (iii), we have

$$y_{n+1} = \beta y_n - \rho$$

for all large n. This implies that $\lim_{n\to\infty} y_n = -\frac{\rho}{1-\beta}$ and so

$$\lim_{n\to\infty} (x_n, y_n) = \lim_{n\to\infty} \left(-\frac{\rho}{1-\beta}, y_n\right) = \left(-\frac{\rho}{1-\beta}, -\frac{\rho}{1-\beta}\right).$$

This completes the proof of the theorem. □

Next, we give necessary and sufficient conditions, in terms of the synaptic delay τ and the initial value Φ, under which the solution $(x_n^\Phi, y_n^\Phi) \to \left(-\frac{\rho}{1-\beta}, -\frac{\rho}{1-\beta}\right)$ and $\left(-\frac{\rho}{1-\beta}, \frac{\rho}{1-\beta}\right)$ respectively.

In the sequel, for a given sequence $\{a_n\}$, we define $\sum_{j=n+1}^{n} a_n = 0$.

Theorem 3.3 *Let* $\sigma = -\frac{\rho}{1-\beta}$ *and let* $\Phi = (\varphi, \psi) \in X$. *Then as* $n \to \infty$, $(x_n^\Phi, y_n^\Phi) \to \left(-\frac{\rho}{1-\beta}, -\frac{\rho}{1-\beta}\right)$ *if and only if one of the following two cases occurs*

(a) $\psi(0) > -\frac{\rho}{1-\beta}$, *and* $\varphi(0)\beta^k + \sum_{j=1}^{k} \beta^{j-1} g(\psi(-j)) \leq -\frac{\rho}{1-\beta}$;

(b) $\psi(0) \leq -\frac{\rho}{1-\beta}$, $\psi(0)\beta^k - \sum_{j=1}^{k} \beta^{j-1} g(\varphi(-j)) > -\frac{\rho}{1-\beta}$, *and*

$$\varphi(0)\beta^k + \sum_{j=1}^{k} \beta^{j-1} g(\psi(-j)) \leq -\frac{\rho}{1-\beta}\left[2\left(\frac{1}{\beta}\right)^{k-k^-+1} - 1\right], \tag{3.1}$$

where k^- *is a unique positive integer in* $N(1,k)$, *such that*

$$\sum_{j=k^-+1}^{k} \beta^{j-1} g(\varphi(-j)) - \frac{\rho}{1-\beta}\beta^{k^-} > \psi(0)\beta^k \geq \sum_{j=k^-}^{k} \beta^{j-1} g(\varphi(-j)) - \frac{\rho}{1-\beta}\beta^{k^- -1} \tag{3.2}$$

and $(x_n^\Phi, y_n^\Phi) \to \left(-\frac{\rho}{1-\beta}, \frac{\rho}{1-\beta}\right)$ *if and only if one of the following three cases occurs*

(c) $\varphi(0)\beta^k + \sum_{j=1}^{k} \beta^{j-1} g(\psi(-j)) > -\frac{\rho}{1-\beta}$;

(d) $\psi(0)\beta^k - \sum_{j=1}^{k} \beta^{j-1} g(\varphi(-j)) \leq -\frac{\rho}{1-\beta}$;

(e) $\psi(0) \leq -\frac{\rho}{1-\beta}$, $\psi(0)\beta^k - \sum_{j=1}^{k} \beta^{j-1} g(\varphi(-j)) > -\frac{\rho}{1-\beta}$, *and*

$$\varphi(0)\beta^k + \sum_{j=1}^{k} \beta^{j-1} g(\psi(-j)) > -\frac{\rho}{1-\beta}\left[2\left(\frac{1}{\beta}\right)^{k-k^-+1} - 1\right], \tag{3.3}$$

where k^- *is as in (3.2).*

Proof Let

$$n^- = \min\left\{n \in N; y_n > -\frac{\rho}{1-\beta}\right\}.$$

Then n^- exists (Lemma 2.3) and $n^- \in N$. For $n \in N(k, k+n^-)$, we have

$$\begin{aligned} x_n &= \beta^n\left[\varphi(0) + \sum_{l=0}^{n-1}\left(\frac{1}{\beta}\right)^{l+1} g(y_{l-k})\right] \\ &= \beta^{n-k}\left[\varphi(0)\beta^k + \sum_{j=1}^{k} \beta^{j-1} g(\psi(-j)) + \frac{\rho}{1-\beta}\left(\frac{1}{\beta}\right)^{n-k} - \frac{\rho}{1-\beta}\right]. \end{aligned} \tag{3.4}$$

For $n \in N(1,k)$, we obtain

$$y_n = \beta^n\left[\psi(0) - \sum_{l=0}^{n-1}\left(\frac{1}{\beta}\right)^{l+1} g(\varphi(l-k))\right]$$

$$= \beta^{n-k}\left[\psi(0)\beta^k - \sum_{j=k-n+1}^{k} \beta^{j-1} g(\varphi(-j))\right]. \tag{3.5}$$

Now we prove the first conclusion of the theorem. In order to prove the sufficiency, let us consider case (a) and (b). In case (a), we have $y_0 > -\frac{\rho}{1-\beta}$. This implies that $n^- = 0$. Thus, it follows from (3.4) that

$$x_{n^-+k} = x_k = \varphi(0)\beta^k + \sum_{j=1}^{k} \beta^{j-1} g(\psi(-j)) \le -\frac{\rho}{1-\beta}.$$

In view of Lemmas 2.1 and 2.3, we must have either $z_n < 0$ or $x_n \equiv -\frac{\rho}{1-\beta}$. Therefore, as in the proof of Theorem 3.2, we see that $(x_n, y_n) \to \left(-\frac{\rho}{1-\beta}, -\frac{\rho}{1-\beta}\right)$ as $n \to \infty$. In case (b), by (3.5) we have

$$y_k = \psi(0)\beta^k - \sum_{j=1}^{k} \beta^{j-1} g(\varphi(-j)) > -\frac{\rho}{1-\beta}.$$

Since $y_0 = \psi(0) \le -\frac{\rho}{1-\beta}$, it follows that $n^- \in N(1,k)$ and

$$\begin{aligned} n^- &= \min\left\{n \in N(1,k); y_n > -\frac{\rho}{1-\beta}\right\} \\ &= k+1-\max\Bigg\{k-n+1 \in N(1,k); \psi(0)\beta^k \\ &\qquad > \sum_{j=k-n+1}^{k} \beta^{j-1} g(\psi(-j)) - \frac{\rho}{1-\beta}\beta^{k-n}\Bigg\}. \end{aligned}$$

This together with (3.2), yields $n^- \le k - k^- + 1$. On the other hand, for $n \in N(0, k-k^-)$, we have (by (3.5))

$$\begin{aligned} y_n &= \beta^{n-k}\left[\psi(0)\beta^k - \sum_{j=k^-+1}^{k} \beta^{j-1} g(\varphi(-j)) + \sum_{j=k^-+1}^{k-n+1} \beta^{j-1} g(\varphi(-j))\right] \\ &< \beta^{n-k}\left[-\frac{\rho}{1-\beta}\beta^{k^-} + \sum_{j=k^-+1}^{k-n+1} \beta^{j-1} g(\varphi(-j))\right] \le -\frac{\rho}{1-\beta}. \end{aligned}$$

This implies that $n^- > k - k^-$. Hence, we have $n^- = k - k^- + 1$. By (3.1) and (3.4), we find that

$$\begin{aligned} x_{n^-+k} &= x_{2k-k^-+1} \\ &= \beta^{k-k^-+1}\left[\varphi(0)\beta^k + \sum_{j=1}^{k} \beta^{j-1} g(\psi(-j)) + \frac{\rho}{1-\beta}\left(\frac{1}{\beta}\right)^{k-k^-+1} - \frac{\rho}{1-\beta}\right] \\ &\le -\frac{\rho}{1-\beta}. \end{aligned}$$

Therefore, as in case (a), we have $(x_n, y_n) \to \left(-\frac{\rho}{1-\beta}, -\frac{\rho}{1-\beta}\right)$. Now we turn to prove the necessity. By the proof of Theorem 3.2, we see that $(x_n, y_n) \to \left(-\frac{\rho}{1-\beta}, -\frac{\rho}{1-\beta}\right)$

implies that either $z_n < 0$ or $x_n \equiv -\frac{\rho}{1-\beta}$. By Lemma 2.3, we get

$$x_{n^-+k} \leq -\frac{\rho}{1-\beta}, \qquad y_k > -\frac{\rho}{1-\beta}. \tag{3.6}$$

From (3.4) and (3.5), it is easy to see that (3.6) is equivalent to

$$\beta^{n^-}\left[\varphi(0)\beta^k + \sum_{j=1}^{k} \beta^{j-1} g(\psi(-j)) + \frac{\rho}{1-\beta}\left(\frac{1}{\beta}\right)^{n^-} - \frac{\rho}{1-\beta}\right] \leq -\frac{\rho}{1-\beta} \tag{3.7}$$

and

$$\psi(0)\beta^k - \sum_{j=1}^{k} \beta^{j-1} g(\varphi(-j)) > -\frac{\rho}{1-\beta},$$

then $n^- = 0$, and (3.7) reduces to

$$\varphi(0)\beta^k + \sum_{j=1}^{k} \beta^{j-1} g(\psi(-j)) \leq -\frac{\rho}{1-\beta}.$$

That is, we have case (a) hold. If $\psi(0) \leq -\frac{\rho}{1-\beta}$, then $n^- = k - k^- + 1$, and (3.7) becomes to

$$\beta^{k-k^-+1}\left[\varphi(0)\beta^k + \sum_{j=1}^{k} \beta^{j-1} g(\psi(-j)) + \frac{\rho}{1-\beta}\left(\frac{1}{\beta}\right)^{k-k^-+1} - \frac{\rho}{1-\beta}\right] \leq -\frac{\rho}{1-\beta},$$

that is, (3.1) is satisfied, and therefore, we have case (b) hold. In the following we prove the second conclusion. First, we consider case (c). We have (by(3.4))

$$\begin{aligned} x_{n^-+k} &= \beta^{n^-}\left[\varphi(0)\beta^k + \sum_{j=1}^{k} \beta^{j-1} g(\psi(-j)) + \frac{\rho}{1-\beta}\left(\frac{1}{\beta}\right)^{n^-} - \frac{\rho}{1-\beta}\right] \\ &> \beta^{n^-}\left[2\frac{\rho}{1-\beta} + \frac{\rho}{1-\beta}\left(\frac{1}{\beta}\right)^{n^-}\right] > -\frac{\rho}{1-\beta}. \end{aligned}$$

According to Lemmas 2.1 and 2.3, we find that $z_n > 0$. As in the proof of Theorem 3.2, we see $(x_n, y_n) \to \left(-\frac{\rho}{1-\beta}, \frac{\rho}{1-\beta}\right)$. Next, we consider case (d). We have

$$y_k = \psi(0)\beta^k - \sum_{j=1}^{k} \beta^{j-1} g(\varphi(-j)) \leq -\frac{\rho}{1-\beta}.$$

By Lemmas 2.1 and 2.3, we see that $z_n > 0$. This implies that $(x_n, y_n) \to \left(-\frac{\rho}{1-\beta}, \frac{\rho}{1-\beta}\right)$. Finally we consider case (e). In this case, one can easily show that, by a similar argument as in case (b), that $x_{n^-+k} > -\frac{\rho}{1-\beta}$. Thus as in case (c), we see that $(x_n, y_n) \to \left(-\frac{\rho}{1-\beta}, \frac{\rho}{1-\beta}\right)$. At last, we prove the necessity part of the second conclusion. To this end, let

$$\varphi(0)\beta^k + \sum_{j=1}^{k} \beta^{j-1} g(\psi(-j)) \leq -\frac{\rho}{1-\beta} \tag{3.8}$$

and let

$$\psi(0)\beta^k - \sum_{j=1}^{k} \beta^{j-1} g(\varphi(-j)) > -\frac{\rho}{1-\beta}.$$

Since $(x_n, y_n) \to \left(-\frac{\rho}{1-\beta}, \frac{\rho}{1-\beta}\right)$, it follows that $z_n > 0$. So we have $x_{n^-+k} > -\frac{\rho}{1-\beta}$, i.e,

$$\beta^{n^-}\left[\varphi(0)\beta^k + \sum_{j=1}^{k} \beta^{j-1} g(\psi(-j)) - \frac{\rho}{1-\beta}\left(\frac{1}{\beta}\right)^{n^-} + \frac{\rho}{1-\beta}\right] > -\frac{\rho}{1-\beta}.$$

If $\psi(0) > -\frac{\rho}{1-\beta}$, then $n^- = 0$ and

$$\varphi(0)\beta^k + \sum_{j=1}^{k} \beta^{j-1} g(\psi(-j)) > -\frac{\rho}{1-\beta}.$$

But this contradicts (3.8). Therefore, we must have $\psi(0) \leq -\frac{\rho}{1-\beta}$. By noting the fact that $n^- = k - k^- + 1$, we see that

$$\beta^{k-k^-+1}\left[\varphi(0)\beta^k + \sum_{j=1}^{k} \beta^{j-1} g(\psi(-j)) + \frac{\rho}{1-\beta}\left(\frac{1}{\beta}\right)^{k-k^-+1} - \frac{\rho}{1-\beta}\right] > -\frac{\rho}{1-\beta}.$$

That is, (3.3) is satisfied and case (e) holds.

This completes the proof of theorem. □

Remark 3.4 Theorems 3.1 and 3.3 improve the following corresponding Theorems in [5] by removing the restriction that $\Phi \in X_\sigma$.

Theorem 1.2[5] Let $\Phi = (\phi, \psi) \in X_\sigma$. Then $\lim_{n\to\infty}(x^\Phi(n), y^\Phi(n)) = \left(-\frac{\rho}{1-\beta}, \frac{\rho}{1-\beta}\right)$ if $\sigma < -\frac{\rho}{1-\beta}$.

Theorem 1.3[5] Let $\sigma = -\frac{\rho}{1-\beta}$. Then $\lim_{n\to\infty}(x^\Phi(n), y^\Phi(n)) = \left(-\frac{\rho}{1-\beta}, \frac{\rho}{1-\beta}\right)$ for $\Phi \in X_\sigma^{+,+} \cup X_\sigma^{+,-} \cup X_\sigma^{-,-}$,and $\lim_{n\to\infty}(x^\Phi(n), y^\Phi(n)) = \left(-\frac{\rho}{1-\beta}, -\frac{\rho}{1-\beta}\right)$ for $\Phi \in X_\sigma^{-,+}$.

References

[1] T. Roska and L. O. Chua. *Cellular neural networks with nonlinear and delay-type template elements*, Int. J. Circuit Theory Appl., **20** (1992), 469–481.

[2] L. Huang and J. Wu. *The role of threshold in preventing delay-induced oscillations of frustrated neural networks with McCulloch-Pitts nonlinearity*, Int. J. Math. Game Theory and Algebra. vol. 11 ,**6** (2001), 71–100.

[3] J. Wu. *Introduction to Neural dynamics and signal transmission delay*, de-Gruyter, 2002.

[4] Z. Zhou. J. S. Yu and L. H. Huang. *Asymptotic behavior of delay difference systems*, Computers Math. Applic., **42** (2001), 283–290.

[5] Z. Zhou and J. Wu. *Attractive periodic orbits in nonlinear discrete-time neural networks with delayed feedback*, J. Diff. Eq. Appl., **8** (2002), 467–483.

Fields Institute Communications
Volume **42**, 2004

Asymptotic Behavior and Periodicity of Solutions for a Difference Equation with Piecewise Constant Nonlinearity

Ye Liu
College of Mathematics and Econometrics
Hunan University
Changsha, Hunan 410082, P. R. China
liuyemaths@sohu.com

Lihong Huang
College of Mathematics and Econometrics
Hunan University
Changsha, Hunan 410082, P. R. China
lhhuang@hnu.net.cn

Zhaohui Yuan
College of Mathematics and Econometrics
Hunan University
Changsha, Hunan 410082, P. R. China
zhaohuiyuan@hotmail.com

Dedicated to Professor Zhicheng Wang on the occasion of his seventieth birthday

Abstract. This paper is concerned with a delay difference equation with piecewise constant nonlinearity of McCulloch-Pitts type. Such a difference equation can be regarded as the discrete analog of a differential equation with piecewise constant argument. Some interesting results are obtained for the asymptotic behavior and periodicity of solutions for the equation. An illustrative example is given.

1 Introduction

Consider the following delay difference equation

$$x_n = ax_{n-1} + bf(x_{n-k}), \qquad n = 0, 1, 2, \cdots, \tag{1.1}$$

2000 *Mathematics Subject Classification.* Primary 39A10; Secondary 39A11 .
This work was supported by the NNSF(10371034) of China, the Key Research Program of Science and Technology of the Ministry of Education of China, the Doctor Program Foundation of the Ministry of Education of China (20010532002) and the Foundation for University Excellent Teacher by the Ministry of Education of China.

where $0 < a \leq \frac{1}{a}$ and $b > 0$ are given constants, k is a given positive integer, $f : R \to R$ is a real function. For such a difference equation, much progress has been made in the study of the oscillation, asymptotic behavior and stability of solutions. See, for example, [1-3, 6-12]. But there are only a limited number of works concerning the periodicity of solutions. On the other hand, to the best of our knowledge, most of the existing works are for the case where f is continuous on R. In this paper, we study the asymptotic behavior and periodicity of solutions of (1.1) when f is of the following McCulloch-Pitts type:

$$f(\xi) = \begin{cases} -1, & \xi > \sigma, \\ 1, & \xi \leq \sigma \end{cases} \tag{1.2}$$

for some $\sigma \in R$. Such a type of function has been introduced into the study of some neural network models described by differential and difference equations (see, for example, [4, 5, 9, 12]).

The difference equation (1.1) can be regarded as the discrete analog of the following differential equation with piecewise constant argument.

$$\frac{dx(t)}{dt} = -\lambda x(t) + \mu f(x([t-\tau])), \tag{1.3}$$

where $\lambda > 0$ and $\mu > 0$ are given constants, τ is a nonnegative integer, $[.]$ denotes the greatest integer function, and f satisfies (1.2). In fact, we can rewrite (1.3) into the following form

$$\frac{d(x(t)e^{\lambda t})}{dt} = \mu e^{\lambda t} f(x([t-\tau])). \tag{1.4}$$

Let n be a nonnegative integer, and integrate (1.4) from $n-1$ to $t \in [n-1, n)$. Then we get

$$x(t)e^{\lambda t} - x(n-1)e^{\lambda(n-1)} = \frac{\mu}{\lambda}(e^{\lambda t} - e^{\lambda(n-1)}) f\big(x([n-1-\tau])\big).$$

Let $t \to n$, and write $x_n = x(n)$, $x_{n-k} = x([n-1-\tau])$. Then we obtain

$$x_n = e^{-\lambda} x_{n-1} + \frac{\mu}{\lambda}(1 - e^{-\lambda}) f\big(x_{n-k}\big),$$

which is of the form of (1.1).

Our purpose is to study the asymptotic behavior and periodicity of solutions of (1.1) when $0 < a \leq \frac{1}{2}$. To simplify our representations, we rescale the variables and parameters in (1.1) by

$$x_n^* = \frac{1-a}{b} x(n), \quad f^*(u) = f(\frac{b}{1-a} u), \quad \sigma^* = \frac{1-a}{b}\sigma,$$

and then drop the $*$ to get

$$x_n = a x_{n-1} + (1-a) f(x_{n-k}) \tag{1.5}$$

with $0 < a \leq \frac{1}{2}$ and f satisfying (1.2). In the remaining part of this paper, we replace (1.1) by (1.5).

For the sake of simplicity, throughout this paper, Z denotes the set of all integers. For any $l, m \in Z$, define $N(l) = \{l, l+1, \cdots\}$, and $N(l, m) = \{l, l+1, \cdots, m\}$ whenever $l \leq m$. In addition, we write $N = N(0)$.

Let $\{x_n\}_{n=-k}^{\infty}$ be a sequence, l, $m \in N(-k)$ and $l \leq m$. If $\{x_l, x_{l+1}, \cdots, x_m\}$ satisfies:

(i) $x_l > \sigma, \quad x_{l+1} > \sigma, \quad \cdots, \quad x_{m-1} > \sigma, \quad x_m > \sigma;$

(ii) either $l = -k$ or $l > -k$ and $x_{l-1} \leq \sigma$;
(iii) either $m = \infty$ or $m < \infty$ and $x_{m+1} \leq \sigma$;

then we call that $\{x_l,\ x_{l+1},\ \cdots,\ x_m\}$ is a *positive semicycle of sequence* $\{x_n\}_{n=-k}^{\infty}$ *relative to* σ. Moreover, we say that the length of the positive semicycle is $m-l+1$. If $\{x_l,\ x_{l+1},\ \cdots,\ x_m\}$ satisfies:

(i) $x_l \leq \sigma,\ x_{l+1} \leq \sigma,\ \cdots,\ x_{m-1} \leq \sigma,\ x_m \leq \sigma$;
(ii) either $l = -k$ or $l > -k$ and $x_{l-1} > \sigma$;
(iii) either $m = \infty$ or $m < \infty$ and $x_{m+1} > \sigma$;

then we call that $\{x_l,\ x_{l+1},\ \cdots,\ x_m\}$ is a *negative semicycle of sequence* $\{x_n\}_{n=-k}^{\infty}$ *relative to* σ. Moreover, we say that the length of the negative semicycle is $m-l+1$.

By a solution of (1.5), we mean a sequence $\{x_n\}$ of points in R that is defined for all $n \in N(-k)$ and satisfies (1.5) for $n \in N$. Clearly, for any initial values $\alpha_{-k},\ \alpha_{-k+1},\ \cdots,\ \alpha_{-1} \in R$, there exists a unique solution of equation (1.5) satisfying $x_n = \alpha_n$ for $n \in N(-k,-1)$. In fact, we can obtain the solution by iteration.

By (1.2) and (1.5), it is clear that equation (1.5) has obvious connection with the following two linear nonhomogeneous difference equations

$$x_n = ax_{n-1} + a - 1, \qquad n \in N(n_0+1) \tag{1.6}$$

and

$$x_n = ax_{n-1} + 1 - a, \qquad n \in N(n_0+1). \tag{1.7}$$

Assuming $n_0 \in N(-1)$, it is not difficult to see that the solutions of difference equations (1.6) and (1.7) with initial value $x(n_0) = x_{n_0}$ are

$$x_n = a^{n-n_0}x_{n_0} + a^{n-n_0} - 1, \qquad n \in N(n_0+1) \tag{1.8}$$

and

$$x_n = a^{n-n_0}x_{n_0} + 1 - a^{n-n_0}, \qquad n \in N(n_0+1), \tag{1.9}$$

respectively.

2 Main results and proofs

Theorem 2.1 *Let $\sigma \geq 1$. If $\{x_n\}_{n=-k}^{\infty}$ is the solution of equation (1.5) with the initial values $\alpha_{-k},\ \alpha_{-k+1},\ \cdots,\ \alpha_{-1} \in R$, then $x_n \to 1$ as $n \to \infty$.*

Proof We first show that there exists an integer $m_1 \in N$ such that $x_{m_1} \leq \sigma$. Otherwise, assume that $x_n > \sigma$ for any $n \in N$. It follows from (1.2) and (1.5) that $x_n = ax_{n-1} + a - 1$ for $n \in N(k)$. By (1.6) and (1.8), we have

$$x_n = a^{n-k+1}x_{k-1} + a^{n-k+1} - 1, \qquad n \in N(k),$$

which implies that $\lim_{n\to\infty} x_n = -1 < \sigma$. This contradicts to the hypothesis.

On the other hand, by (1.2) and (1.5), we can get

$$x_n \leq ax_{n-1} + 1 - a, \qquad n \in N(1).$$

Thus, by iteration, we have

$$x_n \leq a^{n-m_1}x_{m_1} + 1 - a^{n-m_1}, \qquad n \in N(m_1+1). \tag{2.1}$$

This implies that $x_n \leq \sigma$ for $n \in N(m_1)$. It follows from (1.2) and (1.5) that

$$x_n = cx_{n-1} + 1 - a, \qquad n \in N(m_1+k),$$

which, together with (1.9), implies that

$$x_n = a^{n-m_1-k}x_{m_1+k} + 1 - a^{n-m_1-k}, \qquad n \in N(m_1+k).$$

Hence $\lim_{n\to\infty} x_n = 1$. This completes the proof. □

Similarly, we can obtain the following result.

Theorem 2.2 *Let $\sigma \le -1$. If $\{x_n\}_{n=-k}^{\infty}$ is the solution of equation (1.5) with initial values α_{-k}, α_{-k+1}, $\cdots$, $\alpha_{-1} \in R$, then $x_n \to -1$ as $n \to \infty$.*

Now, We consider the case where $|\sigma| < 1$. According to the symmetry of (1.2) and (1.5), we only need to consider the case where $\sigma \in [0, 1)$. We first establish one lemma.

Lemma 2.3 *Let $0 \le \sigma < 1$, and let $\{x_n\}_{n=-k}^{\infty}$ be the solution of equation (1.5) satisfying the initial conditions $x_{-k} > \sigma$, $x_{-k+1} > \sigma$, $\cdots$, $x_{-1} > \sigma$. Then, except perhaps for the first semicycle, the length of every positive semicycle of $\{x_n\}_{n=-k}^{\infty}$ relative to σ has to be k, and the length of every negative semicycle of $\{x_n\}_{n=-k}^{\infty}$ relative to σ has to be not less than k and not more than $\log_a(\frac{\sigma-1}{\sigma+1-2a^{-k}})$.*

Proof By (1.2), (1.5) and the fact that $x_{-k} > \sigma$, $x_{-k+1} > \sigma$, $\cdots$, $x_{-1} > \sigma$, we have

$$x_n = ax_{n-1} + a - 1, \qquad n \in N(0, k-1).$$

In view of (1.8), we get

$$x_n = a^{n+1}x_{-1} + a^{n+1} - 1, \qquad n \in N(0, k-1).$$

Thus, it is certain that there exists an integer $m_1 \in N$ such that $\{x_{-k}, x_{-k+1}, \cdots, x_{m_1-1}\}$ is the first semicycle of $\{x_n\}_{n=-k}^{\infty}$ relative to σ. Moreover, it is a positive semicycle, and hence $x_{m_1} \le \sigma$. It follows that

$$x_n = ax_{n-1} + a - 1, \qquad n \in N(0, m_1 + k - 1).$$

Again by (1.8), we obtain

$$x_n = a^{n-m_1+1}x_{m_1-1} + a^{n-m_1+1} - 1, \qquad n \in N(m_1, m_1 + k - 1). \tag{2.2}$$

Assume that $\{x_{m_1}, x_{m_1+1}, \cdots, x_{m_2-1}\}$ is a negative semicycle of $\{x_n\}_{n=-k}^{\infty}$ relative to σ, where m_2 may be $+\infty$. In fact, we can prove that such an m_2 is a finite positive integer. The proof is as follows.

From (2.2), it is easy to see that $x_{m_1+k-1} \le x_{m_1+k-2} \le \cdots \le x_{m_1} \le \sigma$. Thus, $m_2 - m_1 \ge k$. Since $\{x_{m_1}, x_{m_1+1}, \cdots, x_{m_2-1}\}$ is a negative semicycle, by (1.2) and (1.5), we have

$$x_n = ax_{n-1} + 1 - a, \qquad n \in N(m_1 + k, m_2 + k - 1).$$

From (1.9), it follows that

$$x_n = a^{n-m_1-k+1}x_{m_1+k-1} - a^{n-m_1-k+1} + 1, \quad n \in N(m_1 + k, m_2 + k - 1). \tag{2.3}$$

If $m_2 = +\infty$, then $m_2 + k - 1 = +\infty$ and $x_n \to 1$ as $n \to +\infty$. It follows that for any $\sigma \in [0, 1)$, there exists an integer $n_1 \in N(m_1 + k)$ such that $x_n > \sigma$ for $n > n_1$. This contradicts to the hypothesis that $\{x_{m_1}, x_{m_1+1}, \cdots, x_{m_2-1}\}$ is a negative semicycle of $\{x_n\}_{n=-k}^{\infty}$ relative to σ. As a result, m_2 can not be $+\infty$, which implies $x_{m_2-1} \le \sigma$ and $x_{m_2} > \sigma$. Therefore, by (2.2) and (2.3), we get

$$x_{m_2-1} = a^{m_2-m_1-k}x_{m_1+k-1} - a^{m_2-m_1-k} + 1$$

and

$$\begin{aligned} x_{m_2+k-1} &= a^{m_2-m_1}x_{m_1+k-1} - a^{m_2-m_1} + 1 \\ &\leq a^{m_2-m_1}\sigma - a^{m_2-m_1} + 1 \\ &< 1. \end{aligned} \tag{2.4}$$

On the other hand, according to (2.2), we obtain

$$x_{m_1+k-1} = a^k x_{m_1-1} + a^k - 1.$$

It follows that

$$x_{m_2-1} = a^{m_2-m_1}x_{m_1-1} + a^{m_2-m_1} - 2a^{m_2-m_1-k} + 1.$$

Again since $x_{m_1-1} > \sigma$, we have

$$a^{m_2-m_1}\sigma + a^{m_2-m_1} - 2a^{m_2-m_1-k} + 1 < x_{m_2-1} \leq \sigma.$$

Consequently,

$$m_2 - m_1 \leq \log_a\left(\frac{\sigma - 1}{\sigma + 1 - 2a^{-k}}\right).$$

Assume that $\{x_{m_2}, x_{m_2+1}, \cdots, x_{m_3-1}\}$ is a positive semicycle of $\{x_n\}_{n=-k}^{\infty}$ relative to σ. By (2.3), we can see that $x_{m_2+k-1} \geq x_{m_2+k-2} \geq \cdots \geq x_{m_2} > \sigma$. So, $m_3 - m_2 \geq k$. Again by (2.4) and $x_{m_2} > \sigma$, we have

$$x_{m_2+k} = ax_{m_2+k-1} - 1 + a < a - 1 + a \leq 0.$$

That is to say $x_{m_2+k} \leq \sigma$. Therefore, $m_3 - m_2 \leq k$. This, together with $m_3 - m_2 \geq k$, implies $m_3 - m_2 = k$.

By induction, we can complete the proof of Lemma 2.3. □

Theorem 2.4 *Let $0 \leq \sigma < 1$. For any nonnegative integer m, define*

$$\sigma_m^- = \max\left\{0, 1 - \frac{2a^m(1-a^k)}{1-a^{2k+m}}\right\} \quad \text{and} \quad \sigma_m^+ = 1 - \frac{2a^{m+1}(1-a^k)}{1-a^{2k+m}}.$$

Then, for each $\sigma \in [\sigma_m^-, \sigma_m^+)$, the solution $\{x_n^*\}_{n=-k}^{\infty}$ of equation (1.5) satisfying the initial condition

$$x_{-k}^* > \sigma, x_{-k+1}^* > \sigma, \cdots, x_{-2}^* > \sigma, x_{-1}^* = 1 - \frac{2a^{m+k}(1-a^k)}{1-a^{2k+m}} \geq \sigma_m^+ > \sigma \tag{2.5}$$

is periodic with the minimal period $2k+m$. Moreover, $\{x_0^*, x_1^*, \cdots, x_{k+m-1}^*\}$ is a negative semicycle of $\{x_n^*\}_{n=-k}^{\infty}$ relative to σ, $\{x_{k+m}^*, x_{k+m+1}^*, \cdots, x_{2k+m-1}^*\}$ is a positive semicycle of $\{x_n^*\}_{n=-k}^{\infty}$ relative to σ.

Proof For each $\sigma \in [\sigma_m^-, \sigma_m^+)$. Let $\{x_n^*\}_{n=-k}^{\infty}$ be the solution of equation (1.5) satisfying the initial condition (2.5). By (1.2) and (1.5), we obtain $x_0^* = ax_{-1}^* + a - 1$. It follows that

$$x_0^* = -1 + \frac{2a(1-a^{k+m})}{1-a^{2k+m}}. \tag{2.6}$$

Obviously, $x_0^* \leq \sigma_m^- \leq \sigma$. Moreover, from (1.2), (1.5) and (2.5), it is easy to see that

$$x_n^* = ax_{n-1}^* + a - 1, \qquad n \in N(0, k-1).$$

It follows from (1.8) that

$$x_n^* = a^n x_0^* + a^n - 1, \qquad n \in N(0, k-1). \tag{2.7}$$

Again from (2.6), we can see that $x_0^* > -1$. This, together with (2.7), implies that

$$x_{k-1}^* \leq x_{k-2}^* \leq \cdots \leq x_0^* \leq \sigma_m^- \leq \sigma \tag{2.8}$$

and

$$x_{k-1}^* = a^{k-1}x_0^* + a^{k-1} - 1. \tag{2.9}$$

Noting (1.2) and (1.5), we have

$$x_n^* \leq ax_{n-1}^* + 1 - a, \qquad n \in N(k, k+m-1).$$

By iteration, it is easy to see that

$$x_n^* \leq a^{n-k+1}x_{k-1}^* + 1 - a^{n-k+1}, \qquad n \in N(k, k+m-1).$$

Moreover, from (2.6) and (2.9), it follows that

$$\begin{aligned} x_n^* &\leq a^n x_0^* + a^n - 2a^{n-k+1} + 1 \\ &= \frac{2a^{n+1}(1-a^{-k}) + 1 - a^{2k+m}}{1-a^{2k+m}} \\ &\leq \frac{2a^{k+m}(1-a^{-k}) + 1 - a^{2k+m}}{1-a^{2k+m}} \\ &\leq \sigma_m^- \leq \sigma \end{aligned}$$

for $n \in N(k, k+m-1)$. Thus, we obtain

$$x_k^* \leq \sigma, x_{k+1}^* \leq \sigma, \cdots, x_{k+m-1}^* \leq \sigma. \tag{2.10}$$

This, together with (1.2), (1.5) and (2.8), implies that

$$x_n^* = ax_{n-1}^* + 1 - a, \qquad n \in N(k, 2k+m-1).$$

It follows from (1.9) that

$$x_n^* = a^{n-k+1}x_{k-1}^* + 1 - a^{n-k+1}, \qquad n \in N(k, 2k+m-1), \tag{2.11}$$

which, together with (2.9), implies that

$$x_{k+m}^* = a^{k+m}x_0^* - 2a^{m+1} + a^{k+m} + 1.$$

Again by (2.6), we get

$$x_{k+m}^* = 1 - \frac{2a^{m+1}(1-a^k)}{1-a^{2k+m}} \geq \sigma_m^+ > \sigma,$$

which, together with (2.8) and (2.10), implies that $\{x_0^*, x_1^*, \cdots, x_{k+m-1}^*\}$ is a negative semicycle of $\{x_n^*\}_{n=-k}^{\infty}$ relative to σ.

On the other hand, from (2.11), we can see that

$$1 > x_{2k+m-1}^* \geq x_{2k+m-2}^* \geq \cdots \geq x_{k+m}^* > \sigma. \tag{2.12}$$

Moreover, by (2.9) and (2.11), it is easy to see that

$$x_{2k+m-1}^* = a^{2k+m-1}x_0^* - 2a^{m+k} + a^{2k+m-1} + 1.$$

Again in view of (2.6), we obtain

$$x_{2k+m-1}^* = \frac{1 + a^{2k+m} - 2a^{k+m}}{1-a^{2k+m}}.$$

Since $x_{k+m}^* > \sigma$, according to (1.2), (1.5) and (2.6), we can see that

$$x_{2k+m}^* = ax_{2k+m-1}^* + a - 1 = x_0^* \leq \sigma_m^- \leq \sigma. \tag{2.13}$$

Therefore, by (2.12) and (2.13), we know that $\{x^*_{k+m}, x^*_{k+m+1}, \cdots, x^*_{2k+m-1}\}$ is a positive semicycle of $\{x^*_n\}_{n=-k}^{\infty}$ relative to σ.

By induction, it is easy to prove that $\{x^*_n\}_{n=-k}^{\infty}$ is periodic with the minimal period $2k+m$.

This completes the proof of Theorem 2.4. □

For any $m \in N$, according to the representations of σ_m^- and σ_m^+, it is easy to see that $\sigma_{m+1}^- > \sigma_m^+$. As a result, there exists an interval between $[\sigma_m^-, \sigma_m^+)$ and $[\sigma_{m+1}^-, \sigma_{m+1}^+)$. Next, we will study the existence of periodic solutions for equation (1.5) when $\sigma \in [\sigma_m^+, \sigma_{m+1}^-)$.

Theorem 2.5 *Let $0 \le \sigma < 1$. For any $m \in N$, define the interval $[\sigma_{m,1}^-, \sigma_{m,1}^+) \subset [\sigma_m^+, \sigma_{m+1}^-)$ by*

$$\sigma_{m,1}^- = 1 - \frac{2a^{m+1}(1 - a^k + a^{2k+m} - a^{3k+m})}{1 - a^{4k+2m+1}}$$

and

$$\sigma_{m,1}^+ = 1 - \frac{2a^{m+1}(1 - a^k + a^{2k+m+1} - a^{3k+m+1})}{1 - a^{4k+2m+1}}.$$

Then, for each $\sigma \in [\sigma_{m,1}^-, \sigma_{m,1}^+)$, the solution $\{x^*_n\}_{n=-k}^{\infty}$ of equation (1.5) with the initial values

$$x^*_{-k} > \sigma, x^*_{-k+1} > \sigma, \cdots, x^*_{-1} = \frac{2a^{k+m+1}(-a^{2k+m} + a^k - 1) + 2}{1 - a^{4k+2m+1}} - 1 \ge \sigma_{m,1}^+ > \sigma \tag{2.14}$$

is periodic with the minimal period $(2k+m) + (2k+m+1)$. Moreover, both $\{x^*_0, x^*_1, \cdots, x^*_{k+m-1}\}$ and $\{x^*_{2k+m}, x^*_{2k+m+1}, \cdots, x^*_{3k+2m}\}$ are negative semicycles of $\{x^*_n\}_{n=-k}^{\infty}$ relative to σ, both $\{x^*_{k+m}, x^*_{k+m+1}, \cdots, x^*_{2k+m-1}\}$ and $\{x^*_{3k+2m+1}, x^*_{3k+2m+2}, \cdots, x^*_{4k+2m}\}$ are positive semicycles of $\{x^*_n\}_{n=-k}^{\infty}$ relative to σ.

Proof It is easy to see that $[\sigma_{m,1}^-, \sigma_{m,1}^+) \subset [\sigma_m^+, \sigma_{m+1}^-)$. For any $\sigma \in [\sigma_{m,1}^-, \sigma_{m,1}^+)$, let $\{x^*_n\}_{n=-k}^{\infty}$ be the solution of equation (1.5) satisfying the initial condition (2.14). By (1.2) and (1.5), we obtain $x^*_0 = ax^*_{-1} + a - 1$. It follows from (2.14) that

$$x^*_0 = -1 + \frac{2a^{k-m+2}(-a^{2k+m} + a^k - 1) + 2a}{1 - a^{4k+2m+1}}. \tag{2.15}$$

Obviously, $x^*_0 \le \sigma_{m,1}^- \le \sigma$. Moreover, from (1.2), (1.5) and (2.14), it is easy to see that

$$x^*_n = ax^*_{n-1} + a - 1, \qquad n \in N(1, k-1).$$

Again by (1.8), we have

$$x^*_n = a^n x^*_0 + a^n - 1, \qquad n \in N(1, k-1). \tag{2.16}$$

From (2.15) we can see that $x^*_0 > -1$. Again according to (2.16), we obtain

$$x^*_{k-1} \le x^*_{k-2} \le \cdots \le x^*_0 \le \sigma_{m,1}^- \le \sigma \tag{2.17}$$

and

$$x^*_{k-1} = a^{k-1}x^*_0 + a^{k-1} - 1. \tag{2.18}$$

In view of (1.2) and (1.5), we know

$$x_n^* \le a x_{n-1}^* + 1 - a, \qquad n \in N(k, k+m-1).$$

By iteration, it is easy to show that

$$x_n^* \le a^{n-k+1} x_{k-1}^* + 1 - a^{n-k+1}, \qquad n \in N(k, k+m-1).$$

Therefore, from (2.15) and (2.18), it follows that

$$\begin{aligned} x_n^* &\le \frac{2a^{n+1}(a^{2k+m+1} - a^{k+m+1} + 1 - a^{-k})}{1 - a^{4k+2m+1}} + 1 \\ &\le \frac{2a^{k+m}(a^{2k+m+1} - a^{k+m+1} + 1 - a^{-k})}{1 - a^{4k+2m+1}} + 1 \\ &\le \sigma_{m,1}^- \le \sigma \end{aligned}$$

for $n \in N(k, k+m-1)$. Thus, we have

$$x_k^* \le \sigma, x_{k+1}^* \le \sigma, \cdots, x_{k+m-1}^* \le \sigma, \tag{2.19}$$

which, together with (1.2), (1.5) and (2.17), implies that

$$x_n^* = a x_{n-1}^* + 1 - a, \qquad n \in N(k, 2k+m-1).$$

Again from (1.9), it follows that

$$x_n^* = a^{n-k+1} x_{k-1}^* + 1 - a^{n-k+1}, \qquad n \in N(k, 2k+m-1). \tag{2.20}$$

Noting (2.18) and (2.20), we obtain

$$x_{k+m}^* = a^{k+m} x_0^* - 2a^{m+1} + a^{k+m} + 1.$$

That is

$$x_{k+m}^* = 1 + \frac{2a^{m+1}(-1 + a^k - a^{2k+m+1} + a^{3k+m+1})}{1 - a^{4k+2m+1}}.$$

Clearly, $x_{k+m}^* \ge \sigma_{m,1}^+ > \sigma$, which, together with (2.17) and (2.19), implies that $\{x_0^*, x_1^*, \cdots, x_{k+m-1}^*\}$ is a negative semicycle of $\{x_n^*\}_{n=-k}^{\infty}$ relative to σ. Moreover, in view of (2.20), we know

$$1 > x_{2k+m-1}^* \ge x_{2k+m-2}^* \ge \cdots \ge x_{k+m}^* > \sigma. \tag{2.21}$$

From (2.18) and (2.20), it follows that

$$x_{2k+m-1}^* = a^{2k+m-1} x_0^* - 2a^{k+m} + a^{2k+m-1} + 1. \tag{2.22}$$

Since $x_{k+m}^* > \sigma$, by (1.2), (1.5) and (2.22), it is easy to see that

$$\begin{aligned} x_{2k+m}^* &= a x_{2k+m-1}^* + a - 1 \\ &= a^{2k+m} x_0^* + a^{2k+m} - 2a^{k+m+1} + 2a - 1. \end{aligned}$$

That is

$$x_{2k+m}^* = -1 + \frac{2a(1 - a^{3k+2m+1} + a^{2k+m} - a^{k+m})}{1 - a^{4k+2m+1}}.$$

Obviously, $x_{2k+m}^* \le \sigma_{m,1}^- \le \sigma$, which, together with (2.21), implies that $\{x_{k+m}^*, x_{k+m+1}^*, \cdots, x_{2k+m-1}^*\}$ is a positive semicycle of $\{x_n^*\}_{n=-k}^{\infty}$ relative to σ. Consequently, according to (1.2) and (1.5), we get

$$x_n^* = a x_{n-1}^* + a - 1, \qquad n \in N(2k+m, 3k+m-1).$$

Again by (1.8), we can see that

$$x_n^* = a^{n-2k-m+1}x_{2k+m-1}^* + a^{n-2k-m+1} - 1, n \in N(2k+m, 3k+m-1). \quad (2.23)$$

Noting (2.21) and (2.23), we know that

$$\sigma \geq x_{2k+m}^* \geq x_{2k+m+1}^* \geq \cdots \geq x_{3k+m-1}^*. \quad (2.24)$$

Moreover, according to (2.22) and (2.23), it is easy to see that

$$x_{3k+m-1}^* = a^{3k+m-1}x_0^* + a^{3k+m-1} - 2a^{2k+m} + 2a^k - 1. \quad (2.25)$$

In view of (1.2) and (1.5), we can obtain

$$x_n^* \leq ax_{n-1}^* + 1 - c, \qquad n \in N(3k+m, 3k+2m).$$

By iteration, we know

$$x_n^* \leq a^{n-3k-m+1}x_{3k+m-1}^* + 1 - a^{n-3k-m+1}, \qquad n \in N(3k+m, 3k+2m).$$

Thus, according to (2.15) and (2.25), we obtain

$$\begin{aligned} x_n^* &\leq 1 + \frac{2a^{n+1}(1 - a^{-k} + a^{-2k-m} - a^{-3k-m})}{1 - a^{4k+2m+1}} \\ &\leq 1 + \frac{2a^{2k+2m+1}(1 - a^{-k} + a^{-2k-m} - a^{-3k-m})}{1 - a^{1+4k+2m}} \\ &\leq \sigma_{m,1}^- \leq \sigma \end{aligned}$$

for $n \in N(3k+m, 3k+2m)$. Therefore,

$$x_{3k+m}^* \leq \sigma, x_{3k+m+1}^* \leq \sigma, \cdots, x_{3k+2m}^* \leq \sigma. \quad (2.26)$$

According to (1.2), (1.5), (2.24) and (2.26), it is easy to see that

$$x_n^* = ax_{n-1}^* + 1 - a, \qquad n \in N(3k+m, 4k+2m).$$

Therefore, again from (1.9), it follows that

$$x_n^* = a^{n-3k-m+1}x_{3k+m-1}^* + 1 - a^{n-3k-m+1}, \quad n \in N(3k+m, 4k+2m). \quad (2.27)$$

This, together with (2.25), implies that

$$x_{3k+2m+1}^* = a^{3k+2m+1}x_0^* + a^{3k-2m+1} - 2a^{2k+2m+2} + 2a^{k+m+2} - 2a^{m+2} + 1.$$

By (2.15), it is obvious that $x_{3k+2m+1}^* \geq \sigma_{m,1}^+ > \sigma$, which, together with (2.24) and (2.26),implies that $\{x_{2k+m}^*, x_{2k+m+1}^*, \cdots, x_{3k+2m}^*\}$ is a negative semicycle of $\{x_n^*\}_{n=-k}^{\infty}$ relative to σ.

Noting (2.24) and (2.27), we can see that

$$1 \geq x_{4k+2m}^* \geq x_{4k+2m-1}^* \geq \cdots \geq x_{3k+2m+1}^* > \sigma \quad (2.28)$$

and

$$x_{4k+2m}^* = a^{k+m+1}x_{3k+m-1}^* + 1 - a^{k+m+1}.$$

It follows from (2.25) that

$$x_{4k+2m}^* = a^{4k+2m}x_0^* + a^{4k+2m} - 2a^{3k+2m+1} + 2a^{2k+m+1} - 2a^{k+m+1} + 1.$$

In addition, since $x_{3k+2m+1}^* > \sigma$, by (1.2), (1.5) and (2.15), we obtain

$$\begin{aligned} x_{4k+2m+1}^* &= ax_{4k+2m}^* + a - 1 \\ &= a^{4k+2m+1}(x_0^* + 1) - 2a^{k+m+2}(a^{2k+m} - a^k + 1) + 2a - 1 \\ &= x_0^*. \end{aligned}$$

Again noting $x_0^* \leq \sigma_{m,1}^- \leq \sigma$, we have $x_{4k+2m+1}^* \leq \sigma$, which, together with (2.28), implies that $\{x_{3k+2m+1}^*, x_{3k+2m+2}^*, \cdots, x_{4k+2m}^*\}$ is a positive semicycle of $\{x_n^*\}_{n=-k}^{\infty}$ relative to σ.

By induction, it is easy to prove the periodicity of $\{x_n^*\}_{n=-k}^{\infty}$, and hence the proof of Theorem 2.5 is complete. □

Remark 2.6 Similar to the proofs of Theorems 2.4 and 2.5, when

$$\sigma \in \left[\sigma_m^+, \sigma_{m,1}^-\right) \cup \left[\sigma_{m,1}^+, \sigma_{m+1}^-\right),$$

we can study the existence of periodic solutions with the initial values $x_{-k}^* > \sigma$, $x_{-k+1}^* > \sigma$, $\cdots$, $x_{-1}^* > \sigma$.

3 An illustrative example

In Section 2, for the case where $|\sigma| < 1$, we have only studied the existence of periodic solutions of equation (1.5), but haven't discussed the asymptotic behavior of other solutions of equation (1.5). In fact, we can show that for $\sigma \in [\sigma_m^-, \sigma_m^+)$, any solution $\{x_n\}_{n=-k}^{\infty}$ of equation (1.5) with the initial values $x_{-k}^* > \sigma$, $x_{-k+1}^* > \sigma$, $\cdots$, $x_{-1}^* > \sigma$ converges to a periodic solution $\{x_n^*\}_{n=-k}^{\infty}$ given by Theorem 2.4. For example, take $a = 0.5$ and $k = 1$ in equation (1.5), we have the following numerical simulation figure 1 of solutions for equation (1.5) , which confirms that for each $\sigma \in (0, 1)$, every solution of equation (1.5) converges to a periodic solution.

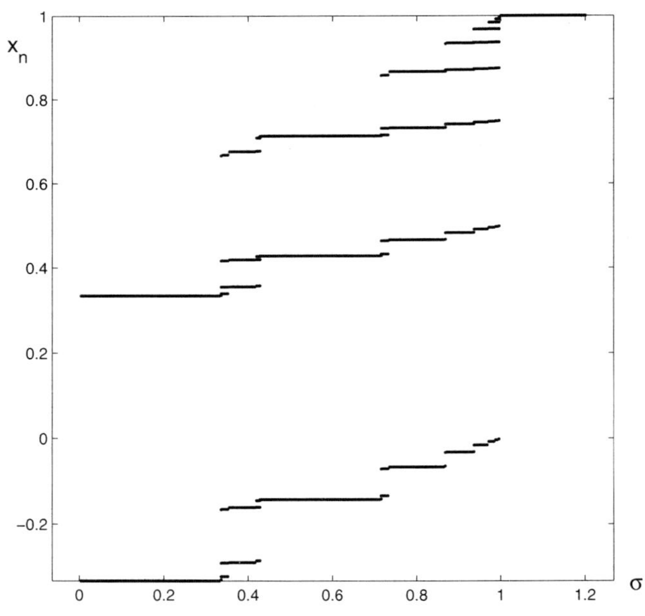

Figure 1. Numberical simulation of solutions of equation (1.5) with a=0.5 and k=1.

References

[1] Agarwal, R. P. and Wong, P. J. Y. *Advanced Topics in Difference Equations*, Kluwer, Dordrecht, 1997.

[2] Agarwal, R. P. *Difference Equations and Inequalities: Theory, Method and Applications*, Marcel Dekker, New York, 1992.

[3] Huang, L. H. and Yu, J. S. *Asymptotic behavior of solutions for a class of retarded difference equations*, Acta Mathematica Sinica **41(6)** (1998), 1215-1218.

[4] Huang, L. H. and Wu, J. *Dynamics of inhibitory artificial neural network with threshold nonlinearity*, Fields Institute Communications, **29** (2001), 235-243.

[5] Huang, L. H. and Wu, J. *The role of threshold in preventing delay-induced oscillations of frustrated neural networks with McCulloch-Pitts nonlinearity*, Int. J. Math. Game Theory and Algebra, **11(6)** (2001), 71-100.

[6] Kocic, V. L. J. and Ladas, G. *Global Behavior of Nonlinear Difference Equations of Higher Order with Applications*, Kluwer Academic, 1993.

[7] Luo, J. W. *Oscillation theorems for a neutral delay difference equation*, Annals of Differential Equations, **13(4)** (1999), 377-384.

[8] Wang, L. *Oscillation of nonlinear neutral difference equations*, Mathematica Applicata, **12(1)** (1999), 97-100.

[9] Yuan, Z. H., Huang, L. H. and Chen, Y. M. *Convergence and periodicity of solutions for a discrete-time network model of two neurons*, Mathematical and Computer Modelling, **35** (2002), 941-950.

[10] Zhang, B. G. and Yang, B. *Oscillation in higher order nonlinear difference equations*, Chinese Annals of Mathematics, **20A(1)** (1999), 71-80.

[11] Zhou, Z. and Yu, J. S. *Stability for nonautonomous neutral delay difference equations*, Acta Mathematica Sinica, **42(6)** (1999), 1093-1102.

[12] Zhou, Z., Yu, J. S. and Huang, L. H. *Asymptotic behavior of delay difference systems*, Computers Math. Applic., **42(3-5)** (2001), 283-290.

Fields Institute Communications
Volume **42**, 2004

Positive Solutions of a Non-autonomous Delay Model of Single Population

Yuji Liu
Department of Mathematics
Beijing Institute of Technology
Beijing 100081, P. R. China
and
Department of Mathematics
Hunan Institute of Technology
Yueyang, Hunan 414000, P. R. China
liuyuji888@sohu.com

Zhan Zhou
Department of Mathematics
Hunan University
Changsha 410082, P. R. China
zzhou0321@hotmail.com

Weigao Ge
Department of Mathematics
Beijing Institute of Technology
Beijing 100081, P. R. China
gew@bit.edu.cn

Abstract. We study the positive solutions to the non-autonomous delay logistic difference equation

$$\Delta x_n = p_n x_n \frac{1 - x_{n-k_n}}{1 - \lambda x_{n-k_n}}, \quad n = 0, 1, 2, \cdots,$$

where $\{p_n\}$ is a sequence of positive real numbers, $\{k_n\}$ a sequence of nonnegative integers such that $\{n - k_n\}$ is nondecreasing, and $\lambda \in [0, 1)$. We obtain some sufficient conditions under which every solution $\{x_n\}$ stays within the interval $(0, 1/\lambda)$, $(0, 1)$ and $(1, 1/\lambda)$, respectively. We also derive some sufficient conditions implying that every solution satisfying $0 < x_n < \frac{1}{\lambda}$ converges to the equilibrium $x = 1$ of the model, or oscillates about $x = 1$. Our results improve and generalize some recent results established by Chen and Yu [1], Zhou and Zhang [16]. The methods used here are new. Some examples and remarks are given to illustrate our results.

2000 *Mathematics Subject Classification.* Primary 34K20, 39A10, 39A11, 34K45; Secondary 92D25.

The first author of this work was supported by the Natural Science Foundation of Educational Committee of Hunan Province and the other authors were supported by NSF of P. R. China.

1 Introduction

Recently, there has an increasing interest in the study of the asymptotic behavior of solutions of the difference equations with delay, especially, of those which have applications in Mathematical Biology. See, for example [5, 6, 11, 14, 15, 16]. In this paper, we consider the non-autonomous logistic type model

$$\Delta x_n = p_n x_n \frac{1 - x_{n-k_n}}{1 - \lambda x_{n-k_n}}, \quad n = 0, 1, 2, \cdots \tag{1.1}$$

where $\{p_n\}$ is a sequence of positive real numbers, $\{k_n\}$ a sequence of nonnegative integers such that $\{n - k_n\}$ is nondecreasing, $\lambda \in [0, 1)$, and $\Delta x_n = x_{n+1} - x_n$.

Eq. (1.1) contains as special case the equation

$$\Delta x_n = p_n x_n (1 - x_{n-k}), \qquad n = 0, 1, 2, \cdots, \tag{1.2}$$

whose global attractivity of the equilibrium $x = 1$ has been well studied by [11, 16]. Eq. (1.2) is also regarded as the analogue of the delay logistic equation

$$N'(t) = r(t)N(t)(1 - N(t - \tau)), \quad t \geq 0$$

which has been studied by many authors, see [7, 9]. In [16], it is proved that if

$$\sum_{n=0}^{+\infty} p_n = +\infty,$$

and for sufficiently large n

$$\sum_{s=n-k_n}^{n} p_s \leq \delta,$$

holds, then every positive solution of (1.2) tends to 1 as n tends to infinity, where $\delta > 0$ satisfies the equation $e^{\frac{x^2}{2}} = 1 + \frac{1}{x}$.

In [1], the authors prove that if

$$\liminf_{n \to +\infty} \sum_{s=n-k_n}^{n} p_n > (\frac{k}{k+1})^{k+1} \tag{1.3}$$

holds, then every positive solution of (1.2) oscillates about 1.

In most results of these papers [1, 16], it is assumed that the solution $\{x_n\}$ satisfies $x_n > 0$. However, we find that this assumption does not always hold, see Remark 2.2 in Section 2.

Four problems appear naturally and are valuable for study. The first one is under what conditions every solution satisfies $0 < x_n < 1/\lambda$. The second one is under what conditions every solution satisfying $0 < x_n < 1/\lambda$ converges to 1. And the third one is under what conditions every solution satisfying $0 < x_n < \frac{1}{\lambda}$ oscillates about 1. The last one is that under what conditions every solution satisfies $0 < x_n < 1$ for all n or $1 < x_n < 1/\lambda$ for all n.

By introducing a new method, we obtain some sufficient conditions under which every solution $\{x_n\}$ stays within the interval $(0, 1/\lambda)$, $(0, 1)$ and $(1, 1/\lambda)$ respectively, see Theorems 2.1 and 4.1. By using an approach different from those in [16], we derive some sufficient conditions that guarantee every solution satisfying $0 < x_n < \frac{1}{\lambda}$ to converge to the equilibrium $x = 1$ of the model (Theorem 3.1 and Corollary 3.2). Finally, we obtain sufficient conditions yielding that every solution satisfying $0 < x_n < 1/\lambda$ oscillates about 1.

This paper is organized as follows: In Section 2, we obtain the sufficient conditions implying that every solution of (1.1) satisfies $0 < x_n < 1/\lambda$. In Section 3, we give sufficient conditions for every solution of (1.1) satisfying $0 < x_n < 1/\lambda$ to converge to 1. In Section 4, sufficient conditions for every solution of (1.1) to satisfy $1/\lambda > x_n > 1$ for all n or $0 < x_n < 1$ for all n are given respectively. In Section 5, we give sufficient conditions for that every solution of (1.1) satisfying $0 < x_n < 1/\lambda$ oscillates about 1 when (1.3) is not valid.

Our results improve and generalize some recent results established by Chen and Yu [1], and Zhou and Zhang [16].

As usual, let $\gamma = -\min\{n - k_n, n \geq 0\}$ be a nonnegative integer, and

$$\sigma_0 = \max\{n : n - k_n < 0\} + 1, \quad \sigma = \max\{n : n - k_n < \sigma_0\} + 1.$$

By a solution of Eq. (1.1) we mean a sequence $\{x_n\}$ which is defined for $n \geq -\gamma$, and satisfies (1.1) for $n \geq 0$. The initial condition of Eq. (1.1) is

$$x_i = a_i, \quad i = -\gamma, -\gamma + 1, \cdots, 0, \tag{1.4}$$

with $a_i \in (0, 1/\lambda)$, for $i = -\gamma, -\gamma + 1, \cdots, 0$. For the general theory of the difference equation, the reader is referred to [6].

By the way, Eq. (1.1) can be regarded as a discrete analogue of the following equation

$$N'(t) = r(t)N(t)\frac{1 - N(t-\tau)}{1 - \lambda N(t-\tau)}, \tag{1.5}$$

which was posed in [12] as a significant population growth model and studied by many authors, see [2]-[13] and the references therein.

2 Conditions for $0 < x_n < 1/\lambda$

In this section, we prove the following theorem.

Theorem 2.1 *Suppose that $\lambda \in [0, 1)$, and there are $\beta > 0$ and $\theta > 1$ such that*

$$\sum_{j=n-k_n}^{n} p_j \leq \alpha, \quad n = \sigma_0, \sigma_0 + 1, \cdots; \quad p_n \leq \beta, \quad n = 0, 1, 2, \cdots, \sigma \tag{2.1}$$

and

$$\frac{\theta\alpha + 1}{\theta\alpha + \lambda}(1 + \beta)^\sigma < \frac{1 + \alpha}{\lambda + \alpha} \tag{2.2}$$

where α is real root of the transcendental equation

$$e^\alpha \left(\frac{1 - \lambda}{1 - \lambda e^{-\alpha}}\right)^{(1-\lambda)/\lambda} = \frac{\alpha + 1}{\alpha + \lambda}, \quad \lambda \in (0, 1), \tag{2.3}$$

or

$$e^{\alpha - 1 + e^{-\alpha}} = 1 + \frac{1}{\alpha}, \quad \lambda = 0.$$

Furthermore, suppose that the initial condition satisfies

$$0 < x_i < \frac{\theta\alpha + 1}{\theta\alpha + \lambda}, \quad i = -r, -r + 1, \cdots, -1, 0. \tag{2.4}$$

Then every solution of Eq. (1.1) with initial condition (2.4) satisfies

$$0 < x_n \leq \frac{\alpha + 1}{\alpha + \lambda}, \quad n = 1, 2, \cdots. \tag{2.5}$$

Proof We first prove the case $\lambda \in (0,1)$. By (1.1), we get

$$x_{n+1} = x_n(1 + p_n \frac{1 - x_{n-k_n}}{1 - \lambda x_{n-k_n}}).$$

Since $0 < x_{-r}, x_{-r+1}, \cdots, x_0 < \frac{\theta\alpha+1}{\theta\alpha+\lambda} < \frac{\alpha+1}{\alpha+\lambda}$, then by (2.1) and (2.2), we find

$$\begin{aligned} x_1 &= x_0\left(1 + p_0 \frac{1 - x_{-k_0}}{1 - \lambda x_{-k_0}}\right) > x_0\left(1 + p_0 \frac{1 - (\alpha+1)/(\alpha+\lambda)}{1 - \lambda(\alpha+1)/(\alpha+\lambda)}\right) \\ &\geq x_0\left(1 + \alpha \frac{1 - (\alpha+1)/(\alpha+\lambda)}{1 - \lambda(\alpha+1)/(\alpha+\lambda)}\right) = 0. \\ x_1 &= x_0\left(1 + p_0 \frac{1 - x_{-k_0}}{1 - \lambda x_{-k_0}}\right) \leq \frac{\theta\alpha+1}{\theta\alpha+\lambda}(1+\beta) < \frac{\alpha+1}{\alpha+\lambda}. \end{aligned}$$

Similarly, one obtain

$$\begin{aligned} x_2 &= x_1\left(1 + p_1 \frac{1 - x_{1-k_1}}{1 - \lambda x_{1-k_1}}\right) > x_1\left(1 + p_1 \frac{1 - (\alpha+1)/(\alpha+\lambda)}{1 - \lambda(\alpha+1)/(\alpha+\lambda)}\right) \\ &\geq x_1\left(1 + \alpha \frac{1 - (\alpha+1)/(\alpha+\lambda)}{1 - \lambda(\alpha+1)/(\alpha+\lambda)}\right) = 0. \\ x_2 &= x_1\left(1 + p_1 \frac{1 - x_{1-k_1}}{1 - \lambda x_{1-k_1}}\right) \leq \frac{\theta\alpha+1}{\theta\alpha+\lambda}(1+\beta)^2 < \frac{\alpha+1}{\alpha+\lambda}. \end{aligned}$$

$$\cdots\cdots$$

Finally, we get

$$\begin{aligned} x_\sigma &= x_{\sigma-1}\left(1 + p_{\sigma-1} \frac{1 - x_{\sigma-1-k_{\sigma-1}}}{1 - \lambda x_{\sigma-1-k_{\sigma-1}}}\right) \\ &> x_{\sigma-1}\left(1 + p_{\sigma-1} \frac{1 - (\alpha+1)/(\alpha+\lambda)}{1 - \lambda(\alpha+1)/(\alpha+\lambda)}\right) \\ &\geq x_{\sigma-1}\left(1 + \alpha \frac{1 - (\alpha+1)/(\alpha+\lambda)}{1 - \lambda(\alpha+1)/(\alpha+\lambda)}\right) = 0. \\ x_\sigma &= x_{\sigma-1}\left(1 + p_{\sigma-1} \frac{1 - x_{\sigma-1-k_{\sigma-1}}}{1 - \lambda x_{\sigma-1-k_{\sigma-1}}}\right) \leq \frac{\theta\alpha+1}{\theta\alpha+\lambda}(1+\beta)^\sigma < \frac{\alpha+1}{\alpha+\lambda}. \end{aligned}$$

Now, it suffices to prove the following result: if $n_0 \geq \sigma$, and

$$0 < x_n \leq \frac{\alpha+1}{\alpha+\lambda}, \quad 0 \leq n \leq n_0, \tag{2.6}$$

then

$$0 < x_{n_0+1} < \frac{\alpha+1}{\alpha+\lambda}. \tag{2.7}$$

By (2.1), (2.5) and $\lambda \in [0,1)$, since the function $f(x) = (x-1)/(1-\lambda x)$ is increasing on $(0, 1/\lambda)$, we have

$$p_{n_0} \frac{x_{n_0-k_{n_0}} - 1}{1 - \lambda x_{n_0-k_{n_0}}} < \alpha \frac{(\alpha+1)/(\alpha+\lambda) - 1}{1 - \lambda(\alpha+1)/(\alpha+\lambda)} = 1$$

which implies by (1.1) that $x_{n_0+1} > 0$. Next, we prove that

$$x_{n_0+1} < \frac{\alpha+1}{\alpha+\lambda}. \tag{2.8}$$

Suppose for the contradiction that $x_{n_0+1} \geq \frac{\alpha+1}{\alpha+\lambda}$. Let $p(t) = p_n$ for $t \in [n, n+1)$, $n = 0, 1, 2, \cdots, n_0$, and

$$x(t) = \begin{cases} x_n, & t = n, \\ x_n(\frac{x_{n+1}}{x_n})^{t-n}, & n \leq t < n+1. \end{cases} \tag{2.9}$$

Then $x(t)$ is a positive continuous function on internal $[0, n_0 + 1]$, and $x(n) = x_n$, $n \geq 0$, $x(t)$ is monotone on $[n, n+1)$. Let $[.]$ denote the maximum integer function, $x'(t)$ stand for the right derivative of function $x(t)$. Then

$$x'(t) = x(t) \ln \left(1 + p(t) \frac{1 - x([t - k_{[t]}])}{1 - \lambda x([t - k_{[t]}])}\right), \quad 0 \leq t < n_0 + 1. \tag{2.10}$$

Since $x([t - k_{[t]}]) > 0$, $0 \leq t \leq n_0 + 1$, then by $(1 - x)/(1 - \lambda x) \leq 1$, $\frac{1}{\lambda} > x \geq 0$, we have

$$x'(t) \leq x(t) \ln(1 + p(t)) \leq p(t)x(t), \quad 0 \leq t < n_0 + 1. \tag{2.11}$$

Again by $\Delta x_{n_0} = x_{n_0+1} - x_{n_0} > 0$ and (1.1), we have $x_{n_0 - k_{n_0}} < 1$. Then there exists $\xi \in [n_0 - k_{n_0}, n_0 + 1)$ such that $x(\xi) = 1$ and $x(t) > 1$ for $t \in (\xi, n_0 + 1]$. When $0 \leq t \leq \xi$, integrating (2.11) from t to ξ, we get

$$x(t) \geq \exp\left(-\int_t^\xi p(s)ds\right), \quad 0 \leq t \leq \xi.$$

If $\xi \leq t < n_0 + 1$ and $[t - k_{[t]}] \leq \xi$, we obtain

$$x([t - k_{[t]}]) \geq \exp\left(-\int_{[t-k_{[t]}]}^\xi p(s)ds\right). \tag{2.12}$$

Substituting (2.12) into (2.10), it follows

$$x'(t) \leq x(t) \ln \left(1 + p(t) \frac{1 - \exp(-\int_{[t-k_{[t]}]}^\xi p(s)ds)}{1 - \lambda \exp(-\int_{[t-k[t]]}^\xi p(s)ds)}\right).$$

Then we have

$$x'(t) \leq x(t)p(t) \frac{1 - \exp(-\int_{[t-k_{[t]}]}^\xi p(s)ds)}{1 - \lambda \exp(-\int_{[t-k_{[t]}]}^\xi p(s)ds)}. \tag{2.13}$$

If $[t - k_{[t]}] > \xi$, since $n - k_n$ is nondecreasing, combining $n_0 + 1 \geq t \geq \xi$, one has $n_0 + 1 - k_{n_0+1} \geq [t] - k_{[t]} = [t - k_{[t]}] > \xi$. This contradicts $\xi \in [n_0 + 1 - k_{n_0+1}, n_0 + 1]$. Hence this case is impossible. Integrating (2.13) from ξ to $n_0 + 1$, noting that $\lambda \in (0, 1)$ and $n_0 \geq \sigma$ implies $n_0 - k_{n_0} \geq \sigma_0$, thus $[t - k_{[t]}] \geq 0$ for $t \in [\xi, n_0 + 1]$, we have

$$\begin{aligned}
&\ln x(n_0+1)\\
&\le \int_{\xi}^{n_0+1} p(t)\frac{1-\exp\left(-\int_{[t-k_{[t]}]}^{\xi} p(s)ds\right)}{1-\lambda\exp\left(-\int_{[t-k_{[t]}]}^{\xi} p(s)ds\right)}dt\\
&= \int_{\xi}^{n_0+1} p(t)\frac{1-\exp\left(-\int_{[t-k_{[t]}]}^{t} p(s)ds+\int_{\xi}^{t} p(s)ds\right)}{1-\lambda\exp\left(-\int_{[t-k_{[t]}]}^{t} p(s)ds+\int_{\xi}^{t} p(s)ds\right)}dt\\
&\le \int_{\xi}^{n_0+1} p(t)\frac{1-e^{-\alpha}\exp(\int_{\xi}^{t} p(s)ds)}{1-\lambda e^{-\alpha}\exp(\int_{\xi}^{t} p(s)ds)}dt\\
&= \int_{\xi}^{n_0+1} p(t)dt-(1-\lambda)\int_{\xi}^{n_0+1} p(t)\frac{e^{-\alpha}\exp(\int_{\xi}^{t} p(s)ds)}{1-\lambda\exp(-\alpha+\int_{\xi}^{t} p(s)ds)}dt\\
&= \int_{\xi}^{n_0+1} p(t)dt+\frac{1-\lambda}{\lambda}\ln\frac{1-\lambda\exp(-\alpha+\int_{\xi}^{n_0+1} p(s)ds)}{1-\lambda e^{-\alpha}}.
\end{aligned}$$

Since $\int_{\xi}^{n_0+1} p(t)dt \le \sum_{j=n_0-k_{n_0}}^{n_0} p_j \le \alpha$, and the function $x+\frac{1-\lambda}{\lambda}\ln\frac{1-\lambda e^{-\alpha+x}}{1-\lambda e^{-\alpha}}$ is increasing in $x\in[0,\alpha]$, we have

$$\ln x(n_0+1) < \alpha+\frac{1-\lambda}{\lambda}\ln\frac{1-\lambda}{1-\lambda e^{-\alpha}},$$

and hence

$$x_{n_0+1}=x(n_0+1)<\exp\left(\alpha+\frac{1-\lambda}{\lambda}\ln\frac{1-\lambda}{1-\lambda e^{-\alpha}}\right)=\frac{\alpha+1}{\alpha+\lambda},$$

which contradicts the assumption $x_{n_0+1}\ge\frac{\alpha+1}{\alpha+\lambda}$. This completes the proof. By a similar method, we can get the theorem in the case $\lambda=0$. □

Remark 2.2 Since $1<\frac{\theta\alpha+1}{\theta\alpha+\lambda}<\frac{\alpha+1}{\alpha+\lambda}<\frac{1}{\lambda}$, Theorem 1 gives sufficient conditions which guarantee $0<x_n<\frac{1}{\lambda}$ for each solution $\{x_n\}$ of Eq. (1.1).

Now, we give examples to show that the solution of equation (1.1) may not satisfy $0<x_n<\frac{1}{\lambda}$ and the solution of (1.2) may not be positive.

Example 2.3 Consider the equation

$$\Delta x_n = p_n x_n\frac{1-x_{n-1}}{1-\lambda x_{n-1}},\quad n=0,1,2,\cdots\quad,$$

where $p_n=\frac{1}{n+6}, k_n\equiv 1, \lambda=\frac{749}{1000}$. The initial condition is $x_{-1}=0.1, x_0=1.21$. It is easy to see by computation that

$$x_1=x_0(1+p_0\frac{1-x_{-1}}{1-\lambda x_{-1}})=1.21\times\frac{4759}{3259}>\frac{1}{\lambda}=\frac{1000}{749}.$$

Hence the solution of the equation (1.1) with this initial condition does not satisfy $0<x_n<\frac{1}{\lambda}$.

Example 2.4 Consider the equation

$$\Delta x_n = 1.3x_n(1-x_{n-1}),\quad n=0,1,2,\cdots\quad,$$

where $p_n \equiv 1.3, k_n \equiv 1, \lambda = 0$. The initial condition is $x_{-1} = 0.23$, $x_0 = 0.2$. Then $x_1 = 0.4002$, $x_2 = 0.8164$, $x_3 = 1.453$, $x_4 = 1.7998$, $x_5 = 0.7399$, $x_6 = -0.62938$, $\cdots$. This shows that solution of Eq. (1.2) with this initial condition is not positive.

3 Global attractivity

In this section,we give the sufficient conditions under which every solution satisfying $0 < x_n < \frac{1}{\lambda}$ converges to 1 as $n \to \infty$.

Theorem 3.1 *Suppose that $\lambda \in (0,1)$, and there is a constant $\delta > 0$ such that for sufficiently large n*

$$\sum_{s=n-k_n}^{n} p_s \le \delta(1-\lambda) \tag{3.1}$$

holds and

$$\sum_{n=1}^{+\infty} p_n = +\infty \tag{3.2}$$

and

$$1 - \delta\left(\delta - \frac{1}{2}\right) e^{(\delta-\frac{1}{2})(1-\lambda)} \ge 0 \tag{3.3}$$

are valid. Then every solution that satisfies $0 < x_n < 1/\lambda$ tends to 1 as $n \to +\infty$.

Proof Suppose that $\{x_n\}$ is a positive solution of Eq. (1.1) satisfying $0 < x_n < \frac{1}{\lambda}$. We divide the proof into four steps as follows.

Step 1. If $\{x_n\}$ is eventually greater than 1, we prove $\lim_{n\to+\infty} x_n = 1$.

Choose N_1 such that $x_{n-k_n} > 1$ for $n \ge N_1$. By (1.1), we see that $\Delta x_n \le 0$ for $n \ge N_1$, hence $\lim_{n\to+\infty} x_n = \mu$ exists. We prove $\mu = 1$. Suppose for contradiction that $\mu > 1$. We choose $\mu > \epsilon > 1$ such that $1 - \lambda\epsilon > 0$ and $1 + \delta(1-\lambda)\frac{1-\epsilon}{1-\lambda\epsilon} > 0$, then there exists a positive integer $N_2 > N_1$ such that $x_{n-k_n} > \epsilon$ for $n > N_2$. Thus

$$\Delta x_n \le p_n x_n \frac{1-\epsilon}{1-\lambda\epsilon}, \quad n \ge N_2,$$

and

$$\ln\frac{x_{n+1}}{x_n} \le \ln\left(1 - p_n\frac{1-\epsilon}{1-\lambda\epsilon}\right) \le p_n\frac{1-\epsilon}{1-\lambda\epsilon}, \quad n \ge N_2.$$

Hence

$$\ln\frac{x_{n+1}}{x_{N_2}} \le \frac{1-\epsilon}{1-\lambda\epsilon}\sum_{s=N_2}^{n} p_s.$$

Let $n \to +\infty$. It follow from (3.2) that $\ln\frac{\mu}{x_{N_2}} \le -\infty$, we get a contradiction.

Step 2. If $\{x_n\}$ is eventually less than 1, we prove that $\lim_{n\to+\infty} x_n = 1$.

Choose N_3 such that $x_{n-k_n} < 1$ for $n \ge N_3$. Then $\Delta x_n \ge 0$, so $\lim_{n\to+\infty} x_n = \mu$ exists. We prove that $\mu = 1$. Suppose for contradiction that $\mu < 1$. We choose $1 > \epsilon > 0$ such that $\delta(1-\lambda)\frac{1-\epsilon}{1-\lambda\epsilon} \le 1$ and choosing N_4 such that $x_{n-k_n} < \epsilon$ for $n \ge N_4$. Again, we have

$$\frac{x_{n+1}}{x_n} \ge 1 + p_n\frac{1-\epsilon}{1-\lambda\epsilon}, \quad n \ge N_4.$$

Since $\ln(1+x) \geq \frac{1}{2}x$ for $x \in [0,1]$, we gat

$$\ln \frac{x_{n+1}}{x_n} \geq \frac{1}{2} p_n \frac{1-\epsilon}{1-\lambda\epsilon}, \quad n \geq N_4.$$

It is easy to get a contradiction.

Step 3. We prove that $\{x_n\}$ is bounded above away from $\frac{1}{\lambda}$. i.e., there exists a number $A > 0$ such that $x_n \leq A < \frac{1}{\lambda}$ for all n.

Let x_{N+1} satisfy that $x_{N+1} \geq x_N$, we will prove $x_{N+1} < \frac{1}{\lambda}$. When $t \in [n, n+1)$, we define that $p(t) = p_n, n = 0, 1, 2, \cdots$, and

$$x(t) = \begin{cases} x_n, & t = n, \\ x_n(\frac{x_{n+1}}{x_n})^{t-n}, & n \leq t < n+1. \end{cases} \tag{3.4}$$

Then $x(n) = x_n, n = -k_0, -k_0+1, \cdots, x(t)$ is monotone in $[n, n+1]$. Denoting $x'(t)$ as the right derivative of $x(t)$, we get

$$x'(t) = x(t) \ln \left(1 + p(t) \frac{1 - x([t - k_{[t]}])}{1 - \lambda x([t - k_{[t]}])} \right). \tag{3.5}$$

If $x_{N+1} \leq 1$, then $x_{N+1} < e^{1-\lambda} < \frac{1}{\lambda}$. If $x_{N+1} > 1$, then by (1.1), $x_{N-k_N} \leq 1$. There exists $\xi \in [N - k_N, N+1)$, such that $x(\xi) = 1$, $x(t) > 1$, for $t \in (\xi, N+1)$. If $t \leq \xi$, by (3.5), we get

$$x'(t) \leq p(t)x(t). \tag{3.6}$$

Integrating (3.6) from t to ξ, we get

$$x(t) \geq \exp\left(-\int_t^{\xi} p(s)ds \right). \tag{3.7}$$

We claim that $(1-x)/(1-\lambda x) \leq (-\ln x)/(1-\lambda)$ for $0 < x \leq \frac{1}{\lambda}$. In fact, Let

$$g(x) = \frac{1-x}{1-\lambda x} + \frac{\ln x}{1-\lambda}.$$

One gets

$$g(1) = 0, \quad g'(x) = \frac{(1-\lambda x)^2 - (1-\lambda)^2 x}{x(1-\lambda x)^2}.$$

Let $h(x) = (1-\lambda x)^2 - (1-\lambda)^2 x$, then $h(1) = 0$ and $h'(x) = -(1 + \lambda^2 - 2\lambda^2 x)$. Since $1 < x < \frac{1}{\lambda}$, one has $h'(x) \leq -(1 + \lambda^2 - 2\lambda) < 0$. Hence $g(x) < g(1) = 0$.

If $[t - k_{[t]}] \leq \xi$, then (3.5) implies

$$x'(t) \leq -\frac{1}{1-\lambda} p(t)x(t) \ln x([t - k_{[t]}]). \tag{3.8}$$

For $t \in [\xi, N+1)$, if $[t - k_{[t]}] \leq \xi$, by (3.7) and (3.8), we get

$$x'(t) \leq \frac{1}{1-\lambda} p(t)x(t) \int_{[t-k_{[t]}]}^{\xi} p(s)ds. \tag{3.9}$$

It is easy to see that the case where $x([t-k_{[t]}])>\xi$ is impossible. Integrating (3.9) from ξ to $N+1$ leads to

$$\begin{aligned}
\ln x(N+1) &\leq \frac{1}{1-\lambda}\int_{\xi}^{N+1} p(t)\int_{[t-k_{[t]}]}^{\xi} p(s)dsdt \\
&= \frac{1}{1-\lambda}\int_{\xi}^{N+1} p(t)\left(\int_{[t-k_{[t]}]}^{t} p(s)ds-\int_{\xi}^{t} p(s)ds\right)dt \\
&\leq \frac{1}{1-\lambda}\int_{\xi}^{N+1} p(t)\left(\delta(1-\lambda)-\int_{\xi}^{t} p(s)ds\right)dt \\
&= \delta\int_{\xi}^{N+1} p(t)dt-\frac{1}{1-\lambda}\int_{\xi}^{N+1} p(t)\int_{\xi}^{t} p(s)dsdt \\
&= \delta\int_{\xi}^{N+1} p(t)dt-\frac{1}{2(1-\lambda)}\left(\int_{\xi}^{N+1} p(t)dt\right)^2.
\end{aligned}$$

Now, we consider two cases as follows.

Case 1. $\int_{\xi}^{N+1} p(t)dt\leq 1-\lambda$. By the increasing property of function $f(x)=\delta x-\frac{1}{2(1-\lambda)}x^2$ for $x<\delta(1-\lambda)$, we have

$$\begin{aligned}
\ln x(N+1) &\leq \delta(1-\lambda)-\frac{1}{2(1-\lambda)}(1-\lambda)^2 \\
&= (\delta-\frac{1}{2})(1-\lambda)\leq 1-\lambda<\ln\frac{1}{\lambda}.
\end{aligned}$$

Noting that (3.3) implies $\delta\geq 1$, we get $x_{N+1}=x(N+1)<e^{1-\lambda}<\frac{1}{\lambda}$.

Case 2. $\int_{\xi}^{N+1} p(t)dt>1-\lambda$. Choose $\eta\in(\xi,N+1)$, such that $\int_{\eta}^{N+1} p(t)dt=1-\lambda$. Integrating (3.6) from ξ to η, integrating (3.9) from η to $N+1$, we have

$$\begin{aligned}
\ln x(N+1) &\leq \int_{\xi}^{\eta} p(t)dt+\frac{1}{1-\lambda}\int_{\eta}^{N+1} p(t)\int_{[t-k_{[t]}]}^{\xi} p(s)dsdt \\
&\leq \int_{\xi}^{\eta} p(t)dt+\frac{1}{1-\lambda}\left(\delta(1-\lambda)\int_{\eta}^{N+1} p(t)dt\right. \\
&\quad\left.-\int_{\eta}^{N+1} p(t)\int_{\xi}^{t} p(s)dsdt\right) \\
&= \int_{\xi}^{\eta} p(t)dt+\frac{1}{1-\lambda}\left[\delta(1-\lambda)\int_{\eta}^{N+1} p(t)dt\right. \\
&\quad\left.-\frac{1}{2}\left((\int_{\xi}^{N+1} p(s)ds)^2-(\int_{\xi}^{\eta} p(s)ds)^2\right)\right]
\end{aligned}$$

$$= \int_\xi^\eta p(t)dt + \frac{1}{1-\lambda}\left[\delta(1-\lambda)\int_\eta^{N+1} p(t)dt - \frac{1}{2}\left(\left(\int_\xi^{N+1} p(t)dt + \int_\xi^\eta p(t)dt\right)\int_\eta^{N+1} p(s)ds\right)\right]$$

$$= (\delta - \frac{1}{2})(1-\lambda) \leq 1-\lambda \leq \ln\frac{1}{\lambda}.$$

Then $x_{N+1} \leq x(N+1) \leq e^{1-\lambda} < \frac{1}{\lambda}$. Hence $\{x_n\}$ is bounded above away from $1/\lambda$, thus we say that $\{x_n\}$ is bounded above away from $\frac{1}{\lambda}$. Step 3 is complete.

Step 4. We prove that if $\{x_n\}$ is oscillatory about 1, then $\{x_n\}$ tends to 1 as n tends to infinity. We suppose that

$$\limsup_{n\to\infty} x_n = u, \quad \liminf_{n\to\infty} x_n = v.$$

Then $0 \leq v \leq 1 \leq u < \frac{1}{\lambda}$. By the same method of the proof of those in Step 3, we get (3.5). Since $x(t)$ is monotone in $[n, n+1]$, we have

$$\limsup_{n\to\infty} x(t) = u, \quad \liminf_{n\to\infty} x(t) = v.$$

For any $\epsilon \in (0,1)$ there is N_5 such that

$$x([t-k_{[t]}]) > \epsilon v = v_1, \quad t \geq N_5. \tag{3.10}$$

Then by (3.5), we have

$$x'(t) \leq x(t)\ln\left(1 + p(t)\frac{1-v_1}{1-\lambda v_1}\right) \leq p(t)x(t)\frac{1-v_1}{1-\lambda v_1},$$

leading to

$$x'(t) \leq p(t)x(t)\frac{1-v_1}{1-\lambda v_1}, \quad t \geq N_5. \tag{3.11}$$

Choose $\{n_i\}$ such that $n_1 < n_2 < \cdots$, and $n_i - k_{n_i} \geq N_5, x_{n_i} \geq \max\{x_{n_i-1}, x_{n_i+1}\}$, and if $x_{n_i} \to u$ if $n_i \to \infty$. By (1.1), we have $x_{n_i - k_{n_i}} \geq 1$ and $x_{n_i-1-k_{n_i-1}} \leq 1$. Noting that $\{n - k_n\}$ is nondecreasing, there is $\xi_i \in [n_i - 1 - k_{n_i-1}, n_i - k_{n_i})$, such that $x(\xi_i) = 1$. If $N_5 \leq t \leq \xi_i$, integrating (3.11) from t to ξ_i, we have that

$$x(t) \geq \exp\left(-\frac{1-v_1}{1-\lambda v_1}\int_t^{\xi_i} p(s)ds\right). \tag{3.12}$$

Similar to those in Step 3, we have

$$x'(t) \leq \frac{1}{1-\lambda}\frac{1-v_1}{1-\lambda v_1}p(t)x(t)\int_{[t-k_{[t]}]}^{\xi_i} p(s)ds, \quad t \in [\xi_i, n_i). \tag{3.13}$$

Using (3.11) and (3.13), similar to those of (3.6) and (3.7) in Step 3, one gets

$$\ln x(n_i) \leq (\delta - \frac{1}{2})(1-\lambda)\frac{1-v_1}{1-\lambda v_1}.$$

Let $i \to \infty$, $\epsilon \to 1$, we get

$$\ln u \leq (\delta - \frac{1}{2})(1-\lambda)\frac{1-v}{1-\lambda v}. \tag{3.14}$$

For any $u_1 \in (u, \frac{1}{\lambda})$, there is N_6, such that if $n \geq N_6$, then $x_{n-k_n} < u_1$. (1.1) implies

$$\Delta x_n = p_n x_n \frac{1 - x_{n-k_n}}{1 - \lambda x_{n-k_n}}.$$

Thus

$$x_{n+1} = x_n \left(1 + p_n \frac{1 - x_{n-k_n}}{1 - \lambda x_{n-k_n}}\right). \tag{3.15}$$

Let $x_{n^*} = \max\{1, x_n\}$, then $x_{(n-k_n)^*} < u_1, n \geq N_6$. Hence

$$x_{n+1} \geq x_n = \left(1 + p_n \frac{1 - x_{(n-k_n)^*}}{1 - \lambda x_{(n-k_n)^*}}\right).$$

Now, we prove that

$$v \geq 1 - \delta(1-\lambda)\frac{u-1}{1-\lambda u}.$$

Choose $\{m_i\}$ such that $m_i - k_{m_i} \geq N_6$, $x_{m_i+1} \leq \min\{x_{m_i}, x_{m_i+2}\}$, and $m_i \to \infty$, $x_{m_i+1} \to v$ as $i \to \infty$. By (1.1), we know that $x_{m_i-k_{m_i}} \geq 1$, then we have

$$\begin{aligned}
x_{m_i+1} &\geq x_{m_i-k_{m_i}} \prod_{n=m_i-k_{m_i}}^{m_i} \left(1 + p_n \frac{1 - x_{(n-k_n)^*}}{1 - \lambda x_{(n-k_n)^*}}\right) \\
&\geq \prod_{n=m_i-k_{m_i}}^{m_i} \left(1 + p_n \frac{1 - x_{(n-k_n)^*}}{1 - \lambda x_{(n-k_n)^*}}\right) \\
&\geq 1 - \sum_{n=m_i-k_{m_i}}^{m_i} p_n \frac{x_{(n-k_n)^*} - 1}{1 - \lambda x_{(n-k_n)^*}} \\
&\geq 1 - \sum_{n=m_i-k_{m_i}}^{m_i} p_n \frac{u_1 - 1}{1 - \lambda u_1} \\
&\geq 1 - \delta(1-\lambda)\frac{u_1 - 1}{1 - \lambda u_1}.
\end{aligned}$$

Let $i \to \infty$, and $u_1 \to u$, we have

$$v \geq 1 - \delta(1-\lambda)\frac{u-1}{1-\lambda u}. \tag{3.16}$$

By (3.14) and (3.16), we get

$$\begin{aligned}
\ln u &\leq (\delta - \frac{1}{2})(1-\lambda)\frac{1-v}{1-\lambda v} \\
&\leq (\delta - \frac{1}{2})(1-\lambda)\frac{\delta(1-\lambda)(u-1)/(1-\lambda u)}{1 - \lambda + \delta\lambda(1-\lambda)(u-1)/(1-\lambda u)} \\
&= \delta(\delta - \frac{1}{2})(1-\lambda)\frac{(u-1)/(1-\lambda u)}{1 + \lambda\delta(u-1)/(1-\lambda u)}.
\end{aligned}$$

Then

$$\ln u \leq \delta(\delta - \frac{1}{2})(1-\lambda)\frac{u-1}{1-\lambda u + \delta\lambda(u-1)}. \tag{3.17}$$

Let

$$F(u) = \ln u - \delta(\delta - \frac{1}{2})(1-\lambda)\frac{u-1}{1-\lambda u + \lambda\delta(u-1)}.$$

It is easy to see that $F(1) = 0$, and

$$F'(u) = \frac{1}{u} - \delta(\delta - \frac{1}{2})(1-\lambda)\frac{1-\lambda}{(1-\lambda u + \lambda\delta(u-1))^2}$$
$$= \frac{(\lambda(\delta-1)u + 1 - \lambda\delta)^2 - \delta(\delta-\frac{1}{2})(1-\lambda)^2 u}{u(1-\lambda u + \lambda\delta(u-1))^2}.$$

Let

$$g(u) = (\lambda(\delta-1)u + 1 - \lambda\delta)^2 - \delta(\delta - \frac{1}{2})(1-\lambda)^2 u.$$

It is easy to see that $\delta > 1$ from (3.3). Again $u \geq 1$, thus $\lambda(\delta-1) \geq 0$. So $\lambda(\delta-1)u + 1 - \lambda\delta \geq \lambda(\delta-1) + 1 - \lambda\delta = 1 - \lambda > 0$. Hence

$$g(u) \geq (1-\lambda)^2\left(1 - \delta(\delta - \frac{1}{2})u\right).$$

By (3.14), we have $u < e^{(\delta-\frac{1}{2})(1-\lambda)}$, thus

$$g(u) > (1-\lambda)^2\left(1 - \delta(\delta - \frac{1}{2})e^{(\delta-\frac{1}{2})(1-\lambda)}\right).$$

From (3.3), $F'(u) > 0$. In this case, if $u \neq 1$, then $u > 1$, so $F(u) > 0$, which contradicts (3.17). Thus $u = 1$, and so $v = 1$, yielding $\lim_{n\to\infty} x_n = 1$. The proof is complete. □

Corollary 3.2 *Suppose that there is $\delta > 0$ such that for sufficiently large n*

$$\sum_{s=n-k_n}^{n} p_s \leq \delta, \tag{3.18}$$

holds and

$$\delta + 1 - \delta e^{\delta - \frac{1}{2}} \geq 0, \tag{3.19}$$

is valid and (3.2) holds. Then every positive solution of the equation (1.2) tends to 1 as n tends to infinity.

Proof Steps 1, 2 and 3 are completely the same as those in the proof of Theorem 3.1 and hence omitted. We give a proof for Step 4:

By a similar method in the proof of Theorem 3.1, one has (3.14) and (3.16), i.e.,

$$\ln u \leq (\delta - \frac{1}{2})(1-v), \quad v \geq 1 - \delta(u-1).$$

Hence

$$\ln u \leq (\delta - \frac{1}{2})\delta(u-1).$$

Let $F(u) = \ln u - \delta(\delta - \frac{1}{2})(u-1)$, one has $F(1) = 0$ and $F'(u) = \frac{1}{u} - \delta(\delta - \frac{1}{2})$ and $F''(u) \leq 0$. Again we have $\ln u < \delta - \frac{1}{2}$. Thus $u < e^{\delta - \frac{1}{2}} = u_0$. So

$$F(u_0) = \ln u_0 - \delta(\delta - \frac{1}{2})(u_0 - 1) = (\delta - \frac{1}{2})(1 + \delta - \delta e^{\delta - \frac{1}{2}}).$$

By (3.19), one get $F(u_0) \geq 0$. Hence $F(u) > 0$ for $1 < u < u_0$. This is a contradiction. The proof is complete. □

Remark 3.3 Corollary 3.2 improves the main result in [6]. In fact, if δ satisfies $\frac{1}{\alpha} + 1 = e^{\frac{\alpha^2}{2}}$ but does not satisfy (3.19), that is $\frac{1}{\delta} + 1 = e^{\frac{\delta^2}{2}}$ and $1 + \delta - \delta e^{\delta - \frac{1}{2}} < 0$, then $e^{\frac{\delta^2}{2}} < e^{\delta - \frac{1}{2}}$, and $(\delta - 1)^2 < 0$, a contradiction.

4 Non-oscillation of positive solutions

In this section, we give sufficient conditions under which the solutions either satisfy $\frac{1}{\lambda} > x_n > 1$ for all n or $0 < x_n < 1$ for all n.

Theorem 4.1 *Suppose that the initial condition (1.4) satisfies*

$$1 < \alpha < x_i < \beta < \frac{1}{\lambda}, \quad i = -r, -r+1, \cdots, 0, \tag{4.1}$$

and

$$0 \leq p_n \leq \frac{1-\lambda\beta}{\beta - 1}(1 - \alpha^{-1/((n+1)(n+2))}), \quad n = 0, 1, \cdots \tag{4.2}$$

holds. Then the solution of Eq. (1.1) with condition (4.1) satisfies $\beta > x_n > 1$ *for all* n.

Proof By (1.1), we have

$$\begin{aligned} x_1 &= x_0\left(1 + p_0\frac{1 - x_{-k_0}}{1 - \lambda x_{-k_0}}\right) \geq \alpha\left(1 + p_0\frac{1-\beta}{1-\lambda\beta}\right) \\ &\geq \alpha\left(1 + \frac{1-\lambda\beta}{\beta-1}(1-\alpha^{-\frac{1}{2}})\frac{1-\beta}{1-\lambda\beta}\right) = \alpha^{\frac{1}{2}}. \end{aligned}$$

Again, we can get $x_1 < \beta$. Now, we suppose that $x_n \geq \alpha^{\frac{1}{n+1}}$ and $x_n < \beta$. Then

$$\begin{aligned} x_{n+1} &= x_n\left(1 + p_n\frac{1 - x_{n-k_n}}{1 - \lambda x_{n-k_n}}\right) \geq \alpha^{\frac{1}{n+1}}\left(1 + p_n\frac{1-\beta}{1-\lambda\beta}\right) \\ &\geq \alpha^{\frac{1}{n+1}}\left[1 + \frac{1-\lambda\beta}{\beta-1}\left(1-\alpha^{-\frac{1}{(n+1)(n+2)}}\right)\frac{1-\beta}{1-\lambda\beta}\right] = \alpha^{\frac{1}{n+2}}. \end{aligned}$$

Together with $x_{n+1} < \beta$. By induction, we have $1 < x_n < \beta$ for all n. □

By a similar method, we get

Theorem 4.2 *Suppose that the initial condition (1.4) satisfies*

$$0 < \alpha < x_i < \beta < 1, \quad i = -r, -r+1, \cdots, 0, \tag{4.3}$$

and

$$0 \leq p_n \leq \frac{1-\lambda\alpha}{1-\alpha}(\beta^{-1/((n+1)(n+2))} - 1), \quad n = 0, 1, \cdots \tag{4.4}$$

holds. Then the solution of Eq. (1.1) with condition (4.3) satisfies $0 < x_n < 1$ *for all* n.

Example 4.3 Consider the equation

$$\Delta x_n = p_n x_n \frac{1 - x_{n-2}}{1 - \frac{1}{2}x_{n-2}}, \quad n = 0, 1, \cdots.$$

If $p_n = \frac{1}{2}\left(1 - (\frac{20}{29})^{1/((n+1)(n+2))}\right)$, $n = 0, 1, \cdots$, by Theorem 4.2, it has solution $\{x_n\}$ satisfying $1 < x_n < \frac{3}{2}$ for all n. In fact, we choose $\alpha = \frac{29}{20}$, $\beta = \frac{3}{2}$, when $\frac{29}{20} < x_{-2}\ x_{-1}, x_0 < \frac{3}{2}$, the solution $\{x_n\}$ satisfies $1 < x_n < \frac{3}{2}$.

5 Oscillation for the equation

In this section, we obtain sufficient conditions by which every solution satisfying $0 < x_n < 1/\lambda$ of (1.1) oscillates about 1.

Theorem 5.1 *Suppose that there is an integer $l \geq 1$ such that*

$$\liminf_{n\to+\infty} p_n^l > \left(\frac{k}{k+1}\right)^{(k+1)l}. \tag{5.1}$$

Then every solution of (1.1) satisfying $0 < x_n < 1/\lambda$ oscillates about 1, where

$$p_n^1 = (1-\lambda)\sum_{s=n+1}^{n+k} p_s, \quad p_n^{i+1} = (1-\lambda)\sum_{s=n+1}^{n+k} p_s p_s^i, \quad i = 1, 2, \cdots.$$

Proof By (1.1), let $y_n = (1 - x_n)/1 - \lambda x_n)$ for $n = 0, 1, \cdots$, we get

$$\Delta y_n + (1-\lambda)p_n(1+\lambda y_n)(1+y_n)y_{n-k} = 0, \quad n = 0, 1, \cdots. \tag{5.2}$$

It suffices to show that every solution of (5.2) oscillates about zero. For the sake of contradiction, assume that $\{y_n\}$ is a non-oscillatory solution of (5.2). We consider two cases.

Case 1. $\{y_n\}$ is eventually positive. Combine with (5.1), choose $N_1 > 0$ and $c > 0$ such that

$$p_n^l > \left[\left(\frac{k}{k+1}\right)^l (1+c)\right]^{k+1}, \quad n \geq N_1, \tag{5.3}$$

and

$$y_{n-k} \geq y_n \geq y_{n+1} > 0, \quad \Delta y_n \leq 0, \quad n \geq N_1. \tag{5.4}$$

Let

$$z_n = 1 - \frac{y_{n+1}}{y_n}, \quad n \geq N_1.$$

Thus $0 \leq z_n < 1$ for $n \geq N_1$ and

$$\Delta y_n + (1-\lambda)p_n y_{n-k} \leq 0, \quad n \geq N_1. \tag{5.5}$$

Hence, we get

$$z_n \geq (1 - \lambda p_n \prod_{s=n-k}^{n-1} (1-z_s)^{-1}, \quad n \geq N_1 + k.$$

By employing the arithmetic-geometric mean inequality, we have

$$z_n \geq (1-\lambda)p_n \left(1 - \frac{1}{k}\sum_{s=n-k}^{n-1} z_s\right)^{-k}, \quad n \geq N_1 + k. \tag{5.6}$$

It is easy to show that

$$(1-\frac{x}{k})^{-k} \geq (\frac{k+1}{k})^{k+1} x \quad for \quad 0 \leq x < k. \tag{5.7}$$

Then

$$z_n \geq (\frac{k+1}{k})^{k+1}(1-\lambda)p_n \sum_{s=n-k}^{n-1} z_s, \quad n \geq N_1 + k.$$

Let

$$\begin{aligned} z_n^0 &= z_n & n &\geq N_1, \\ z_n^i &= (1-\lambda)p_n \textstyle\sum_{s=n-k}^{n-1} z_n^{i-1} & n &\geq N_1 + ik, \quad i = 1, 2, \cdots, l. \end{aligned} \tag{5.8}$$

From (5.6) and (5.8), we can easily obtain by simple induction

$$z_n \geq \left(\frac{k+1}{k}\right)^{(k+1)(l-1)} z_n^{l-1}, \quad n \geq N_1 + (l-1)k. \tag{5.9}$$

So by (5.6), (5.9) and $0 \leq z_n < 1$ we find that

$$0 \leq \left(\frac{k+1}{k}\right)^{(k+1)(l-1)} z_n^{l-1} < k, \quad n \geq N_1 + lk, \tag{5.10}$$

and

$$z_n \geq (1-\lambda)p_n \left(1 - \frac{1}{k}\left(\frac{k+1}{k}\right)^{(k+1)(l-1)} \sum_{s=n-k}^{n-1} z_s^{l-1}\right)^{-k}, \quad n \geq N_1 + lk. \tag{5.11}$$

It is easy to prove that

$$r\left(1 - \frac{x}{k}\right)^{-k} \geq x + k\left(\frac{k+1}{k} r^{1/(k+1)} - 1\right), \quad r \geq 0, \ 0 \leq x < k.$$

From (5.10) and (5.11), we have

$$\begin{aligned}
z_n p_n^l \ \geq\ & \left(\frac{k}{k+1}\right)^{(k+1)(l-1)} (1-\lambda)p_n \left[\left(\frac{k+1}{k}\right)^{(k+1)(l-1)} p_n^l\right] \\
& \times \left[1 - \frac{1}{k}\left(\frac{k+1}{k}\right)^{(k+1)(l-1)} \sum_{s=n-k}^{n-1} z_s^{l-1}\right]^{-k} \\
\geq\ & \left(\frac{k}{k+1}\right)^{(k+1)(l-1)} (1-\lambda)p_n \left\{\left(\frac{k+1}{k}\right)^{(k+1)(l-1)} \sum_{s=n-k}^{n-1} z_s^{l-1}\right. \\
& \left. + k\left[\frac{k+1}{k}\left(\frac{k+1}{k}\right)^{(k+1)(l-1)} (p_n^l)^{1/(k+1)} - 1\right]\right\} \\
=\ & (1-\lambda)p_n \left\{\sum_{s=n-k}^{n-1} z_s^{l-1} + k\left(\frac{k+1}{k}\right)^{(k+1)(1-l)}\right. \\
& \left. \times \left[\left(\frac{k+1}{k}\right)^{l} (p_n^l)^{1/(k+1)} - 1\right]\right\}, \qquad n \geq N_1 + lk.
\end{aligned}$$

It follows that

$$\begin{aligned}
& z_n p_n^l - (1-\lambda)p_n \sum_{s=n-k}^{n-1} z_s^{l-1} \\
\geq\ & k\left(\frac{k+1}{k}\right)^{(k+1)(1-l)} (1-\lambda)p_n \left[\left(\frac{k+1}{k}\right)^{l} (p_n^l)^{1/(k+1)} - 1\right],
\end{aligned}$$

for $n \geq N_1 + lk$. Set $N_2 = N_1 + lk$, then for $N > N_2 + lk$, we have

$$\begin{aligned}
&\sum_{n=N_2}^{N} z_n p_n^l - (1-\lambda) \sum_{n=N_2}^{N} p_n \sum_{s=n-k}^{n-1} z_s^{l-1} \\
&\quad \geq k \left(\frac{k+1}{k}\right)^{(k+1)(1-l)} (1-\lambda) \sum_{N=n_2}^{N} p_n \left[\left(\frac{k+1}{k}\right)^{l} (p_n^l)^{1/(k+1)} - 1 \right] \\
&\quad \geq (1-\lambda) k \left(\frac{k+1}{k}\right)^{(k+1)(1-l)} \sum_{n=N_2}^{N} p_n.
\end{aligned}$$

By interchanging the order of summation, we find that

$$\begin{aligned}
\sum_{n=N_2}^{N} (1-\lambda) p_n \sum_{s=n-k}^{n-1} z_s^{l-1} &\geq \sum_{n=N_2}^{N} z_n^{l-1} \sum_{s=n+1}^{n+k} (1-\lambda) p_s \\
&= \sum_{n=N_2}^{N-k} (1-\lambda) p_n^1 p_n \sum_{s=n-k}^{n-1} z_s^{l-2} \geq \sum_{n=N_2}^{N-2k} z_n^{l-2} \sum_{s=n+1}^{n+k} (1-\lambda) p_s^1 p_s \\
&= \sum_{n=N_2}^{N-2k} (1-\lambda) p_n^2 p_n \sum_{s=n-k}^{n-1} z_s^{l-3} \\
&\quad \cdots\cdots \\
&\geq \sum_{n=N_2}^{N-lk} z_n^0 \sum_{s=n+1}^{n+k} (1-\lambda) p_s^{l-1} p_s \\
&= \sum_{n=N_2}^{N-lk} z_n p_n^l.
\end{aligned}$$

From this and the above inequality, we get

$$\sum_{n=N-lk+1}^{N} z_n p_n^l \geq c(1-\lambda) k \left(\frac{k+1}{k}\right)^{(k+1)(1-l)} \sum_{n=N_2}^{N} p_n.$$

Thus

$$\sum_{n=N-lk+1}^{N} p_n^l \geq c(1-\lambda) k \left(\frac{k+1}{k}\right)^{(k+1)(1-l)} \sum_{n=N_2}^{N} p_n. \tag{5.12}$$

On the other hand, from the definition of $\{p_n^i\}$, we see that

$$\sum_{n=0}^{+\infty} p_n = +\infty.$$

In fact, if $\sum_{n=0}^{+\infty} p_n < +\infty$, then p_n tends to 0 as n tends to infinity. Thus p_n^1 tends to 0 as n tends to infinity. $\cdots$. So p_n^l tends to 0 as n tends to infinity. This contradicts (5.1). From (5.12), one gets

$$\lim_{n \to +\infty} \sum_{n=N-lk+1}^{N} p_n^l = +\infty. \tag{5.13}$$

Now, we are going to derive a fact that contradicts (5.13). By (5.5), we obtain

$$y_{n+k+1} - y_{n+1} + (1-\lambda) \sum_{s=n+1}^{n+k} p_s y_{s-k} \le 0, \quad n \ge N_1. \tag{5.14}$$

By omitting the first term in (5.14), and using the non-increasing nature of $\{y_n\}$, we have

$$y_{n+1} \ge y_n (1-\lambda) \sum_{s=n+1}^{n+k} p_s = y_n p_n^1, \quad n \ge N_1, \tag{5.15}$$

and so

$$p_n^1 = (1-\lambda) \sum_{s=n+1}^{n+k} p_s \le \frac{z_{n+1}}{z_n} \le 1. \tag{5.16}$$

Hence by (5.14), (5.15), (5.16) and the definition of p_n^i, we have $p_n^i \le 1$ for $n \ge N_1$ and $i = 1, 2, \cdots, l$. It follows that

$$\sum_{n=N-lk+1}^{N} p_n^l \le lk, \quad N \ge N_1 + lk.$$

This contradicts (5.13) and completes the proof of Case 1.

Case 2. $\{y_n\}$ is eventually negative. It is easy to see that $\sum_{n=0}^{+\infty} p_s = +\infty$. Hence, we know that y_n tends to zero as n tends to infinity. Choose $1 > \epsilon > 0$, $c > 0$ and $N_4 > 0$ such that

$$(1-\lambda\epsilon)^l (1-\epsilon)^l p_n^l > \left[\left(\frac{k}{k+1} \right)^l (1+c) \right]^{k+1}, \quad n \ge N_4,$$

and

$$0 > y_{n+1} > y_n > y_{n-k} > -\epsilon, \quad n \ge N_4.$$

Hence, we get

$$\Delta y_n + (1-\lambda)(1-\lambda\epsilon)(1-\epsilon) p_n y_{n-k} \ge 0, \quad n \ge N_4.$$

Let $z_n = -y_n$. We have

$$\Delta z_n + (1-\lambda) r_n z_{n-k} \le 0, \quad n \ge N_4,$$

where $r_n = (1-\lambda\epsilon)(1-\epsilon)p_n$. By a similar method in Case 1, we derive a contradiction. The proof is complete. □

Example 5.2 *Consider the difference equation*

$$\Delta x_n = p_n x_n (1 - x_{n-3}), \quad n = 0, 1, 2, \cdots \tag{5.17}$$

where $k = 3$, $\lambda = 0$, *and*

$$p_n = \begin{cases} 0.1, & 2 \le j \le 4, \\ 0.24, & 5 \le j \le 13, \end{cases}$$

with $p_n = p_{n+12}$ *for every integer* n.

It is easy to know that

$$p_n^1 = \begin{cases} 0.30, & j = 1, \\ 0.44, & j = 2, \\ 0.58, & j = 3, \\ 0.72, & 4 \le j \le 10, \\ 0.58, & j = 11, \\ 0.44, & j = 12. \end{cases}$$

$$p_n^2 = \begin{cases} 0.132, & j = 0, \\ 0.2748, & j = 1, \\ 0.4036, & j = 2, \\ 0.5184, & 3 \le j \le 7, \\ 0.4848, & j = 8, \\ 0.4176, & j = 9, \\ 0.2748, & j = 10, \\ 0.1796, & j = 11. \end{cases}$$

Hence,

$$\liminf_{n \to +\infty} p_n^2 > \left(\frac{3}{3+1}\right)^{(3+1)\times 2} \approx 0.100113.$$

By Theorem 5.1, every positive solution of (5.17) oscillates about 1. However, since

$$\liminf_{n \to +\infty} p_n^1 = 0.3 < \left(\frac{3}{4}\right)^4 \approx 0.3164,$$

the Theorem in [1, 16] can not be applied.

References

[1] Chen M.P. and Yu J.S., Oscillation and global attractivity in a delay logistic difference equation. J. Differ. Equations Appl., 1995, 1: 227-237.

[2] Cui Q. and Lawson G.J., Study on model of single population: an expansion of the logistic and exponential equation. J. Theor. Boil., 1982, 98: 645-659.

[3] Cui Q., et al., Study of models of single populations: development of equations used in microorganism cultures. Biotechnology and Bioengineering, 1984, 24: 682-686.

[4] Cui Q., et al., Mathematical model in chemistry system, Proceeding of international biology conference, Xian, P.R. China. 1987.

[5] Erbe L.H., Xia H. and Yu J.S., Global attractivity in nonlinear delay difference equations. J. Differ. Equations Appl., 1995, 1:151-161.

[6] Kocic V.L.J. and Ladas G., Global Behavior of Nonlinear Difference Equations of Higher Order with Applications. Kluwer Academic, Boston. 1993, 75-80.

[7] Kuang Y., Delay Differential Equations with Applications in Population Dynamics. Academic Press, Boston. 1993.

[8] Kuang Y., Global stability for a class of nonlinear non-autonomous delay equations. Nonl., Anal. TAM., 1991, 17:627-634.

[9] Kuang Y., Zhang B. G. and Zhao T.,Qualitative analysis of a non-autonomous differential equations. Tohoku Math. J., 1991, 43:509-528.

[10] Liu Y.J., Global attractivity for population model. J. of Biomath., 2000, 15(1):65-69.

[11] Philos Ch.G., Oscillations in a non-autonomous delay logistic difference equation. Proc. of the Edinburgh Math. Soc., 1992, 35:121-131.

[12] Thompson D.W., On growth and form. Cambrige Univ. Press. 1952.

[13] Yu J.S., Wu J. and Zou X., On the hyper-logistic delay equation. Glasgow Math. J., 1996, 38:255-261.

[14] Zhang B.G. and Gopalsamy K., Global attractivity in a delay logistic equation with variable parameters. Math. Proc. Camb. Phil. Soc., 1990, 107:579-590.

[15] Zhang B.G. and Tian C.J., Nonexistence and existence of positive solutions for difference equations with unbounded delay. Computers Math. Applic., 1998, 36(1):1-8.

[16] Zhou Z. and Zhang Q.Q., Global attractivity of a non-autonomous logistic equation with delays. Computers And Mathematics With Applications. 1999, 38:57-64.

Fields Institute Communications
Volume **42**, 2004

Stability Regions for a Class of Delay Difference Systems

Hideaki Matsunaga
Department of Mathematical Sciences
Osaka Prefecture University
Sakai 599-8531, Japan
hideaki@ms.osakafu-u.ac.jp

Abstract. We give some new necessary and sufficient conditions for the asymptotic stability of a linear delay difference system

$$x_{n+1} - ax_n - Bx_{n-k} = 0, \quad n = 0, 1, 2, \ldots, \tag{L}$$

by using root-analysis for its characteristic equation. Here, a is a real number, B is a 2×2 real constant matrix, and k is a positive integer. In case

$$B = b\begin{pmatrix} \cos\theta & -\sin\theta \\ \sin\theta & \cos\theta \end{pmatrix}, \quad b \in \mathbf{R}, \ |\theta| \le \frac{\pi}{2},$$

the stability region for (L), that is, the region of (a, b) for which the zero solution of (L) with fixed k and θ is asymptotically stable, is symmetric with respect to the b-axis (resp. origin) if k is odd (resp. even).

1 Introduction

In this paper we study the asymptotic stability of a linear delay difference system with two coefficients and one delay

$$x_{n+1} - ax_n + Bx_{n-k} = 0, \quad n = 0, 1, 2, \ldots, \tag{1.1}$$

where a is a real number, B is a 2×2 real constant matrix, and k is a positive integer.

In the scalar case, Kuruklis [5] has shown that *the zero solution of the difference equation*

$$x_{n+1} - ax_n + bx_{n-k} = 0, \quad n = 0, 1, 2, \ldots, \tag{1.2}$$

where a is nonzero and b is a real number, is asymptotically stable if and only if $|a| < (k+1)/k$, and

$$|a| - 1 < b < (a^2 + 1 - 2|a|\cos\phi)^{1/2} \quad \text{if } k \text{ is odd}$$

$$|b - a| < 1 \quad \text{and} \quad |b| < (a^2 + 1 - 2|a|\cos\phi)^{1/2} \quad \text{if } k \text{ is even,}$$

2000 *Mathematics Subject Classification.* Primary 39A10, 39A11; Secondary 34K20.

The author was supported in part by Sasakawa Grants for Science Fellows of The Japan Science Society.

where ϕ is the solution in $(0, \pi/(k+1))$ of $\sin(k\phi)/\sin((k+1)\phi) = 1/|a|$. Note that in case $a = 0$ the necessary and sufficient condition for the asymptotic stability of (1.2) is $|b| < 1$. (Some other proofs of this result are given in [2] and [7].)

Also, the author and Hara [6] have recently shown that *the zero solution of the difference system*

$$x_{n+1} - x_n + bR(\theta)x_{n-k} = 0, \quad n = 0, 1, 2, \dots, \tag{1.3}$$

where b is a real number and $R(\theta)$ represents a 2×2 matrix

$$\begin{pmatrix} \cos\theta & -\sin\theta \\ \sin\theta & \cos\theta \end{pmatrix}$$

with $|\theta| \leq \pi/2$, is asymptotically stable if and only if

$$0 < b < 2\cos\frac{k\pi + |\theta|}{2k+1}. \tag{1.4}$$

The purpose of this paper is to give some new necessary and sufficient conditions for the asymptotic stability of the zero solution of (1.1). It is an extension of the previous results above. Our main results are proved by using the fact that the zero solution of (1.1) is asymptotically stable if and only if all the roots of its associated characteristic equation

$$F(\lambda) \equiv \det(\lambda^{k+1}I - a\lambda^k I + B) = 0 \tag{1.5}$$

are inside the unit disk. Here, I is the 2×2 identity matrix. For the general background of delay difference systems, one can refer to recent books [1], [3] and [4, Chapter 7].

By the transfomation $x_n = Py_n$ with a nonsingular matrix P, the system (1.1) can be written as

$$y_{n+1} - ay_n + P^{-1}BPy_{n-k} = 0, \quad n = 0, 1, 2, \dots.$$

Consequently, we have only to consider (1.1) where the matrix B is either of the following two matrices in Jordan form:

$$\text{(I)} \quad B = \begin{pmatrix} b_1 & d \\ 0 & b_2 \end{pmatrix}, \qquad \text{(II)} \quad B = bR(\theta) \equiv b\begin{pmatrix} \cos\theta & -\sin\theta \\ \sin\theta & \cos\theta \end{pmatrix},$$

where b_1, b_2, d, b and θ are real numbers with $0 < |\theta| \leq \pi/2$.

2 Main results

2.1 Triangular matrix. We consider the difference system (1.1) where B is given by (I), that is,

$$x_{n+1} - ax_n + \begin{pmatrix} b_1 & d \\ 0 & b_2 \end{pmatrix} x_{n-k} = 0. \tag{2.1}$$

Then (1.5) becomes

$$F(\lambda) = (\lambda^{k+1} - a\lambda^k + b_1)(\lambda^{k+1} - a\lambda^k + b_2) = 0.$$

It is obvious that for $j = 1, 2$, the equation $\lambda^{k+1} - a\lambda^k + b_j = 0$ is the characteristic equation of (1.2) with $b = b_j$, and so one can immediately get the following theorem from the previous result given by Kuruklis [5].

Theorem 2.1 *Suppose that* $a \neq 0$. *Then the zero solution of* (2.1) *is asymptotically stable if and only if* $|a| < (k+1)/k$, *and for* $j = 1, 2$

$$|a| - 1 < b_j < (a^2 + 1 - 2|a| \cos\phi)^{1/2} \quad \text{if } k \text{ is odd}$$

$$|b_j - a| < 1 \quad \text{and} \quad |b_j| < (a^2 + 1 - 2|a| \cos\phi)^{1/2} \quad \text{if } k \text{ is even,}$$

where ϕ *is the solution in* $(0, \pi/(k+1))$ *of* $\sin(k\phi)/\sin((k+1)\phi) = 1/|a|$.

In this case if $a = 0$, the necessary and sufficient condition for the asymptotic stability of (2.1) is $|b_1| < 1$ and $|b_2| < 1$. We also notice that for $j = 1, 2$, the region of (a, b_j) for which the zero solution of (2.1) with fixed k is asymptotically stable is symmetric with respect to the b_j-axis (resp. origin) if k is odd (resp. even).

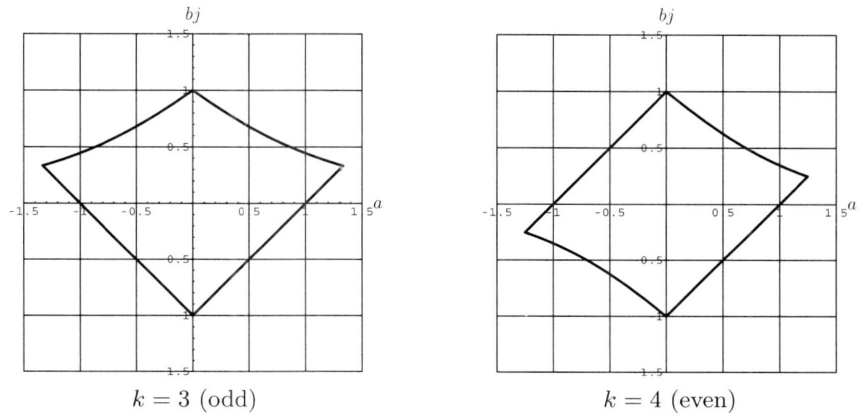

Figure 1 The stability regions for (2.1)

2.2 Rotation matrix. We consider the difference system (1.1) where B is given by (II), that is,

$$x_{n+1} - a x_n + b R(\theta) x_{n-k} = 0. \tag{2.2}$$

Here, for the sake of convenience, we set

$$S(\omega) = \frac{\sin(k\omega - |\theta|)}{\sin((k+1)\omega - |\theta|)}, \tag{2.3}$$

$$C_l(a) = (a^2 + 1 - 2|a| \cos\omega_l)^{1/2}, \quad 1 \leq l \leq 4, \tag{2.4}$$

where ω_l $(1 \leq l \leq 4)$ are the solutions of $S(\omega) = 1/|a|$ with

$$\frac{|\theta| - \pi}{k+1} < \omega_1 \leq \frac{2|\theta| - \pi}{2k+1} < \omega_2 < \omega^* < \omega_3 < 0 < \omega_4 < \frac{|\theta|}{k+1},$$

and ω^* is the solution in $((2|\theta| - \pi)/(2k+1), 0)$ of

$$k \sin\omega - \sin(k\omega - |\theta|) \cos((k+1)\omega - |\theta|) = 0. \tag{2.5}$$

Note that the square root in (2.4) is valid because $a^2 + 1 - 2|a| \cos\omega_l \geq (|a| - 1)^2 \geq 0$.

Theorem 2.2 *Suppose that* $a \neq 0$, $0 < |\theta| < \pi/2$, *and* k *is odd. Then the zero solution of* (2.2) *is asymptotically stable if and only if any one of the following conditions holds.*

(i) $|a| \leq 1$ *and* $0 < b < C_1(a)$.
(ii) $1 < |a| < 1/S(\omega^*)$ *and* $C_3(a) < b < C_2(a)$.

(iii) $|a| < 1$ *and* $-C_4(a) < b \le 0$.

Theorem 2.3 *Suppose that* $a \neq 0$, $0 < |\theta| < \pi/2$, *and* k *is even. Then the zero solution of* (2.2) *is asymptotically stable if and only if any one of the following conditions holds.*

(i) $ab > 0$, $|a| \le 1$, *and* $|b| < C_1(a)$.
(ii) $ab > 0$, $1 < |a| < 1/S(\omega^*)$, *and* $C_3(a) < |b| < C_2(a)$.
(iii) $ab \le 0$, $|a| < 1$, *and* $|b| < C_4(a)$.

Theorem 2.4 *Suppose that* $a \neq 0$ *and* $|\theta| = \pi/2$. *Then the zero solution of* (2.2) *is asymptotically stable if and only if*

$$|a| < 1 \quad \text{and} \quad |b| < (a^2 + 1 - 2|a| \cos\psi)^{1/2},$$

where ψ *is the solution in* $(0, \pi/2(k+1))$ *of* $\cos(k\psi)/\cos((k+1)\psi) = 1/|a|$.

In this case if $a = 0$, the necessary and sufficient condition for the asymptotic stability of (2.2) is $|b| < 1$. We also notice that the region of (a, b) for which the zero solution of (2.2) with fixed k and θ is asymptotically stable is symmetric with respect to the b-axis (resp. origin) if k is odd (resp. even).

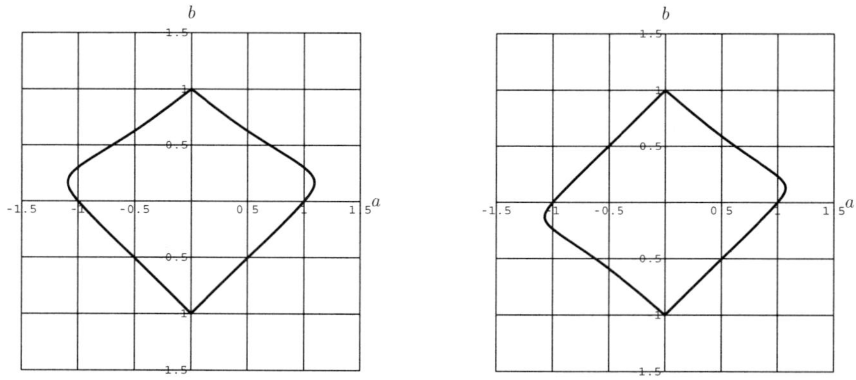

$k = 3$ (odd) and $\theta = \pi/6$ $\qquad$ $k = 4$ (even) and $\theta = \pi/6$

Figure 2 The stability regions for (2.2)

Remark 2.5 All the boundary of the stability region for (2.2) consists of parametric curves. Indeed, in case $0 < |\theta| < \pi/2$, the curve in the right half plane is given parametrically by the equation

$$a = \frac{\sin((k+1)\omega - |\theta|)}{\sin(k\omega - |\theta|)}, \quad b = \frac{\sin\omega}{\sin(k\omega - |\theta|)}, \quad \frac{|\theta| - \pi}{k+1} < \omega < \frac{|\theta|}{k+1}$$

(see (2.3) and (2.4)).

Remark 2.6 In case $a = 1$, Theorems 2.2 and 2.3 show that the zero solution of (2.2) is asymptotically stable if and only if $0 < b < C_1(1)$ where

$$C_1(1) = \sqrt{2(1 - \cos\omega_1)} = 2\sin\frac{|\omega_1|}{2} = 2\sin\frac{\pi - 2|\theta|}{2(2k+1)} = 2\cos\frac{k\pi + |\theta|}{2k+1},$$

which coincides with (1.4).

In order to prove our main theorems, we will deal with the roots of (1.5). Let $f(\lambda) = \lambda^{k+1} - a\lambda^k + be^{i\theta}$. Then, by (1.5), we have

$$\begin{aligned} F(\lambda) &= (\lambda^{k+1} - a\lambda^k + b\cos\theta)^2 + b^2\sin^2\theta \\ &= (\lambda^{k+1} - a\lambda^k + b\cos\theta + ib\sin\theta)(\lambda^{k+1} - a\lambda^k + b\cos\theta - ib\sin\theta) \\ &= f(\lambda)\overline{f(\bar{\lambda})}, \end{aligned}$$

where $\bar{\lambda}$ is the complex conjugate of any complex λ. Hence it suffices to investigate the roots of the equation

$$f(\lambda) = \lambda^{k+1} - a\lambda^k + be^{i\theta} = 0. \tag{2.6}$$

We may also assume that $\theta > 0$. Furthermore, for $a \neq 0$, the equation (2.6) can be equivalently as

$$\mu^{k+1} - \mu^k + ce^{i\theta} = 0, \tag{2.7}$$

where $\mu = \lambda/a$ and $c = b/a^{k+1}$. Thus, all the roots of (2.6) are inside the unit disk if and only if all the roots of (2.7) are inside the disk $|\mu| < 1/|a|$, and therefore, the following lemma holds.

Lemma 2.7 *Let $a \neq 0$ and $c = b/a^{k+1}$. Then the zero solution of (2.2) is asymptotically stable if and only if all the roots of (2.7) are inside the disk $|\mu| < 1/|a|$.*

Consequently, we will now discuss the location of the roots of (2.7) as c varies. Note that, for $c = 0$, the roots of (2.7) are 0 (multiplicity k) and 1 (simple), and for $c \neq 0$, the equation (2.7) has no real roots.

Let μ be expressed in the polar form, that is, $\mu = re^{i\omega}$. Then we obtain $r^{k+1}e^{i((k+1)\omega-\theta)} - r^k e^{i(k\omega-\theta)} + c = 0$, namely

$$r^k\{\cos(k\omega - \theta) - r\cos((k+1)\omega - \theta)\} = c, \tag{2.8}$$

and

$$r^k\{\sin(k\omega - \theta) - r\sin((k+1)\omega - \theta)\} = 0. \tag{2.9}$$

It is obvious that $\sin((k+1)\omega - \theta) \neq 0$, so (2.9) becomes

$$r = \frac{\sin(k\omega - \theta)}{\sin((k+1)\omega - \theta)}, \tag{2.10}$$

which, together with (2.8), yields

$$c = r^k \frac{\sin\omega}{\sin((k+1)\omega - \theta)}. \tag{2.11}$$

Hence, we have the following proposition on the existence region of the arguments of complex roots of (2.7), by examining the sign of the right-hand side in both (2.10) and (2.11).

Proposition 2.8 *Assume that $0 < \theta \leq \pi/2$. Let $re^{i\omega}$ be a complex root of (2.7), where $r > 0$ and $0 < |\omega| < \pi$. Then the following statements hold.*

(a) *$c > 0$ and $\omega > 0$ imply $\omega \in I_m^+ = ((\theta + 2m\pi)/k, (\theta + (2m+1)\pi)/(k+1))$, $m = 0, 1, \ldots, [(k-1)/2]$.*

(b) *$c > 0$ and $\omega < 0$ imply $\omega \in I_0^- = ((\theta - \pi)/(k+1), 0)$ or $\omega \in I_m^- = ((\theta - (2m+1)\pi)/(k+1), (\theta - 2m\pi)/k)$, $m = 1, 2, \ldots, [k/2]$.*

(c) *$c < 0$ and $\omega > 0$ imply $\omega \in J_0^+ = (0, \theta/(k+1))$ or $\omega \in J_m^+ = ((\theta + (2m-1)\pi)/k, (\theta + 2m\pi)/(k+1))$, $m = 1, 2, \ldots, [k/2]$.*

(d) $c<0$ *and* $\omega<0$ *imply* $\omega\in J_m^-=((\theta-(2m+2)\pi)/(k+1),(\theta-(2m+1)\pi)/k)$, $m=0,1,\ldots,[(k-1)/2]$.

Here, $[\cdot]$ *denotes the greatest-integer function.*

The following proposition determines the arguments of complex roots of (2.7) on the unit circle.

Proposition 2.9 *Assume that* $0<\theta\le\pi/2$. *Then the arguments of complex roots of* (2.7) *on the unit circle are given by*

$$\omega=\xi_p\equiv\frac{2\theta+(2p-1)\pi}{2k+1},\quad p=0,\pm1,\ldots,\pm k. \tag{2.12}$$

Proof Substituting $r=1$ into (2.10), we get

$$\sin((k+1)\omega-\theta)-\sin(k\omega-\theta)=2\sin(\omega/2)\cos((k+1/2)\omega-\theta)=0.$$

By $\sin(\omega/2)\ne0$, we have $\cos((k+1/2)\omega-\theta)=0$, which implies (2.12). □

Next, we will observe how the roots of (2.7) move in the complex plane when c varies.

Proposition 2.10 *Assume that* $0<\theta<\pi/2$. *Then the simple root* $\mu=1$ *of* (2.7) *with* $c=0$ *moves inside the unit disk (resp. outside the unit disk) as* c *increases from* 0 *(resp. decreases from* 0*).*

Proof Let $re^{i\omega}$ be a root of (2.7). Then it follows from (2.10) and (2.11) that

$$\begin{aligned}\frac{dr}{d\omega}&=\frac{k\cos(k\omega-\theta)\sin((k+1)\omega-\theta)-(k+1)\sin(k\omega-\theta)\cos((k+1)\omega-\theta)}{\sin^2((k+1)\omega-\theta)}\\&=\frac{k\sin\omega-\sin(k\omega-\theta)\cos((k+1)\omega-\theta)}{\sin^2((k+1)\omega-\theta)}\end{aligned}$$

and

$$\begin{aligned}\frac{dc}{d\omega}&=kr^{k-1}\frac{dr}{d\omega}\cdot\frac{\sin\omega}{\sin((k+1)\omega-\theta)}\\&\quad+r^k\frac{\cos\omega\sin((k+1)\omega-\theta)-(k+1)\sin\omega\cos((k+1)\omega-\theta)}{\sin^2((k+1)\omega-\theta)}\\&=kr^k\frac{\{k\sin\omega-\sin(k\omega-\theta)\cos((k+1)\omega-\theta)\}\sin\omega}{\sin(k\omega-\theta)\sin^2((k+1)\omega-\theta)}\\&\quad+r^k\frac{\sin(k\omega-\theta)-k\sin\omega\cos((k+1)\omega-\theta)}{\sin^2((k+1)\omega-\theta)}\\&=r^k\frac{k^2\sin^2\omega-2k\sin\omega\sin(k\omega-\theta)\cos((k+1)\omega-\theta)+\sin^2(k\omega-\theta)}{\sin(k\omega-\theta)\sin^2((k+1)\omega-\theta)}.\end{aligned}$$

By using the relations above, we thus obtain

$$\frac{dr}{dc}=\frac{dr}{d\omega}\left(\frac{dc}{d\omega}\right)^{-1}=\frac{\sin(k\omega-\theta)G(\omega)}{r^kH(\omega)}, \tag{2.13}$$

where

$$\begin{aligned}G(\omega)&=k\sin\omega-\sin(k\omega-\theta)\cos((k+1)\omega-\theta),\\H(\omega)&=k^2\sin^2\omega-2k\sin\omega\sin(k\omega-\theta)\cos((k+1)\omega-\theta)+\sin^2(k\omega-\theta).\end{aligned}$$

In view of (2.11), $c = 0$ is equivalent to $\omega = 0$, and hence, we have

$$\left.\frac{dr}{dc}\right|_{\substack{r=1\\c=0}} = \frac{\sin(-\theta)G(0)}{H(0)} = -\cos\theta < 0.$$

The proof is complete. □

Proposition 2.11 *Assume that $0 < \theta \leq \pi/2$. Then the absolute values of complex roots of (2.7) increase as $|c|$ increases except for the root $\tilde{\mu}$ satisfying $\omega^* < \arg\tilde{\mu} < 0$, where ω^* is the solution in $(\xi_0, 0)$ of $G(\omega) = 0$ (see (2.5)). Precisely, the absolute value of $\tilde{\mu}$ decreases as the positive number c increases.*

Proof Let $re^{i\omega}$ be a root of (2.7). Proposition 2.8 then implies

$$\omega \in \Omega \equiv \left(\bigcup_{m=0}^{[(k-1)/2]} I_m^+\right) \cup \left(\bigcup_{m=0}^{[k/2]} I_m^-\right) \cup \left(\bigcup_{m=0}^{[k/2]} J_m^+\right) \cup \left(\bigcup_{m=0}^{[(k-1)/2]} J_m^-\right).$$

Note that $G(\omega)$ is strictly increasing for $\omega \in \Omega$, since, by (2.10),

$$G'(\omega) = (2k+1)\sin(k\omega - \theta)\sin((k+1)\omega - \theta) > 0.$$

We consider the case $c > 0$. (In case $c < 0$, the proof is similar.)

Case (a): $c > 0$ and $\omega > 0$. For $\omega \in I_m^+$, $m = 0, 1, \ldots, [(k-1)/2]$, it is enough to show that $dr/dc > 0$. From the increasing property of $G(\omega)$, the inequality $G((\theta + 2m\pi)/k) = k\sin((\theta + 2m\pi)/k) > 0$ yields $G(\omega) > 0$. Proposition 2.8 also implies that $\sin(k\omega - \theta) > 0$ and $H(\omega) > (k\sin\omega - \sin(k\omega - \theta))^2 \geq 0$. Hence, together with (2.13), we obtain $dr/dc > 0$.

Case (b): $c > 0$ and $\omega < 0$. For $\omega \in I_m^-$, $m = 0, 1, \ldots, [k/2]$, we see that $\sin(k\omega - \theta) < 0$ and $H(\omega) > 0$ in a similar way to Case (a). Also, from the increasing property of $G(\omega)$, the inequality $G((\theta - 2m\pi)/k) = -k\sin((\theta + 2m\pi)/k) < 0$ yields that $G(\omega) < 0$ for $\omega \in I_m^-$, $m = 1, 2, \ldots, [k/2]$, while the inequalities $G(0) = (\sin 2\theta)/2 > 0$ and $G(\xi_0) = (k + 1/2)\sin((2\theta - \pi)/(2k+1)) < 0$ yield that $G(\omega) > 0$ for $\omega \in (\omega^*, 0)$ and $G(\omega) < 0$ for $\omega \in ((\theta - \pi)/(k+1), \omega^*)$ by the definition of ω^*. Thus, together with (2.13), the assertion of this proposition holds. □

Propositions 2.10 and 2.11 show that as $|c|$ increases, the roots of (2.7) are moving continuously and monotonically away from 0 except for the root $\tilde{\mu}$.

Furthermore, we will find the minimum value of $|c|$ for which (2.7) has a root on the circle $|\mu| = 1/|a|$ as $|c|$ increases from 0. The following proposition provides the value of $|c|$ in terms of $|a|$ and the argument ω when a root of (2.7) lies on the circle $|\mu| = 1/|a|$.

Proposition 2.12 *Let $re^{i\omega}$ be a root of (2.7) on the circle $|\mu| = 1/|a|$. Then*

$$|c| = \frac{(a^2 + 1 - 2|a|\cos\omega)^{1/2}}{|a|^{k+1}}. \tag{2.14}$$

Proof By squaring both sides of (2.8) and (2.9), and adding them together, we have

$$c^2 = r^{2k}(r^2 + 1 - 2r\cos\omega),$$

which implies (2.14) from $r = 1/|a|$. □

We notice that the value of $|c|$ given by (2.14) is increasing with respect to $|\omega|$, and hence, the minimum value of $|c|$ corresponds to the minimum value of $|\omega|$ satisfying $r = 1/|a|$, or equivalently

$$S(\omega) = \frac{\sin(k\omega - \theta)}{\sin((k+1)\omega - \theta)} = \frac{1}{|a|} \tag{2.15}$$

(see (2.3) and (2.10)). Therefore, by virtue of Propositions 2.8 and 2.11, one can easily see that $S(\omega)$ is strictly increasing (resp. decreasing) with respect to $|\omega|$, $\omega \in \Omega \setminus (\omega^*, 0)$ (resp. $\omega \in (\omega^*, 0)$), and so $S(\omega^*)$ is only a local minimum of $S(\omega)$. In fact, the graph of $r = S(\omega)$ becomes as follows:

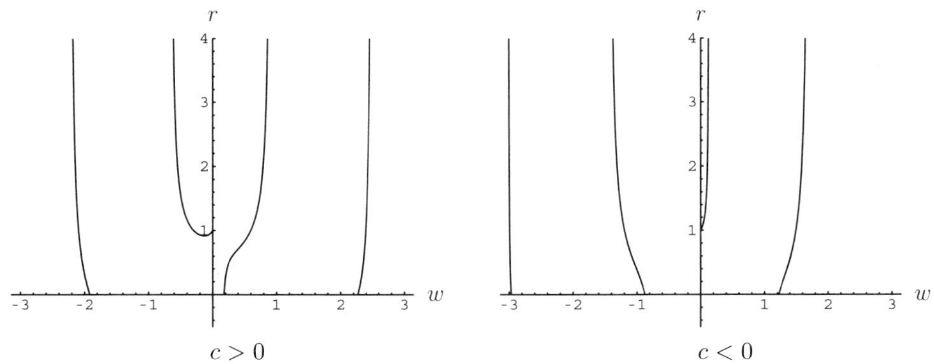

Figure 3 The graphs of $r = S(\omega)$ with $k = 3$ and $\theta = \pi/6$

Now, we are in a position to present necessary and sufficient conditions for the roots of (2.7) to be inside the disk $|\mu| < 1/|a|$ (Lemmas 2.13–2.15).

Lemma 2.13 *Suppose that* $a \neq 0$, $0 < \theta < \pi/2$, *and* $c > 0$. *Then all the roots of* (2.7) *are inside the disk* $|\mu| < 1/|a|$ *if and only if*

$$0 < |a| \leq 1 \ \text{ and } \ 0 < c < C_1(a)/|a|^{k+1}$$

or

$$1 < |a| < 1/S(\omega^*) \ \text{ and } \ C_3(a)/|a|^{k+1} < c < C_2(a)/|a|^{k+1},$$

where $S(\omega)$ *and* $C_l(a)$ $(l = 1, 2, 3)$ *are given by* (2.3) *and* (2.4), *respectively, and* ω^* *is the soluition in* $(\xi_0, 0)$ *of* $G(\omega) = 0$.

Proof We first note that $0 < C_3(a) < C_2(a) < C_1(a)$ holds and the values of ω satisfying $r = S(\omega) = 1$ in I_0^- and I_0^+ are ξ_0 and ξ_1, respectively, by the definitions. Let us investigate the location of the roots of (2.7) in detail. Recall that the roots of (2.7) with $c = 0$ are 0 (multiplicity k) and 1 (simple). Also, the branch of the simple root $\mu = 1$ stays in the region $\{\mu \in \mathbf{C} : \arg\mu \in I_0^-\}$ as c increases from 0, by the continuity of the roots with respect to c. Our argument is divided into three cases.

Case (I): $1/|a| \geq 1$, i.e., $0 < |a| \leq 1$. By the graph of $r = S(\omega)$, the equation (2.15) has only one solution $\omega_1 \in ((\theta - \pi)/(k+1), \xi_0]$ and $\hat{\omega}_1 \in [\xi_1, (\theta + \pi)/(k+1))$, respectively. Then, Propositions 2.10–2.12 show that the trace of the branch of $\mu = 1$ is the following: it stays inside the disk $|\mu| = 1/|a|$ as c increases form 0 to $C_1(a)/|a|^{k+1}$, it reaches the circle $|\mu| = 1/|a|$ as $c = C_1(a)/|a|^{k+1}$, and it moves outside the disk $|\mu| = 1/|a|$ as c increases from $C_1(a)/|a|^{k+1}$. Therefore, we assert that all the roots of (2.7) are inside the disk $|\mu| = 1/|a|$ if and only if $0 < c < C_1(a)/|a|^{k+1}$. To this end, it suffices to verify

$$|\omega_1| < \hat{\omega}_1 \tag{2.16}$$

from the increasing property of c with respect to $|\omega|$.

Subcase (Ia): $(\pi - \theta)/(k+1) \leq \xi_1$, i.e., $k\pi/(4k+3) \leq \theta < \pi/2$. It is obvious that $|\omega_1| = -\omega_1 < (\pi - \theta)/(k+1) \leq \xi_1 \leq \hat{\omega}_1$ and so (2.16) holds.

Subcase (Ib): $(\pi-\theta)/(k+1) > \xi_1$, i.e., $0 < \theta < k\pi/(4k+3)$. We prove (2.16) by contradiction. Suppose that $|\omega_1| \geq \hat{\omega}_1$. Then there exists $\eta \in (\xi_1, (\pi - \theta)/(k+1))$ such that $S(\eta) \geq S(-\eta)$, that is,

$$\frac{\sin(k\eta - \theta)}{\sin((k+1)\eta - \theta)} \geq \frac{\sin(k\eta + \theta)}{\sin((k+1)\eta + \theta)}. \tag{2.17}$$

For simplicity, let $\alpha = k\eta + \theta$ and $\beta = k\eta - \theta$. Then (2.17) becomes

$$\frac{\sin\beta\sin(\alpha + \eta) - \sin\alpha\sin(\beta + \eta)}{\sin(\alpha + \eta)\sin(\beta + \eta)} \geq 0. \tag{2.18}$$

We see that $\sin(\alpha+\eta) > 0$ and $\sin(\beta+\eta) > 0$, since $((4k+3)\theta+(k+1)\pi)/(2k+1) < \alpha + \eta < \pi$ and $(\theta + (k+1)\pi)/(2k+1) < \beta + \eta < \pi - 2\theta$, respectively. Thus (2.18) yields $\sin\beta\sin(\alpha + \eta) - \sin\alpha\sin(\beta + \eta) \geq 0$, namely

$$\sin\eta\sin(\beta - \alpha) \geq 0.$$

By using $\sin\eta > 0$, we therefore obtain $\sin(\beta-\alpha) = -\sin 2\theta \geq 0$, which contradicts the choice of θ.

Case (II): $S(\omega^*) < 1/|a| < 1$, i.e., $1 < |a| < 1/S(\omega^*)$. By the graph of $r = S(\omega)$, the equation (2.15) has only one solution $\omega_2 \in (\xi_0, \omega^*)$, $\omega_3 \in (\omega^*, 0)$, and $\hat{\omega}_2 \in (\theta/k, \xi_1)$, respectively. Then, Propositions 2.10–2.12 show that the trace of the branch of $\mu = 1$ is the following: it is outside the disk $|\mu| = 1/|a|$ as c increases from 0 to $C_3(a)/|a|^{k+1}$, it reaches the circle $|\mu| = 1/|a|$ as $c = C_3(a)/|a|^{k+1}$ or $c = C_2(a)/|a|^{k+1}$, it stays inside the disk $|\mu| = 1/|a|$ as c increases from $C_3(a)/|a|^{k+1}$ to $C_2(a)/|a|^{k+1}$, and it moves outside the disk $|\mu| = 1/|a|$ as c increases from $C_2(a)/|a|^{k+1}$. Therefore, we assert that all the roots of (2.7) are inside the disk $|\mu| = 1/|a|$ if and only if $C_3(a)/|a|^{k+1} < c < C_2(a)/|a|^{k+1}$. It is enough to verify

$$|\omega_2| < \hat{\omega}_2 \tag{2.19}$$

from the increasing property of c with respect to $|\omega|$.

Subcase (IIa): $\theta/k \geq -\xi_0$, i.e., $k\pi/(4k+1) \leq \theta < \pi/2$. It is obvious that $|\omega_2| = -\omega_2 < -\xi_0 \leq \theta/k < \hat{\omega}_2$ and so (2.19) holds.

Subcase (IIb): $\theta/k < -\xi_0$, i.e. $0 < \theta < k\pi/(4k+1)$. We prove (2.19) by contradiction. Suppose that $|\omega_2| \geq \hat{\omega}_2$. Then there exists $\zeta \in (\theta/k, -\xi_0)$ such that $S(\zeta) \geq S(-\zeta)$, that is,

$$\frac{\sin(k\zeta - \theta)}{\sin((k+1)\zeta - \theta)} \geq \frac{\sin(k\zeta + \theta)}{\sin((k+1)\zeta + \theta)}. \tag{2.20}$$

For simplicity, let $\alpha = k\zeta + \theta$ and $\beta = k\zeta - \theta$ again. Then (2.20) becomes

$$\frac{\sin\beta\sin(\alpha + \zeta) - \sin\alpha\sin(\beta + \zeta)}{\sin(\alpha + \zeta)\sin(\beta + \zeta)} \geq 0. \tag{2.21}$$

We see that $\sin(\alpha + \zeta) > 0$ and $\sin(\beta + \zeta) > 0$, since $(2k+1)\theta/k < \alpha + \zeta < ((k+1)\pi - \theta)/(2k+1)$ and $\theta/k < \beta + \zeta < ((k+1)\pi - (4k+3)\theta)/(2k+1)$, respectively. Thus (2.21) yields $\sin\beta\sin(\alpha+\zeta)-\sin\alpha\sin(\beta+\zeta)\geq 0$, namely

$$\sin\zeta\sin(\beta-\alpha)\geq 0.$$

By using $\sin\zeta>0$, we therefore obtain $\sin(\beta-\alpha)=-\sin 2\theta\geq 0$, which contradicts the choice of θ.

Case (III): $0<1/|a|\leq S(\omega^*)$, i.e., $|a|\geq 1/S(\omega^*)$. By the graph of $r=S(\omega)$, we always have $S(\omega)\geq 1/|a|$ for $\omega\in I_0^-$. This means that there exists a root μ of (2.7) which corresponds to the branch of $\mu=1$ such that $|\mu|\geq 1/|a|$.

This completes the proof. □

Lemma 2.14 *Suppose that $a\neq 0$, $0<\theta<\pi/2$, and $c<0$. Then all the roots of (2.7) are inside the disk $|\mu|<1/|a|$ if and only if*

$$0<|a|<1 \text{ and } -C_4(a)/|a|^{k+1}<c<0,$$

where $C_4(a)$ is given by (2.4).

Proof Note that the branch of $\mu=1$ stays in the region $\{\mu\in\mathbf{C}:\arg\mu\in J_0^+\}$ as c decreases from 0 by the continuity of the roots with respect to c. There are two possible cases to consider.

Case (I): $1/|a|>1$, i.e., $0<|a|<1$. By the graph of $r=S(\omega)$, the equation (2.15) has only one solution $\omega_4\in(0,\theta/(k+1))$ and $\hat{\omega}_4\in((\theta-2\pi)/(k+1),(\theta-\pi)/k)$, respectively. Then, Propositions 2.10–2.12 show that the trace of the branch of $\mu=1$ is the following: it stays inside the disk $|\mu|=1/|a|$ as c decreases form 0 to $-C_4(a)/|a|^{k+1}$, it reaches the circle $|\mu|=1/|a|$ as $c=-C_4(a)/|a|^{k+1}$, and it moves outside the disk $|\mu|=1/|a|$ as c decreases from $-C_4(a)/|a|^{k+1}$. Therefore, we assert that all the roots of (2.7) are inside the disk $|\mu|=1/|a|$ if and only if $-C_4(a)/|a|^{k+1}<c<0$. To this end, we have only to verify

$$\omega_4<|\hat{\omega}_4|$$

from the decreasing property of c with respect to $|\omega|$. Indeed, it is valid because $(\pi-\theta)/k-\theta/(k+1)=(k(\pi-2\theta)+\pi-\theta)/k(k+1)>0$.

Case (II): $0<1/|a|\leq 1$, i.e., $|a|\geq 1$. By the graph of $r=S(\omega)$, we always have $S(\omega)\geq 1/|a|$ for $\omega\in J_0^+$. This means that there exists a root μ of (2.7) which corresponds to the branch of $\mu=1$ such that $|\mu|\geq 1/|a|$.

This completes the proof. □

Lemma 2.15 *Suppose that $ac\neq 0$ and $\theta=\pi/2$. Then all the roots of (2.7) are inside the disk $|\mu|<1/|a|$ if and only if*

$$0<|a|<1 \text{ and } 0<|c|<(a^2+1-2|a|\cos\psi)^{1/2}/|a|^{k+1},$$

where ψ is the solution in $(0,\pi/2(k+1))$ of $\cos(k\psi)/\cos((k+1)\psi)=1/|a|$.

The proof is carried out in an argument similar to that above with $\theta=\pi/2$ and will be omitted.

Finally, we will prove our main theorems. In case $b=0$, the system (2.2) becomes $x_{n+1}-ax_n=0$, and hence, the zero solution of (2.2) is asymptotically stable if and only if $|a|<1$. Consequently, we have only to consider (2.2) with $b\neq 0$.

Proof of Theorem 2.2 We note that $a^{k+1}>0$ because $a\neq 0$ and k is odd.

If $b > 0$ then $c = b/a^{k+1} > 0$, and thus, Lemmas 2.7 and 2.13 imply that the zero solution of (2.2) is asymptotically stable if and only if

$$|a| \leq 1 \quad \text{and} \quad b < C_1(a)$$

or

$$1 < |a| < 1/S(\omega^*) \quad \text{and} \quad C_3(a) < b < C_2(a).$$

Also if $b < 0$ then $c = b/a^{k+1} < 0$, and hence, Lemmas 2.7 and 2.14 imply that the zero solution of (2.2) is asymptotically stable if and only if

$$|a| < 1 \quad \text{and} \quad -C_4(a) < b.$$

This completes the proof of Theorem 2.2. □

Proof of Theorem 2.3 We note that $a^k > 0$ because $a \neq 0$ and k is even.
If $ab > 0$ then $c = b/a^{k+1} > 0$, and thus, Lemmas 2.7 and 2.13 imply that the zero solution of (2.2) is asymptotically stable if and only if

$$|a| \leq 1 \quad \text{and} \quad |b| < C_1(a)$$

or

$$1 < |a| < 1/S(\omega^*) \quad \text{and} \quad C_3(a) < |b| < C_2(a).$$

Also if $ab < 0$ then $c = b/a^{k+1} < 0$, and hence, Lemmas 2.7 and 2.14 imply that the zero solution of (2.2) is asymptotically stable if and only if

$$|a| < 1 \quad \text{and} \quad |b| < C_4(a).$$

This completes the proof of Theorem 2.3. □

Proof of Theorem 2.4 Since $b = c/a^{k+1}$, Lemmas 2.7 and 2.15 imply Theorem 2.4. □

References

[1] R. P. Agarwal, "Difference Equations and Inequalities, Second Edition, Revised and Expanded: Theory, Methods and Applications," Marcel Dekker, New York, 2000.

[2] S. S. Cheng and Y.-Z. Lin. *Exact stability regions for linear difference equations with two constant coefficients and one delay* (to appear).

[3] S. N. Elaydi, "An Introduction to Difference Equations, Second Edition," Springer-Verlag, New York, 1999.

[4] I. Györi and G. Ladas, "Oscillation Theory of Delay Differential Equations," Clarendon Press, Oxford, 1991.

[5] S. A. Kuruklis, *The asymptotic stability of $x_{n+1} - ax_n + bx_{n-k} = 0$*, J. Math. Anal. Appl., **188** (1994), 719–731.

[6] H. Matsunaga and T. Hara, *The asymptotic stability of a two-dimensional linear delay difference equation*, Dynam. Contin. Discrete Impuls. Systems, **6** (1999), 465–473.

[7] V. G. Papanicolaou, *On the asymptotic stability of a class of linear difference equations*, Math. Magazine, **69** (1996), 34–43.

Fields Institute Communications
Volume **42**, 2004

Oscillation and Nonoscillation Results for Higher-Order Nonlinear Difference Equations

Małgorzata Migda
Institute of Mathematics
Poznań University of Technology
Piotrowo 3A, 60-965 Poznań, Poland
mmigda@math.put.poznan.pl

Abstract. The nonlinear difference equations

$$L_m x(n) + \delta(-1)^m a(n) f(x(\sigma(n))) = 0, \quad m \geq 2, \quad \delta = \pm 1$$

are considered. Some criteria for the oscillation of bounded solutions and sufficient conditions for existence of certain types of nonoscillatory solutions of the above equations are presented.

1 Introduction

In this paper we consider the m-th order difference equations

$$L_m x(n) + \delta(-1)^m a(n) f(x(\sigma(n))) = 0, \qquad n \in N \qquad (E\delta)$$

where $m \geq 2$, $\delta = \pm 1$, $N = \{1, 2, ...\}$ and

$$L_m x(n) = \Delta(r_{m-1}(n)\Delta(...r_2(n)\Delta(r_1(n)\Delta x(n))...)),$$

a, r_i $(i = 1, 2..., m-1)$ are positive real sequences, $f : R \to R$, $\sigma : N \to Z$ (Z denote the set of integers) and $\lim_{n\to\infty} \sigma(n) = \infty$. We define the difference operator Δ as usual

$$\Delta x(n) = x(n+1) - x(n), \quad \Delta^m x(n) = \Delta(\Delta^{m-1} x(n)) \quad \text{for} \quad n \in N, \quad m \geq 2.$$

For $k \in N$ we use the factorial notation

$$n^{(k)} = n(n-1)..(n-k+1) \quad \text{with} \quad n^{(0)} = 1.$$

For given sequences r_i $(i = 1, 2, ..., m-1)$ and any real sequence x we denote

$$L_0 x(n) = x(n), \quad L_i x(n) = r_i(n)\Delta L_{i-1} x(n), \quad i = 1, 2, ...m, \qquad (1.1)$$

where $r_m \equiv 1$. The sequences $L_i x$ (i=0,1,...,m) are called quasi-differences of x.

By a solution of equation $(E\delta)$ we mean a real sequence x which is defined for $n \geq \min_{i\geq 0} \sigma(i)$ and satisfies the equation for sufficiently large n. A solution of

2000 *Mathematics Subject Classification.* Primary 39A10.

equation $(E\delta)$ is called oscillatory if for every $n_0 \in N$ there exists an $n \geq n_0$ such that $x(n)x(n+1) \leq 0$. Otherwise it is called nonoscillatory.

We will obtain sufficient conditions for every bounded solution of equation $(E+)$ to be oscillatory and for every bounded solution of equation $(E-)$ to be either oscillatory or such that its quasi-differences to approach zero monotonically as $n \to \infty$. Moreover we will give sufficient conditions under which for every constant $c \in R$, there exists a solution x of equation $(E\delta)$ convergent to c. Motivation on the first part of the present study was the work of Lovelady [11] in which the oscillatory behaviour of bounded solutions of the linear differential equations

$$(p_{n-1}(...p_2(p_1u')'...)' + (-1)^n qu = 0$$

and

$$(p_{n-1}(...p_2(p_1u')'...)' + (-1)^{n+1} qu = 0$$

was studied.

For $m = 2$ and $\sigma(n) = n + 1$ equation $(E-)$ takes the form

$$\Delta(r_n \Delta x_n) = a_{n+1} f(x_{n+1}).$$

In last few years this (or every similar) equation has been considered by a number of authors, see for example [3]-[6],[9],[13],[14],[16]-[17],[19] and the monograph by Agarwal [1]. The oscillatory and nonoscillatory behaviour of solutions of difference equations of the form $(E\delta)$ in the case $m > 2$ were discussed in [2], [7], [8],[10], and [18].

2 Oscillation results

In this part we will use the following conditions:

(H1) $\sum\limits_{n=1}^{\infty} \frac{1}{r_k(n)} = \infty, \quad k = 1, 2, ..., m-1$

(H2) $xf(x) > 0$ for $x \neq 0$

(H3) $|f(x)|$ is bounded away from zero if $|x|$ is bounded away from zero.

Let

$$\omega_1(n) = \sum_{i_1=1}^{n} \frac{1}{r_1(i_1)}$$

$$\omega_2(n) = \sum_{i_2=1}^{n} \frac{1}{r_2(i_2)} \sum_{i_1=1}^{i_2} \frac{1}{r_1(i_1)} = \sum_{i_2=1}^{n} \frac{\omega_1(i_2)}{r_2(i_2)}$$

$$\vdots$$

$$\omega_k(n) = \sum_{i_k=1}^{n} \frac{\omega_{k-1}(i_k)}{r_k(i_k)} \quad \textit{for} \quad k = 1, 2, ..., m-1 \quad \textit{and} \quad n \in N.$$

Our first theorem gives sufficient conditions for all bounded solutions of equation (E+) to be oscillatory.

Theorem 2.1 *Assume (H1)-(H3) and*

$$\sum_{n=1}^{\infty} \omega_{m-1}(n)a(n) = \infty. \tag{2.1}$$

Then every bounded solution of equation $(E+)$ *is oscillatory.*

Proof Let x be a bounded and nonoscillatory solution of equation $(E+)$. We can assume without loss of generality that x is eventually positive. Therefore there is an $n_0 \in \mathrm{N}$ such that $x(n) > 0$ for $n \geq n_0$. From $(E+)$ and (1.1) we have

$$\begin{aligned} \Delta L_0 x(n) &= \frac{L_1 x(n)}{r_1(n)} \\ \Delta L_1 x(n) &= \frac{L_2 x(n)}{r_2(n)} \\ &\vdots \\ \Delta L_{m-1} x(n) &= \frac{L_m x(n)}{r_m(n)} \\ L_m x(n) &= (-1)^{m-1} a(n) f(x(\delta(n))). \end{aligned} \tag{2.2}$$

Hence, by (H2) $L_m x(n)$ is of one sign for $n \geq n_0$. So, $\Delta L_{m-1} x$ is eventually of one sign. Therefore $L_{m-1} x$ is eventually of one sign. Continuing in this way we see that there exists $n_1 \geq n_0$ such that all $L_k x$ $(k = 0, 1, 2, ..., m)$ are of one sign for $n \geq n_1$.

We will show that if $k \geq 1$ then $L_k x(n) \Delta L_k x(n) \leq 0$ for $n \geq n_1$. For $k = 1, 2, ..., m$ and $n \geq n_0$ we have

$$\Delta L_{k-1} x(n) = \frac{L_k x(n)}{r_k(n)}$$

Summing up both sides of the above equality from n_1 to $n-1$ we get

$$L_{k-1} x(n) = L_{k-1} x(n_1) + \sum_{i=n_1}^{n-1} \frac{L_k x(i)}{r_k(i)}, \quad k = 1, 2, ..., m. \tag{2.3}$$

Suppose that for some $k \geq 1$, $L_k x(n) \Delta L_k x(n) > 0$, $n \geq n_1$. Thus $L_k x$ is either eventually positive and increasing or eventually negative and decreasing. In either case from (2.3) and (H1) we get that $L_{k-1} x$ is unbounded and has the same eventual sign as $L_k x$. Repeating this process k times we see that x is unbounded, which is a contradiction to the boundedness of x. So, for every $k = 1, 2, ..., m-1$ we have

$$L_k x(n) \Delta L_k x(n) \leq 0, \quad n \geq n_1. \tag{2.4}$$

By (2.2) $L_m x(n) \leq 0$ $[\geq 0]$ if m is even [odd], $n \geq n_1$. Therefore, by (2.4) we see that, irrespectively of the parity of m, if $k \geq 1$ then $L_k x(n) \geq 0$ $[\leq 0]$ for odd [even] k. By mathematical induction we can show that if $n \geq n_1$ and $k = 1, 2, ..., m-1$ then

$$\begin{aligned} x(n) =& x(n_1) + \sum_{j=1}^{k} (-1)^{j+1} \omega_j(n-1) L_j x(n) - \sum_{j=1}^{k} (-1)^{j+1} \omega_j(n_1 - 1) L_j x(n_1) \\ &+ (-1)^k \sum_{j=n_1}^{n-1} \omega_k(j) \Delta L_k x(j). \end{aligned}$$

In particular

$$x(n) = x(n_1) + \sum_{j=1}^{m-1}(-1)^{j+1}\omega_j(n-1)L_jx(n) - \sum_{j=1}^{m-1}(-1)^{j+1}\omega_j(n_1-1)L_jx(n_1)$$
$$+(-1)^{m-1}\sum_{j=n_1}^{n-1}\omega_{m-1}(j)\Delta L_{m-1}x(j) =$$
$$= x(n_1) + \sum_{j=1}^{m-1}(-1)^{j+1}\omega_j(n-1)L_jx(n) - \sum_{j=1}^{m-1}(-1)^{j+1}\omega_j(n_1-1)L_jx(n_1)$$
$$+\sum_{j=n_1}^{n-1}\omega_{m-1}(j)a(j)f(x(\sigma(j))). \tag{2.5}$$

By our above remarks on the sign of $L_1x, L_2x, ..., L_{m-1}x$ we get

$$\sum_{j=1}^{m-1}(-1)^{j+1}\omega_j(n-1)L_jx(n) \geq 0 \quad \text{for} \quad n \geq n_1.$$

However, $L_1x(n) \geq 0$ for $n \geq n_1$, and so x is nondecreasing for $n \geq n_1$. Therefore, there is $n_2 \geq n_1$ such that $x(\sigma(n)) > c > 0$ for all $n \geq n_2$ and by (H3) there exists a positive constant M such that $f(x(\sigma(n)) \geq M$ for $n \geq n_2$. Hence, by (2.5)

$$x(n) \geq x(n_1) - \sum_{j=1}^{m-1}(-1)^{j+1}\omega_j(n_1-1)L_jx(n_1) + M\sum_{j=n_1}^{n-1}\omega_{m-1}(j)a(j).$$

In view of (2.1) we see that $x(n) \longrightarrow \infty$ as $n \longrightarrow \infty$ which is a contradiction to the boundness of x. Thus the proof is complete. □

Example 2.2 The equation

$$\Delta\left(\left(\frac{3}{4}\right)^{n+1}\Delta\left(\left(\frac{2}{3}\right)^{n+1}\Delta x(n)\right)\right) - \frac{5}{2^{n+1}}x(n+1) = 0$$

satisfies the hypotheses of Theorem 2.1, so all bounded solutions of this equation are oscillatory. One such solution is $x(n) = (-\frac{1}{2})^n$.

For equation (E–) we have the following theorem

Theorem 2.3 *Assume (H1)-(H3) and*

$$\sum_{n=1}^{\infty}\omega_{m-1}(n)a(n) = \infty.$$

Then every bounded solution of equation (E–) is either oscillatory or such that L_kx are eventually monotonic and

$$\lim_{n\to\infty} L_kx(n) = 0 \quad \text{for} \quad k = 0, 1, ..., m-1.$$

Proof Let x be a bounded and eventually positive solution of equation (E–). As before, we find $n_1 \in N$ such that each L_kx is of one sign for $n \geq n_1$ and $L_kx(n)\Delta L_kx(n) \leq 0$ for $n \geq n_1$, $k = 1, 2, ..., m$. This time $L_kx(n) \leq 0$ [≥ 0] if k is odd [even] for $n \geq n_1$. Therefore each L_kx ($k = 0, 1, ..., m-1$) is either nonnegative and nonincreasing or nonpositive and nondecreasing, and so L_kx is eventually monotonic and $\lim_{n\to\infty} L_k x(n)$ exists. By (H1) and (2.3) for $k \geq 1$ if $\lim_{n\to\infty} L_k x(n) \neq 0$ then $L_{k-1}x$ is unbounded, and so $\lim_{n\to\infty} L_k x(n) = 0$, for $k = 1, 2, ..., m-1$. Hence, it suffices to show that $\lim_{n\to\infty} x(n) = 0$. We know $\lim_{n\to\infty} x(n) = d \geq 0$. Suppose $d > 0$. Then there is $n_2 \geq n_1$ such that $x(\sigma(n)) > d$ for $n \geq n_2$. If $n \geq n_2$ then

$$\sum_{j=1}^{m-1} (-1)^{j+1} \omega_j(n-1) L_j x(n) \leq 0$$

and by (H3) there exists a positive constant M such that $f(x(\sigma(n))) \geq M$. Hence

$$\begin{aligned} x(n) &= x(n_1) + \sum_{j=1}^{m-1} (-1)^{j+1} \omega_j(n-1) L_j x(n) - \sum_{j=1}^{m-1} (-1)^{j+1} \omega_j(n_1-1) L_j x(n_1) \\ &\quad - \sum_{j=n_1}^{n-1} \omega_{m-1}(j) a(j) f(x(\sigma(j))) \\ &\leq x(n_1) - \sum_{j=1}^{m-1} (-1)^{j+1} \omega_j(n_1-1) L_j x(n_1) - M \sum_{j=n_1}^{n-1} \omega_{m-1}(j) a(j) \end{aligned}$$

and by (2.1) $x(n) \to -\infty$ as $n \to \infty$, and we get a contradiction since x is bounded. Hence $\lim_{n\to\infty} x(n) = 0$ which completes the proof. □

Example 2.4 The equation

$$L_m x(n) - (-1)^m \frac{1}{n} x(n+1) = 0,$$

where $r_1(n) = r_2(n) = \cdots = r_{n-1}(n) = n+1$, satisfies all the hypotheses of Theorem 2.3 and has a nonoscillatory solution $x(n) = \frac{1}{n} \to 0$ as $n \longrightarrow \infty$. For this solution we have

$$L_k x(n) = \frac{(-1)^{k+1}}{n} \to 0 \quad \text{as} \quad n \to \infty, \quad k = 1, 2, ..., m-1.$$

The equation

$$\Delta(\frac{1}{2^n} \Delta(\frac{1}{2^n} \Delta x(n))) + \frac{15}{4^{n+1}} x^3(n) = 0$$

satisfies the hypotheses of Theorem 2.3. This equation has an oscillatory solution $x(n) = (-1)^n$.

Remark 2.5 It is easy to calculate that if $r_k(n) \equiv 1$ for all $k = 1, 2, \ldots, m$ than $\omega_{m-1}(n) = \frac{(n+m-2)^{(m-1)}}{(m-1)!}$. So, our above results extend Theorems 1 and 2 of [14].

Similar results for the equation

$$\Delta[a_n \Delta^{m-1}(x_n - p_n x_{n-k})] + \delta q_n f(x_{\sigma_n}) = 0$$

can be found in [20].

3 Existence of nonoscillatory solutions

In this section we assume

(H4) $\sum_{n=1}^{\infty} a(n) < \infty$

(H5) $\sum_{n=1}^{\infty} \frac{1}{r_k(n)} < \infty, \quad k = 1, 2, ..., m-1.$

We will use the following notations

$$s_1(n) = \sum_{j_1=n}^{\infty} a(j_1)$$

$$s_2(n) = \sum_{j_2=n}^{\infty} \frac{1}{r_{m-1}(j_2)} \sum_{j_1=j_2}^{\infty} a(j_1) = \sum_{j_2=n}^{\infty} \frac{s_1(j_2)}{r_{m-1}(j_2)}$$

$$s_3(n) = \sum_{j_3=n}^{\infty} \frac{1}{r_{m-2}(j_3)} \sum_{j_2=j_3}^{\infty} \frac{1}{r_{m-1}(j_2)} \sum_{j_1=j_2}^{\infty} a(j_1) = \sum_{j_3=n}^{\infty} \frac{s_2(j_3)}{r_{m-2}(j_3)}$$

$$\vdots$$

$$s_k(n) = \sum_{j_k=n}^{\infty} \frac{s_{k-1}(j_k)}{r_{m-k+1}(j_k)}, \quad for \quad k = 2, 3, ..., m.$$

Obviously, by (H4) and (H5)

$$\sum_{n=1}^{\infty} \frac{s_i(n)}{r_{m-i}(n)} < \infty \quad for \quad i = 1, 2, ..., m-1. \tag{3.1}$$

Using a fixed point theorem we will give sufficient conditions under which for every real constant there exist a solution x of equation $(E\delta)$ convergent to this constant. Note that we assume only the continuity of the nonlinear function f.

Theorem 3.1 *Assume (H4),(H5) and the function f is continuous. Then for every $c \in R$ there exists a solution x of equation $(E\delta)$ such that*

$$\lim_{n\to\infty} x(n) = c.$$

Proof Let $c \in R$ and let us choose a real number $a > 0$. Then there exists a constant $M > 1$ such that

$$|f(t)| \leq M \quad \text{for every} \quad t \in [c-a, c+a].$$

From (3.1) there exists an index $n_0 \in N$ such that

$$Ms_m(n) < a \quad \text{for} \quad \text{every} \quad n \geq n_0.$$

Let l_∞ denote the Banach space of all real bounded sequences with norm $\|x\| = \sup_{n\geq n_0} |x(n)|$. We define the subset T of l_∞ by

$$T = \{x \in l_\infty : x(0) = x(1) = ... = x(n_0 - 1) = c \quad \text{and} \quad |x(n) - c| < Ms_m(n) \quad \text{for} \quad n \geq n_0\}.$$

If $x \in T$, then $x(n) \in [c-a, c+a]$ for each $n \in N$. Hence $|f(x(n))| < M$ for every $x \in T, n \in N$. Let $x \in T$. Since $|f(x(n))| < M$ for every $n \in N$ the series

$\sum\limits_{j=1}^{\infty} a(j)f(x(\sigma(j))$ is absolutely convergent. Denoting

$$u_1(n) = \sum_{j=n}^{\infty} a(j)f(x(\sigma(j)), \quad n \in N$$

we have

$$|u_1(n)| \leq M \sum_{j=n}^{\infty} a(j) = Ms_1(n). \tag{3.2}$$

By (3.1) the series $\sum\limits_{j=1}^{\infty} \frac{|u_1(j)|}{r_{m-1}(j)}$ is convergent. Let $u_2(n) = \sum\limits_{j=n}^{\infty} \frac{u_1(j)}{r_{m-1}(j)}$. From (3.2) we have $|u_2(n)| \leq M \sum\limits_{j=n}^{\infty} \frac{s_1(j)}{r_{m-1}(j)} = Ms_2(n)$ and by (3.1) the series $\sum\limits_{j=1}^{\infty} \frac{|u_2(j)|}{r_{m-2}(j)}$ is corvergent, too. Similary we get

$$|u_k(n)| \leq Ms_k(n), \quad k = 2, 3, ..., m, \tag{3.3}$$

where $u_k(n) = \sum\limits_{j=n}^{\infty} \frac{u_{k-1}(j)}{r_{m-k+1}(j)}$, and the convergence of the series $\sum\limits_{j=1}^{\infty} \frac{|u_k(j)|}{r_{m-k+1}(j)}$ for $k = 1, 2, ..., m-1$.

Now, we define the sequence Ax by

$$Ax(n) = \begin{cases} c & \text{for} \quad n < n_0 \\ c - \delta u_m(n) & \text{for} \quad n \geq n_0. \end{cases}$$

Next we show that the hypotheses of Schauder's fixed point theorem are satisfied.

(a) Obviously, T is a convex and closed subset of the space l_∞. It is easy to construct a finite ε-net for the set T. Hence T is compact.

(b) A maps T into itself. In fact, for $n \geq n_0$, by (3.3) we have $|Ax(n) - c| = |u_m(n)| \leq Ms_m(n)$. Hence $Ax \in T$ for every $x \in T$.

(c) A is a continuous map. Let $x \in T, \varepsilon > 0$. Since the function f is uniformly continuous on the interval $[c-a, c+a]$, there exists $\delta > 0$ such that for $s, t \in [c-a, c+a]$ and $|t-s| < \delta$ we have

$$|f(t) - f(s)| < \varepsilon. \tag{3.4}$$

Now, let $z \in T$ be such that $\|x - z\| < \delta$. Then $|x(n) - z(n)| < \delta$ for every $n \in N$. Hence $|f(x(n)) - f(z(n))| < \varepsilon$ for each $n \in N$.

Let us denote

$$v_1(n) = \sum_{j=n}^{\infty} a(j)f(z(\sigma(j)))$$

and

$$v_k(n) = \sum_{j=n}^{\infty} \frac{v_{k-1}(j)}{r_{m-k+j}(j)}, \quad k = 2, 3, ..., m, \ n \in N.$$

Then

$$\|Ax - Az\| \le \sup_{n \ge n_0} |u_m(n) - v_n(n)| \le \sup_{n \ge n_0} \sum_{j=n}^{\infty} \frac{|u_{m-1}(j) - v_{m-1}(j)|}{r_1(j)}$$

$$\le \sum_{j=n_0}^{\infty} \frac{|u_{m-1}(j) - v_{m-1}(j)|}{r_1(j)}.$$

By (3.4) we have

$$|u_{m-1}(n) - v_{m-1}(n)| \le \sum_{j_{m-1}=n}^{\infty} \frac{1}{r_2(j_{m-1})} \sum_{j_{m-2}=j_{m-1}}^{\infty} \frac{1}{r_3(j_{m-2})} \cdots$$

$$\cdots \sum_{j_1=j_2}^{\infty} a(j_1) \, |f(x(\sigma(j_1))) - f(z(\sigma(j_1)))| \le \varepsilon s_{m-1}(n).$$

Hence $\|Ax - Az\| \le \varepsilon \sum_{j=n_0}^{\infty} \frac{s_{m-1}(j)}{r_1(j)} = \varepsilon s_m(n).$

This shows that A is a continuous map.

By Schandauer's fixed point theorem there exists $x \in T$ such that $Ax = x$. So, we have

$$x(n) = c - \delta u_m(n) \quad \text{for} \quad n \ge n_0.$$

Hence

$$\Delta x(n) = \delta \frac{u_{m-1}(n)}{r_1(n)}$$

and therefore

$$L_1 x(n) = \delta u_{m-1}(n) \qquad for \quad n \ge n_0.$$

Taking the difference of the above equation and multiplying by $r_2(n)$ we get

$$L_2 x(n) = (-1)\delta u_{m-1}(n)$$

and after m steps

$$L_m x(n) = (-1)^{m-1} \delta u_1(n) = (-1)^{m-1} \delta \sum_{j=n}^{\infty} a(j) f(x(\sigma(j))).$$

So, we see that the sequence x fulfils equation $(E\delta)$ for $n \ge n_0$. From the convergence of the series $\sum_{j=1}^{\infty} \frac{u_{m-1}(j)}{r_1(j)}$ it follows that $\lim_{n\to\infty} u_m(n) = 0$ and therefore $\lim_{n\to\infty} x(n) = c$. This completes the proof of the theorem. □

Remark 3.2 Theorem 3.1 extends Theorem 7 of [16].

If condition (H5) is not satisfied, then we need a stronger assumption of the sequence a then (H4).

Theorem 3.3 *Assume*

$$\sum_{n=1}^{\infty} \frac{s_{m-1}(n)}{r_1(n)} < \infty \tag{3.5}$$

and the function f is continuous. Then for every $c \in R$ there exists a solution x of equation $(E\delta)$ such that

$$\lim_{n\to\infty} x(n) = c.$$

Remark 3.4 If $r_k(n) \equiv 1$ for all $k = 1, 2, \ldots, m-1$ then (3.5) is equivalent to $\sum_{n=1}^{\infty} n^{m-1} a(n) < \infty$ and Theorem 3.3 extends Theorem 1 of [12].

Example 3.5 Consider the equation

$$\Delta[\frac{1}{n}\Delta(n\Delta x_n)] + \frac{1}{n(n+1)^2} x_{n+1} = 0, \quad n \geq 1. \tag{3.6}$$

It is easy to check that condition (3.5) is satisfied. Hence, from Theorem 3.3 there exist nonoscillatory, bounded solutions of equation (3.6). One such solution is $x_n = 1 + \frac{1}{n}$, (see also [2, Th.3.1].)

Agarwal, Grace and O'Regan in their recent paper [2] studied the equation

$$\Delta([\frac{1}{a_{n-1}(k)}\Delta(\frac{1}{a_{n-2}(k)}\Delta(\ldots \frac{1}{a_1(k)}\Delta x(k)\ldots))]^{\alpha} + q(k)f(x[g(k)]) = 0,$$

where α is the ratio of two positive odd integers, $a_i(k)$, $q(k) : N \longrightarrow R^+$, $i = 1, 2, \ldots, n-1$, $g(k) : N \longrightarrow N$, $\lim_{k\longrightarrow\infty} g(k) = \infty$ and $\{g(k)\}$ is a nondecreasing sequence, $f(x) : R \longrightarrow R$ is continuous and $xf(x) > 0$ for $x \neq 0$ and $f'(x) \geq 0$ for $x \neq 0$. The result obtained by them for even n are similar but for odd n are different from ours.

References

[1] R.P.Agarwal, *Difference Equations and Inequalities*, Marcel Dekker, New York, (1992).

[2] R.P.Agarwal, S.R.Grace, D.O'Regan, *Oscillation crtiteria of comparison type for higher order difference equations*, Nonlinear Analysis Forum 7,1 73-91 (2002).

[3] M.Cecchi, Z.Došlá, M.Marini, *Positive decreasing solutions of quasilinear difference equations*, Comput.Math.Appl. 42, 10-11, 1401-1410 (2001).

[4] S.S.Cheng, H.J.Li *Bounded solutions of nonlinear difference equations*, Tamkang Journal of Math. 21,2,137-142 (1990).

[5] S.S.Cheng, T.Ch.Yan, H.J.Li, *Oscillation criteria for second order difference equation*, Funkcialaj Ekvacioj 34, 223-239 (1991).

[6] S.R.Grace, H.A.El-Morshedy, *Oscillation and nonoscillation theorems for certain second-order difference equations with forcing term*, J.Math. Anal. Appl. 216, 614-625 (1997).

[7] J.R.Graef, J.Jaroš, A.Miciano, P.W.Spikes and E.Thandapani, *Oscillation and nonoscillation results for general nonlinear difference equations*, Proc.of Dynamic Systems and Appl. 2 , 199-206 (1996).

[8] J.R,Graef, E.Thandapani, *Oscillation and asymptotic behaviour of fourth order nonlinear delay difference equations*, Fasciculi Math. 31, 23-36 (2001).

[9] H.J.Li, S.S Cheng, *Asymptotically monotone solutions of a nonlinear difference equations*, Tamkang Journal of Math. 24,3, 269-282 (1993).

[10] B.Liu, S.S.Cheng, *Monotone solutions of a higher order nonlinear difference equation with advancement*, (to appear).

[11] D.L.Lovelady, *On the oscillatory behaviour of bounded solutions of higher order differential equations*, J.Diff.Eq. 19,167-175 (1975).

[12] M.Migda, J.Migda, *On the asymtotic behaviour of solutions of higher order nonlinear diference equations*, Nonlinear Analysis 47, 4687-4695 (2001).

[13] M.Migda, *Asymptotic behaviour of solutions of nonlinear delay difference equations*, Fasciculi Math. 31, 57-62 (2001).

[14] Z.Szafrański, B.Szmanda, *Oscillation theorems for some nonlinear difference equations*, Appl.Math. and Comp. 83:43-52, 43-52 (1997).

[15] B.Szmanda, *Note on the oscillation of certain difference equations*, Glasnik Math. 3,115-121 (1996).

[16] E.Thandapani, M.M.S.Manuel, J.R.Graef, W.Spikes, *Monotone properties of certain classes of solutions of second-order difference equations*, Comp.Math.Appl. 36:10-12,291-297 (1998).

[17] E.Thandapani, I.Gyori, B.S.Lalli, *An application of discrete inequality to second order non-linear oscillation*, J.Math. Anal. Appl. 186, 200-208 (1994).

[18] E.Thandapani, I.M.Arockiasamy, *Some oscillation and nonoscillation theorems for fourth order difference equations*, Z.Aual.Anwendungen 19,3, 863-872 (2000).

[19] P.J.Y.Wong, R.P.Agarwal, *Oscillation theorems for certain second order nonlinear difference equations*, J.Math. Anal. Appl. 204, 813-829 (1996).

[20] A.Zafer, R.S.Dahiya, *Oscillation of a neutral difference equation*, Appl. Math. Lett. 2,71-74 (1993).

Fields Institute Communications
Volume **42**, 2004

Stability of Center Fiber Bundles for Nonautonomous Difference Equations

Christian Pötzsche
Department of Mathematics
University of Augsburg
D-86135 Augsburg, Germany
christian.poetzsche@math.uni-augsburg.de

Abstract. Center fiber bundles are the generalization of center manifolds to nonautonomous difference equations. Following closely an idea of Palmer, we deduce the essential stability properties of such fiber bundles, namely their asymptotic phase and a reduction principle.

1 Introduction and preliminaries

With the aid of elementary tools like the variation of constants formula and Gronwall's inequality, Palmer [11] proved a useful lemma in the theory of center manifolds for finite dimensional autonomous ordinary differential equations. It means that as long as a solution remains in the vicinity of an equilibrium with given center manifold, it must be close to some solution on the mentioned manifold. This lemma is helpful, since such stability properties as asymptotic phase and Pliss's reduction principle are derived simply and directly. In this paper we show that the same result also holds in the situation of nonautonomous difference equations in arbitrary Banach spaces, with the restriction that we need a finite dimensional center manifold to deduce its asymptotic phase and the reduction principle, since a compactness argument is involved. The importance of a nonautonomous theory is due to the fact that the investigation of nonconstant solutions canonically leads to time dependent problems in form of the equation of perturbed motion. To our best knowledge, related results under global assumptions on the nonlinearities can be found solely in Wanner [14] and Janglajew [8], whereas our situation is more realistic and — concerning the spectrum of the linear part — more general (cf. Hypothesis 2.1). We only refer to Aulbach [2], Carr [7] and Palmer [11] for further references on ordinary differential equations.

2000 *Mathematics Subject Classification.* Primary 37D10, 39A11; Secondary 37C60.

Research supported by the "Graduiertenkolleg: Nichtlineare Probleme in Analysis, Geometrie und Physik" (GRK 283) financed by the Deutsche Forschungsgemeinschaft and the State of Bavaria.

Now we introduce our notation. $\mathbb{Z}$ stands for the integers and $\mathbb{N}$ is the set of positive integers. A *discrete interval* is the intersection of a real interval with the integers. For $\kappa, K \in \mathbb{Z}$ we write $\mathbb{Z}_\kappa^+ := \{k \in \mathbb{Z} : \kappa \leq k\}$ and $[\kappa, K]_\mathbb{Z} := [\kappa, K] \cap \mathbb{Z}$. The Banach spaces X, Y are real or complex throughout this paper and their norm is denoted by $\|\cdot\|$. $L(X)$ is the Banach space of continuous endomorphisms on X, I_X the identity map on X and $GL(X)$ the multiplicative group of bijective mappings in $L(X)$. On the Cartesian product $X \times Y$ we always use the norm

$$\|(x, y)\| := \max\{\|x\|, \|y\|\}. \tag{1.1}$$

In a normed space, $B_\epsilon(x)$ is the ball with center x and radius $\epsilon > 0$. In case a mapping $F : \mathbb{Z} \times Y \to X$ depends differentiably on the second variable, then its partial derivative is denoted by $D_2 F$.

We write

$$x' = f(k, x) \tag{1.2}$$

to denote the difference equation $x(k+1) = f(k, x(k))$, with the *right-hand side* $f : \mathbb{Z} \times U \to X$, where U is a subset of the Banach space X. A sequence $\nu : I \to X$ is said to solve (1.2) on a discrete interval I, if $\nu(k+1) = f(k, \nu(k))$ as long as ν exists, i.e., as long as $\nu(k) \in U$ holds for $k \in I$. Let φ denote the *general solution* of equation (1.2), i.e., $\varphi(\cdot; \kappa, \xi)$ solves (1.2) and satisfies the initial condition $\varphi(\kappa; \kappa, \xi) = \xi$ for $\kappa \in I$, $\xi \in U$. In forward time, $\varphi(\cdot; \kappa, \xi)$ can be defined recursively

$$\varphi(k; \kappa, \xi) := \begin{cases} \xi & \text{for } k = \kappa \\ f(k-1, \varphi(k-1; \kappa, \xi)) & \text{for } k > \kappa \end{cases},$$

as long as $\varphi(k-1; \kappa, \xi) \in U$. Given an operator sequence $A : \mathbb{Z} \to L(X)$ we define the *evolution operator* $\Phi(k, \kappa) \in L(X)$ of the linear difference equation

$$x' = A(k)x$$

as the mapping

$$\Phi(k, \kappa) := \begin{cases} I_X & \text{for } k = \kappa \\ A(k-1) \cdots A(\kappa) & \text{for } k > \kappa \end{cases}$$

and if $A(k)$ is invertible (in $L(X)$) for $k < \kappa$, then

$$\Phi(k, \kappa) := A(k)^{-1} \cdots A(\kappa - 1)^{-1} \quad \text{for } k < \kappa.$$

2 Existence of center fiber bundles

In this section we repeat and summarize some basic facts about center fiber bundles. For the autonomous situation, Carr [7, pp. 33–36, Section 2.8] is still a good reference. However, our nonautonomous setting is as follows:

Hypothesis 2.1 Let $U \subseteq X \times Y$ be an open neighborhood of $(0, 0)$, $m \in \mathbb{N}$ and consider a system of nonautonomous difference equations

$$\begin{cases} x' = A(k)x + F(k, x, y) \\ y' = B(k)y + G(k, x, y) \end{cases}, \tag{2.1}$$

where X, Y are arbitrary Banach spaces, $A : \mathbb{Z} \to L(X)$, $B : \mathbb{Z} \to GL(Y)$ and the mappings $F : \mathbb{Z} \times U \to X$, $G : \mathbb{Z} \times U \to Y$ are m-times continuously differentiable with respect to (x, y). Moreover, we assume:

(i) Hypothesis on linear part: The evolution operators Φ and Ψ of the systems $x' = A(k)x$ and $y' = B(k)y$, respectively, satisfy for all $k, l \in \mathbb{Z}$ the estimates

$$\|\Phi(k,l)\| \le K_1\alpha^{k-l} \quad \text{for } k \ge l, \qquad \|\Psi(k,l)\| \le K_2\beta^{k-l} \quad \text{for } l \ge k, \tag{2.2}$$

with real constants $K_1, K_2 \ge 1$ and α, β with $0 < \alpha < 1$, $\alpha < \beta$.

(ii) Hypothesis on nonlinearities: We have

$$F(k,0,0) \equiv 0, \qquad G(k,0,0) \equiv 0 \quad \text{on } \mathbb{Z},$$

and the partial derivatives of F and G satisfy

$$\lim_{(x,y)\to(0,0)} D_{(2,3)}(F,G)(k,x,y) = 0 \quad \text{uniformly in } k \in \mathbb{Z}. \tag{2.3}$$

Remark 2.2 Let U be an open neighborhood of 0 in a Hilbert space Z. Suppose that the mapping $f : \mathbb{Z}\times U \to Z$ is of class C^m, $m \in \mathbb{N}$, in the state space variable and satisfies the following assumptions:

(i)′ $D_2 f(k,0) \in GL(Z)$ for $k \in \mathbb{Z}$, and the variational equation

$$z' = D_2 f(k,0)z \tag{2.4}$$

possesses an exponential dichotomy on $\mathbb{Z}$ with growth rates $\alpha \in (0,1)$, $\beta > \alpha$,

(ii)′ $f(k,0) \equiv 0$ on $\mathbb{Z}$ and $\lim_{z\to 0} D_2 f(k,z) = 0$ holds uniformly in $k \in \mathbb{Z}$.

Then any nonautonomous difference equation

$$z' = f(k,z)$$

can be brought into the "decoupled" form (2.1) such that Hypothesis 2.1 is fulfilled. This can be shown using methods from [4, Theorem 5] via a Lyapunov transformation. If, moreover, Z is finite dimensional, then the assumption $D_2 f(k,0) \in GL(Z)$, $k \in \mathbb{Z}$, can be dropped (cf. [12, p. 33, Satz 1.5.7]) and instead of (i)′ we can assume that the dichotomy spectrum (cf. [5]) of (2.4) is disjoint from $[\alpha, \beta]$. Finally, in case of an arbitrary Banach space Z, and if $C := D_2 f(k,0)$ does not depend on $k \in \mathbb{Z}$, it is sufficient to assume that the spectrum $\sigma(C) \subseteq \mathbb{C}$ of the operator $C \in L(Z)$ can be separated into a "stable" spectral part $\sigma_1 \subseteq B_\alpha(0)$, $0 < \alpha < 1$, and a disjoint "pseudo-unstable" spectral part σ_2 outside a circle with center 0 and radius $\beta > \alpha$ in the complex plane.

Our next aim is to introduce a nonautonomous counterpart of invariant manifolds. Thereto let $U_1 \subseteq X$, $U_2 \subseteq Y$ be open neighborhoods of 0 with $U_1 \times U_2 \subseteq U$ and let $\varphi = (\varphi_1, \varphi_2)$ denote the general solution of (2.1). Then the graph

$$C := \{(\kappa, c(\kappa,\eta), \eta) \in \mathbb{Z}\times X \times Y : \kappa \in \mathbb{Z}, \eta \in U_2\}$$

of a function $c : \mathbb{Z}\times U_2 \to U_1$ is called a *locally invariant fiber bundle* of the difference equation (2.1), if the implication

$$(\kappa, \xi, \eta) \in C \quad \Rightarrow \quad (k, \varphi(k;\kappa,\xi,\eta)) \in C$$

holds for all $k \in \mathbb{Z}_\kappa^+$ with the property $\varphi_2(k;\kappa,\xi,\eta) \in U_2$. Locally invariant fiber bundles evidently satisfy the following nonlinear functional equation, denoted as *invariance equation*

$$c(\kappa+1, B(\kappa)\eta + G(\kappa, c(\kappa,\eta), \eta)) = A(\kappa)c(\kappa,\eta) + F(\kappa, c(\kappa,\eta), \eta) \tag{2.5}$$

for all $(\kappa, \eta) \in \mathbb{Z}\times U_2$ such that $B(\kappa)\eta + G(\kappa, c(\kappa, \eta), \eta) \in U_2$. Finally, a locally invariant fiber bundle C is called a *center fiber bundle* of (2.1), if the corresponding mapping c is continuously differentiable in $\eta \in U_2$ and if the relations

$$c(\kappa, 0) \equiv 0 \quad \text{on } \mathbb{Z}, \qquad \lim_{\eta\to 0} D_2 c(\kappa, \eta) = 0 \quad \text{uniformly in } \kappa \in \mathbb{Z} \tag{2.6}$$

are satisfied. We remark that the conditions (2.6) imply that C contains the zero solution of (2.1), and that C is fiber-wise tangent to the "pseudo-unstable" vector bundle $\mathbb{Z}\times \{0\} \times Y$. Such center fiber bundles are not unique in general, which can be seen from the succeeding

Example 2.3 Consider the two-dimensional autonomous difference equation

$$\begin{cases} x' = & e^{-1}x \\ y' = y + \frac{y^2}{1-y} \end{cases}, \tag{2.7}$$

satisfying Hypothesis 2.1 with $U = \mathbb{R}\times (-\infty, 1)$, $K_1 = K_2 = 1$, $\alpha = e^{-1}$ and $\beta = 1$. Now it is easy to verify that

$$C_\gamma := \left\{(\kappa, \xi, \eta) \in \mathbb{Z}\times\mathbb{R}\times (-\infty, 1) : \xi = \gamma e^{1/\eta} \text{ for } \eta < 0, \xi = 0 \text{ for } \eta \geq 0\right\}$$

is a center fiber bundle of (2.7) for any parameter $\gamma \in \mathbb{R}$.

Nevertheless, center fiber bundles do exist under reasonable assumptions. In fact — and for purely technical reasons — we only need the additional assumption in Hypothesis 2.1 that X, Y are C^n-Banach spaces, $n \in \mathbb{N}$; i.e., the norms on X, Y have to be of class C^n away from 0. A characterization of such spaces, as well as examples, can be found in Kriegl and Michor [9, pp. 127–152, Section 13].

Theorem 2.4 (existence of center fiber bundles) *Suppose that the above Hypothesis 2.1 is satisfied under the gap condition $\alpha < \beta^n$ for some $n \in \{1, \dots, m\}$, and that X, Y are C^n-Banach spaces. Then there exists a real number $\rho_0 > 0$ and a function $c : \mathbb{Z}\times B_{\rho_0}(0) \subseteq \mathbb{Z}\times Y \to B_{\rho_0}(0) \subseteq X$ with the following properties:*

(a) *$c : \mathbb{Z}\times B_{\rho_0}(0) \to X$ is continuous and $c(\kappa, \cdot) : B_{\rho_0}(0) \to X$ is n-times continuously differentiable for any $\kappa \in \mathbb{Z}$,*
(b) *the graph $C := \{(\kappa, c(\kappa, \eta), \eta) \in \mathbb{Z}\times X\times Y : \kappa \in \mathbb{Z}, \eta \in B_{\rho_0}(0)\}$ is a center fiber bundle of (2.1),*
(c) *if the mappings A, B and F, G are periodic in k with period $\theta \in \mathbb{N}$, then*

$$c(\kappa + \theta, \eta) = c(\kappa, \eta) \quad \textit{for } \kappa \in \mathbb{Z},\ \eta \in B_{\rho_0}(0),$$

and if equation (2.1) is autonomous, then c is independent of $\kappa \in \mathbb{Z}$, i.e., the subset $C_0 := \{(c(\eta), \eta) \in X\times Y : \eta \in B_{\rho_0}(0)\}$ of the state space $X\times Y$ is an invariant manifold of (2.1).

Proof One shows the existence of the mentioned mapping c by extending the nonlinearities F, G smoothly such that they are defined on $\mathbb{Z}\times X\times Y$ and have globally bounded derivatives there — in this cut-off technique the fact that X, Y are C^n-Banach spaces plays a decisive role, and the explicit construction can be found in [12, p. 73, Lemma 2.3.2]. Then it is possible to apply a general theorem on invariant fiber bundles (cf. [6, Theorem 4.1(b)]), where the above smoothness assertion follows from [13, Theorem 5.1(b)]. The fact that c satisfies the limit relation in (2.6) can be seen as in [12, p. 64, Korollar 2.2.15]. After all, the assertion (c) follows from [3, Corollary 4.2]. □

3 Stability properties of center fiber bundles

From now on we assume that Hypothesis 2.1 is satisfied with $U = B_{\rho_0}(0)\times B_{\rho_0}(0)$ for some $\rho_0 > 0$, and that the difference equation (2.1) possesses a center fiber bundle C given by the mapping $c : \mathbb{Z}\times B_{\rho_0}(0)\to B_{\rho_0}(0)$. First of all, we observe that the mappings $\omega_F, \omega_G, \omega_c : [0, \rho_0) \to\mathbb{R}$,

$$\begin{aligned} \omega_F(\rho) &:= \sup_{(k,x,y)\in\mathbb{Z}\times B_\rho(0)\times B_\rho(0)} \left\|D_{(2,3)}F(k,x,y)\right\|, \\ \omega_G(\rho) &:= \sup_{(k,x,y)\in\mathbb{Z}\times B_\rho(0)\times B_\rho(0)} \left\|D_{(2,3)}G(k,x,y)\right\|, \\ \omega_c(\rho) &:= \sup_{(\kappa,\eta)\in\mathbb{Z}\times B_\rho(0)} \left\|D_2 c(\kappa,\eta)\right\| \end{aligned}$$

are well-defined, increasing, and obtain by Hypothesis 2.1(ii) (cf. (2.3)) and (2.6) the limit relations

$$\lim_{\rho\searrow 0} \omega_F(\rho) = 0, \qquad \lim_{\rho\searrow 0} \omega_G(\rho) = 0, \qquad \lim_{\rho\searrow 0} \omega_c(\rho) = 0.$$

Lemma 3.1 *For any $\rho \in [0, \rho_0)$ we have*

$$\|c(\kappa,\eta)\| \le \omega_c(\rho)\,\|\eta\| \quad \text{for } \kappa\in\mathbb{Z},\ \eta\in B_\rho(0). \tag{3.1}$$

Proof Using the mean value theorem (cf. [10, p. 341, Theorem 4.2]) we get

$$\|c(\kappa,\eta)\| \overset{(2.6)}{=} \|c(\kappa,\eta) - c(\kappa,0)\| \le \left\|\int_0^1 D_2c(\kappa,h\eta)\,dh\right\| \|\eta\| \le \omega_c(\rho)\,\|\eta\|$$

for $\kappa\in\mathbb{Z}$ and $\eta\in B_\rho(0)$. □

Lemma 3.2 (Palmer's lemma) *Let $\kappa\le K$ be integers, let the real constant $\gamma\in(0, \min\{1,\beta\}-\alpha)$ be fixed and choose $\rho\in\left(0,\frac{\rho_0}{2}\right)$ so small that the estimates*

$$\begin{gathered} \max\{2\omega_G(\rho), \omega_c(2\rho)\} < 1, \\ K_1\left(\omega_F(\rho) + \omega_c(\rho)\omega_G(\rho)\right) \le \gamma, \\ K_2\left[\omega_G(2\rho)(1+\omega_c(2\rho)) + 2K_1\omega_G(\rho)\right] < \beta-\alpha-\gamma \end{gathered} \tag{3.2}$$

hold. If $\nu = (\nu_1,\nu_2)$ is a solution of (2.1) defined on a discrete interval $[\kappa,K]_\mathbb{Z}$ and $\|\nu(k)\| < \rho$ for $k\in[\kappa,K]_\mathbb{Z}$, and if $\tilde\nu$ denotes a solution of the difference equation

$$y' = B(k)y + G(k, c(k,y), y), \tag{3.3}$$

with $\tilde\nu(K) = \nu_2(K)$, then the following holds:

(a) *$\tilde\nu$ is defined on the discrete interval $[\kappa,K]_\mathbb{Z}$,*
(b) *$\|\tilde\nu(k)\| < 2\rho$ for all $k\in[\kappa,K]_\mathbb{Z}$, and*
(c) *$\tilde\nu : [\kappa,K]_\mathbb{Z}\to Y$ satisfies the estimate*

$$\left\|\nu(k) - \begin{pmatrix} c(k,\tilde\nu(k)) \\ \tilde\nu(k)\end{pmatrix}\right\| \le \left(K_1 + \tfrac12\right)\|\nu_1(\kappa) - c(\kappa,\nu_2(\kappa))\|\,(\alpha+\gamma)^{k-\kappa} \tag{3.4}$$

for all $k\in[\kappa,K]_\mathbb{Z}$.

Proof (I) We begin the present proof by deriving a preparatory estimate. Due to Lemma 3.1 the inequality

$$\|c(k,\nu_2(k))\| \overset{(3.1)}{\le} \omega_c(\rho)\,\|\nu_2(k)\| \le \omega_c(\rho)\rho \overset{(3.2)}{<} \rho \quad \text{for } k\in[\kappa,K]_\mathbb{Z} \tag{3.5}$$

holds, and taking the solution property of ν_2 into account, one has

$$\begin{aligned}
&\|B(k)\nu_2(k)+G(k,c(k,\nu_2(k)),\nu_2(k))\| \\
&\stackrel{(2.1)}{\leq} \|\nu_2(k+1)\|+\|G(k,c(k,\nu_2(k)),\nu_2(k))-G(k,\nu(k))\| \\
&\leq \|\nu_2(k+1)\|+\omega_G(\rho)\,\|c(k,\nu_2(k))-\nu_1(k)\| \\
&\leq \rho+\omega_G(\rho)\left(\|c(k,\nu_2(k))\|+\|\nu_1(k)\|\right) \\
&\stackrel{(3.5)}{\leq} \rho+2\omega_G(\rho)\rho \stackrel{(3.2)}{<} 2\rho<\rho_0 \quad \text{for } k\in[\kappa,K-1]_{\mathbb{Z}}
\end{aligned} \tag{3.6}$$

by the mean value inequality (cf. [10, p. 342, Corollary 4.3]) applied to $G(k,\cdot)$. Using the mean value theorem (cf. [10, p. 341, Theorem 4.2]) and the invariance equation (2.5) we find the identity

$$\begin{aligned}
&A(k)\nu_1(k)+F(k,\nu(k))-c\big(k+1,B(k)\nu_2(k)+G(k,c(k,\nu_2(k)),\nu_2(k))\big) \\
&\qquad +c\big(k+1,B(k)\nu_2(k)+G(k,c(k,\nu_2(k)),\nu_2(k))\big) \\
&\qquad -c\big(k+1,B(k)\nu_2(k)+G(k,\nu(k))\big) \\
&\stackrel{(2.5)}{=} A(k)\big(\nu_1(k)-c(k,\nu_2(k))\big)+F(k,\nu(k))-F(k,c(k,\nu_2(k)),\nu_2(k)) \\
&\qquad +\int_0^1 D_2c(k+1,\eta_k(h))\,dh\,[G(k,c(k,\nu_2(k)),\nu_2(k))-G(k,\nu(k))]
\end{aligned}$$

for all $k\in[\kappa,K-1]_{\mathbb{Z}}$, and here we have abbreviated

$$\eta_k(h):=B(k)\nu_2(k)+G(k,\nu(k))+h\,[G(k,c(k,\nu_2(k)),\nu_2(k))-G(k,\nu(k))]\,.$$

Notice that the convexity of the ball $B_{\rho_0}(0)\subseteq Y$ and (3.6) implies $\eta_k(h)\in B_{\rho_0}(0)$ for $k\in[\kappa,K-1]_{\mathbb{Z}}$, $h\in[0,1]$. Then $\Delta_1(k):=\nu_1(k)-c(k,\nu_2(k))$ solves the linear inhomogeneous difference equation

$$\begin{aligned}
x' &= A(k)x+F(k,\nu(k))-F(k,c(k,\nu_2(k)),\nu_2(k)) \\
&\qquad +\int_0^1 D_2c(k+1,\eta_k(h))\,dh\,[G(k,c(k,\nu_2(k)),\nu_2(k))-G(k,\nu(k))]
\end{aligned}$$

on $[\kappa,K]_{\mathbb{Z}}$. Now the variation of constants formula (cf. [1, pp. 57–58]) gives us

$$\begin{aligned}
\Delta_1(k) &= \Phi(k,\kappa)\Delta_1(\kappa)+\sum_{n=\kappa}^{k-1}\Phi(k,n+1)\Big[F(n,\nu(n))-F(n,c(n,\nu_2(n)),\nu_2(n)) \\
&\qquad +\int_0^1 D_2c(n+1,\eta_n(h))\,dh\,[G(n,c(n,\nu_2(n)),\nu_2(n))-G(n,\nu(n))]\Big],
\end{aligned}$$

so that applying the mean value inequality (cf. [10, p. 342, Corollary 4.3]) yields

$$\|\Delta_1(k)\|\stackrel{(2.2)}{\leq} K_1\alpha^{k-\kappa}\|\Delta_1(\kappa)\|+K_1\left(\omega_F(\rho)+\omega_c(\rho)\omega_G(\rho)\right)\sum_{n=\kappa}^{k-1}\alpha^{k-n-1}\|\Delta_1(n)\|$$

and consequently

$$\frac{\|\Delta_1(k)\|}{\alpha^k}\stackrel{(2.2)}{\leq} K_1\alpha^{-\kappa}\|\Delta_1(\kappa)\|+\frac{K_1\left(\omega_F(\rho)+\omega_c(\rho)\omega_G(\rho)\right)}{\alpha}\sum_{n=\kappa}^{k-1}\frac{\|\Delta_1(n)\|}{\alpha^n}$$

for $k\in[\kappa,K]_{\mathbb{Z}}$. By Gronwall's lemma (cf. [1, p. 183, Corollary 4.1.2]) we obtain

$$\|\Delta_1(k)\|\leq K_1\left[\alpha+K_1(\omega_F(\rho)+\omega_c(\rho)\omega_G(\rho))\right]^{k-\kappa}\|\Delta_1(\kappa)\|$$

and using (3.2) it follows that

$$\|\nu_1(k) - c(k, \nu_2(k))\| \leq K_1 (\alpha + \gamma)^{k-\kappa} \|\nu_1(\kappa) - c(\kappa, \nu_2(\kappa))\| \tag{3.7}$$

for $k \in [\kappa, K]_{\mathbb{Z}}$.

(II) Now let $k_0 \in [\kappa, K]_{\mathbb{Z}}$ be the least integer such that $\tilde{\nu}(k)$ is defined on the discrete interval $[k_0, K]_{\mathbb{Z}}$ and $\|\tilde{\nu}(k)\| < 2\rho$ there. Then $\Delta_2(k) := \nu_2(k) - \tilde{\nu}(k)$ solves the linear-inhomogeneous difference equation

$$y' = B(k)y + G(k, \nu(k)) - G(k, c(k, \tilde{\nu}(k)), \tilde{\nu}(k))$$

on $[k_0, K]_{\mathbb{Z}}$ and satisfies the initial condition $\Delta_2(K) = 0$. By the variation of constants formula in backward time (cf. [1, p. 58]) it follows

$$\begin{aligned} \Delta_2(k) &= -\sum_{n=k}^{K-1} \Psi(k, n+1)\big[G(n, \nu(n)) - G(n, c(n, \nu_2(n)), \nu_2(n)) \\ &\quad + G(n, c(n, \nu_2(n)), \nu_2(n)) - G(n, c(n, \tilde{\nu}(n)), \nu_2(n)) \\ &\quad + G(n, c(n, \tilde{\nu}(n)), \nu_2(n)) - G(n, c(n, \tilde{\nu}(n)), \tilde{\nu}(n))\big], \end{aligned}$$

so that

$$\begin{aligned} \|\Delta_2(k)\| &\stackrel{(3.7)}{\leq} \frac{K_1 K_2 \omega_G(\rho)}{\beta - \alpha - \gamma} (\alpha + \gamma)^{k-\kappa} \|\nu_1(\kappa) - c(\kappa, \nu_2(\kappa))\| \\ &\quad + K_2 \omega_G(2\rho)(1 + \omega_c(2\rho)) \sum_{n=k+1}^{K} \beta^{k-n} \|\Delta_2(n-1)\| \end{aligned} \tag{3.8}$$

for $k \in [k_0, K]_{\mathbb{Z}}$. This yields

$$\begin{aligned} \frac{\|\Delta_2(k)\|}{\beta^k} &\leq \frac{K_1 K_2 \omega_G(\rho)}{\beta - \alpha - \gamma} (\alpha + \gamma)^{-\kappa} \left(\frac{\alpha + \gamma}{\beta}\right)^k \|\nu_1(\kappa) - c(\kappa, \nu_2(\kappa))\| \\ &\quad + \frac{K_2 \omega_G(2\rho)(1 + \omega_c(2\rho))}{\beta} \sum_{n=k}^{K-1} \frac{\|\Delta_2(n)\|}{\beta^n} \end{aligned}$$

and because of (3.2) we can apply Gronwall's lemma in backward time (cf. e.g. [14, pp. 68–69, Satz 2.1.3(b)]) to obtain

$$\|\Delta_2(k)\| \leq \frac{K_1 K_2 \omega_G(\rho)}{\beta - \alpha - \gamma - K_2 \omega_G(2\rho)(1 + \omega_c(2\rho))} (\alpha + \gamma)^{k-\kappa} \|\nu_1(\kappa) - c(\kappa, \nu_2(\kappa))\|$$

for $k \in [k_0, K-1]_{\mathbb{Z}}$; it is easy to see from (3.8) that the above estimate also holds in case $k = K$, which implies

$$\|\nu_2(k) - \tilde{\nu}(k)\| \stackrel{(3.2)}{\leq} \tfrac{1}{2} (\alpha + \gamma)^{k-\kappa} \|\nu_1(\kappa) - c(\kappa, \nu_2(\kappa))\| \tag{3.9}$$

for $k \in [k_0, K]_{\mathbb{Z}}$. This, in turn, leads to

$$\begin{aligned} \|\tilde{\nu}(k)\| &\leq \|\nu_2(k)\| + \|\tilde{\nu}(k) - \nu_2(k)\| \leq \\ &\stackrel{(3.9)}{\leq} \|\nu_2(k)\| + \tfrac{1}{2} (\|\nu_1(\kappa)\| + \|c(\kappa, \nu_2(\kappa))\|) \leq \\ &\stackrel{(3.5)}{\leq} \rho + \tfrac{1}{2} (\rho + \omega_c(\rho)\rho) < 2\rho \quad \text{for } k \in [k_0, K]_{\mathbb{Z}}, \end{aligned} \tag{3.10}$$

in particular, $\|\tilde{\nu}(k_0)\| < 2\rho$ and so $k_0 = \kappa$. Hence, $\tilde{\nu}$ is defined on $[\kappa, K]_{\mathbb{Z}}$ and the inequalities (3.9), (3.10) hold for $k \in [\kappa, K]_{\mathbb{Z}}$. So we proved assertions (a) and (b).

(III) In order to show the remaining estimate (3.4) we get from (3.10), as well as from [10, p. 342, Corollary 4.3]

$$\begin{aligned}
&\left\| \nu(k) - \begin{pmatrix} c(k,\tilde{\nu}(k)) \\ \tilde{\nu}(k) \end{pmatrix} \right\| \\
&\overset{(1.1)}{=} \max\left\{ \|\nu_1(k) - c(k,\tilde{\nu}(k))\|, \|\nu_2(k) - \tilde{\nu}(k)\| \right\} \\
&\leq \max\left\{ \|\nu_1(k) - c(k,\nu_2(k))\| + \|c(k,\nu_2(k)) - c(k,\tilde{\nu}(k))\|, \|\nu_2(k) - \tilde{\nu}(k)\| \right\} \\
&\overset{(3.7)}{\leq} \max\big\{ K_1 (\alpha+\gamma)^{k-\kappa} \|\nu_1(\kappa) - c(\kappa,\nu_2(\kappa))\| + \omega_c(2\rho) \|\nu_2(k) - \tilde{\nu}(k)\|, \\
&\qquad \|\nu_2(k) - \tilde{\nu}(k)\| \big\} \\
&\overset{(3.9)}{\leq} \max\left\{ K_1 + \tfrac{1}{2}\omega_c(2\rho), \tfrac{1}{2} \right\} (\alpha+\gamma)^{k-\kappa} \|\nu_1(\kappa) - c(\kappa,\nu_2(\kappa))\| \\
&\overset{(3.2)}{\leq} \left(K_1 + \tfrac{1}{2}\right) (\alpha+\gamma)^{k-\kappa} \|\nu_1(\kappa) - c(\kappa,\nu_2(\kappa))\|
\end{aligned}$$

for all $k \in [\kappa, K]_{\mathbb{Z}}$. This concludes the proof. □

For the rest of the paper we fix $\gamma \in (0, \min\{1,\beta\} - \alpha)$ and choose $\rho \in \left(0, \frac{\rho_0}{2}\right)$ so small that the estimates (3.2) are fulfilled. As a first corollary of Lemma 3.2 we show that small bounded solutions of (2.1) must lie on the center fiber bundle.

Theorem 3.3 (asymptotic description of C) (a) *If the equation (2.1) has a solution $\nu = (\nu_1, \nu_2) : \mathbb{Z} \to X \times Y$ satisfying $\|\nu(k)\| < \rho$ for all $k \in \mathbb{Z}$, then $\nu_1(k) = c(k, \nu_2(k))$ holds for all $k \in \mathbb{Z}$.*

(b) *If $S \subseteq \mathbb{Z} \times B_\rho(0) \times B_\rho(0) \subseteq \mathbb{Z} \times X \times Y$ is a set such that for any $(\kappa, \xi, \eta) \in S$ the solution $\varphi(\cdot; \kappa, \xi, \eta)$ of (2.1) exists on $\mathbb{Z}$ and satisfies $\|\varphi(k; \kappa, \xi, \eta)\| < \rho$ there, then $S \subseteq C$.*

Proof (a) Putting $k = K$ in (3.4) we obtain from Lemma 3.2

$$\begin{aligned}
\|\nu_1(K) - c(K,\nu_2(K))\| &\leq \left(K_1 + \tfrac{1}{2}\right) \|\nu_1(\kappa) - c(\kappa,\nu_2(\kappa))\| (\alpha+\gamma)^{K-\kappa} \\
&\overset{(3.1)}{\leq} 2\rho \left(K_1 + \tfrac{1}{2}\right) (\alpha+\gamma)^{K-\kappa}
\end{aligned}$$

and due to $\alpha + \gamma < 1$, letting $\kappa \to -\infty$ gives us $\nu_1(K) = c(K, \nu_2(K))$. Since $K \in \mathbb{Z}$ was arbitrary, we get the assertion.

(b) This follows from the above assertion (a) applied to the individual solutions $\varphi(\cdot; \kappa, \xi, \eta)$ of (2.1) with $(\kappa, \xi, \eta) \in S$. □

The next result enables us to relate the asymptotic behavior of small solutions of equation (2.1) to solutions of (3.3), and guarantees that center fiber bundles are exponentially attractive.

Theorem 3.4 (asymptotic phase of C) *Let Y be finite dimensional, $\kappa \in \mathbb{Z}$, and let $\nu : \mathbb{Z}_\kappa^+ \to X \times Y$ be a solution of (2.1) satisfying $\|\nu(k)\| < \rho$ for all $k \in \mathbb{Z}_\kappa^+$. Then there exists a solution $\tilde{\nu}_* : \mathbb{Z}_\kappa^+ \to Y$ of (3.3) such that*

$$\left\| \nu(k) - \begin{pmatrix} c(k,\tilde{\nu}_*(k)) \\ \tilde{\nu}_*(k) \end{pmatrix} \right\| \leq \left(K_1 + \tfrac{1}{2}\right) \|\nu_1(\kappa) - c(\kappa,\nu_2(\kappa))\| (\alpha+\gamma)^{k-\kappa} \qquad (3.11)$$

for all $k \in \mathbb{Z}_\kappa^+$.

Proof Let $\tilde{\nu}_m$, $m \in \mathbb{Z}_\kappa^+$, denote the solution of (3.3) satisfying $\tilde{\nu}_m(m) = \nu_2(m)$. Then Lemma 3.2 implies that $\tilde{\nu}_m$ is defined on $[\kappa, m]_{\mathbb{Z}}$ and $\|\tilde{\nu}_m(k)\| < 2\rho$ there. Moreover, we simply set $\tilde{\nu}_m(k) := 0$ for $k \in \mathbb{Z}_\kappa^+$, $k > m$. Hence, $(\tilde{\nu}_m(\kappa))_{m \in \mathbb{Z}_\kappa^+}$

is bounded and since the space Y is finite dimensional, there exists a convergent subsequence $(\tilde{\nu}_{m_n^{(\kappa)}}(\kappa))_{n\in\mathbb{Z}_\kappa^+}$ in Y with $\kappa+1\le m_\kappa^{(\kappa)}$. In addition, $(\tilde{\nu}_{m_n^{(\kappa)}}(\kappa+1))_{n\in\mathbb{Z}_\kappa^+}$ is bounded and there exists a convergent subsequence $(\tilde{\nu}_{m_n^{(\kappa+1)}}(\kappa+1))_{n\in\mathbb{Z}_\kappa^+}$ in Y with $\kappa+2\le m_\kappa^{(\kappa+1)}$. Iterating this construction, we obtain a sequence $(\tilde{\nu}_{m_n^{(l+1)}})_{n\in\mathbb{Z}_\kappa^+}$, which is a subsequence of $(\tilde{\nu}_{m_n^{(l)}})_{n\in\mathbb{Z}_\kappa^+}$, $l\in\mathbb{Z}_\kappa^+$, such that $l+1\le m_\kappa^{(l)}$ and accordingly $(\tilde{\nu}_{m_n^{(l)}}(k))_{n\in\mathbb{Z}_\kappa^+}$ converges for $k\in[\kappa,l]_{\mathbb{Z}}$. Now we define $\bar{\nu}_n(k):=\tilde{\nu}_{m_n^{(n)}}(k)$, $k\in\mathbb{Z}_\kappa^+$, and $(\bar{\nu}_n(k))_{n\in\mathbb{Z}_\kappa^+}$ converges for any $k\in\mathbb{Z}_\kappa^+$, because beginning with the index $n=k$, $(\bar{\nu}_n(k))_{n\ge k}$ is a subsequence of the convergent sequence $(\tilde{\nu}_{m_n^{(k)}}(k))_{n\in\mathbb{Z}_\kappa^+}$. Now, by definition, $(\bar{\nu}_n)_{n\in\mathbb{Z}_\kappa^+}$ satisfies for any $n\in\mathbb{Z}_\kappa^+$

$$\begin{aligned}\bar{\nu}_n(k+1) &\overset{(3.3)}{=} B(k)\bar{\nu}_n(k)+G(k,c(k,\bar{\nu}_n(k)),\bar{\nu}_n(k)) \quad \text{for } k\in\left[\kappa,m_n^{(n)}-1\right]_{\mathbb{Z}},\\ \left\|\nu(k)-\begin{pmatrix}c(k,\bar{\nu}_n(k))\\ \bar{\nu}_n(k)\end{pmatrix}\right\| &\overset{(3.4)}{\le} \left(K_1+\tfrac{1}{2}\right)\|\nu_1(\kappa)-c(\kappa,\nu_2(\kappa))\|\,(\alpha+\gamma)^{k-\kappa}\end{aligned}$$

for $k\in\left[\kappa,m_n^{(n)}\right]_{\mathbb{Z}}$, and passing over to the limit $n\to\infty$ in these relations for k fixed, we see that $\tilde{\nu}_*(k):=\lim_{n\to\infty}\bar{\nu}_n(k)$, $k\in\mathbb{Z}_\kappa^+$, is a solution of the difference equation (3.3) on $\mathbb{Z}_\kappa^+$ satisfying (3.11). □

Having Theorem 3.4 available, it is not difficult to maintain stability properties (cf. e.g. [1, p. 240, Definition 5.4.1]) for (2.1) from the corresponding properties of the zero solution of the finite dimensional difference equation (3.3).

Theorem 3.5 (reduction principle) *Let Y be finite dimensional. If the zero solution of* (3.3) *is stable (uniformly stable, asymptotically stable, uniformly asymptotically stable, exponentially stable with rate $\tilde{\gamma}$, unstable, respectively), then also the zero solution of* (2.1) *is stable (uniformly stable, asymptotically stable, uniformly asymptotically stable, exponentially stable with rate* $\max\{\alpha+\gamma,\tilde{\gamma}\}$, *unstable, respectively).*

Proof If the zero solution of (3.3) is unstable, then by invariance of C, the zero solution of equation (2.1) is unstable, too. Let $\epsilon>0$ and $\kappa\in\mathbb{Z}$ be given arbitrarily, but w.l.o.g. $\epsilon\le 2\rho_0$. We suppose now that the zero solution 0 is stable for (3.3); then there exists a $\delta>0$ such that

$$\|\tilde{\nu}(k)\|<\tfrac{\epsilon}{2}\quad \text{for } k\in\mathbb{Z}_\kappa^+ \tag{3.12}$$

and any solution $\tilde{\nu}:\mathbb{Z}_\kappa^+\to Y$ of the reduced equation (3.3) with $\|\tilde{\nu}(\kappa)\|<\delta$. Henceforth let $\nu=(\nu_1,\nu_2):\mathbb{Z}_\kappa^+\to X\times Y$ be an arbitrary solution of (2.1) with $\|\nu(k)\|<\rho$ for $k\in\mathbb{Z}_\kappa^+$ and $\|\nu(\kappa)\|<\min\left\{\frac{\delta}{2K_1+2},\frac{\epsilon}{4K_1+3}\right\}$. Then due to Theorem 3.4 there exists a solution $\tilde{\nu}_*:\mathbb{Z}_\kappa^+\to Y$ of (3.3) satisfying (3.11). This, in particular, yields

$$\begin{aligned}\|\tilde{\nu}_*(\kappa)\| &\le \|\nu_2(\kappa)-\tilde{\nu}_*(\kappa)\|+\|\nu_2(\kappa)\|\\ &\overset{(3.11)}{\le} \left(K_1+\tfrac{1}{2}\right)\|\nu_1(\kappa)-c(\kappa,\nu_2(\kappa))\|+\|\nu_2(\kappa)\|\\ &\overset{(3.1)}{\le} \left(K_1+\tfrac{1}{2}\right)(\|\nu_1(\kappa)\|+\|\nu_2(\kappa)\|)+\|\nu_2(\kappa)\|<\delta\end{aligned}$$

and consequently $\|\tilde{\nu}_*(k)\| < \frac{\epsilon}{2}$ for $k \in \mathbb{Z}_\kappa^+$ by (3.12). On the other hand, we have the estimate

$$\begin{aligned} \|\nu(k)\| &\leq \left\| \nu(k) - \begin{pmatrix} c(k, \tilde{\nu}_*(k)) \\ \tilde{\nu}_*(k) \end{pmatrix} \right\| + \left\| \begin{pmatrix} c(k, \tilde{\nu}_*(k)) \\ \tilde{\nu}_*(k) \end{pmatrix} \right\| \\ &\overset{(3.11)}{\leq} \left(K_1 + \tfrac{1}{2}\right) \|\nu_1(\kappa) - c(\kappa, \nu_2(\kappa))\| \, (\alpha + \gamma)^{k-\kappa} + \left\| \begin{pmatrix} c(k, \tilde{\nu}_*(k)) \\ \tilde{\nu}_*(k) \end{pmatrix} \right\| \\ &\overset{(3.1)}{\leq} \left(K_1 + \tfrac{1}{2}\right) (\|\nu_1(\kappa)\| + \|\nu_2(\kappa)\|) \, (\alpha + \gamma)^{k-\kappa} + \|\tilde{\nu}_*(k)\| < \epsilon \end{aligned}$$

for all $k \in \mathbb{Z}_\kappa^+$ and therefore, the solution ν is stable. Finally, one can show the assertions about the remaining stability properties along the same lines. □

References

[1] Agarwal, R.P. *Difference Equations and Inequalities.* Marcel Dekker, New York, 1992.

[2] Aulbach, B. *A reduction principle for nonautonomous differential equations.* Archiv der Mathematik, **39** (1982), 217–232.

[3] Aulbach, B. *The fundamental existence theorem on invariant fiber bundles.* Journal of Difference Equations and Applications, **3** (1998), 501–537.

[4] Aulbach, B., Van Minh, N. and Zabreiko, P.P. *The concept of spectral dichotomy for linear difference equations.* Journal of Math. Analysis and Applications, **185** (1994), 275–287.

[5] Aulbach, B. and Siegmund, S. *The dichotomy spectrum for noninvertible systems of linear difference equations.* Journal of Difference Equations and Applications, **7** (2001), 895–913.

[6] Aulbach, B., Pötzsche, C. and Siegmund, S. *A smoothness theorem for invariant fiber bundles.* Journal of Dynamics and Differential Equations, **14** (2002), 519-547.

[7] Carr, J. *Applications of Centre Manifold Theory.* Applied Mathematical Sciences, 35, Springer-Verlag, New York, 1981.

[8] Janglajew, K. *On the reduction principle of difference equations.* Dynamics of Continuous, Discrete and Impulsive Systems, **6** (1999), 381–388.

[9] Kriegl, A. and Michor, P.W. *The Convenient Setting of Global Analysis.* Mathematical Surveys and Monographs, 53, American Mathematical Society, Providence, 1997.

[10] Lang, S. *Real and Functional Analysis.* Graduate Texts in Mathematics, 142, Springer-Verlag, New York, 1993.

[11] Palmer, K. *On the stability of the center manifold.* Journal of Applied Mathematics and Physics (ZAMP), **38** (1987), 273–278.

[12] Pötzsche, C. *Nichtautonome Differenzengleichungen mit stationären und invarianten Mannigfaltigkeiten* (in german). Diploma thesis, University of Augsburg, 1998.

[13] Pötzsche, C. and Siegmund, S. *C^m-Smoothness of invariant fiber bundles.* To appear, Topological Methods in Nonlinear Analysis, 2004.

[14] Wanner, T. *Invariante Faserbündel und topologische Äquivalenz bei dynamischen Prozessen* (in german). Diploma thesis, University of Augsburg, 1991.

Fields Institute Communications
Volume **42**, 2004

Bifurcation of Maps and Cycling in Genetic Systems

Robert J. Sacker
Department of Mathematics
University of Southern California
Los Angeles, CA 90089-1113 U.S.A.
rsacker@math.usc.edu

Hubertus F. von Bremen
Department of Mathematics
University of Southern California
Los Angeles, CA 90089-1113 U.S.A.
vonbrem@usc.edu

Abstract. We study a first order system of difference equations describing the evolution under random mating of an infinite population of gametes in a 2-locus, 2-allele genetic model with recombination and selection acting. For a certain choice of fitness coefficients it is proved that as the recombination coefficient moves through a critical value, a fixed point loses its stability and an invariant curve bifurcates due to a Neimark-Sacker bifurcation. This gives rise to a recurrent or cycling orbit and verifies an earlier observation of Alan Hastings. Cycling guarantees a continual change in the distribution of genotypes in a given population.

1 Introduction

In this paper we will study a greatly simplified model of reproduction in which a certain recurrence or "cycling" behavior takes place due to the effects of *natural election* and *recombination* only. The first simplifications will be to assume an infinite population of individuals and random mating in order to arrive at a deterministic model. We take as the beginning of reproduction the process known as *gametogenesis*, the formation of *gametes* or reproductive cells such as ova or sperm. This takes place during an involved process called *meiosis* where chromosome pairs (one member of the pair from "Mom" the other from "Dad") are replicated . There are 23 distinct types of chromosome pairs in a human and we consider just one

2000 *Mathematics Subject Classification.* Primary 39A10, 39A11; Secondary 92D10, 92D25.

The authors thank Peter Calabrese Magnus Nordborg and especially Simon Tavare for help in understanding various biological concepts.

The first author was supported by University of Southern California, Letters Arts and Sciences Faculty Development Grant.

type. After replication, each original chromosome contributes 4 gametes. Ignoring, only for the moment, *crossover*, 2 of these gametes carry only genetic information from "Mom", the other 2 only from "Dad".

With this scenario (no crossover) when Mom and Dad become grandparents only one of them will contribute genetic information to their grandchild. Before discussing crossover, which requires some assumptions as to what a *chromosome* is, let us carry this scenario to completion of one generational cycle. Upon mating, a gamete (sperm, say) unites with an egg to form a *zygote*, the earliest form of an embryo. The *fitness* of this zygote is a number which measures its (a) ability to survive to reproductive age, (b) fertility and (c) ability to successfully mate. The fitness will be a function of the gametes which formed the zygote.

A *chromosome* is a linear helical chain consisting of thousands of loci, each locus being itself a shorter helical chain of DNA, called a *gene* which acts as a genetic marker or determining factor for some trait (Rothwell)[15]. Genes come in different forms called *alleles* and for simplicity we assume just 2 loci. At locus A there are 2 alleles, "A" and "a" and at locus B, "B" and "b". We shall refer to "A" as the *dominant* allele and "a", *recessive.* Thus a typical chromosome pair before meiosis will be designated by $\frac{A\,b}{a\,B}$ giving rise (no crossover) to gametes Ab and aB. Crossover occurs when this chromosome pair breaks between the A and B loci and recombines to form gametes AB and ab.

Letting x_1, x_2, x_3, x_4 denote the frequencies (and the names) of the genotypes AB, Ab, aB, ab, one readily obtains the *recombination mapping* (Roughgarden [16]).

$$x \;\rightarrow\; x^* = R(x) \qquad x = (x_1, x_2, x_3, x_4)^T$$

where

$$\begin{aligned} x_1^* &= x_1 - r(x_1x_4 - x_2x_3) \\ x_2^* &= x_2 + r(x_1x_4 - x_2x_3) \\ x_3^* &= x_3 + r(x_1x_4 - x_2x_3) \\ x_4^* &= x_4 - r(x_1x_4 - x_2x_3) \end{aligned} \tag{1.1}$$

and r, the *recombination coefficient*, satisfies $0 \le r \le 1/2$. The coefficient r is a measure of the degree to which crossover takes place: $r \approx 0$ represents *tight linkage* whereas $r \approx 1/2$ represents *loose linkage* between the loci. The quantity in parentheses is called *linkage disequilibrium* and is a measure of the statistical dependency between the loci. In statistical terms it is the *covariance* between the loci [16, p.113].

The next mapping to discuss is the *selection* mapping. After random mating the gametes form zygotes and each zygote is assigned a fitness coefficient $F_{ij} = fitness\ of\ \frac{x_i}{x_j}$, e.g.,$F_{13} = fitness\ of\ \frac{AB}{a\,B}$. The selection mapping is given by

$$x \;\rightarrow\; x^* = S(x) \qquad x = (x_1, x_2, x_3, x_4)^T$$

where

$$x_i^* = x_i\left(\frac{F_i}{\overline{F}}\right) \tag{1.2}$$

and

$$F_i = \sum_{j=1}^{4} F_{ij}x_j \qquad and \qquad \overline{F} = \sum_{i,j=1}^{4} F_{ij}x_ix_j = \sum_{i=1}^{4} F_ix_i$$

The quantity F_i is the *marginal mean fitness* of x_i and $\overline{F}$ is the *mean fitness* of the population. Dividing by the quantity $\overline{F}$ in (1.2) preserves the relation $\sum x_i = 1$. Composing these maps we obtain the *generational map*

$$x^* = G(x) \qquad G = S \circ R. \tag{1.3}$$

It is this mapping that we will study.

A general treatment of the equations of gametic frequency change with linkage and selection in the 2-locus model was given by Kimura [9] in the continuous time case and by Lewontin and Kojima [11] in the discrete time case. In the discrete case they obtain the mapping

$$x_i^* = \frac{1}{\overline{F}}\Big\{x_i F_i + \epsilon_i r(F_{14}x_1x_4 - F_{23}x_2x_3)\Big\} \tag{1.4}$$

where $\epsilon_2 = \epsilon_3 = -\epsilon_1 = -\epsilon_4 = 1$ ($F_{14} = F_{23}$ in their derivation). Upon close examination of their derivation one can see that (1.4) is just $x^* = R \circ S(x)$. The difference between this map and G is minor since if x is a fixed point for $R \circ S$ then $S(x)$ is a fixed point for $S \circ R$ and similarly for a cycling point–the dynamics are equivalent.

A fixed point of the mapping G is a frequency distribution of the genotypes that remains the same from one generation to the next. Our interest will be in studying a fixed point which is stable for certain values of r but loses its stability due to a complex conjugate pair of eigenvalues of the linearized mapping leaving the unit circle as r increases (or decreases) through some critical value.

Along the lines of an earlier result of E. Hopf [7] for ordinary differential equations, Neimark [13] conjectured that an invariant curve would bifurcate from the fixed point provided a certain non-linear term has the right sign. Later Sacker [18] independently stated and proved, for the first time, the bifurcation theorem for maps and discovered that certain low order "resonances" had to be avoided, i.e., the eigenvalues must avoid 3rd and 4th roots of unity as they leave the unit circle. The result has come to be known as the Neimark-Sacker bifurcation theorem, e.g. Kuznetsov [10]. The additional requirement referred to as "nonlinear damping" is needed but its verification depends on first putting the mapping in a certain normal form. See Section 3 for a statement of the theorem using the notation of Section 2. See Iooss and Joseph [8, p.254] for further discussion and references. The result was later rediscovered by Ruelle and Takens [17]. Also see Marsden and McCracken [12] and Hassard, Kazarnoff and Wan [4].

In the article by Hastings [6] a fixed point for (1.4) with complex eigenvalues is found and "cycling" or recurrence is observed using numerical methods. This was done using a linear programming technique developed in Hastings [5]. The nonlinear damping condition is not verified but rather inferred from the fact that cycling does indeed occur. Although in Akin [1], [2] the continuous time model is treated and the nonlinear damping is computed using a formula of [12], the discrete time model seems to be the most straightforward to analyze both numerically and analytically.

For a discussion of bifurcations (not including cycling) in the case in which the coupling (*cis*) and repulsion (*trans*) double heterozygotes have different fitness, i.e., $F_{14} \neq F_{23}$, see Nordborg, Franklin and Feldman [14].

Regarding terminology, the term "cycling", borrowed from the literature, is known as "recurrence" in the area of Dynamical Systems. Specifically, it does *not* imply periodicity.

2 Discussion of mapping

It is clear from the previous discussion that the map G preserves the unit simplicial face in R^4:

$$\sum_{i=1}^{4} x_i = 1, \quad x_i \geq 0.$$

One of the four equations in (1.3) is therefore redundant and can be ignored. We take the first three equations and make the obvious substitution $x_4 = 1-x_1-x_2-x_3$ in the remaining equations. We then obtain a mapping $T : \triangle \to \triangle$ where $\triangle$ is the unit simplex in R^3

$$\triangle : \qquad \sum_{i=1}^{3} x_i \leq 1, \quad x_i \geq 0.$$

We now make the following choice of fitness coefficients and recombination coefficient:

$$\begin{array}{llll} F_{11} = 0.83301, & F_{12} = 0.6585664, & F_{13} = 0.5057075, & F_{14} = 1.000, \\ F_{22} = 1.33068, & F_{23} = 1.000, & F_{24} = 0.22458, & F_{33} = 0.80724, \\ F_{34} = 0.46357, & F_{44} = 1.41881, & r_0 = 0.180223678355 & \end{array} \tag{2.1}$$

For these choices, closely related to those studied in [6], we find (using MATLAB) one fixed point (out of the 13 found) which is of particular interest:

$$x_{fix} : \quad x_1 \approx 0.797844, \quad x_2 \approx 0.0883542, \quad x_3 \approx 0.0280991 \tag{2.2}$$

For the values (2.1) the linearized mapping at the fixed point(2.2) has eigenvalues:

$$\begin{array}{rcl} \lambda & \approx & 0.999578 + 0.0290354i, \quad \bar{\lambda} \\ \sigma & \approx & 0.695037 \end{array} \tag{2.3}$$

The eigenvalue λ lies on the unit circle and since $\Im\lambda \neq 0$, the implicit function theorem guarantees the fixed point will vary smoothly as r varies on a small neighborhood of r_0. Thus if we let $r = r_0 - \mu$ and denote the smooth family of fixed points by $x_{fix}(\mu)$ we have $x_{fix}(0)$ given by (2.2). Let v and w be eigenvectors(columns) corresponding to $\lambda(\mu)$ and $\sigma(\mu)$ respectively and define

$$M(\mu) = [\Re \mathrm{v}, \Im \mathrm{v}, w]$$

Then the change of variables

$$x = x_{fix}(\mu) + M(\mu)y, \qquad r = r_0 - \mu \tag{2.4}$$

reduces the mapping to

$$y^* = \begin{bmatrix} \Re\lambda(\mu) & \Im\lambda(\mu) & 0 \\ -\Im\lambda(\mu) & \Re\lambda(\mu) & 0 \\ 0 & 0 & \sigma(\mu) \end{bmatrix} y + f(y, \mu). \tag{2.5}$$

Using the procedure introduced in [18] we define $z = y_1 + iy_2$ and $W = y_3$. Then the mapping takes the form

$$\begin{aligned} z^* &= \lambda(\mu)z + P(z, \bar{z}, W, \mu) \\ W^* &= \sigma(\mu)W + Q(z, \bar{z}, W, \mu) \end{aligned} \tag{2.6}$$

where P and Q are rational functions in $z, \bar{z}$ and W whose power series begin with quadratic terms while P and $Q(0, 0, 0, \mu) = 0$ identically for μ in some small neighborhood of zero. Since all the conditions for bifurcation are verified at $\mu = 0$ we shall suppress the μ dependence until it is actually needed. Thus we consider

$$\begin{aligned} z^* &= \lambda z + P(z, \bar{z}, W) \\ W^* &= \sigma W + Q(z, \bar{z}, W) \end{aligned} \tag{2.7}$$

and proceed to put this into the desired canonical where the bifurcation conditions can be verified.

The first step is to "remove", from the equations (2.7), all the quadratic terms of the form $z^k\bar{z}^l$, $k + l = 2$ by means of the transformations

$$\begin{aligned} z &= \zeta + a\zeta^2 + b\zeta\bar{\zeta} + c\bar{\zeta}^2 \\ W &= Y + \alpha\zeta^2 + \beta\zeta\bar{\zeta} + \gamma\bar{\zeta}^2 \end{aligned}$$

in a sufficiently small neighborhood of the origin. If $\lambda^3 \neq 1$ and $|\sigma| < 1$, then $a, \ldots, \gamma$ can be uniquely chosen so that in the new mapping

$$\begin{aligned} \zeta^* &= \lambda\zeta + \ldots \\ Y^* &= \sigma Y + \ldots \end{aligned} \tag{2.8}$$

the equations are devoid of these *specific* quadratic terms.

The exclusion of the fourth roots of unity (together with $|\sigma| < 1$) also allows us to remove terms of the form

$$\zeta^k\bar{\zeta}^l, \quad k + l = 3, \quad k \neq 2; \qquad \zeta Y \textit{ and } \bar{\zeta}Y \tag{2.9}$$

from the first equation of (2.8) to obtain the canonical form

$$\begin{aligned} \zeta^* &= \lambda\zeta + b\zeta^2\bar{\zeta} + \mathcal{F}_4(\zeta, \bar{\zeta}, Y) \\ Y^* &= \sigma Y + \mathcal{F}_3(\zeta, \bar{\zeta}, Y) \end{aligned} \tag{2.10}$$

where the Taylor expansion of $\mathcal{F}_j$ has no monomials of "weight" $< j$. The *weight* τ of the monomial $Y^s\zeta^t\bar{\zeta}^u$ is defined to be $\tau = 2s + t + u$. See [18] for more details.

Thus we will work with (2.10) which we rewrite

$$\begin{aligned} \zeta^* &= \lambda\zeta + b\zeta^2\bar{\zeta} + \mathcal{F}_4(\zeta, \bar{\zeta}, Y) \\ &= \lambda(1 + \beta|\zeta|^2)\zeta + \mathcal{F}_4 \\ &= e^{\alpha}e^{\beta|\zeta|^2}\zeta + \hat{\mathcal{F}}_4 \\ &= e^{\alpha(\mu)+\beta(\mu)|\zeta|^2}\zeta + \hat{\mathcal{F}}_4(\zeta, \bar{\zeta}, Y, \mu) \\ Y^* &= \sigma Y + \mathcal{F}_3(\zeta, \bar{\zeta}, Y) \end{aligned} \tag{2.11}$$

where in the last two equalities we have returned the μ dependence. Note, however, that the μ dependence in, e.g., $\hat{\mathcal{F}}_4$ comes about only through coefficients of monomials of weight 4 and higher.

3 Statement of bifurcation theorem

In this section we state the theorem directly from [18] as it applies to the above equations

Theorem 3.1 *In (2.6),viz.,*

$$\begin{aligned} z^* &= \lambda(\mu)z + P(z,\bar{z},W,\mu) \\ W^* &= \sigma(\mu)W + Q(z,\bar{z},W,\mu) \end{aligned} \tag{3.1}$$

where $P(0,0,0,\mu) = Q(0,0,0,\mu) = 0$ *identically for* μ *in some small neighborhood of zero. Assume further that*

1. $|\lambda(0)| = 1, \quad \lambda^3(0) \neq 1,$ *and* $\lambda^4(0) \neq 1$ *(non resonance)*
2. $\Re\lambda'(0) > 0$ *(loss of stability)*
3. $|\sigma(0)| < 1$ *(stability in "normal" direction) and*
4. *In the normal form* (2.11), $\Re\beta(0) < 0$ *(nonlinear damping)*

Then in a sufficiently small μ *neighborhood,* $0 < \mu < \mu_*$, *(3.1) has an asymptotically stable invariant curve parametrized by* θ, $0 \leq \theta \leq 2\pi$

$$\begin{aligned} z &= a_0\sqrt{\mu}\exp(i\theta) + \mu f(\theta,\mu) \\ W &= \mu g(\theta,\mu) \end{aligned} \tag{3.2}$$

where $a_0 = \sqrt{-\Re\alpha'(0)/\Re\beta(0)}$.

In (2.11) the pertinent constants are:

$$\begin{aligned} \alpha(\mu) &= \ln\lambda(\mu) = \alpha(0) + \alpha'(0)\mu + \ldots \\ &= -0.0290395i + (10.545392 + 10.718922i)\mu + \ldots \\ \beta(0) &= b(0)/\lambda(0) = -2.936877 - 6.270699i \end{aligned} \tag{3.3}$$

The calculation of $\alpha'(0)$ is done numerically using MATLAB while the quantity $b(0)$ is computed using MAPLE by actually transforming the mapping into the desired canonical form.

Thus the existence of an invariant curve for the generational map (1.3) is established. The curve is most easily seen (analytically) in the coordinates ζ and Y in which the mapping (2.11), in the first approximation, is

$$\begin{aligned} \zeta^* &= \zeta\exp[\Re\alpha'(0)\mu + \Re\beta(0)|\zeta|^2]\exp i[\Im\alpha(0) + \Im\beta(0)|\zeta|^2] \\ Y^* &= \sigma Y. \end{aligned} \tag{3.4}$$

where it is clear that the circle $|\zeta| = a_0\sqrt{\mu}$, $\quad Y = 0$ is invariant. Further it can be seen from (3.4) that on this circle points are moved by the mapping, in the first approximation, by $\Im\alpha(0) = -0.0290395$ radians and therefore it takes 217 iterations of the map to return approximately to the initial point. This agrees with numerical simulations to within one iteration. See Bürger[3, p.53] for additional discussion and Sacker and von Bremen [19] for further details.

References

[1] Ethan Akin. Cycling in simple genetic systems. *J. Math. Biol.*, 13:305–324, 1982.

[2] Ethan Akin. Cycling in simple genetic systems (part 2. the symmetric cases). *Lecture Notes in Econ. and Math. Sci. 287*, pages 139–153, 1987.

[3] Reinhard Bürger. *The Mathematical Theory of Selection, Recombination and Mutation.* John Wiley, West Sussex, PO19 IUD, UK, 2000.

[4] B. Hassard, N. Kazarnoff, and Y.-H. Wan. *Theory and Application of Hopf Bifurcation.* London Math. Society Lecture Note Ser., 41. Cambridge Univ. Press, London, 1981.

[5] Alan Hastings. Disequilibrium, selection and recombination: limits in two locus, two allele models. *Genetics*, 98:659–668, July 1981.
[6] Alan Hastings. Stable cycling in discrete-time genetic models. *Proc. Natl. Acad. Sci. USA*, 78(11):7224–7225, Nov. 1981.
[7] E. Hopf. Abzweigung einer periodischen lösung von einer stationären lösung eines differentialsystems. *Sächs. Akad. Wiss. Leipzig, Math.-Phys. Klasse, Berichten*, 94:667–693, Jan. 1942.
[8] Gerard Iooss and Daniel D. Joseph. *Elementary Stability and Bifurcation Theory*. Undergraduate Texts in Mathematics. Springer, New York, second edition, 1990.
[9] M. Kimura. A model of a genetic system which leads to closer linkage by natural selection. *Evolution*, 10:278–287, 1956.
[10] Yuri A. Kuznetsov. *Elements of Applied Bifurcation Theory*. Applied Mathematical Sciences, v. 112. Springer, New York, second edition, 1998.
[11] R.C. Lewontin and K. Kojima. The evolutionary dynamics of complex polymorphisms. *Evolution*, 14:458–472, 1960.
[12] J. Marsden and M. McCracken. *The Hopf Bifurcation and its Applications*. Applied Mathematical Sciences, v.19. Springer, New York, 1976.
[13] Ju. Neimark. On some cases of periodic motions depending on parameters. *Dokl. Akad. Nauk SSSR*, 129:736–739, 1959.
[14] Magnus Nordborg, Ian R. Franklin, and Marcus W. Feldman. Effects of cis-trans viability selection on some two-locus models. *Theor. Pop. Biol.*, 47(3):365–392, Jun. 1995.
[15] Norman V. Rothwell. *Understanding Genetics, a Molecular Approach*. Wiley-Liss, New York, NY, 1993.
[16] Jonathan Roughgarden. *Theory of Population Genetics and Evolutionary Ecology*. MacMillan Publ. Co., 1979. Reprinted by Prentice Hall, 1996.
[17] D. Ruelle and F. Takens. On the nature of turbulence. *Com. Math. Phys.*, 20:167–192, 1971.
[18] Robert J. Sacker. On invariant surfaces and bifurcation of periodic solutions of ordinary differential equations. IMM-NYU 333, Courant Institute - New York University, October 1964. Downloadable(pdf) from http://math.usc.edu/~rsacker.
[19] Robert J. Sacker and Hubertus F. von Bremen. A new approach to cycling in a 2-locus 2-allele genetic model. *J. Difference Eq. and Appl.*, 9(5): 441–448, 2003.

Fields Institute Communications
Volume **42**, 2004

Nonoscillation and Oscillation Theorems for a Fourth Order Nonlinear Difference Equation

Ewa Schmeidel
Institute of Mathematics
Poznań University of Technology
Piotrowo 3A, 60-965 Poznań, Poland
eschmeid@math.put.poznan.pl

Abstract. We study a fourth order nonlinear difference equation

$$\Delta^2(r_n\Delta^2 y_n) = f(n, y_n).$$

Two types of nonoscillatory solutions are considered. We divide the set of solutions of this equation into two types: F_+ and F_--solutions. Relations between these types of solutions and their nonoscillatory behavior are the main purpose of this paper.

1 Introduction

We will study fourth order difference equations of the form

$$\Delta^2(r_n\Delta^2 y_n) = f(n, y_n). \tag{E}$$

We denote by N the set of positive integers, by R the set of real numbers, by R_+ the set of real positive numbers and by R_- the set of negative numbers. For a function $x : N \to R$, the forward difference operator are define as follows:

$$\Delta x_n = x_{n+1} - x_n \quad \text{and} \quad \Delta^k x_n = \Delta(\Delta^{k-1} x_n), \quad \text{for} \quad k = 2, 3, \ldots.$$

The sequence y is the trivial sequence if there exists $n_0 \in N$, such that $y_n = 0$, for all $n \geq n_0$. By a solution of (E) we mean any nontrivial sequence y satisfying equation (E) for all $n \in N$.

A solution is oscillatory if for every $m \in N$, there exists $n \geq m$, such that $y_n y_{n+1} \leq 0$. Therefore a nonoscillatory solution of equation (E) is an eventually positive or eventually negative. We assume that void sum is equal to zero.
In this paper we assume that the function $f : N \times R \to R$ satisfies the condition

$$xf(n, x) < 0, \quad \text{for} \quad n \in N, \; x \in R \setminus \{0\}. \tag{$*$}$$

Throughout this paper we assume also that $r : N \to R_+$ and $\sum\limits_{k=1}^{\infty} \frac{k}{r_k} = \infty$.

2000 *Mathematics Subject Classification.* Primary 39A10.

The study of higher order difference equations has received a great deal of attention in last few years (see [1], [3], [6] and [7]). Fourth order difference equations was studied for example by Cheng [2], Graef and Thandapani [4], Hooker and Patula [5], Liu and Yan [8] and Schmeidel and Szmanda [12]. Results obtained in this paper are motivated by some results obtained by Smith and Taylor [13], [14] and [15], Zhang and Cheng [16], and Popenda and Schmeidel [9].

We define operator F as

$$F(x_n) = x_{n-1}\Delta(r_n\Delta^2 x_n) - r_n(\Delta x_{n-1})(\Delta^2 x_n), \qquad n \in N.$$

We use the F operator to classify solutions of equation (E).

Definition 1.1 If $F(y_n) \geq 0$ for all $n \in N$, then a solution y of equation (E) is called F_+-solution. If $F(y_n) < 0$ for some n, then a solution y of equation (E) is called F_--solution.

Operator F divides the set of solutions into two disjoint subsets F_+ and F_--solutions.

The following properties (see [11]) of nonoscillatory solutions of equation (E) are known.

Every nonoscillatory solution $\{y_n\}$ of equation (E) is one of four types:

$$y_n > 0,\ \Delta y_n > 0,\ \Delta^2 y_n > 0,\ \Delta(r_n\Delta^2 y_n) > 0,\ \Delta^2(r_n\Delta^2 y_n) < 0, \qquad (A4+)$$

$$y_n < 0,\ \Delta y_n < 0,\ \Delta^2 y_n < 0,\ \Delta(r_n\Delta^2 y_n) < 0,\ \Delta^2(r_n\Delta^2 y_n) > 0, \qquad (A4-)$$

$$y_n > 0,\ \Delta y_n > 0,\ \Delta^2 y_n < 0,\ \Delta(r_n\Delta^2 y_n) > 0,\ \Delta^2(r_n\Delta^2 y_n) < 0, \qquad (A2+)$$

$$y_n < 0,\ \Delta y_n < 0,\ \Delta^2 y_n > 0,\ \Delta(r_n\Delta^2 y_n) < 0,\ \Delta^2(r_n\Delta^2 y_n) > 0, \qquad (A2-)$$

eventually.

In this paper we consider relationship between F_+, F_--solutions and $(A4+)$, $(A4-)$, $(A2+)$, $(A2-)$ types.

Lemma 1.2

$$\Delta F(x_n) = x_n\Delta^2(r_n\Delta^2 x_n) - r_{n+1}\Delta^2 x_{n-1}\Delta^2 x_{n+1}. \qquad (1.1)$$

Proof The proof is evident. □

For $r_n \equiv 1$ we get operator F using by Schmeidel in [10].

Lemma 1.3 *If $\{y_n\}$ is a nonoscillatory F_+-solution of equation (E), then*

$$\sum_{n=1}^{\infty} r_{n+1}(\Delta^2 y_{n-1})(\Delta^2 y_{n+1}) < \infty, \qquad (1.2)$$

and

$$\lim_{n\to\infty} r_{n+1}(\Delta^2 y_{n-1})(\Delta^2 y_{n+1}) = 0. \qquad (1.3)$$

Proof Let $\{y_n\}$ be a F_+-solution of equation (E).

From (1.1), we get

$$\Delta F(y_k) = y_k\, f(k, y_k) - r_{k+1}\Delta^2 y_{k-1}\Delta^2 y_{k+1}.$$

Now by summation, we obtain

$$F(y_n) = F(y_1) + \sum_{k=1}^{n-1} y_k\, f(k, y_k) - \sum_{k=1}^{n-1} r_{k+1}\Delta^2 y_{k-1}\, \Delta^2 y_{k+1}.$$

Since $F(y_n) \geq 0$, then

$$0 \leq F(y_1) + \sum_{k=1}^{n-1} y_k f(k, y_k) - \sum_{k=1}^{n-1} r_{k+1}\Delta^2 y_{k-1}\,\Delta^2 y_{k+1},$$

and from $(*)$

$$F(y_1) \geq \sum_{k=1}^{n-1} r_{k+1}\Delta^2 y_{k-1}\,\Delta^2 y_{k+1}.$$

Hence

$$\sum_{k=1}^{\infty} r_{k+1}\Delta^2 y_{k-1}\,\Delta^2 y_{k+1} < \infty.$$

Condition (1.3) follows directly from (1.2). □

Remark 1.4 Assume that $\{r_n\}$ is nondecreasing sequence. If $\{y_n\}$ is a nonoscillatory F_+-solution of equation (E), then

$$\sum_{n=1}^{\infty} (\Delta^2 y_{n-1})(\Delta^2 y_{n+1}) < \infty,$$

and

$$\lim_{n\to\infty} (\Delta^2 y_{n-1})(\Delta^2 y_{n+1}) = 0.$$

2 Existence theorems

The following theorem is proved by Bin and Jurang in [8] (Theorem 3, Remark 1).

Theorem 2.1 *Let $f(n, y)$ be monotone in y for each fixed n. Then a necessary and sufficient condition for (E) to have a solution $\{y_n\}$ such that $\lim_{n\to\infty} y_n = b \neq 0$ is that*

$$\sum_{k=n_0}^{\infty} \sum_{i=m}^{k-1} \frac{(k-1-i)i}{r_i} |f(k, c)| < \infty$$

for some $c \neq 0$.

It is easy to see that this solution of (E) is an $(A2)$-solution. Then there exists $(A2)$-solution of equation (E).

In linear case the following existence theorem holds.

Theorem 2.2 *Assume that $f(n, y_n) = b_n y_n$ in (E), where $b : N \to R$. Then there exists F_+-solution of equation (E).*

Proof See proof of Theorem 2.1 in [11]. □

Theorem 2.3 *If $\{r_n\}$ is a nondecreasing sequence and $\sum_{i=1}^{\infty} |b_i| = \infty$, $b : N \to R_-$, then all solutions of equation (E) are oscillatory.*

Proof See proof of Theorem 2.3 in [11]. □

Hence, from Theorem 2.2 and Theorem 2.3, there exists oscillatory F_+-solution of equation (E).

Existence of nonoscillatory F_+-solution is proved in part 3 of this paper.

Assume that $f : N \times R \to R$. To prove that there exists an F_--solution of equation (E) we can choose initial conditions such that $F(x_2) < 0$. For example, let $x_1 = 0$, $x_2 = r_1$, $x_3 = 3r_1$ and $x_4 = 6r_1$, then $F(x_2) < 0$.

3 Main results

Theorem 3.1 *If $\{y_n\}$ is a nonoscillatory solution of equation (E), then $F(y_n)$ is an eventually decreasing function.*

Proof Let $\{y_n\}$ is an eventually positive solution of equation (E). (For a eventually negative solution the proof is similar.) Then $\{y_n\}$ is $(A4+)$-solution or $(A2+)$-solution, implies $\Delta^2 y_{n-1}\Delta^2 y_{n+1} > 0$. From (1.1) and $(*)$, we have

$$\begin{aligned} \Delta F(y_n) &= y_n \Delta^2(r_n \Delta^2 y_n) - r_{n+1}\Delta^2 y_{n-1}\,\Delta^2 y_{n+1} \\ &= y_n\, f(n, y_n) - r_{n+1}\Delta^2 y_{n-1}\,\Delta^2 y_{n+1} < 0. \end{aligned}$$

Hence the proof is completed. □

Remark 3.2 Assume that $\{y_n\}$ is a positive (negative) solution of equation (E), then $F(y_n)$ is decreasing function on N.

Theorem 3.3 *If $\{y_n\}$ is a nonoscillatory F_+-solution of equation (E) then*

$$\sum_{n=1}^{\infty} r_n(\Delta^{2+j} y_n)^2 < \infty, \tag{3.1}$$

and

$$\lim_{n\to\infty} r_n(\Delta^{2+j} y_n)^2 = 0, \qquad \text{for} \quad j = 0, 1, 2, \ldots. \tag{3.2}$$

Proof We prove (3.1) for $j = 0$, first. Let $\{y_n\}$ is eventually positive solution of equation (E). (For a eventually negative solution the proof is similar.) Then $\{y_n\}$ is $(A4+)$-solution or $(A2+)$-solution. Let $\{y_n\}$ be a $(A4+)$-solution. Then $r_{n+1}\Delta^2 y_{n+1} > r_{n-1}\Delta^2 y_{n-1} > 0$. Hence $r_{n+1}\Delta^2 y_{n-1}\,\Delta^2 y_{n+1} > r_{n-1}(\Delta^2 y_{n-1})^2$, and

$$\sum_{n=1}^{\infty} r_{n+1}\Delta^2 y_{n-1}\,\Delta^2 y_{n+1} > \sum_{n=1}^{\infty} r_{n-1}(\Delta^2 y_{n-1})^2.$$

From Lemma 1.3 we get

$$\sum_{n=1}^{\infty} r_{n-1}(\Delta^2 y_{n-1})^2 < \infty, \quad \text{then} \quad \sum_{n=1}^{\infty} r_n(\Delta^2 y_n)^2 < \infty.$$

If $\{y_n\}$ is $(A2+)$-solution the proof is analogous. □

We will prove (3.1) by induction. Let $\sum\limits_{n=1}^{\infty} r_n(\Delta^{2+j}\, y_n)^2 < \infty$. We will prove that $\sum\limits_{n=1}^{\infty} r_n(\Delta^{2+j+1}\, y_n)^2 < \infty$.

Using inequality $-2ab \le a^2 + b^2$, we have

$$\begin{aligned} (\Delta^{2+j+1} y_n)^2 &= (\Delta^{2+j} y_{n+1})^2 - 2\Delta^{2+j} y_{n+1}\,\Delta^{2+j} y_n + (\Delta^{2+j} y_n)^2 \\ &\le 2(\Delta^{2+j} y_{n+1})^2 + 2(\Delta^{2+j} y_n)^2. \end{aligned}$$

So,

$$\begin{aligned}\sum_{n=1}^{\infty} r_n(\Delta^{2+j+1}y_n)^2 &\le 2\sum_{n=1}^{\infty} r_n(\Delta^{2+j}y_{n+1})^2 + 2\sum_{n=1}^{\infty} r_n(\Delta^{2+j}y_n)^2 \\ &= 2\sum_{n=2}^{\infty} r_n(\Delta^{2+j}y_n)^2 + 2\sum_{n=1}^{\infty} r_n(\Delta^{2+j}y_n)^2 < \infty.\end{aligned}$$

Hence (3.1) holds, for $j = 0, 1, 2, \dots$.

Condition (3.2) follows directly from (3.1).

Remark 3.4 Assume that $\{r_n\}$ is nondecreasing sequence. If $\{y_n\}$ is a nonoscillatory F_+-solution of equation (E) then

$$\sum_{n=1}^{\infty} (\Delta^{2+j}y_n)^2 < \infty,$$

and

$$\lim_{n\to\infty} \Delta^{2+j}y_n = 0, \qquad for \quad j = 0, 1, 2, \dots.$$

Theorem 3.5 *If there exists $\varepsilon > 0$ such that*

$$|f(n,x)| \ge \varepsilon, \qquad for \quad n \in N, \quad x \in R\backslash\{0\}, \tag{3.3}$$

then equation (E) does not have a nonoscillatory F_+-solution.

Proof Suppose that there exists nonoscillatory F_+-solution $\{y_n\}$ of equation (E).

By Theorem 3.3, we have

$$\lim_{n\to\infty} r_n\Delta^2 y_n = 0.$$

Then

$$\lim_{n\to\infty} \Delta^2(r_n\Delta^2 y_n) = 0.$$

Hence, from equation (E), $\lim_{n\to\infty} f(n, y_n) = 0$. So $\lim_{n\to\infty} |f(n,y_n)| = 0$. On the other hand, by (3.3), if there exists $\lim_{n\to\infty} |f(n,y_n)|$, then $\lim_{n\to\infty} |f(n,y_n)| \ge \varepsilon$. The proof is completed by contradiction. □

Solution $\{y_n\}$ of equation (E) is called (A2)-solution if $\{y_n\}$ is (A2+)-solution or (A2-)-solution. Solution $\{y_n\}$ of equation (E) is called (A4)-solution if $\{y_n\}$ is (A4+)-solution or (A4-)-solution.

Theorem 3.6 *Each positive (negative) F_+-solution of equation (E) is (A2)-solution. Each (A2)-solution is F_+-solution.*

Proof Let $\{y_n\}$ be a positive solution. (The proof for a negative solution is similar.)

Suppose to the contrary that $\{y_n\}$ is (A4+)-solution. Then from $\Delta(r_n\Delta^2 y_n) > 0$ we get $r_n\Delta^2 y_n > r_M\Delta^2 y_M > 0$ for $n > M$. So,

$$\lim_{n\to\infty} r_n\Delta^2 y_n \ge r_M\Delta^2 y_M > 0.$$

But, from Theorem 3.3, we have $\lim_{n\to\infty} r_n\Delta^2 y_n = 0$. This contradiction shows that $\{y_n\}$ is (A2+)-solution.

Assume that $\{y_n\}$ is (A2+)-solution. Then we have: $y_n > 0$, $\Delta(r_n\Delta^2 y_n) > 0$, $\Delta y_n > 0$ and $\Delta^2 y_n < 0$, so $F(y_n) > 0$. □

From Theorem 2.1 and Theorem 3.6 we get that there exists nonoscillatory F_+-solution of equation (E).

Theorem 3.7 *Let $\{y_n\}$ be a positive solution of equation (E). If*

$$\sum_{n=1}^{\infty} r_n(\Delta^2 y_n)^2 < \infty, \tag{3.4}$$

then $\{y_n\}$ is F_+-solution.

Proof Let $\{y_n\}$ be a positive solution of equation (E), for which condition (3.1) holds. Suppose that $\{y_n\}$ is F_--solution. Then, for some $m \in N$, we have $F(y_m) < 0$. So, by Remark 3.2, $F(y_n) < F(y_m) < 0$, for $n \geq m$. Since $\{y_n\}$ is positive solution of equation (E) then $\{y_n\}$ is (A4+)-solution or (A2+)-solution. We exclude both of the cases.

For (A4+)-solution we have $\Delta(r_n\Delta^2 y_n) > 0$, then $r_n(\Delta^2 y_n)^2$ is increasing also. Hence $\sum\limits_{n=1}^{\infty} r_n(\Delta^2 y_n)^2 = \infty$.

By Theorem 3.6, (A2+)-solution is F_+-solution. Thus $\{y_n\}$ is a F_+-solution. □

Remark 3.8 Assume that $\{r_n\}$ is nondecreasing sequence. Let $\{y_n\}$ be a positive solution of equation (E). If

$$\sum_{n=1}^{\infty} (\Delta^2 y_n)^2 < \infty,$$

then $\{y_n\}$ is F_+-solution.

Remark 3.9 Let $\{y_n\}$ be a negative solution of equation (E). If condition (3.4) holds, then $\{y_n\}$ is F_+-solution.

Remark 3.10 Let $\{y_n\}$ be a positive (negative) solution of equation (E). The $\{y_n\}$ is F_+-solution if and only if condition

$$\sum_{n=1}^{\infty} r_n(\Delta^2 y_n)^2 < \infty$$

holds.

Remark 3.11 Assume that $\{r_n\}$ is nondecreasing sequence. Let $\{y_n\}$ be a positive (negative) solution of equation (E). The $\{y_n\}$ is F_+-solution if and only if condition

$$\sum_{n=1}^{\infty} (\Delta^2 y_n)^2 < \infty$$

holds.

Theorem 3.12 *If $\{y_n\}$ is a positive (negative) F_--solution of equation (E), then*

$$\lim_{n\to\infty} r_n\Delta y_n = \infty.$$

Proof By Theorem 3.6, the positive F_--solution has to be an (A4+)-solution. (For negative solution the proof is analogous.) Then for positive solution we have

$$r_n\Delta^2 y_n > r_m\Delta^2 y_m > 0, \qquad \text{for} \quad n > m,$$

so by summation we get

$$r_n \Delta y_n \geq r_m \Delta y_m + r_m (n-m) \Delta^2 y_m, \qquad n \geq m.$$

Hence $\lim_{n \to \infty} r_n \Delta y_n = \infty$. □

Remark 3.13 Assume that $\{r_n\}$ is a bounded sequence. Every positive (negative) F_--solution of equation (E) has unbounded first difference.

Remark 3.14 Assume that $\{r_n\}$ is a bounded sequence. Every positive (negative) F_--solution of equation (E) is unbounded.

Proof Suppose that $\{y_n\}$ is F_--solution of equation (E) and $\{y_n\}$ is bounded, then there exists C such that

$$|y_n| \leq C, \qquad \text{for} \quad n \in N.$$

So, $|\Delta y_n| \leq |y_{n+1}| + |y_n| \leq 2C$. It is impossible by Theorem 3.6. □

Remark 3.15 Assume that $\{r_n\}$ is a bounded sequence. Every positive (negative) bounded solution of equation (E) is a F_+-solution.

Remark 3.16 Assume that $\{r_n\}$ is a bounded sequence. Every positive (negative) F_+-solution of equation (E) has a bounded first difference.

Theorem 3.17 *If there exists a positive constant δ such that*

$$x f(n, x) \leq -\frac{\delta}{n}, \qquad for \quad n \in N, \quad x \in R \backslash \{0\}, \tag{3.5}$$

then equation (E) does not have a nonoscillatory F_+-solution.

Proof Suppose that there exists a F_+-solution of equation (E). Let $\{y_n\}$ be such a solution. Then

$$0 \leq F(y_m) = F(y_1) + \sum_{j=1}^{m-1} y_j f(j, y_j) - \sum_{j=1}^{m-1} r_{n+1} \Delta^2 y_{j-1} \Delta^2 y_{j+1}.$$

Hence

$$-\sum_{j=1}^{m-1} y_j f(j, y_j) \leq F(y_1) - \sum_{j=1}^{m-1} r_{n+1} \Delta^2 y_{j-1} \Delta^2 y_{j+1}.$$

From Lemma 1.3

$$-\sum_{j=1}^{\infty} y_j f(j, y_j) \leq F(y_1) - \sum_{j=1}^{\infty} r_{n+1} \Delta^2 y_{j-1} \Delta^2 y_{j+1} < \infty.$$

On the other hand, by (3.5)

$$-y_j f(j, y_j) \geq \frac{\delta}{j}.$$

Then

$$-\sum_{j=1}^{\infty} y_j f(j, y_j) \geq \delta \sum_{j=1}^{\infty} \frac{1}{j} = \infty.$$

The proof is complete by contradiction. □

Theorem 3.18 *If for arbitrary positive constant ε there exists $\delta = \delta(\varepsilon) > 0$, such that*

$$|f(n,x)| > \frac{\delta}{n^2}, \qquad \text{for} \quad n \in N \text{ and } |x| > \varepsilon,$$

then every F_+-solution of equation (E) is oscillatory.

Proof Let $\{y_n\}$ be a nonoscillatory F_+-solution of equation (E). Then, from Theorem 3.3

$$\lim_{n\to\infty} r_n\Delta^2 y_n = \lim_{n\to\infty} \Delta(r_n\Delta^2 y_n) = 0.$$

Suppose that $\{y_n\}$ is an eventually positive solution of equation (E).(For an eventually negative solution the proof is similar.) Then there exists $m \in N$ such that $y_n > 0$, for $n > m$. Summing equation (E), we obtain

$$\Delta(r_n\Delta^2 y_n) - \Delta(r_k\Delta^2 y_k) = \sum_{j=k}^{n-1} f(j, y_j).$$

Then

$$-\Delta(r_k\Delta^2 y_k) = \sum_{j=k}^{\infty} f(j, y_j). \tag{3.6}$$

Therefore series $\sum\limits_{j=k}^{\infty} f(i, y_j)$ converges.

From (3.6) we have

$$-r_n\Delta^2 y_n + r_m\Delta^2 y_m = \sum_{k=m}^{n-1}\sum_{j=k}^{\infty} f(j, y_j). \tag{3.7}$$

Then $r_m\Delta^2 y_m = \sum\limits_{k=m}^{\infty}\sum\limits_{j=k}^{\infty} f(j, y_j)$. Therefore series $\sum\limits_{k=m}^{\infty}\sum\limits_{j=k}^{\infty} f(j, y_j)$ converges. Also

$$\begin{aligned}
\sum_{k=m}^{n-1}\sum_{j=k}^{\infty} f(j, y_j) &= \sum_{k=m}^{n-2}(k+1-m)f(k, y_k) + (n-m)\sum_{k=n+1}^{\infty} f(k, y_k) \\
&\le \sum_{k=m}^{n-2}(k+1-m)f(k, y_k),
\end{aligned}$$

for $n > m$.

From (3.7) we get

$$r_n\Delta^2 y_n - r_m\Delta^2 y_m \ge -\sum_{k=m}^{n-2}(k+1-m)f(k, y_k).$$

So,

$$\begin{aligned}
-r_m\Delta^2 y_m &= \lim_{n\to\infty}\left[\Delta^2 y_n - \Delta^2 y_m\right] \\
&\ge \sum_{k=m}^{\infty}(k+1-m)\left[-f(k, y_k)\right] \\
&\ge \sum_{k=m}^{\infty}(k+1-m)\frac{\delta}{k^2} = \infty.
\end{aligned}$$

Because $-r_m\Delta^2 y_m$ is finite, the obtained contradiction proves our theorem. □

References

[1] R.P. Agarwal, *Difference equations and inequalities. Theory, methods and applications*, Marcel Dekker, Inc. New York 1992.

[2] S.S. Cheng, On a class of fourth order linear recurrence equations, *Internat. J. Math. and Math. Sci*, 7(1) (1984), 131-149.

[3] S.N. Elaydi, *An introduction to difference equation*, Springer-Verlag, New York 1996.

[4] J.R. Graef, E.Thandapani, Oscillatory and asymptotic behaviour of fourth order nonlinear difference equation, *Fasciculi Mathematici*, 31(2001), 23-36.

[5] J.W. Hooker, W.T. Patula, Growth and oscillation properties of solutions of a fourth order linear difference equation, *J. Austral. Math. Soc.* Ser. B 26(1985), 310-328.

[6] W.G. Kelley, A.C. Peterson, *Difference Equations: An Introduction with Applications*, Harcourt/Academic Press, Boston-San Diego, Second Edition, 2001.

[7] V.L. Kocic, G. Ladas, *Global behavior of nonlinear difference equations of higher order with applications*, Kulver Publ. Dordrecht, 1993.

[8] B. Liu, J. Yan, Oscillatory and asymptotic behavior of fourth order nonlinear difference equations, *Acta Mathematica Sinica*, 13(1)(1997), 105-115.

[9] J. Popenda, E. Schmeidel, On the solution of fourth order difference equations, *Rocky Mountain Journal of Mathematics*, 25(4)(1995), 1485-1499.

[10] E. Schmeidel, Oscillation and nonoscillation theorems for fourth order difference equations, *Rocky Mountain Journal of Mathematics*, 33(3)(2003).

[11] E. Schmeidel, Nonoscillation and oscillation properties of fourth order difference equations, *Procedineengs of 7th ICDEA Conference*, (to appear).

[12] E. Schmeidel, B. Szmanda, Oscillatory and asymptotic behavior of certain difference equation, *Nonlinear Analysis*, 47(2001), 4731-4742.

[13] B. Smith, W.E. Taylor, Oscillatory and asymptotic behavior of certain fourth order difference equations, *Rocky Mountain Journal of Mathematics*, 16(2)(1986), 403-406.

[14] B. Smith, W.E. Taylor, Oscillation properties of fourth order linear difference equations, *Tamkang Journal of Mathematics*, 18(4)(1987), 89-95.

[15] W.E. Taylor, Oscillation properties of fourth order difference equations, *Portugaliae Mathematica*, 45(1)(1988), 105-144.

[16] B.G. Zhang, S.S. Cheng, On a class of nonlinear difference equations, *Journal of Difference Equations and Applications*, 1(1995), 391-411.

Fields Institute Communications
Volume **42**, 2004

On the Equation $x_{n+1} = cx_n + f(x_n - x_{n-1})$

Hassan Sedaghat
Department of Mathematics
Virginia Commonwealth University
Richmond, VA 23284-2014, U.S.A.
hsedagha@vcu.edu

Abstract. In this note we present two results about the equation in the title. The first concerns oscillatory behavior and the non-existence of period-2 solutions. The second gives a condition for the existence of monotonic solutions.

In this paper, we consider the second order difference equation

$$x_{n+1} = cx_n + f(x_n - x_{n-1}), \quad c \in [0, 1) \tag{0.1}$$

with real initial values x_0, x_{-1}. We assume that f is continuous on the real line. Variants of this equation have appeared in macroeconomic models of the business cycle for over half a century. One of the earliest cases was Samuelson's linear model in [4] in which the function f was linear-affine of type $at + b$. Nonlinear models appeared subsequently when it was noticed that persistent oscillatory behavior (bounded, non-convergent) in linear models is not structurally stable. The model by Hicks in [2] used a piecewise linear form for f that incorporated known economic mechanisms. Goodwin proposed (in a continuous time model in [1]) that the "investment function" f must be a non-decreasing sigmoid, similar to the inverse tangent. More recently, we have seen variations of this sigmoid in the work of Puu [3]. For more details see [5], [6] and [8].

In the preceding models, conditions leading to persistent oscillatory behavior were of particular interest, since data always indicated the existence of such a behavior in the actual business cycle; see [5]- [8]. Stochastic terms could then be added to a viable nonlinear equation for more realistic modeling. Strange and unusual behavior, such as the existence of an unstable, yet globally attracting fixed point is seen to occur for (0.1); see [6] and [8]. Further, in [8] Puu's variation is discussed in detail and it is shown how chaotic behavior may arise from a version of (0.1) in the case where savings are fully consumed in each period. In this note two additional new results about solutions of (0.1) are presented. The first is about the oscillatory behavior of solutions and the second gives a condition that implies the monotonicity of all solutions.

2000 *Mathematics Subject Classification.* Primary 39A10, 39A11; Secondary 39A12.

1 The main results

In general, oscillatory solutions of (0.1) need not oscillate about the equilibrium. For example, if

$$f(t) = \min\{1, |t|\}, \quad c = 0$$

then $f(0) = 0$ so that the origin is the unique fixed point or equilibrium of (0.1). However, (0.1) has a period-3 solution $\{0, 1, 1\}$ which is clearly non-negative and oscillatory. The condition in the following lemma forces all oscillations to occur about the origin.

Lemma 1.1. *If $tf(t) \geq 0$ for all t, then every eventually non-negative and every eventually non-positive solution of* (0.1) *is eventually monotonic.*

Proof Suppose that $\{x_n\}$ is a solution of (0.1) that is eventually non-negative, i.e., there is $k > 0$ such that $x_n \geq 0$ for all $n \geq k$. Either $x_n \geq x_{n-1}$ for all $n > k$ in which case $\{x_n\}$ is eventually monotonic, or there is $n > k$ such that $x_n \leq x_{n-1}$. In the latter case,

$$x_{n+1} = cx_n + f(x_n - x_{n-1}) \leq cx_n \leq x_n$$

so that by induction, $\{x_n\}$ is eventually non-increasing, hence monotonic. The argument for an eventually non-positive solution is similar and omitted. □

Theorem 1.1. *If $tf(t) \geq 0$ for all t, then* (0.1) *has no solutions that are eventually of period two.*

Proof Let $\{x_n\}$ be a solution of (0.1). We claim that if $c > 0$, then for all $k \geq 1$

$$x_k > 0 > x_{k+1} \Rightarrow x_{k+2} < 0 \quad \text{and} \quad x_k < 0 < x_{k+1} \Rightarrow x_{k+2} > 0.$$

For suppose that $x_k > 0 > x_{k+1}$ for some $k \geq 1$. Then

$$x_{k+2} = cx_{k+1} + f(x_{k+1} - x_k) \leq cx_{k+1} < 0.$$

The argument for the other case is similar and omitted. Now by Lemma 1, if a solution $\{x_n\}$ eventually has period 2, then for all sufficiently large n, there is $x_n > 0$, $x_{n+1} < 0$ and $x_{n+2} = x_n > 0$. If $c > 0$, then this contradicts the above claim. If $c = 0$ then

$$0 < x_n = x_{n+2} = f(x_{n+1} - x_n) \leq 0$$

which is again a contradiction. Hence, no solution of (0.1) can eventually have period two. □

Theorem 1.2. *Assume that $tf(t) \geq 0$ and there is $a \in (0,1)$ such that $|f(t)| \leq a|t|$ for all t. If*

$$c \geq 2\sqrt{a} - a \tag{1.1}$$

then every solution of (0.1) *is eventually monotonic and converges to zero.*

Proof Inequality (1.1) is equivalent to $(a+c)^2 \geq 4a$ so that

$$p = \frac{a + c - \sqrt{(a+c)^2 - 4a}}{2}$$

is a real number (the significance of p becomes clear below). First, assume that

$$0 \leq px_{-1} \leq x_0 \leq x_{-1} \tag{1.2}$$

and let $\{x_n\}$ be the solution generated by the initial values x_{-1}, x_0. Suppose that there is an integer $m \geq 1$ such that $x_m \leq 0$ but $x_{m-1} > 0$. Then $x_{-1} \geq x_0 > \cdots > x_m$ and

$$0 \geq x_m = cx_{m-1} + f(x_{m-1} - x_{m-2}) \geq cx_{m-1} + a(x_{m-1} - x_{m-2})$$

so that

$$x_{m-1} \leq \frac{a}{a+c} x_{m-2}. \tag{1.3}$$

Further,

$$\frac{a}{a+c} x_{m-2} \geq x_{m-1} = cx_{m-2} + f(x_{m-2} - x_{m-3}) \geq cx_{m-2} + a(x_{m-2} - x_{m-3})$$

which yields

$$x_{m-2} \leq \frac{a}{a+c-a/(a+c)} x_{m-3}. \tag{1.4}$$

The development of coefficients in inequalities (1.3) and (1.4) follows a pattern that is described next. Define the mapping

$$\phi(t) = \frac{a}{a+c-t}, \qquad 0 \leq t < a+c.$$

In terms of ϕ, inequalities (1.3) and (1.4) may be written as

$$x_{m-1} \leq \phi(0) x_{m-2}$$

and

$$x_{m-2} \leq \phi^2(0) x_{m-3}$$

respectively. This pattern continues inductively, since if for any $n = 2, \ldots, m$ we have

$$x_{m-n+1} \leq \phi^{n-1}(0) x_{m-n}$$

then

$$\phi^{n-1}(0) x_{m-n} \geq cx_{m-n} + f(x_{m-n} - x_{m-n-1}) \geq cx_{m-n} + a(x_{m-n} - x_{m-n-1})$$

which yields

$$x_{m-n} \leq \frac{a}{a+c-\phi^{n-1}(0)} x_{m-n-1} = \phi^n(0) x_{m-n-1}.$$

In particular, for $n = m$, we obtain

$$x_0 \leq \phi^m(0) x_{-1}.$$

But, upon iterating the mapping ϕ above, it is easy to see that the real sequence $\{\phi^n(0)\}$ is strictly increasing towards p, which is a fixed point of ϕ. Therefore, $\phi^m(0) < p$ and we obtain

$$x_0 \leq \phi^m(0) x_{-1} < p x_{-1}$$

which contradicts (1.2). It follows that $x_n > 0$ for all n if (1.2) holds so by Lemma 1 $\{x_n\}$ is monotonic, and in fact, non-increasing. Since the origin is the only fixed point of (0.1), it follows that $\{x_n\}$ converges to zero.

A similar argument shows that if $x_{-1} \leq x_0 \leq px_{-1} \leq 0$, then $x_n < 0$ for all $n \geq 1$ and thus again the solution is non-decreasing and converges to zero.

In the general case, the solution starting from an arbitrary pair of initial values x_{-1}, x_0 if not monotonic, will have a term x_k which is either a positive maximum or a negative minimum, i.e., $x_k \geq 0$ and $x_{k-1}, x_{k+1} \leq x_k$ or $x_k \leq 0$ and $x_{k-1}, x_{k+1} \geq x_k$. Consider the positive maximum case. Then

$$x_k \geq x_{k+1} = cx_k + f(x_k - x_{k-1}) \geq cx_k. \tag{1.5}$$

Now, since

$$(c-a)^2 = c^2 - 2ac + a^2 > c^2 - 2a + a^2 = (c+a)^2 - 4a$$

it follows that

$$c - p = \frac{c - a - \sqrt{(a+c)^2 - 4a}}{2} > 0.$$

From (1.5) it follows that $x_k \geq x_{k+1} > px_k \geq 0$ at which point the argument for case (1.2) holds and establishes that the solution $\{x_n\}$ is monotonically decreasing for $n > k$. A similar argument for the negative minimum case completes the proof. □

An interesting conjecture that might naturally follow Theorem 2 is that all solutions of (0.1) converge to zero in Theorem 2 *even if* (1.1) *does not hold.* Condition (1.1) may be re-written equivalently as

$$a \leq \left(1 - \sqrt{1-c}\right)^2 = b. \tag{1.6}$$

It may be conjectured that *if $b < a < 1$ then all solutions converge to zero in an oscillatory fashion.* This claim is true for the linear version of (0.1) where $f(t) = at$. In the linear case, the number b is the demarkation line where solutions change from monotonic to oscillatory as a exceeds b. Indeed, if (1.6) holds, then the linear version of (0.1) has two positive eigenvalues the smaller of which is the number p in the proof of Theorem 2. If (1.6) does not hold, then there are two complex eigenvalues whose common modulus is a.

References

[1] Goodwin, R.M. *The Nonlinear Accelerator and The Persistence of Business Cycles*, Econometrica, **19** (1951) 1–17

[2] Hicks, J.R. *A Contribution to the Theory of the Trade Cycle*, (2nd ed.) Clarendon Press, Oxford, 1965 [First edition: Oxford University Press, Oxford, 1950]

[3] Puu, T. *Nonlinear Economic Dynamics*, (3rd ed.) Springer-Verlag, 1993

[4] Samuelson, P.A. *Interaction between the Multiplier Analysis and the Principle of Acceleration*, The Review of Economic Statistics, **21** (1939) 75–78

[5] Sedaghat, H. *A Class of Nonlinear Second Order Difference Equations from Macroeconomics*, Nonlinear Analysis, **29** (1997) 593–603

[6] Sedaghat, H. *Bounded Oscillations in the Hicks Business Cycle Model and Other Delay Equations*, J. Difference Equations and Applications, **4** (1998) 325–341

[7] Sedaghat, H. *Geometric Stability Conditions for Higher Order Difference Equations*, J. Mathematical Analysis and Applications, **224** (1998) 255–272

[8] Sedaghat, H. *Nonlinear Difference Equations: Theory with Applications to Social Science Models*, Kluwer Academic Publishers, Dordrecht, 2003

Fields Institute Communications
Volume **42**, 2004

Stability Criteria for a Kind of Delay Functional Equation

Chuanjun Tian
College of Information Engineering
ShenZhen University
ShenZhen 518060, P. R. China
tiancj@sina.com.cn

Jihong Zhang
College of Information Engineering
ShenZhen University
ShenZhen 518060, P. R. China
zhangjh@szu.edu.cn

Zhen Ji
College of Information Engineering
ShenZhen University
ShenZhen 518060, P. R. China
jizhen@szu.edu.cn

Abstract. This paper is concerned with the delay functional equation of the form

$$u(x, y+t) = a(x,y)u(x+s,y) + b(x,y)u(x,y) + p(x,y)u(x-\sigma, y-\tau),$$

where $s > 0$, $t > 0$, σ and τ are nonnegative constants, $a(x,y)$, $b(x,y)$ and $p(x,y)$, $x, y \in [0,\infty)$, are three real functions. Some sufficient conditions for this equation to be stable and unstable are derived.

1 Introduction

Recently, there have been many investigations into the study of delay difference equations, see for example, [1-8]. In particular, Zhang and Yan [2], Shen [3], Zhang [4-5] discussed the oscillation of the functional equations. In this paper, we consider the stability of the two-variables functional equation of the form

$$u(x, y+t) = a(x,y)u(x+s,y) + b(x,y)u(x,y) + p(x,y)u(x-\sigma, y-\tau), \tag{1.1}$$

where $s > 0$, $t > 0$, σ and τ are nonnegative constants, $a(x,y)$, $b(x,y)$ and $p(x,y)$ are real functions defined on $x \geq 0$ and $y \geq 0$.

By a solution of (1.1) we mean a real function $u(x,y)$ which is defined for $x \geq -\sigma$ and $y \geq -\tau$, and satisfies (1.1) for $x \geq 0$ and $y \geq t$.

2000 *Mathematics Subject Classification.* Primary 39A10; Secondary 34K15.

The first author was supported by NNSF Grant #60172065 and #60374017.

Let h be a real number, $R = (-\infty, \infty)$, $R_h = [h, \infty)$ and $\Omega = R_{-\sigma} \times R_{-\tau} \backslash R_0 \times R_t$. For any function $\varphi = \varphi(x, y)$ defined on Ω, it is easy to construct by induction a function $u(x, y)$ which equals $\varphi(x, y)$ on Ω and satisfies (1.1) on $R_0 \times R_t$. Obviously, the solution of the initial value problem of (1.1) is unique.

For any function $\varphi(x, y)$ on Ω, let $\|\varphi\| = \sup_{(x,y)\in\Omega} |\varphi(x, y)|$. For any positive real number $H > 0$, let $S_H = \{\varphi | \|\varphi\| < H\}$.

Stability and exponential asymptotic stability are defined as follows.

Definition 1.1 Equation (1.1) is said to be stable if there exists a constant $M > 0$ such that every solution of (1.1) with the initial function φ defined on Ω satisfies

$$|u(x, y)| \leq M\|\varphi\|, \quad x, y \in R_0.$$

From the above definition, it is obvious that equation (1.1) is stable implies that for every $\varepsilon > 0$, there exists a $\delta > 0$ such that for every $\varphi \in S_\delta$, the corresponding solution $u = u(x, y)$ of (1.1) satisfies $|u(x, y)| \leq \varepsilon, \quad x, y \in R_0$.

Definition 1.2 Equation (1.1) is said to be exponentially asymptotically stable if for any $\delta > 0$, there exist a positive constant M_δ and a real number $\xi \in (0, 1)$ such that $\varphi \in S_\delta$ implies that

$$|u(x, y)| \leq M_\delta \xi^y, \quad x, y \in R_0, \tag{1.2}$$

where $u(x, y)$ is a solution of (1.1) with the initial function $\varphi(x, y)$.

2 Main results

Theorem 2.1 *Assume that*

$$|a(x, y)| + |b(x, y)| + |p(x, y)| \leq 1 \quad \textit{for all} \quad x, y \in R_0. \tag{2.1}$$

Then equation (1.1) is stable.

Proof For a given solution $u(x, y)$ of equation (1.1), let

$$w_u(y) = V(u, x, y) = \max_{x \geq 0} |u(x, y)| \quad \textit{for} \quad y \geq 0. \tag{2.2}$$

Obviously, for any $x \geq 0$ and $t \leq y < 2t$, $(x+s, y-t)$, $(x, y-t)$, $(x-\sigma, y-t-\tau) \in \Omega$. Thus, from (1.1), for any $x \in R_0$ and $y \in [t, 2t)$, we get

$$\begin{aligned} |u(x, y)| &\leq |a(x, y-t)||u(x+s, y-t)| + |b(x, y-t)||u(x, y-t)| \\ &\quad + |p(x, y-t)||u(x-\sigma, y-t-\tau)| \\ &\leq (|a(x, y-t)| + |b(x, y-t)| + |p(x, y-t)|) \cdot \|\varphi\| \leq \|\varphi\|. \end{aligned}$$

Hence $|u(x, y)| \leq w_u(y) \leq \|\varphi\|$ for any $x \in R_0$ and $y \in [0, 2t)$. Assume that for some fixed integer $n > 1$,

$$u(x, y) \leq w_u(y) \leq \|\varphi\|, \quad \textit{for any} \quad x \in R_0 \quad \textit{and} \quad y \in [0, nt).$$

Then for any $y \in [nt, (n+1)t)$, $y - t \in [0, nt)$, $y - t - \tau \in [-\tau, nt)$, and

$$\begin{aligned} |u(x, y)| &\leq (|a(x, y-t)| + |b(x, y-t)| + |p(x, y-t)|) \\ &\quad \times \max\{|u(x+s, y-t)|, |u(x, y-t)|, |u(x-\sigma, y-t-\tau)|\} \leq \|\varphi\|. \end{aligned}$$

By induction, $|u(x, y)| \leq w_u(y) \leq \|\varphi\|$ for any $y \geq 0$. Hence (1.1) is stable. □

Example 2.2 Consider the functional equation

$$u(x, y+3) = a(x,y)u(x+2,y) + b(x,y)u(x,y) + p(x,y)u(x-1,y-1), \qquad (2.3)$$

where

$$a(x,y) = -\frac{x+y}{3(x+y+1)}, \quad b(x,y) = -\frac{x^2+y^2}{3(x^2+y^2+1)}, \quad p(x,y) = \frac{1}{3}, \quad (x,y) \in R_0^2.$$

It is easy to see that $|a(x,y)| + |b(x,y)| + |p(x,y)| \leq 1$ for any $(x,y) \in R_0^2$. Hence by Theorem 2.1, (2.3) is stable.

If (2.1) does not hold, then we can obtain the following result.

Let $A(x,y) = |a(x,y)| + |b(x,y)| + |p(x,y)|$ for any $(x,y) \in R_0^2$, and

$$\overline{a}(x,y) = a(x,y)A(x+s, y-t), \quad \overline{b}(x,y) = b(x,y)A(x, y-t).$$

Theorem 2.3 *Assume that there exists a constant $C > 1$ such that*

$$|a(x,y)| + |b(x,y)| + |p(x,y)| \leq C \;\; \textit{for any} \;\; x \in R_0 \;\; \textit{and} \;\; y \in [0,t),$$

$$|\overline{a}(x,y)| + |\overline{b}(x,y)| + |p(x,y)| \leq 1 \;\; \textit{for any} \;\; x \in R_0 \;\; \textit{and} \;\; y \in [t, +\infty). \qquad (2.4)$$

Then equation (1.1) is stable.

Proof For a given solution $u(x,y)$ of (1.1), let $V(u,x,y)$ and $w_u(y)$ be defined in (2.2). From (1.1), for any $x \in R_0$ and $y \in [t, 2t)$, we have $y - t \in [0,t)$ and

$$\begin{aligned} u(x,y) \;\; &\leq \;\; (|a(x,y-t)| + |b(x,y-t)| + |p(x,y-t)|) \\ &\quad \times \max\{|u(x+s,y-t)|, |u(x,y-t)|, |u(x-\sigma, y-t-\tau)|\} \leq C\|\varphi\|. \end{aligned}$$

Hence $w_u(y) \leq C\|\varphi\|$ for $y \in [0, 2t)$. From (1.1), we have

$$\begin{aligned} |u(x,y)| \;\; &\leq \;\; |a(x,y-t)||u(x+s,y-t)| + |b(x,y-t)||u(x,y-t)| \\ &\quad + |p(x,y-t)||u(x-\sigma, y-t-\tau)| \\ &\leq \;\; |a(x,y-t)|(|a(x+s,y-2t)||u(x+2s,y-2t)| \\ &\quad + |b(x+s\; y-2t)||u(x+s,y-2t)| \\ &\quad + |p(x+s,y-2t)||u(x+s-\sigma, y-2t-\tau)|) \\ &\quad + |b(x,y-t)|(|a(x,y-2t)||u(x+s,y-2t)| \\ &\quad + |b(x,y-2t)||u(x,y-2t)| \\ &\quad + |p(x,y-2t)||u(x-\sigma,y-2t-\tau)|) \\ &\quad + |p(x,y-t)||u(x-\sigma,y-t-\tau)|. \end{aligned} \qquad (2.5)$$

Hence, from (2.5), for any $x \in R_0$ and $y \in [2t, 3t)$,

$$|u(x,y)| \leq (|\overline{a}(x,y-t)| + |\overline{b}(x,y-t)| + |p(x,y-t)|) \cdot C\|\varphi\| \leq C\|\varphi\|.$$

Thus $w_u(y) \leq C\|\varphi\|$ for $y \in [0\; 3t)$. Assume that for some fixed integer $n > 2$

$$w_u(y) \leq C\|\varphi\|, \quad y \in [0, nt).$$

Then in view of (2.5), for any $x \in R_0$ and $y \in [nt, (n+1)t)$, we can obtain

$$|u(x,y)| \leq (|\overline{a}(x,y-t)| + |\overline{b}(x,y-t)| + |p(x,y-t)|) \cdot C\|\varphi\| \leq C\|\varphi\|.$$

By induction, $w_u(y) \leq C\|\varphi\|$ for $y \geq 0$. Hence $|u(x,y)| \leq w_u(y) \leq C\|\varphi\|$ for all $(x,y) \in R_0^2$, i.e., equation (1.1) is stable. □

Example 2.4 Consider the functional equation

$$u(x, y+2) = a(x, y)u(x+\pi/2, y) + b(x, y)u(x, y) + p(x, y)u(x-2, y-1), \quad (2.6)$$

where

$$a(x, y) = sin(x), \quad b(x, y) = 0.1 \ \ and \ \ p(x, y) = 0.1 \ \ for \ \ (x, y) \in R_0^2.$$

It is easy to see that $|a(x,y)| + |b(x,y)| + |p(x,y)| = |sin(x)| + 0.2$ for any $(x,y) \in R_0^2$, then by Theorem 2.1, it is difficult to assert that equation (2.6) is stable. But it is obvious that

$$A(x, y) = |sin(x)| + 0.2 \quad for \ any \ \ (x, y) \in R_0^2 \ \ and \ \ y \in [0, t),$$

$$|\overline{a}(x, y+2)| = \frac{|sin(x)|}{5} + \frac{|sin(2x)|}{2} \qquad x \in R_0 \ \ and \ \ y \geq t,$$

$$|\overline{b}(x, y+2)| = \frac{1}{50} + \frac{|sin(x)|}{10} \qquad x \in R_0 \ \ and \ \ y \in (t, \infty).$$

Then for $x \in R_0$ and $y \geq 0$,

$$|\overline{a}(x, y-2)| + |\overline{b}(x, y-2)| + |p(x, y-2)| = \frac{3}{25} + \frac{3|sin(x)|}{10} + \frac{|sin(2x)|}{2} \leq \frac{23}{25}.$$

By Theorem 2.3, we can conclude that equation (2.6) is stable.

Let $\widetilde{A}(x,y) = |a(x,y)| + |b(x,y)| + \xi^{-\tau}|p(x,y)|$ for any $(x,y) \in R_0^2$, and

$$\widetilde{a}(x, y) = a(x, y)\widetilde{A}(x+s, y-t), \quad \widetilde{b}(x, y) = b(x, y)\widetilde{A}(x, y-t). \quad (2.7)$$

Theorem 2.5 *Assume that $\sigma = 0$, $\tau \geq t$, and there exist a constant $C > 1$ and a constant $\xi \in (0,1)$ such that either*

$$|a(x, y)| + |b(x, y)| + \xi^{-\tau}|p(x, y)| \leq \xi^t \quad for \ \ all \ \ x, y \in R_0 \quad (2.8)$$

or

$$|a(x, y)| + |b(x, y)| + |p(x, y)| \leq C \quad for \ \ any \ \ x \in R_0 \ \ and \ \ y \in [0, t),$$

$$|\widetilde{a}(x, y)| + |\widetilde{b}(x, y)| + \xi^{-(\tau-t)}|p(x, y)| \leq \xi^{2t} \quad for \ \ all \ \ x \in R_0 \ \ and \ \ y \geq t. \quad (2.9)$$

Then equation (1.1) is exponentially asymptotically stable.

Proof For a given solution $u(x,y)$ of (1.1), let $V(u,x,y)$ and $w_u(y)$ be defined in (2.2).

If (2.8) holds, then from (1.1), for any $\delta > 0$ and $\varphi \in S_\delta$, there exists a constant $M_\delta \geq C\xi^{-(2t+\tau)}\|\varphi\|$ such that for $x \in R_0$ and $y \in [t, 2t)$,

$$|u(x, y)| \leq (|a(x, y-t)| + |b(x, y-t)| + |p(x, y-t)|)\|\varphi\| \leq M_\delta \xi^y. \quad (2.10)$$

Hence $w_u(y) \leq M_\delta \xi^y$ for any $y \in [0, 2t)$. In general, we can obtain

$$w_u(y) \leq M_\delta \xi^y \ \ for \ \ any \ \ y \in [0, t+\tau).$$

Assume that for any positive integer $n \geq 0$,

$$w_u(y) \leq M_\delta \xi^y \ \ for \ \ any \ \ y \in [0, t+\tau+nt).$$

Then from (1.1), for any $y \in [t+\tau+nt, t+\tau+(n+1)t)$, we have

$$|u(x, y)| \leq (|a(x, y-t)| + |b(x, y-t)| + \xi^{-\tau}|p(x, y-t)|)M_\delta \xi^{y-t} \leq M_\delta \xi^y.$$

By induction, $u(x,y) \leq w_u(y) \leq M_\delta \xi^y$ for any $y \geq 0$. Hence equation (1.1) is exponentially asymptotically stable.

If (2.9) holds, then for any $\delta > 0$ and $\varphi \in S_\delta$, (2.10) holds for any $x \in R_0$ and $y \in [0,t)$, and then $u(x,y) \le w_u(y) \le M_\delta \xi^y$ for $y \in [0,2t)$. From (1.1), for any $y \ge 2t$, we have

$$\begin{aligned} |u(x,y)| \le\ & |a(x,y-t)|(|a(x+s,y-2t)||u(x+2s,y-2t)| \\ & +|b(x+s,y-2t)||u(x+s,y-2t)| \\ & +|p(x+s,y-2t)||u(x+s,y-2t-\tau)|) \\ & +|b(x,y-t)|(|a(x,y-2t)||u(x+s,y-2t)| \\ & +|b(x,y-2t)||u(x,y-2t)| + |p(x,y-2t)||u(x,y-2t-\tau)|) \\ & +|p(x,y-t)||u(x,y-t-\tau)|. \end{aligned} \tag{2.11}$$

Hence, for any $x \in R_0$ and $y \in [2t,3t)$, from (2.11), we obtain

$$|u(x,y)| \le (|\widetilde{a}(x,y-t)| + |\widetilde{b}(x,y-t)| + \xi^{-(\tau-t)}|p(x,y-t)|) \cdot \|\varphi\| \le M_\delta \xi^y.$$

Thus $w_u(y) \le M_\delta \xi^y$ for $y \in [0,3t)$. In general, from (2.11), we have

$$w_u(y) \le M_\delta \xi^y, \quad x \in R_0 \ \ and \ \ y \in [0, 3t+\tau).$$

Assume that for some fixed integer $n \ge 0$,

$$w_u(y) \le M_\delta \xi^y, \ \ for \ \ x \in R_0 \ \ and \ \ y \in [0, 3t+\tau+nt).$$

Then in view of (2.11), for any $x \in R_0$ and $y \in [3t+\tau+nt, 3t+\tau+(n+1)t)$, we can obtain

$$|u(x,y)| \le (|\widetilde{a}(x,y-t)| + |\widetilde{b}(x,y-t)| + \xi^{-(\tau-t)}|p(x,y-t)|) M_\delta \xi^{y-2t} \le M_\delta \xi^y.$$

By induction, $u(x,y) \le M_\delta \xi^y$ for any $x \ge 0$ and $y \ge 0$. □

Let $B(x,y) = |a(x,y)| + |b(x,y)|$ for any $x, y \in R_0$, and

$$\widetilde{A}(x,y) = a(x,y)B(x+s,y-t), \qquad \widetilde{B}(x,y) = b(x,y)B(x,y-t).$$

Theorem 2.6 *Assume that $a(x,y) \ge 0$, $b(x,y) \ge 0$, $p(x,y) \ge 0$ for $x, y \in R_0$, and there exists a positive constant $r > 1$ such that either*

$$a(x,y) + b(x,y) \ge r^{2t} \quad for \ \ all \ \ x, y \in R_0 \tag{2.12}$$

or

$$\begin{aligned} a(x,y) + b(x,y) \ge r^{2t} \quad & for \ \ x \in R_0 \ \ and \ \ y \in [0,t), \\ \widetilde{A}(x,y) + \widetilde{B}(x,y) \ge r^{2t} \quad & for \ \ x \in R_0 \ \ and \ \ y \ge t. \end{aligned} \tag{2.13}$$

Then equation (1.1) is unstable.

Proof If (2.12) holds, we take $\varphi(x,y) = \delta > 0$ for all $(x,y) \in \Omega$, where δ is a positive constant. In view of (1.1), for any $x \in R_0$ and $y \in [t,2t)$, we have

$$u(x,y) = \delta(a(x,y-t) + b(x,y-t) + p(x,y-t)) \ge \delta r^{2t} \ge \delta r^y > 0. \tag{2.14}$$

Assume that for some fixed integer $n > 1$,

$$u(x,y) \ge \delta \cdot r^y > 0 \quad for \ \ x \in R_0 \ \ and \ \ y \in [0, nt).$$

Then from (1.1), for any $x \in R_0$ and $y \in [nt, (n+1)t)$, we have

$$u(x,y) \ge \delta r^{y-t}(a(x,y-t) + b(x,y-t)) \ge \delta r^{y+t} \ge \delta r^y > 0.$$

By induction, we have

$$u(x,y) \ge \delta r^y > 0 \quad for \ \ x, y \in R_0.$$

Obviously, $u(x,y) \to \infty$ as $y \to \infty$ for any constant $\delta > 0$, then equation (1.1) is unstable.

If (2.13) holds, then we also take $\varphi(x,y) = \delta > 0$ for all $(x,y) \in \Omega$. In view of (1.1), for any $x \in R_0$ and $y \in [t, 2t)$, (2.14) holds. From (1.1), for any $x \in R_0$ and $y \geq 2t$, we obtain

$$\begin{aligned} u(x,y) \quad = \quad & a(x,y-t)(a(x+s,y-2t)u(x+2s,y-2t) \\ & +b(x+s,y-2t)u(x+s,y-2t) \\ & +p(x+s,y-2t)u(x+s-\sigma,y-2t-\tau)) \\ & +b(x,y-t)(a(x,y-2t)u(x+s,y-2t)+b(x,y-2t)u(x,y-2t) \\ & +p(x,y-2t)u(x-\sigma,y-2t-\tau)) \\ & +p(x,y-t)u(x-\sigma,y-t-\tau). \end{aligned} \tag{2.15}$$

Hence from (2.13) and (2.15), for any $x \in R_0$ and $y \in [2t, 3t)$, we get

$$u(x,y) \geq \delta(\widetilde{A}(x,y-t) + \widetilde{B}(x,y-t)) \geq \delta r^{2t} \geq \delta r^y > 0.$$

Assume that for some fixed integer $n > 2$,

$$u(x,y) \geq \delta r^y > 0 \quad for \;\; x \in R_0 \;\; and \;\; y \in [0, nt).$$

Then from (1.1), (2.13) and (2.15), for any $x \in R_0$ and $y \in [nt, (n+1)t)$, we have

$$u(x,y) \geq \delta r^{y-2t}(\widetilde{A}(x,y-t) + \widetilde{B}(x,y-t)) \geq \delta r^y > 0.$$

By induction, we have

$$u(x,y) \geq \delta r^y > 0 \quad for \;\; any \;\; x, y \in R_0.$$

Obviously, $u(x,y) \to \infty$ as $y \to \infty$ for any $\delta > 0$, then equation (1.1) is unstable. □

Example 2.7 Consider the functional equation

$$u(x,y+2) = a(x,y)u(x+1,y) + b(x,y)u(x,y) + p(x,y)u(x-2,y-1), \tag{2.16}$$

where

$$a(x,y) = \frac{y^2+1}{y^2+2}, \;\; b(x,y) = \frac{1}{y^2+1} + \frac{1}{5}, \;\; p(x,y) = \frac{1}{y^2+1}, \;\; (x,y) \in R_0^2.$$

It is easy to see that $|a(x,y)| + |b(x,y)| \geq 1.1 = (\sqrt[4]{1.1})^4$ for any $(x,y) \in R_0^2$. Hence by Theorem 2.6, equation (2.16) is unstable.

References

[1] Agarwal,R.P *Difference Equations and Inequalities, Theory, Methods and Applications*, Second Edition, Marcel Dekker, New York, 2000.

[2] Zhang, Y.Z, Yan, J. R. *Oscillation criteria for difference equations with continuous arguments (in Chinese)*, Acta Math. SINICA. **38** (1995), 406-411.

[3] Shen, J.H. *Oscillation and comparison theorems of difference equations with continuous arguments and applications*, Chinese Sci.Bull. **41**(1996), 1441-1444.

[4] Zhang, B.G. *Oscillation criteria for partial difference equations with continuous arguments (in Chinese)*, Acta Math. SINICA . **42**(1999), 487-494.

[5] Zhang, B. G., Zhou, Y. *Oscillation of a kind of two-variables functional equation*, Computers Math. Applic. **42**(2001), 369-378.

[6] Lin, Y. Z., Cheng, S. S. *Stability criteria for two partial difference equations*, Computers Math. Applic. **32**(1996), 87-103.

[7] Zhang, B. G., Tian, C. J. *Stability criteria for a class of linear delay partial difference equations*, Computers Math. Applic. **38**(1999), 37-43.

[8] Xie, S. L., Cheng, S. S. *Stability criteria for parabolic type partial difference equations*, J. Computer. Math. Appl. **75**(1996), 57-66.

Fields Institute Communications
Volume **42**, 2004

Exponential Stability of Delay Difference Equations with Applications to Neural Networks

Lin Wang
Department of Mathematics and Statistics
Memorial University of Newfoundland
St. John's, NF, Canada A1C 5S7
lin@math.mun.ca
Current address: Department of Mathematics and Statistics
McMaster University
Hamilton, ON, Canada L8S 4K1
wanglin@math.mcmaster.ca

Xingfu Zou
Department of Mathematics and Statistics
Memorial University of Newfoundland
St. John's, NF, Canada A1C 5S7
Current address: Department of Applied Mathematics
University of Western Ontario
London, ON, Canada N6A 5B7
xzou@math.mun.ca

Dedicated to Professor Zhicheng Wang on the occasion of his seventieth birthday

Abstract. Global exponential stability for a class of delay difference equations is obtained in this paper via constructing a discrete Liapunov functional. Applications to some discrete-time neural networks with delays show that our results improve some existing ones.

1 Introduction

In this paper, we shall study the exponential stability of the following system

$$x_i(n+1) = a_i(x_i(n)) + \sum_{j=1}^{m} w_{ij} g_j(x_j(n)) + \sum_{j=1}^{m} b_{ij} f_j(\sum_{p=1}^{r_{ij}} k_{ij}(p) x_j(n-p)), \quad (1.1)$$

where $n = 0, 1, \cdots$, r_{ij} are positive integers, $a_i(0) = 0$, $g_i(0) = 0$, and $f_i(0) = 0$ and $i = 1, 2, \cdots, m$. When it comes to neural network models, in (1.1), $m \geq 2$ denotes the number of neurons in the network; x_i describes the *activation* of the

2000 *Mathematics Subject Classification.* Primary 39A10, 39A11; Secondary 92B20.

Research partially supported by NSERC of Canada and by a Petro-Canada Young Innovator Award.

ith neuron; the $m \times m$ connection matrices $W = (w_{ij})$ and $B = (b_{ij})$ tell how the neurons are connected in the network at present states and the past states, respectively; the activation functions $g_j, f_j, j = 1, 2, \cdots, m$ show how the neurons react to one another. It is seen that the discrete kernels k_{ij} in (1.1) show the neurons states depend on some kind of average over periods of past time.

Clearly, if one assumes $B = 0$, $a_i(x) = -a_i x$, for $i = 1, \cdots, m$, then (1.1) reduces to

$$x_i(n+1) = -a_i x_i(n) + \sum_{j=1}^{m} w_{ij} g_j(x_j(n)), \tag{1.2}$$

which represents a discrete-time neural network model whose global asymptotic stability was investigated in [10] via Liapunov function method. The following special form of (1.1)

$$x_i(n+1) = (1 - a_i h) x_i(n) + \sum_{j=1}^{m} h w_{ij} g_j(x_j(n)), \tag{1.3}$$

is the discrete analog of the Hopfield neural network model

$$\dot{x}_i(t) = -a_i x_i(t) + \sum_{j=1}^{m} w_{ij} g_j(x_j), i = 1, \cdots, m. \tag{1.4}$$

In fact, (1.3) can be obtained via Euler method with the uniform step size h. Using a different way, Mohamad and Gopalsamy [13] obtained another discrete version for (1.4), which is described by

$$x_i(n+1) = e^{-a_i h} x_i(n) + \frac{1 - e^{-a_i h}}{a_i} \sum_{j=1}^{m} w_{ij} g_j(x_j(n)), i = 1, 2, \ldots, m \tag{1.5}$$

For the continuous time Hopfield neural network (1.4), there have been a large number of papers addressing its dynamics and applications in literature since it was proposed by Hopfield in 1984 [7]. For example, stability and instability were discussed in [6], [13], [23]; applications to solve optimization problems such as linear and quadratic programming problems, variational inequalities, can be found in [21] and [22]. There have been lots of evidences showing that time delays should be taken into consideration when modelling the neural networks (for details, see the recent book, [19]). With appearance of time delays, much more complicated dynamic behavior may occur in the delayed neural networks. For impact of delays on neural dynamics, we refer to [1], [14] and the references therein. For studies on stability and applications of delayed Hopfield neural networks, we refer to [2], [5], [12], [15], [16], [17], [18] and [20]. However, for discrete-time neural networks, the results on stability and its applications are much less than that of continuous time neural networks (see, [8], [9], [10] and [11]). Especially, for the models with delays, the results are scarce (for references, see, [13]). The aim of this paper is to investigate the global exponential stability (GES) of system (1.1) and apply our results to some discrete-time neural networks. As we shall see, by using Liapunov function method, we can obtain a GES result which improved the known results in the literature.

The rest of the paper is organized as follows. In Section 2 we introduce our basic notations and assumptions. Section 3 is devoted to the globally exponential stability results and the proofs for (1.1). Applications and numerical simulations are given in Section 4.

2 Preliminaries

In this paper, we use the following notations: $N(a) = \{a, a+1, \cdots,\}$; $Z^+ = N(0)$; $N(a,b) = \{a, a+1, \cdots, b-1, b\}$; $\rho(W)$: the spectrum radius of the matrix W; W^T: the transpose of the matrix W and $|W| = (|w_{ij}|)$. The initial conditions associated with (1.1) are of the form

$$x_i(s) = \phi_i(s),\ i = N(1,m),\ s \in N(-k,0) \tag{2.1}$$

where $k = \max\{r_{ij}, i, j \in N(1,m)\}$. We are now ready to give some assumptions.

(H_1) For each $i \in N(1,m)$, a_i is globally Lipschitz continuous with

$$\sup_{u,v\in\mathbb{R},u\neq v} \frac{|a_i(u) - a_i(v)|}{|u-v|} \le \alpha_i.$$

(H_2) For each $i \in N(1,m)$, $g_i, f_i : \mathbb{R} \to \mathbb{R}$ are globally Lipschitz continuous with

$$\sup_{u,v\in\mathbb{R},u\neq v} \frac{|g_i(u) - g_i(v)|}{|u-v|} = \beta_i, \quad \sup_{u,v\in\mathbb{R},u\neq v} \frac{|f_i(u) - f_i(v)|}{|u-v|} = l_i.$$

(H_3) The discrete kernels k_{ij} are nonnegative and bounded with

$$\sum_{p=1}^{r_{ij}} k_{ij}(p) = \kappa_{ij}.$$

3 Exponential stability for (1.1)

The following is the main result.

Theorem 3.1 *Suppose that $(H_1)-(H_3)$ hold. If there exist positive real numbers q_i, $i \in N(1,m)$ such that*

$$\alpha_i + \beta_i \sum_{j=1}^{m} q_j q_i^{-1} |w_{ji}| - l_i \sum_{j=1}^{m} q_j q_i^{-1} |b_{ji}| \kappa_{ji} < 1,\ i \in N(1,m), \tag{3.1}$$

then the trivial solution of (1.1) is globally exponentially stable in the sense that the following inequality holds.

$$\sum_{i=1}^{m} |x_i(n)| \le C \left(\frac{1}{\gamma^*}\right)^n \max_{i\in N(1,m)} \left\{ \sup_{s\in N(-k,0)} |\phi_i(s)| \right\} \tag{3.2}$$

where $C > 0$ and $\gamma^* > 1$ will be specified below.

Proof Combining (H_1) with (H_2), we have

$$|x_i(n+1)| \le \alpha_i |x_i(n)| + \sum_{j=1}^{m} |w_{ij}| \beta_j |x_j(n)| + \sum_{j=1}^{m} \left(|b_{ij}| l_j \sum_{p=1}^{r_{ij}} k_{ij}(p) |x_j(n-p)| \right). \tag{3.3}$$

Let

$$\mu_i(\gamma) := 1 - \alpha_i \gamma - \beta_i \sum_{j=1}^{m} q_j q_i^{-1} \gamma |w_{ji}| - l_i \sum_{j=1}^{m} q_j q_i^{-1} |b_{ji}| \sum_{p=1}^{r_{ji}} k_{ji}(p) \gamma^{p+1} \tag{3.4}$$

for $i \in N(1,m)$. By virtue of (3.1), we see that for each $i \in N(1,m)$, $\mu_i(1) > 0$ and as a function of γ, $\mu_i(\gamma)$ is decreasing, hence there is a $\gamma_i > 1$ such that $\mu_i(\gamma) > 0$ for $\gamma \in (1, \gamma_i]$. Let

$$\gamma^* = \min\{\gamma_i, i \in N(1,m)\}. \tag{3.5}$$

Then, for $i \in N(1,m)$, $\mu_i(\gamma^*) > 0$. Letting

$$|x_i(n)| = q_i^{-1}(\gamma^*)^{-n} y_i(n), \tag{3.6}$$

and using (3.3), we can obtain

$$\begin{aligned} y_i(n+1) \quad &\leq \quad \gamma^*|\alpha_i| y_i(n) + \sum_{j=1}^{m} \gamma^* \beta_j q_i q_j^{-1} |w_{ij}| y_j(n) \\ &\quad + \sum_{j=1}^{m} \left(l_j q_i q_j^{-1} |b_{ij}| \sum_{p=1}^{r_{ij}} (\gamma^*)^{p+1} k_{ij}(p) y_j(n-p) \right). \end{aligned}$$

Define $V(n) = V(y)(n)$ by

$$V(n) = \sum_{i=1}^{m} \left\{ y_i(n) + \sum_{j=1}^{m} |b_{ij}| l_j q_i q_j^{-1} \left[\sum_{p=1}^{r_{ij}} (\gamma^*)^{p+1} k_{ij}(p) \sum_{s=n-p}^{n-1} y_j(s) \right] \right\}. \tag{3.7}$$

We can estimate $\Delta V(n)$ along the solution of (1.1) as below:

$$\begin{aligned} \Delta V(n) \quad &= \quad V(n+1) - V(n) \\ &= \quad \sum_{i=1}^{m} \left\{ \Delta y_i(n) + \Delta \sum_{j=1}^{m} |b_{ij}| l_j q_i q_j^{-1} \left[\sum_{p=1}^{r_{ij}} (\gamma^*)^{p+1} k_{ij}(p) \sum_{s=n-p}^{n-1} y_j(s) \right] \right\} \\ &= \quad \sum_{i=1}^{m} \left\{ y_i(n+1) - y_i(n) + \sum_{j=1}^{m} |b_{ij}| l_j q_i q_j^{-1} \right. \\ &\quad \left. \sum_{p=1}^{r_{ij}} (\gamma^*)^{p+1} k_{ij}(p) (y_j(n) - y_j(n-p)) \right\} \\ &\leq \quad \sum_{i=1}^{m} \left[(\gamma^* \alpha_i - 1) y_i(n) + \sum_{j=1}^{m} \gamma^* \beta_j q_i q_j^{-1} |w_{ij}| y_j(n) \right. \\ &\quad \left. + \sum_{j=1}^{m} l_j q_i q_j^{-1} |b_{ij}| \sum_{p=1}^{r_{ij}} k_{ij}(p) (\gamma^*)^{p+1} y_j(n) \right] \\ &= \quad - \sum_{i=1}^{m} \mu_i(\gamma^*) y_i(n) \\ &\leq \quad - \min_{i \in N(1,m)} \{\mu_i(\gamma^*)\} \sum_{i=1}^{m} y_i(n) \\ &\leq \quad 0. \end{aligned}$$

Therefore, the nonnegative function $V(y)(n)$ is decreasing in n, then we have

$$\begin{aligned}
&V(y)(n)\\
&\quad\le\quad V(y)(0)\\
&\quad=\quad \sum_{i=1}^{m}\left(y_i(0)+\sum_{j=1}^{m}|b_{ij}|l_jq_iq_j^{-1}\sum_{p=1}^{r_{ij}}k_{ij}(p)(\gamma^*)^{p+1}\sum_{s=-p}^{-1}y_j(s)\right)\\
&\quad\le\quad \sum_{i=1}^{m}\left(1+\sum_{j=1}^{m}|b_{ij}|l_jq_iq_j^{-1}\sum_{p=1}^{r_{ij}}k_{ij}(p)(\gamma^*)^{p+1}\right)\max_{i\in N(1,m)}\left\{\sup_{s\in N(-k,0)}y_i(s)\right\}\\
&\quad:=\quad C_0\max_{i\in N(1,m)}\left\{\sup_{s\in N(-k,0)}y_i(s)\right\}.
\end{aligned}$$

It follows from (3.6) that

$$\begin{aligned}
\max_{i\in N(1,m)}\left\{\sup_{s\in N(-k,0)}y_i(s)\right\} &= \max_{i\in N(1,m)}\left\{\sup_{s\in N(-k,0)}|q_i(\gamma^*)^s x_i(s)|\right\}\\
&\le \max_{i\in N(1,m)}\left\{q_i\sup_{s\in N(-k,0)}|\phi_i(s)|\right\}.
\end{aligned}$$

Thus, we have

$$\sum_{i=1}^{m}y_i(n) \quad\le\quad V(n)\le C_0\max_{i\in N(1,m)}\left\{q_i\sup_{s\in N(-k,0)}|\phi_i(s)|\right\}.$$

Put

$$C:=C_0\frac{\max\{q_i, i\in N(1,m)\}}{\min\{q_i, i\in N(1,m)\}}. \tag{3.8}$$

Then, using (3.6), we finally obtain

$$\sum_{i=1}^{m}|x_i(n)|\le C\left(\frac{1}{\gamma^*}\right)^n\max_{i\in N(1,m)}\left\{\sup_{s\in N(-k,0)}|\phi_i(s)|\right\}.$$

This completes the proof. □

If we denote $A=diag(\alpha_1,\alpha_2,\cdots,\alpha_m)$, $W=(w_{ij})$, $\Lambda=diag(\beta_1,\beta_2,\cdots,\beta_m)$, $B=(b_{ij})$, $\hat{B}=(b_{ij}\kappa_{ij})$ and $L=diag(l_1,l_2,\cdots,l_m)$. Then we have

Corollary 3.2 *Assume that* $(H_1)-(H_3)$ *hold. If one of the following condition holds, then (1.1) is exponentially stable.*

(1) *the matrix* $\left(I-(A+|W|\Lambda+|\hat{B}|L)\right)^T$ *is diagonally dominant;*

(2) *the matrix* $\left(I-(A+|W|\Lambda+|\hat{B}|L)\right)$ *is diagonally dominant;*

(3)

$$\rho\left(I-(A+|W|\Lambda+|\hat{B}|L)\right)<1; \tag{3.9}$$

(4)

$$Re\left(\lambda\left(I-(A+|W|\Lambda+|\hat{B}|L)\right)\right)>0; \tag{3.10}$$

(5) *there exist some real positive numbers* $p_i, i \in N(1,m)$ *such that*

$$\alpha_i + \sum_{j=1}^{m} \beta_j |w_{ij}| p_j p_i^{-1} + \sum_{j=1}^{m} |b_{ij}| l_j p_j p_i^{-1} < 1, \quad i \in N(1,m); \tag{3.11}$$

(6)

$$\alpha_i + \sum_{j=1}^{m} \beta_i |w_{ji}| + \sum_{j=1}^{m} |b_{ji}| l_i < 1, \quad i \in N(1,m); \tag{3.12}$$

(7)

$$\alpha_i + \sum_{j=1}^{m} \beta_j |w_{ij}| + \sum_{j=1}^{m} |b_{ij}| l_j < 1, \quad i \in N(1,m). \tag{3.13}$$

Proof Note that (1) $\Longleftrightarrow$ (3.1) $\Longleftrightarrow$ (2) $\Longleftrightarrow$ (5). Moreover, since the matrix $A + |W|\Lambda + |\hat{B}|L$ is a positive matrix, from [4], we know that (2) $\Longleftrightarrow$ (3) $\Longleftrightarrow$ (4). Therefore, the conclusion is true if each of (1) – (5) holds. Let the positive real numbers $q_i, p_i, i \in N(1,m)$ be 1 in (3.1) and (3.11), respectively, we immediately have (3.12) and (3.13). Hence the proof is completed. □

4 Applications to discrete-time neural networks

Incorporating time delays and adding a fixed input $J = (J_1, \cdots, J_m)^T$ from outside of the network into (1.2), we have the following discrete-time neural network model with delays

$$x_i(n+1) = -a_i x_i(n) + \sum_{j=1}^{m} b_{ij} g_j(x_j(n - r_{ij})) + J_i. \tag{4.1}$$

Here $g_j(0), j \in N(1,m)$ are not necessarily be zero. Instead, we assume that

(H_4) for each $i \in N(1,m)$, g_i is bounded with $|g_i| \le M_i$.

Now we first establish an existence result for the equilibrium of system (4.1).

Theorem 4.1 *Assume that* (H_2) *and* (H_4) *hold. If* $a_i \neq -1, i \in N(1,m)$, *then there exists an equilibrium for system (4.1).*

Proof We know that x^* is an equilibrium of (4.1) if and only if $x^* = (x_1^*, \cdots, x_m^*)^T$ is a solution of the following system

$$-a_i x_i + \sum_{j=1}^{n} b_{ij} g_j(x_j) + J_i = x_i, i \in N(1,m). \tag{4.2}$$

From (H_4), we have

$$|\sum_{j=1}^{m} b_{ij} g_j(x_j) + J_i| \le \sum_{j=1}^{m} |b_{ij}| M_j + |J_i| =: P_i.$$

Consider

$$x_i = h_i(x_1, x_2, \cdots, x_m) = \frac{1}{a_i + 1} \left(\sum_{j=1}^{m} b_{ij} g_j(x_j) + J_i \right)$$

for $i \in N(1,m)$. We have

$$|h_i(x_1, x_2, \cdots, x_n)| \le \max\{\frac{P_i}{1+a_i}, -\frac{P_i}{1+a_i}\} =: D_i, \text{ for } i \in N(1,m).$$

It follows that $(h_1, h_2, \cdots, h_m)^T$ maps a bounded set $D := [-D_1, D_1] \times [-D_2, D_2] \times \cdots \times [-D_m, D_m]$ into itself. Then the existence of the equilibrium follows from the Brouwer's fixed point theorem (Theorem 3.2, [3]). The proof is thus completed. □

Let x^* be an equilibrium of (4.1) and substitute $x(n)$ with $u(n) + x^*$ into (4.1) resulting

$$u_i(n+1) = -a_i u_i(n) + \sum_{j=1}^{m} b_{ij} g_j^*(u_j(n - r_{ij})), i \in N(1, m) \tag{4.3}$$

where $g_j^*(u_j(n)) = g_j(u_j(n) + x_j^*) - g_j(x_j^*)$.

Note that (4.3) is a special case of (1.1) with $a_i(x) = -a_i x, W = 0, k_{ij}(p) = 0, p \in N(1, r_{ij} - 1), k_{ij}(r_{ij}) = 1$ and thus $\alpha_i = |a_i|, \quad \kappa_{ij} = 1$ for $i, j \in N(1, m)$. Applying Theorem 3.1 to system (4.3), we have

Theorem 4.2 *Assume that* (H_2) *and* (H_4) *hold. If there exist* $q_i > 0, i \in N(1, m)$ *such that*

$$|a_i| + \sum_{j=1}^{m} \beta_i |b_{ji}| \frac{q_j}{q_i} < 1, \text{ for } i \in N(1, m), \tag{4.4}$$

then, for any input J*, the equilibrium* x^* *of (4.1) is globally exponentially stable in the sense that the following inequality holds:*

$$\sum_{i=1}^{m} |x_i(n) - x_i^*| \le C \left(\frac{1}{\lambda^*}\right)^n \max_{i \in N(1,m)} \left\{ \sup_{s \in N(-k,0)} |\phi_i(s) - x_i^*| \right\}, \tag{4.5}$$

where $C > 0$ *and* $\lambda^* > 1$ *can be similarly given as in Theorem 3.1.*

Proof The proof can be easily completed by applying Theorem 3.1 to (4.3). □

Consequently, applying Corollary 3.2 to (4.3), we have

Corollary 4.3 *Assume that* (H_2) *and* (H_4) *hold. If one of the following condition is satisfied*

(i) *the matrix* $(I - (A + |B|\Lambda))^T$ *or* $(I - (A + |B|\Lambda))$ *is an M-matrix;*

(ii)

$$\rho(A + |B|\Lambda) < 1; \tag{4.6}$$

(iii)

$$Re(\lambda(I - (A + |B|\Lambda))) > 0; \tag{4.7}$$

(iv) *there exist real numbers* $p_i, i \in N(1, m)$ *such that*

$$|a_i| + \sum_{j=1}^{m} \beta_j |b_{ij}| \frac{q_j}{q_i} < 1, \quad i \in N(1, m); \tag{4.8}$$

(v)

$$|a_i| + \sum_{j=1}^{m} |b_{ij}| \beta_j < 1, \quad i \in N(1, m); \tag{4.9}$$

or

$$|a_i| + \sum_{j=1}^{m} |b_{ji}| \beta_i < 1, \quad i \in N(1, m), \tag{4.10}$$

then every solution of (4.1) exponentially converges to its unique equilibrium.

Remark 4.4 Note that Jin and Gupta [9] obtained similar results for the global stability of the corresponding differentiable system without delays. Here we extend their results to not necessarily differentiable delayed difference system and indeed we can obtain global exponential stability.

As in [13], we next consider

$$x_i(n+1) = e^{-a_i h} x_i(n) + \psi_i(h) \sum_{j=1}^{m} b_{ij} f_j(x_j(n)) + \psi_i(h) J_i, \tag{4.11}$$

and

$$x_i(n+1) = e^{-a_i h} x_i(n) + \psi_i(h) \sum_{j=1}^{m} b_{ij} f_j(x_j(n - r_{ij})) + \psi_i(h) J_i, \tag{4.12}$$

with $\psi_i(h) = \frac{1-e^{-a_i h}}{a_i}$, for $i \in N(1,m), n \in N(0)$. Let $x_i(n) = \psi_i(h) y_i(n)$, then (4.11) and (4.12) reduce to

$$y_i(n+1) = e^{-a_i h} y_i(n) + \sum_{j=1}^{m} b_{ij} f_j(\psi_j(h) y_j(n)) + J_i, \tag{4.13}$$

and

$$y_i(n+1) = e^{-a_i h} y_i(n) + \sum_{j=1}^{m} b_{ij} f_j(\psi_j(h) y_j(n - r_{ij})) + J_i. \tag{4.14}$$

Using a similar way as in the above theorem and employing Corollary 4.3, we have

Corollary 4.5 *Assume that $a_i > 0, h > 0, f_i$ is globally Lipschitz continuous and bounded satisfying (H_2) and (H_4) for each $i \in N(1,m)$. If one of the following conditions is satisfied, then every solution of (4.11) (or (4.12)) exponentially converges to its unique equilibrium x^*.*

(I) *there exist real numbers $p_i, i \in N(1,m)$ such that*

$$a_i q_i > \sum_{j=1}^{m} l_j |b_{ij}| q_j, \quad i \in N(1,m); \tag{4.15}$$

or

$$a_i q_i > \sum_{j=1}^{m} l_i |b_{ji}| q_j, \quad i \in N(1,m); \tag{4.16}$$

(II)

$$\rho\left(diag(a_1, a_2, \cdots, a_m) - |B|L\right) < 1; \tag{4.17}$$

(III)

$$a_i > \sum_{j=1}^{m} |b_{ij}| l_j, \quad i \in N(1,m); \tag{4.18}$$

or

$$a_i > \sum_{j=1}^{m} |b_{ji}| l_i, \quad i \in N(1,m). \tag{4.19}$$

Indeed, we have

$$\sum_{i=1}^{m} \frac{|x_i(n) - x_i^*|}{\psi_i(h)} \leq C \left(\frac{1}{\lambda^*}\right)^n \max_{i \in N(1,m)} \left\{ \sup_{s \in N(-k,0)} \frac{|x_i(s) - x_i^*|}{\psi_i h} \right\}. \tag{4.20}$$

Proof The existence of equilibrium x^* of (4.11) ((4.12)) is equivalent to that of (4.13) ((4.14)), which can be similarly proved. Suppose that y^* is an equilibrium of (4.13)((4.14)). The transformation $u_i(n) = y_i(n) - y_i^*, i \in N(1,m)$ leads to

$$u_i(n+1) = e^{-a_i h} u_i(n) + \sum_{j=1}^{m} b_{ij} f_j^*(\psi_j(h) u_j(n)), \tag{4.21}$$

with $f_i^*(u_i(n)) = f_i(u_i(n) + y_i^*) - f_i(y_i^*), i \in N(1,m)$. The proof can be easily completed by employing Corollary 4.3. For example, (4.16) implies

$$((1 - e^{-a_i h})(a_i q_i - l_i \sum_{j=1}^{m} |b_{ji}| q_j) > 0,$$

which gives

$$e^{-a_i h} + \psi_i(h) l_i \sum_{j=1}^{m} |b_{ji}| q_j q_i^{-1} < 1.$$

Hence the conclusion immediately follows from Corollary 4.3 and thus the proof is complete. □

Remark 4.6 Note that condition (4.19) was derived in [13] for the global exponential stability and which is just a special case of (4.16).

Now we are ready to give some examples to demonstrate our results.

Example 4.7 Consider

$$\begin{cases} x_1(n+1) &= a_1 x_1(n) + b_{11} g_1(x_1(n - r_{11})) + b_{12} g_2(x_2(n - r_{12}) + J_1 \\ x_2(n+1) &= a_2 x_2(n) + b_{21} g_1(x_1(n - r_{21})) + a_{22} g_2(x_2(n - r_{22}) + J_2 \end{cases} \tag{4.22}$$

$n \in N(0)$, where g_1 and g_2 are bounded Lipschitz continuous with $\beta_1 = \beta_2 = 1$.

If we let $a_1 = -1/2, a_2 = 1/2$, and $b_{11} = 1/4, b_{12} = 1/8, b_{21} = 1/4, b_{22} = 1/16$. Then it is easy to check that

$$|a_1| + |b_{11}| + |b_{12}| = 7/8 < 1, \quad |a_2| + |b_{21}| + |b_{22}| = 13/16 < 1.$$

Therefore, v) in Corollary 4.3 holds and thus for any given fixed input $J = (J_1, J_2)^T$, (4.22) has a unique equilibrium which is exponentially stable independent of delays r_{ij}, $i, j = 1, 2$. A numeric simulation is given in **Fig. 1**.

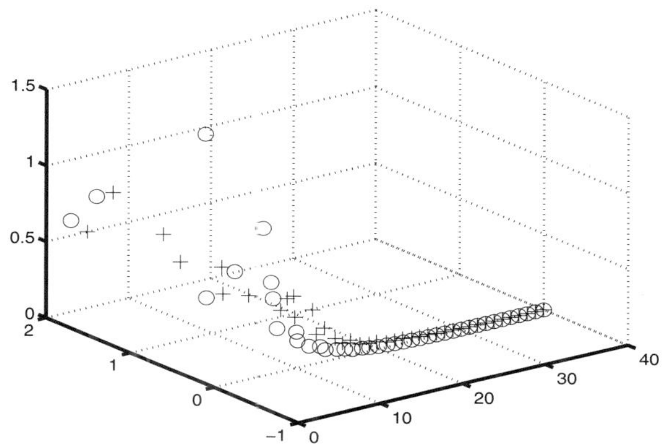

Figure 1 Numeric solutions of (4.22).

If we let $a_1 = a_2 = a, b_{11} = b_{22} = 0, b_{12} = b_{21} = b, r_{12} = r_{21}$ with $a + b < 1$, then Corollary 4.3 shows that the corresponding system has a unique equilibrium to which every solution converges. In this case if $a + b > 1$, the system may admit periodic solutions, for instance, if $a = 1/4$ and $b = 2$, $g_1(x) = g_2(x) = 1/2(|x+1| - |x-1|)$, there always exists a period two solution, say, $\{(1.6, -1.6)^T, (-1.6, 1.6)^T\}$.

Example 4.8 Consider

$$\left.\begin{array}{rcl} x_i(n+1) & = & e^{-a_i}x_i(n) + \psi_i(1)\sum_{j=1}^{3} b_{ij}\tanh(x_j(n-r_{ij}) \\ & & +\psi_i(1)J_i, \quad i=1,2,3, \quad n \in N(0) \end{array}\right\} \tag{4.23}$$

with

$$\left.\begin{array}{lllll} a_1 = 1.0, & b_{11} = 0.2, & b_{12} = -0.4, & b_{13} = 1.4, & l_1 = 1 \\ a_2 = 2.5, & b_{21} = 0.5, & b_{22} = -0.5, & b_{23} = -1.5, & l_2 = 1 \\ a_3 = 3.0, & b_{31} = -0.3, & b_{32} = 0.6, & b_{33} = 0.1, & l_3 = 1 \end{array}\right\} \tag{4.24}$$

In this example, as we can see,

$$a_1 = |b_{11}| + |b_{21}| + |b_{31}| = 1, \quad a_3 = |b_{13}| + |b_{23}| + |b_{33}| = 3.0,$$

which shows that (4.19) does not hold, that is, the main condition in [13] can not be satisfied and the result (Theorem 4.2, [13]) is not applicable. However, if we let $q_1 = 1, q_2 = 0.5$ and $q_3 = 0.8$, then

$$\left.\begin{array}{l} a_1q_1 = 1.0 > \sum_{j=1}^{3}|b_{j1}|q_j = 0.69 \\ a_2q_2 = 1.25 > \sum_{j=1}^{3}|b_{j2}|q_j = 1.13 \\ a_3q_3 = 2.4 > \sum_{j=1}^{3}|b_{j3}|q_j = 2.23 \end{array}\right\} \tag{4.25}$$

which shows that (4.16) holds and thus (4.23) admits a unique equilibrium x^* for any fixed input J and every solution of it exponentially converges to x^* independent of choice of delays $r_{ij}, i, j \in N(1,3)$. The corresponding numeric simulations are given in **Fig. 2**.

Example 4.9 Consider a discrete-time neural network model with distributed delays

$$\left.\begin{array}{rl} x_i(n+1) = & a_ix_i(n) + \sum_{j=1}^{2} b_{ij}tanh\left(\sum_{p=1}^{r_{ij}} k_{ij}(p)x_j(n-p)\right) \\ & i = 1,2, \ n \in N(0) \end{array}\right\} \tag{4.26}$$

where

$$\left.\begin{array}{llll} a_1 = 1/2, & b_{11} = 1/4, & b_{12} = 1/8, & l_1 = 1 \\ a_2 = -1/4, & b_{21} = 1/4, & b_{22} = -1/4, & l_2 = 1 \end{array}\right\} \tag{4.27}$$

$$r_{11} = 2, \quad r_{12} = 4, \quad r_{21} = 5, \quad r_{22} = 6 \tag{4.28}$$

$$\left.\begin{array}{llll} k_{11}(1) = 0.9, & k_{11}(2) = 0.1; & & \\ k_{12}(1) = 0.6, & k_{12}(2) = 0.3, & k_{12}(3) = 0.2, & k_{12}(4) = 0.1; \\ k_{21}(1) = 0.5, & k_{21}(2) = 0.4, & k_{21}(3) = 0.1, & k_{21}(4) = 0.05, \\ k_{21}(5) = 0.05; & & & \\ k_{22}(1) = 0.6, & k_{22}(2) = 0.3, & k_{22}(3) = 0.2, & k_{22}(4) = 0.2, \\ k_{22}(5) = 0.1, & k_{22}(6) = 0.1. & & \end{array}\right\} \tag{4.29}$$

It is easy to check that

$$\kappa_{11} = 1, \ \kappa_{12} = 1.2, \ \kappa_{21} = 1.1, \ \kappa_{22} = 1.5,$$

and

$$a_1 + |b_{11}|\kappa_{11}l_1 + |b_{12}|\kappa_{12}l_2 = 7/8 < 1,$$
$$a_2|b_{21}|\kappa_{21}l_1 + |b_{22}|\kappa_{22}l_2 = 9/10 < 1.$$

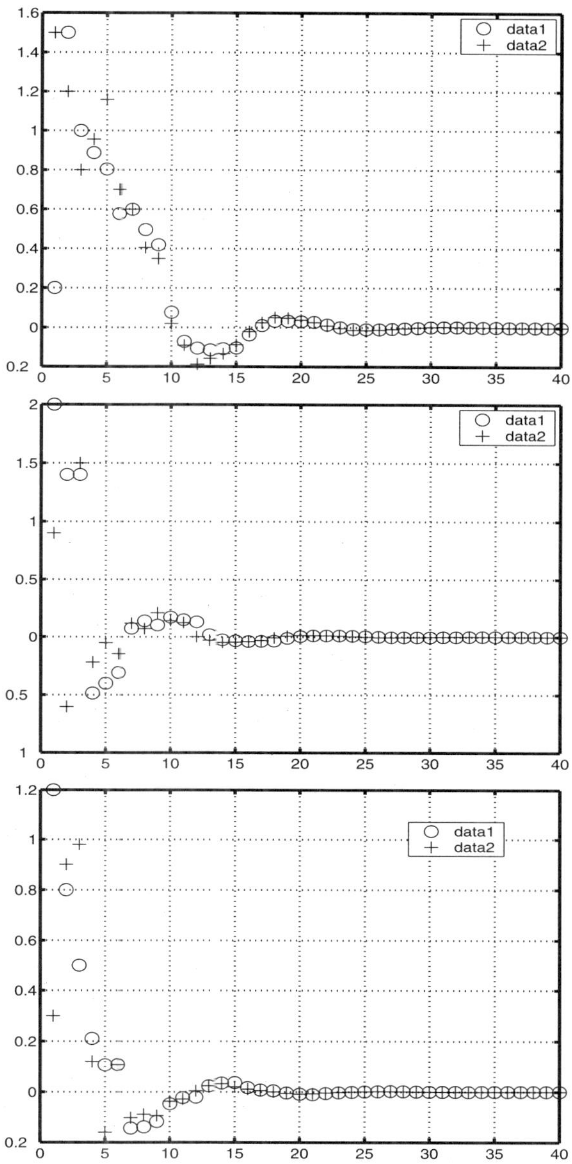

Figure 2 Exponential sta๖ility of discrete-time neural network (4.23)

Thus, (3.11) is satisfied and Corollary 3.2 shows that all solutions of (4.26) exponentially converge to zero. This conclusion is confirmed in our numerical simulation **Fig. 3**.

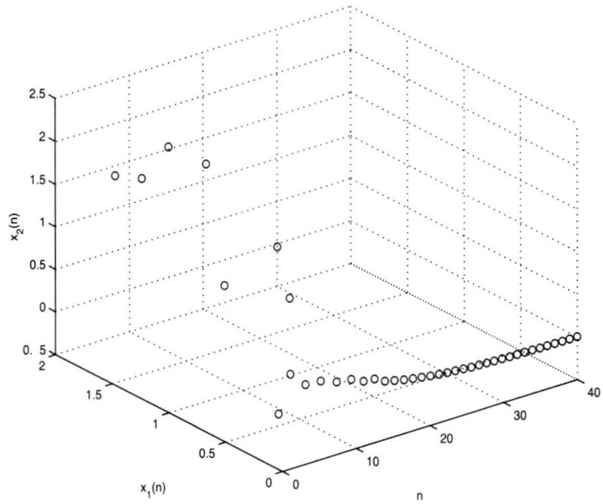

Figure 3 Numeric solutions of (4.26).

References

[1] P. Baldi and A. F. Atiya, *How delays affect Neural Dynamics and Learning* , IEEE Trans. Neural Networks, 5(1994), 612–621.

[2] Y. Chen and S. Fang, Neurocomputing with time delay analysis for solving convex quadratic programming problems, IEEE Trans. Neural Networks. 11(2000), 230–240.

[3] K. Deimling, *Nonlinear functional analysis*, Springer, New York, 1985.

[4] M. Fielder, Special matrices and their applications in numerical mathematics.

[5] K. Gopalsamy and X. He, *Stability in asymmetric Hopfield nets with transmission delays*, Phys. D 76(1994), 344–358.

[6] Z. Guan, G. Chen and Y. Qin, *On equilibria, stability, and instability of Hopfield neural networks*, IEEE Trans. Neural Networks, 2(2000), 534–540.

[7] J. Hopfield, *Neurons with graded response have collective computational properties like those of two-stage neurons*, Proc. Nat. Acad. Sci. U.S.A. 81(1984), 3088–3092.

[8] S. Jagannathan and F. L. Lewis, *Multilayer discrete-time neural-net controller with guaranteed performance*, IEEE Trans. Neural Networks, 7(1996), 107–130.

[9] L. Jin and M. M. Gupta, *Globally asymptotical stability of discrete time analog neural networks, IEEE Trans*, Neural Networks, 7(1996), 1024–1031.

[10] L. Jin, P. N. Nikiforuh and M. M. Gupta, *Absolute stability conditions for discrete-time recurrent neural networks*, IEEE Trans. Neural Networks, 5(1994), 945–955.

[11] D. Liu and A. N. Michel, *Asymptotic stability of discrete-time systems with saturation nonlinearities with applications to digital filters*, IEEE Trans. Circuits Syst., 39(1992), 798–807.

[12] H. Lu, *On stability of nonlinear continuous-time neural networks with delays,* Neural Networks, 13(2000), 1135–1143.

[13] S. Mohamad and K. Gopalsamy, *Dynamics of a class of discrete-time neural networks and their continuous-time counter parts*, Math. Comput. Simulation, 53(2000), 1–39.

[14] K. Pakdaman, C. Grotta-Ragazzo, C. P. Malta, O. Arino and J. -F. Vibert, *Effect of delay on the boundary of the basin of attraction in a system of two neurons*, Neural Networks, 11(1998), 509–519.

[15] K. Pakdaman, C. Grotta-Ragazzo and C. P. Malta, *Transient region duration in continuous-time neural networks with delay*, Physical Review E. 58(1998), 3623–3627.

[16] K. Pakdaman, and C. P. Malta, *A note on convergence under dynamical thresholds with delays, IEEE Trans. Neural Networks*, 9(1998), 231–233.

[17] P. van den Driessche, J. Wu and X. Zou, *Stablization role of inhibitory self-connections in a delayed neural network*, Physica D., 150(2001), 84–90.

[18] P. van den Driessche and X. Zou, *Global attractivity in delayed Hopfield neural network models*, SIAM J. Appl. Math. 58(1998), 1878–1890.
[19] J. Wu, *Introduction to neural dynamics and signal transmission delay,* New York: Walter de Gruyter, 2001.
[20] J. Wu, *Symmetric functional-differential equations and neural networks with memory,* Trans. Amer. Math. Soc. 350(1999), 4799–4838.
[21] Y. Xia,*A new neural network for solving linear and quadratic programming problems*, IEEE Trans. Neural Networks, 7(1996), 1544–1547.
[22] Y. Xia, *A new neural network for solving linear programming problems and its applications*, IEEE Trans. Neural Networks, 7(1996), 525–529.
[23] H. Yang and T. S. Dillon, *Exponential Stability and oscillation of hopfield graded response neural network*, IEEE Trans. Neural Networks, 5(1994), 719–729.

Fields Institute Communications
Volume **42**, 2004

Asymptotic Behavior of Solutions of Linear Difference-Differential Equations with Oscillating Coefficients

Xiaoping Wang
Department of Mathematics
Xiangnan University
Chenzhou, Hunan 423000, P. R. China
wangxp123@hotmail.com

Liusheng Liao
Department of Mathematics
Xiangnan University
Chenzhou, Hunan 423000, P. R. China

Abstract. In this note, we investigate asymptotic behavior of solutions of linear difference-differential equations with oscillating coefficients

$$x'(t) + \sum_{i=1}^{m} p_i(t)x(t-\tau_i) = 0 \qquad (t \geq t_0),$$

where $p_i \in C([t_0, \infty), R), i = 1, 2, \cdots, m;\ 0 \leq \tau_1 \leq \tau_2 \leq \cdots \leq \tau_m$. A new convergence result is established which improves some known results.

Consider the linear difference-differential equation

$$x'(t) + \sum_{i=1}^{m} p_i(t)x(t-\tau_i) = 0, \qquad t \geq t_0, \tag{1}$$

where $p_i \in C([t_0, \infty), R), i = 1, 2, \cdots, m;\ 0 \leq \tau_1 \leq \tau_2 \leq \cdots \leq \tau_m$.

The asymptotic behavior of solutions of (1) has been studied by a number of authors, for example, see [1-15]. However, most of them only consider the case when $p_i(t) \geq 0$. When the coefficients $p_i(t)$ are allowed to oscillate, there are only few papers [2,7,9] dealing with this case. In 1987, [2] proved the following theorem.

2000 *Mathematics Subject Classification.* Primary 34C10, 34K15.
Key words and phrases. Asymptotic behavior, delay, oscillating coefficients.

Theorem 1 *Assume that there are positive constants A, B, Q_1, and Q_2 such that*

$$|p_i(t)| \le A, \quad i = 1, 2, \cdots, m; \quad \sum_{i=1}^{m} p_i(t+\tau_i) \ \ge \ B, \tag{2}$$

$$\sum_{i=1}^{m} \int_{t-\tau_m}^{t-\tau_i} |p_i(s+\tau_i)|ds \ \le \ Q_1, \tag{3}$$

$$\sum_{i=1}^{m} \int_{t-\tau_m}^{t} |p_i(s-\tau_m+\tau_i)|ds \ \le \ Q_2, \tag{4}$$

and

$$2Q_1 + Q_2 < 1. \tag{5}$$

Then every solution of (1) tends to zero as $t \to \infty$.

In 1996, [7] improved the above result by weakening conditions (3) –(5). In fact they obtained the following

Theorem 2 *Assume that there are positive constants A, B, Q_1, Q_2 and $r \in [0, \tau_m]$ such that (2) holds, and that*

$$\sum_{i=1}^{m} \int_{t-\tau_i}^{t} |p_i(s+\tau_i)|ds \ \le \ Q_1, \tag{6}$$

$$\sum_{i=1}^{m} \int_{t-r}^{t-\tau_i} sign(r-\tau_i)|p_i(s+\tau_i)|ds \ \le \ Q_2, \tag{7}$$

and

$$Q_1 + Q_2 < 1. \tag{8}$$

Then every solution of (1) tends zero as $t \to \infty$.

In 2000, [9] proved that the conclusion of Theorem 2 still holds if (2) is replaced by the following weaker condition

$$\sum_{i=1}^{m} p_i(t+\tau_i) > 0, \quad \int_0^{\infty} \sum_{i=1}^{m} p_i(s+\tau_i)ds = \infty. \tag{9}$$

Moreover, [9] obtained the following result on non-oscillatory solutions of (1)

Theorem 3 *Assume that (9) holds, and that*

$$\sum_{\tau_i<r} \int_{t-r}^{t-\tau_i} p_i^-(s+\tau_i)ds + \sum_{\tau_i>r} \int_{t-\tau_i}^{t-r} p_i^+(s+\tau_i)ds \le \alpha < 1, \tag{10}$$

where $r \in [0, \tau_m]$, $p_i^+(s) = (|p_i(s)| + p_i(s))/2, p_i^-(s) = (|p_i(s)| - p_i(s))/2$. Then every nonoscillatory solution of (1) tends zero as $t \to \infty$.

In this note, we eastablish the following new convergence result for (1)

Theorem 4 *Assume that there are positive constants Q_2, Q_3 and $r \in [0, \tau_m]$ such that (7) and (9) hold. If*

$$\sum_{i=1}^{m} \min_{t-\tau_i \le \eta_i \le t} \left[\int_{t-\tau_i}^{\eta_i} |p_i(s+\tau_i)|ds + \int_{\eta_i}^{t} |p_i(s+\tau_i)| \int_s^t \sum_{j=1}^{m} |p_j(v)|dvds \right] \le Q_3 \tag{11}$$

and

$$Q_2 + Q_3 < 1 \tag{12}$$

hold, then every solution of (1) tends to zero as $t \to \infty$.

Proof It is easy to see that (7) and (12) imply (10). Thus, by Theorem 3, every nonoscillatory solution of (1) tends to zero as $t \to \infty$. Therefore, in the sequel, we only consider the case when $x(t)$ is oscillatory. Set

$$y(t) = x(t) - \sum_{i=1}^{m} \int_{t-\tau_i}^{t-r} p_i(s+\tau_i)x(s)ds, \tag{13}$$

then

$$y'(t) + \left[\sum_{i=1}^{m} p_i(t-r+\tau_i)\right] x(t-r) = 0, \qquad t \geq t_0. \tag{14}$$

We first prove that $x(t)$ is bounded. Otherwise, $\mu = \limsup_{t\to\infty} |x(t)| = \infty$. Note that

$$\begin{aligned} |y(t)| &\geq |x(t)| - \sum_{i=1}^{m} \int_{t-r}^{t-\tau_i} \operatorname{sign}(r-\tau_i)|p_i(s+\tau_i))||x(s)|ds \\ &\geq |x(t)| - Q_2 \max_{t_0 \leq s \leq t} |x(s)|, \end{aligned}$$

which implies

$$\max_{t_0 \leq s \leq t} |y(s)| \geq (1-Q_2) \max_{t_0 \leq s \leq t} |x(s)|. \tag{15}$$

It follows that $\limsup_{t\to\infty} y(t) = \infty$. Notice that $x(t)$ is oscillatory, so is $y'(t)$, and hence there exists a sequence $\{t_n\}$ such that

$$|y(t_n)| \to \infty, \quad y'(t_n) = 0, \quad |y(t_n)| > |y(s)|, \quad t_0 \leq s < t_n.$$

Without loss of generality, we can assume that $y(t_n) = |y(t_n)|$. Then by (9) and (14) we know $x(t_n - r) = 0$. Integrating (1) from t to $t_n - r$, we have

$$-x(t) = -\sum_{i=1}^{m} \int_{t}^{t_n-r} p_i(s)x(s-\tau_i)ds.$$

Then

$$|x(t)| \leq \max_{t_0 \leq s \leq t_n} |x(s)| \sum_{i=1}^{m} \int_{t}^{t_n-\tau} |p_i(s)|ds, \quad t_n - r - \tau \leq t \leq t_n - r \tag{16}$$

By (11), there are $\eta_i = \eta_i(t_n - r) \in [t_n - r - \tau_i, t_n - r]$ such that

$$\sum_{i=1}^{m} \left[\int_{t_n-r-\tau_i}^{\eta_i} |p_i(s+\tau_i)|ds + \int_{\eta_i}^{t_n-r} |p_i(t+\tau_i)| \int_{t}^{t_n-r} \sum_{i=1}^{m} |p_j(s)|dsdt \right] \leq Q_3.$$

Thus, integrating (14) again, we have

$$\begin{aligned}
y(t_n) &= y(t_n-r)-\int_{t_n-r}^{t_n}\sum_{i=1}^{m}p_i(s-r+\tau_i)x(s-r)ds\\
&= -\sum_{i=1}^{m}\int_{t_n-r-\tau_i}^{t_n-2r}p_i(s+\tau_i)x(s)ds-\sum_{i=1}^{m}\int_{t_n-2r}^{t_n-r}p_i(s+\tau_i)x(s)ds\\
&= -\sum_{i=1}^{m}\int_{t_n-r-\tau_i}^{t_n-r}p_i(t+\tau_i)x(t)dt\\
&\le \sum_{i=1}^{m}\int_{t_n-r-\tau_i}^{t_n-r}|p_i(t+\tau_i)||x(t)|dt\\
&= \sum_{i=1}^{m}\left[\int_{t_n-r-\tau_i}^{\eta_i}|p_i(t+\tau_i)||x(t)|dt+\int_{\eta_i}^{t_n-r}|p_i(t+\tau_i)||x(t)|dt\right]\\
&\le \max_{t_0\le s\le t_n}|x(s)|\sum_{i=1}^{m}\left[\int_{t_n-r-\tau_i}^{\eta_i}|p_i(s+\tau_i)|ds\right.\\
&\qquad\left.+\int_{\eta_i}^{t_n-r}|p_i(t+\tau_i)|\int_{t}^{t_n-r}\sum_{j=1}^{m}|p_j(s)dsdt\right]\\
&\le Q_3\max_{t_0\le s\le t_n}|x(s)|.
\end{aligned}$$

On the other hand, similar to (15), we have

$$y(t_n)\ge(1-Q_2)\max_{t_0\le s\le t_n}|x(s)|,$$

It follows that

$$(1-Q_2-Q_3)\max_{t_0\le s\le t_n}|x(s)|\le 0,$$

which contradicts (12), and so $\mu<\infty$.

Next we prove that $\mu=0$. Set $\lambda=\limsup_{t\to\infty}|y(t)|$. Then by (13), $\lambda<\infty$. Similar to (15), we can show that

$$\lambda\ge(1-Q_2)\mu. \tag{17}$$

Let $\varepsilon>0$ be any positive number and $T>0$ such that

$$|x(t-\tau_m)|<\mu+\varepsilon,\quad t\ge T.$$

Choose a sequence $\{T_n\}$ such that

$$|y(T_n)|\to\lambda,\quad y'(T_n)=0,\quad n=1,2,\cdots.$$

Without loss of generality, we can assume that $y(T_n)=|y(T_n)|$. Then by (9) and (14) we know $x(T_n-r)=0$. Similar to (16), we have

$$|(x(t)|\le(\mu+\varepsilon)\sum_{i=1}^{m}\int_{t}^{T_n-r}|p_i(s)|ds,\quad T_n-r-\tau_m\le t\le T_n-r. \tag{18}$$

By (11), there are $\eta_i=\eta_i(T_n-r)\in[T_n-r-\tau_i,T_{n-r}]$ such that

$$\sum_{i=1}^{m}\left[\int_{T_n-r-\tau_i}^{\eta_i}|p_i(s+\tau_i)|ds+\int_{\eta_i}^{T_n-r}|p_i(t+\tau_i)|\int_{t}^{T_n-r}\sum_{i=1}^{m}|p_j(s)|dsdt\right]\le Q_3.$$

Thus, integrating (14) again, we have

$$\begin{aligned}
y(T_n) &= y(T_n - r) - \int_{T_n - r}^{T_n} \sum_{i=1}^{m} p_i(s - r + \tau_i)x(s - r)ds \\
&= -\sum_{i=1}^{m} \int_{T_n - r - \tau_i}^{T_n - 2r} p_i(s + \tau_i)x(s)ds - \sum_{i=1}^{m} \int_{T_n - 2r}^{T_n - r} p_i(s + \tau_i)x(s)ds \\
&= -\sum_{i=1}^{m} \int_{T_n - r - \tau_i}^{T_n - r} p_i(t + \tau_i)x(t)dt \\
&\leq \sum_{i=1}^{m} \int_{T_n - r - \tau_i}^{T_n - r} |p_i(t + \tau_i)||x(t)|dt \\
&= \sum_{i=1}^{m} \left[\int_{T_n - r - \tau_i}^{\eta_i} |p_i(t + \tau_i)||x(t)|dt + \int_{\eta_i}^{T_n - r} |p_i(t + \tau_i)||x(t)|dt \right] \\
&\leq (\mu + \varepsilon) \sum_{i=1}^{m} \left[\int_{T_n - r - \tau_i}^{\eta_i} |p_i(s + \tau_i)|ds \right. \\
&\qquad \left. + \int_{\eta_i}^{T_n - r} |p_i(t + \tau_i)| \int_{t}^{T_n - r} \sum_{j=1}^{m} |p_j(s)|dsdt \right] \\
&\leq Q_3(\mu + \varepsilon).
\end{aligned}$$

Let $n \to \infty$ and $\varepsilon \to 0$ in the above, we obtain

$$\lambda \leq Q_3 \mu.$$

Combining this and (17), we have

$$(1 - Q_2 - Q_3)\mu \leq 0.$$

It follows from (12) that $\mu = 0$. The proof is complete. □

Applying Theorem 4 to the delay differential with constant coefficients

$$x'(t) + \sum_{i=1}^{m} p_i x(t - \tau_i) = 0, \quad t \geq t_0, \tag{19}$$

where $p_i \in R$ and τ_i are the same as in (1), then we have the following

Corollary 1 *Assume that there is a $r \in [0, \tau_m]$ such that*

$$\sum_{i=1}^{m} p_i > 0, \quad P = \sum_{i=1}^{m} |p_i| \tag{20}$$

and

$$\sum_{i=1}^{m} |p_i||r - \tau_i| + \sum_{P\tau_i \geq 1} \left(|p_i|\tau_i - \frac{|p_i|}{2P} \right) + \frac{P}{2} \sum_{P\tau_i < 1} |p_i|\tau_i^2 < 1. \tag{21}$$

Then every solution of (19) tends to zero as $t \to \infty$.

Proof First, we establish the following identity:

$$\begin{aligned}&\int_{t-\tau_i}^{\eta_i}|p_i(s+\tau_i)|ds+\int_{\eta_i}^{t}|p_i(s+\tau_i)|\int_s^t\sum_{j=1}^m|p_j(v)|dvds\\ =\;& |p_i|(\eta_i+\tau_i-t)+\frac{1}{2}\left(\sum_{j=1}^m|p_i|\right)|p_i|(t-\eta_i)^2\\ =\;& |p_i|\tau_i-|p_i|(t-\eta_i)+\frac{1}{2}P|p_i|(t-\eta_i)^2\end{aligned}$$

Note that $t-\tau_i\le\eta_i\le t$ implies $0\le t-\eta_i\le\tau_i$. Thus, if $\tau_i\ge P^{-1}$, then

$$\min_{t-\tau_i\le\eta_i\le t}\left[\int_{t-\tau_i}^{\eta_i}|p_i(s+\tau_i)|ds+\int_{\eta_i}^{t}|p_i(s+\tau_i)|\int_s^t\sum_{j=1}^m|p_j(v)|dvds\right]=|p_i|\tau_i-\frac{|p_i|}{2P};$$

If $\tau_i<P^{-1}$, then

$$\min_{t-\tau_i\le\eta_i\le t}\left[\int_{t-\tau_i}^{\eta_i}|p_i(s+\tau_i)|ds+\int_{\eta_i}^{t}|p_i(s+\tau_i)|\int_s^t\sum_{j=1}^m|p_j(v)|dvds\right]=\frac{P}{2}|p_i|\tau_i^2.$$

By Theorem 4, (20) and (21) imply that every solution of (19) tends to zero as $t\to\infty$. The proof is complete. □

If $(\sum_{i=1}^m|p_i|)\tau_1\ge 1$, then (21) reduces to

$$\sum_{i=1}^m|p_i||r-\tau_i|+\sum_{i=1}^m|p_i|\tau_i<\frac{3}{2}.$$

So, we have the following.

Corollary 2 *Assume that there is a* $r\in[0,\tau_m]$ *such that*

$$\sum_{i=1}^m p_i>0,\quad\left(\sum_{i=1}^m|p_i|\right)\tau_1\ge 1,\tag{22}$$

and

$$\sum_{i=1}^m|p_i||r-\tau_i|+\sum_{i=1}^m|p_i|\tau_i<\frac{3}{2}.\tag{23}$$

Then every solution of (19) tends to zero as $t\to\infty$.

Remark 1 For equation (19), condition (8) reduces to

$$\sum_{i=1}^m|p_i||r-\tau_i|+\sum_{i=1}^m|p_i|\tau_i<1.\tag{24}$$

Obviously, (21) is better than (24) and (12) is also better than (8). These show that Theorem 4 improves Theorem 2.

Remark 2 Theorem 4 fits to the case when the coefficients $p_i(t)$ is oscillatory, as long as conditions (7) and (12) hold.

References

[1] G. Ladas, Y. G. sficas and I.P. Stavroulakis, Asymptotic behavior of solutions of retarded differential equations, Proc. Amer. Math. Soc. 88(1983), 247-253.

[2] M. R. S. Kulenovic, G. Ladas and A. Meimaridou, Stability of solutions of linear delay differential equations, Proc. Amer. Math. Soc. 100(1987), 433-441.

[3] G. Ladas and Y. G. Sficas, Asymptotic behavior of oscillatory solutions, Hiroshima Math. J. 18(1988), 351-359.

[4] J. A. Yorke, Asymptotic stability for one dimensional differential delay equations, J. Differential Equations 7(1970), 189-202.

[5] T. Yoneyama, On the 3/2 stability theorem for one-dimensional delay-differential equations, J. Math. Anal. Appl. 125(1987), 161-173.

[6] T. Yoneyama, On the 3/2 stability theorem for one-dimensional delay-differential equations with unbounded delay, J. Math. Anal. Appl. 165(1992), 133-143.

[7] A. Zhao and J. Yan, Asymptotic behavior of solutions of impulsive delay differential equations, J. Math. Anal. Appl. 201(1996), 943-954.

[8] J. W. H. So, J. S. Yu and M. P. Chen, Asymptotic stability for scalar delay differential equations, Funkcial. Ekvac. 39 (1996),1-17.

[9] J. Luo and J. Shen, Asymptotic behavior of forced nonlinear delay differential equations with impulses, Acta Math. Sinica, English Series, 16(2000), 565-572.

[10] T. Krisztin, Stability for functional differential equations and some variational problems, Tohoku Math. J. 42(1990), 407-417.

[11] T. Krisztin, On the stability properties for one-dimensional functional differential equations, Funkcial. Ekvac. 34(1991), 241-256.

[12] K. Gopalsamy, Stability and Oscillations in Delay Differential Equations of Population Dynamics, Kluwer Academic Publishers, Boston, 1992.

[13] Y. Kuang, Delay Differential Equations with Applications in Population Dynamics, Academic Press, Boston, 1993.

[14] I. Győri and G. Ladas, Oscillation Theory of Delay Differential Equations with Applications, clarendon Press, Oxford, 1991.

[15] J. K. Hale, Theory of Functional Differential Equations, Springer-Verlag, New York, 1997.

Fields Institute Communications
Volume **42**, 2004

Analysis of a Discrete Predator-prey Model

Wang Wendi
Department of Mathematics
Southwest Normal University
Chongqing 400715, P. R. China
wendi@swnu.edu.cn

Abstract. A discrete predator-prey model is analyzed one equation of which is not of Lotka-Volterra type. Sufficient conditions are established that ensure the persistence of the prey and the extinction of the predator.

1 The Model

Predator-prey systems are important population models and there have been many researches for the persistence and extinction of populations in these models(see [1, 5, 6, 8, 9]). Papers [4, 10, 11, 12] have studied the persistence and extinction of population models of Lotka-Volterra type. Few papers consider the persistence and extinction of population models that are not of Lotka-Volterra type because it is difficult to construct suitable persistence functions and is difficult to analyze the structures of ω limit sets for such models.

Motivated by [2], we consider the following model

$$\begin{cases} x(n+1) = x(n)e^{r_1 - a_1 x(n) - a_2 y(n)}, \\ y(n+1) = k\,x(n)\,(1 - e^{-a_2 y(n)}) + y(n)\,e^{-b_1 - b_2 y(n)}. \end{cases} \tag{1.1}$$

where $x(n)$ is the number of a prey at generation n, $y(n)$ is the number of a predator at generation n r_1 is the intrinsic growth rate of the prey, a_1 is the density-dependent coefficient of the prey, $1 - e^{-a_2 y_a(n)}$ is the probability that an individual of the prey is eaten by the predator through generation n, the number of prey eaten during the generation is $x(n)(1 - e^{-a_2 y_a(n)})$, k is the conversion coefficient that the predator converts the eaten prey into offsprings of the predator, it is assumed that the reproduction process mainly occurs following generation n and before generation $n-1$, $e^{-b_1 - b_2 y(n)}$ is the probability that an individual of the predator survives through the generation. We assume that all the parameters are positive constants. This model is a modification to the model proposed by

---

2000 *Mathematics Subject Classification.* Primary 92D25; Secondary 39A11.

The author is supported by the National Natural Science Fund of China (No. 10271096).

Gyllenberg, et al.([2]). If we let $b_2 = 0$ and let the reproduction of predator take place during generation, (1.1) reduces to the model in [2].

2 Analysis

Let us begin from equilibria of (1.1). It has equilibria $E_0 = (0,0)$ and $E_1 = (r_1/a_1, 0)$ for all parameters. It is easy to see that E_0 is a saddle. The Jacobian matrix at E_1 is

$$\begin{bmatrix} 1-r_1 & -a_2r_1/a_1 \\ 0 & kr_1a_2/a_1 + e^{-b_1} \end{bmatrix}$$

Hence, we have the following results about the stability of the equilibria.

Theorem 2.1 *E_0 is a saddle point. E_1 is asymptotically stable if $kr_1a_2 < a_1(1-e^{-b_1})$ and is unstable if $kr_1a_2 > a_1(1-e^{-b_1})$.*

Let us now discuss the persistence and the extinction of the populations. Set

$$f_1(y) = \frac{1-e^{-a_2y}}{y}, \qquad y > 0.$$

Lemma 2.2 *For any $y > 0$, we have $f_1(y) < a_2$.*

Proof Note that $f_1(0+) = a_2$ and that

$$f_1'(y) = \frac{a_2\, e^{-a_2\, y} y - 1 + e^{-a_2\, y}}{y^2}.$$

If we define

$$f_2(y) = a_2\, e^{-a_2\, y} y - 1 + e^{-a_2\, y},$$

since $f_2(0) = 0$ and $f_2'(y) = -{a_2}^2 e^{-a_2\, y} y < 0$, we see that $f_1'(y) < 0$, and therefore, $f_1(y)$ is monotonically decreasing. Consequently, we have $f_1(y) < a_2$ for any $y > 0$. □

Theorem 2.3 *Suppose that $kr_1a_2 < a_1(1-e^{-b_1})$. Then there are positive constants ϵ_1 and M_1 such that for any positive solution $(x(n), y(n))$ of (1.1), we have $\lim_{n\to\infty} y(n) = 0$ and*

$$\epsilon_1 \le x(n) \le M_1, \quad \textit{for all large} \quad n. \tag{2.1}$$

Proof First, we show that population x is permanent, i.e., (2.1) is valid. For any positive solution $(x(n), y(n))$ of (1.1), observe that

$$x(n+1) < x(n)\exp(r_1 - a_1x(n)). \tag{2.2}$$

Let us define $f_3(x) = xe^{r-ax}$, where r and a are positive constants. It is easy to see that

$$\max_{0\le x<\infty} f_3(x) = \exp(r-1)/a.$$

It follows from (2.2) that

$$x(n+1) < e^{r_1-1}/a_1 \triangleq M_1. \tag{2.3}$$

Next, we consider the following function

$$f_4(y) = kM_1(1-e^{-a_2y}) + y\, e^{-b_1-b_2y}.$$

Similarly, we have

$$\max_{0\le y<\infty} f_4(y) = kM_1 + \exp(-b_1-1)/b_2 \triangleq M_2. \tag{2.4}$$

Define

$$X = \{(x, y) : x \geq 0, y \geq 0\},$$
$$X_{10} = \{(x, y) \in X : x > 0\},$$
$$\partial X_{10} = X \setminus X_{10}.$$

It then suffices to show that (1.1) is uniformly persistent with respect to $(X_{10}, \partial X_{10})$. First, by the form of (1.1), it is easy to see that both X and X_{10} are positively invariant. Clearly, ∂X_{10} is relatively closed in X. Furthermore, (2.3) and (2.4) imply that system (1.1) is point dissipative. Set

$$M_\partial = \{(x(0), y(0)) : (x(n), y(n)) \quad \text{satisfies (1.1) and } (x(n), y(n)) \in \partial X_{10}, \forall n \geq 0\}.$$

Then it is easy to see that

$$M_\partial = \{(0, y) : y \geq 0\} = \partial X_{10}. \tag{2.5}$$

It is clear that $(0, 0)$ is the unique equilibrium in M_∂. We now claim $W^s((0,0)) \cap X_{10} = \emptyset$. Assume the contrary. Then there exists a solution $(x(n), y(n))$ of system (1.1) with $x(n) > 0$ such that

$$(x(n), y(n)) \to (0, 0) \text{ as } n \to +\infty.$$

Then for large n, by the first equation of (1.1) we have

$$x(n+1) > x(n)e^{r_1/2}.$$

Since $r_1 > 0$, it follows that $x(n) \to \infty$ as $n \to +\infty$. We reach a contradiction.

Note that every orbit in M_∂ tends to $(0, 0)$ as $n \to \infty$. This result, together with the arguments to show $W^s(E_i) \cap X_{10} = \emptyset$, shows that $(0, 0)$ is an isolated invariant set in X. Clearly $(0, 0)$ is acyclic in M_∂. By [7, Theorem 4.6](see also [3, Theorem 4.3 and Remark 4.3] for a stronger repelling property of ∂X_{10}), we conclude that ∂X_{10} repels uniformly the solutions of (1.1) with positive $x(n)$. It follows that there is an $\epsilon_1 > 0$ such that (2.1) holds.

Now, we show that $\lim_{n\to\infty} y(n) = 0$. Let us rewrite the second equation of (1.1) as

$$y(n+1) = y(n)\,[k\,x(n)\,f_1(y(n)) + g_1(y(n))] \triangleq y(n)F_n, \tag{2.6}$$

where

$$f_1(y(n)) = \frac{1 - e^{-a_2 y(n)}}{y(n)}, \qquad g_1(y(n)) = e^{-b_1 - b_2 y(n)}.$$

For any positive integer m, (2.6) implies that

$$\frac{y(n+m)}{y(n)} = F_n F_{n+1} \cdots F_{n+m-1}. \tag{2.7}$$

By Lemma 2.2, we have

$$F_{n+i} < ka_2 x(n+i) + e^{-b_1}.$$

Note that inequality property implies that

$$\begin{aligned} \prod_{i=0}^{m-1} F_{n+i} &\leq \left(\sum_{i=0}^{m-1} \frac{F_{n+i}}{m} \right)^m \\ &\leq \left[e^{-b_1} + ka_2 \sum_{i=0}^{m-1} \frac{x(n+i)}{m} \right]^m. \end{aligned} \tag{2.8}$$

By the first equation of (1.1), we obtain

$$\begin{aligned} a_1 \sum_{i=0}^{m-1} \frac{x(n+i)}{m} &\le a_1 \sum_{i=0}^{m-1} \frac{x(n+i)}{m} + a_2 \sum_{i=0}^{m-1} \frac{y(n+i)}{m} \\ &= r_1 - \frac{1}{m} \ln(\frac{x(n+m)}{x(n)}). \end{aligned} \tag{2.9}$$

Suppose that (2.1) holds for $n \ge N_0$. Then for all $n \ge N_0$, (2.1) implies that we can choose m (independent of n) such that

$$\sum_{i=0}^{m-1} \frac{x(n+i)}{m} < \frac{r_1}{a_1} + \epsilon_0 \tag{2.10}$$

where ϵ_0 is a sufficiently small number such that

$$\eta := e^{-b_1} + ka_2(r_1/a_1 + \epsilon_0) < 1.$$

Then (2.8) and (2.10) lead to

$$\prod_{i=0}^{m-1} F_{n+i} < \left[e^{-b_1} + ka_2(r_1/a_1 + \epsilon_0)\right]^m = \eta^m. \tag{2.11}$$

This inequality and (2.7) imply that $\liminf_{i\to\infty} y(i) = 0$. Now, we show that $\limsup_{i\to\infty} y(i) = 0$. Set

$$A = \max_{1 \le i \le m-1} \left[e^{-b_1} + ka_2 M_1\right]^i.$$

For any $\epsilon > 0$, choose $\delta > 0$ small enough such that $\delta A < \epsilon$. By (2.11) and (2.7), we see that there is a large $n_0 > N_0$ such that $y(n_0) < \delta$. By means of (2.3) and Lemma 2.2 , from the second equation of (1.1) we obtain

$$y(n_0+1) < y(n_0)\left[e^{-b_1} + ka_2 M_1\right] \le A\,\delta < \epsilon.$$

Similarly, we have

$$y(n_0+i) \le y(n_0)\left[e^{-b_1} + ka_2 M_1\right]^i \le A\,\delta < \epsilon, \qquad \text{for } 1 \le i \le m-1. \tag{2.12}$$

Note that (2.11) and (2.7) imply that

$$y(n_0+m) < y(n_0) < \delta.$$

By repeating the above procedures and using induction, we see that

$$y(n_0+i) < \epsilon, \qquad \text{for all } i \ge n_0 + m.$$

This means that $\limsup_{i\to\infty} y(i) = 0$. This means that predator will become extinct. □

References

[1] K. Gopalsamy, *Stability and oscillations in delay differential equations of population dynamics.* Kluwer Academic Publishers Group, Dordrecht, 1992.

[2] M. Gyllenberg; I. Hanski and T. Linsdtrom, *A predator-prey model with optimal suppression of reproduction in the prey.* Math. Biosci. **134** (1996), 119-152.

[3] W. M. Hirsch, H. L. Smith and X.-Q. Zhao, *Chain transitivity, attractivity, and strong repellors for semidynamical systems.* J. Dynam. Differential Equations. **13** (2001), 107-131.

[4] J. Hofbauer; V. Hutson and W. Jansen, *Coexistence for systems governed by difference equations of Lotka-Volterra type.* J. Math. Biol. **25** (1987), 553-570.

[5] Y. Kuang, *Delay differential equations with applications in population dynamics.* Academic Press, Inc. Boston. 1993.

[6] N. Madras;J. Wu and X. Zou, *Local-nonlocal interaction and spatial-temporal patterns in single species population over a patchy environment.* Canadian Applied Mathematics Quarterly. **4** (1996), 109-134.

[7] H. R. Thieme, *Persistence under relaxed point-dissipativity (with application to an endemic model).* SIAM J. Math. Anal. **24** (1993), 407-435.

[8] H. R. Thieme and X. Q. Zhao, *A non-local delayed and diffusive predator-prey model.* Nonlinear Analysis: Real World Applications. **2** (2001), 145-160.

[9] Wang Wendi; Chen Lansun, *A predator-prey system with stage-structure for predator.* Computers Math. Applic. **33** (1997), 83-91.

[10] Wang Wendi and Lu Zhengyi, *Global stability of discrete models of Lotka-Volterra type.* Nonlinear. Analysis. **35** (1999), 1019-1030.

[11] Wang Wendi; G. Mulone; F. Salemi; V. Salone, *Global Stability of Discrete Population Models With Time Delays and Fluctuating Environment.* J. Math. Anal. Appl. **264** (2001), 147-167.

[12] Lu Zhengyi and Wang Wendi, *Permanence and global attractivity for Lotka-Volterra difference systems.* J. Math. Biol. **39** (1999), 269-282.

Fields Institute Communications
Volume **42**, 2004

Oscillation for Nonautonomous Neutral Difference Equations with Variable Coefficients

Hong-Wu Wu
Department of Mathematics
Sun Yat-Sen University
Guangzhou, 510275, P.R. China
wuhongwu007@163.com

Yuan-Tong Xu
Department of Mathematics
Sun Yat-Sen University
Guangzhou, 510275, P. R. China
xyt@zsu.edu.cn

Abstract. This paper is concerned with oscillation of nonautonomous neutral difference equation with variable coefficients of the form

$$\Delta(x_n - c_n x_{n-l_n}) + H(n, x_{n-h_n}) + G(n, x_{n-g_n}) = 0,$$

where $n \in N = \{0, 1, 2, \dots\}$, $l_n \in N\backslash\{0\}$, $c_n \in R$, $h_n, g_n \in N$, $H(n,x), G(n,x) : N \times R \to R$ are continuous in x for each $n \in N$. The delays may be unbounded provided all of $n - l_n$, $n - h_n$, and $n - g_n \to \infty$ as $n \to \infty$. Sufficient conditions are established for the existence of bounded positive solutions and for oscillation of all solutions of this equation.

1 Introduction

The oscillatory behavior of neutral type differential equations has received a great deal of attention in the last few decades. See [1]-[7]. Recently, there has been increasing interest in studying oscillation of difference equations. Among numerous papers dealing with the subject we refer in particular to [8]–[14] and the references cited therein. We note that these equations occur naturally in applications to problems in numerical calculus, automation, economics, biology, etc.

In this paper we study oscillation of nonautonomous neutral difference equation with variable coefficients of the form

$$\Delta(x_n - c_n x_{n-l_n}) + H(n, x_{n-h_n}) + G(n, x_{n-g_n}) = 0, \tag{1.1}$$

2000 *Mathematics Subject Classification.* Primary 39A10; Secondary 34C10.

Project supported by NNSF of China and the Foundation of Sun Yat-Sen University Advanced Research Center.

where $n \in N = \{0, 1, 2, \ldots\}$, $l_n \in N \backslash \{0\}$, $c_n \in R$, $h_n, g_n \in N$, $H, G : N \times R \to R$ and $H(n, \cdot)$ and $G(n, \cdot)$ are continuous for each $n \in N$. The operator Δ is the forward difference operator defined by $\Delta x_n = x_{n+1} - x_n$.

We do not assume that any of the delays l_n, h_n, and g_n is bounded. Rather we assume that

$$n - l_n,\ n - h_n, \text{ and } n - g_n \text{ all} \to \infty \text{ as } n \to \infty. \tag{1.2}$$

Hence for each $n_0 \in N$, $r_{n_0} := \min\{n - l_n,\ n - h_n,\ n - g_n : n \geq n_0\} > -\infty$ and $r_{n_0} \to \infty$ as $n_0 \to \infty$. So we say that the real sequence $\{x_n\}_{n=n_0}^{\infty}$ is a solution to Eq. (1.1) provided $\{x_n\}_{n=m_0}^{\infty}$ satisfies Eq. (1.1) if $r_{m_0} \geq n_0$.

It is easy to see that Eq. (1.2) has a unique solution for any initial value conditions $x_n = a_n$, $r_{n_0} \leq n \leq n_0 - 1$. As is customary, a solution $\{x_n\}_{n=n_0}^{\infty}$ of Eq. (1.1) is said to be an eventually positive solution if $x_n > 0$ when n sufficiently large. A solution of Eq. (1.1) is said to be oscillatory if it is neither eventually positive nor eventually negative; otherwise, it is called nonoscillatory.

The paper is organized as follows. Section 2 deals with the existence of positive solutions of Eq. (1.1). We extend the results in Theorem 1 of [4] to more general equations. In Section 3 we tackle the problem of oscillation of all solutions of Eq. (1.1) and establish, with help of Lemma 1 in [1], new sufficient conditions for oscillation. Also, by some examples, we illustrate that the results of this paper can be applied to cases to which criteria in [11]–[14] fail to apply.

2 Existence of positive solutions

Suppose that there exist nonnegative real sequences $\{\xi_n\}$ and $\{\eta_n\}$ and there is an $n_0 \in N$ such that for all $x, y \in R^+$, $n \geq n_0$ we have

$$\begin{cases} H(n,0) = 0, & |H(n,x) - H(n,y)| \leq \xi_n |x - y|, \\ G(n,0) = 0, & |G(n,x) - G(n,y)| \leq \eta_n |x - y|. \end{cases} \tag{2.1}$$

Theorem 2.1 *Assume $0 < a < b$, $-1 < c \leq 0$ and $a/b < (1+c)/(1-3c)$. Also assume that (1.2) and (2.1) holds and there is a $n_1 \in N$ such that $-1 < c \leq c_n \leq 0$ for all $n \geq n_1 \geq n_0$. If both the series $\sum\limits_{n=0}^{\infty} \xi_n$ and $\sum\limits_{n=0}^{\infty} \eta_n$ are convergent, then Eq. (1.1) has a bounded eventually positive solution.*

Proof Since $\sum\limits_{n=0}^{\infty} \xi_n$ and $\sum\limits_{n=0}^{\infty} \eta_n$ are both convergent, there exists a sufficiently large integer $m \in N$, $m \geq n_1 \geq n_0$, such that

$$\sum_{n=m}^{\infty} \xi_n < \frac{\beta}{b} \qquad \text{and} \qquad \sum_{n=m}^{\infty} \eta_n < \frac{\beta}{b}, \tag{2.2}$$

where $\beta = (b-a)(1-|c|)/8$. For $r = r_m$, let Y denote the Banach space consisting of all real bounded sequences $x = \{x_n\}_{n=m-r}^{\infty}$ with norm defined by $\|x\| = \sup\limits_{n \geq m-r} |x_n|$. Set $X = \{x \in Y : a \leq x_n \leq b \ \forall\ n \geq m - r\}$. Then X is a closed subset of Y. Define an operator $T : X \to Y$ by

$$(Tx)_n = \begin{cases} \alpha + c_n x_{n-l_n} + \sum\limits_{s=n}^{\infty} [H(s, x_{s-h_s}) + G(s, x_{s-g_s})], & n \geq m \\ (Tx)_m, & m - r \leq n \leq m - 1 \end{cases} \tag{2.3}$$

where $\alpha = (b + a)(1 - c)/2$.

We shall show that T is a contraction mapping on X. It is easy to see that T maps X into itself. In fact, for each $x \in X$, $n \geq m$, and $j \in N$ we have

$$\begin{cases} |H(n,x_j)| = |H(n,x_j) - H(n,0)| \leq \xi_n|x_j| \leq \xi_n|x|, \\ |G(n,x_j)| = |G(n,x_j) - G(n,0)| \leq \eta_n|x_j| \leq \eta_{n_n}|x| \end{cases} \tag{2.4}$$

since (2.1) holds and $a > 0$. Using (2.2) it follows that

$$(Tx)_n \geq \alpha + c_n b - \beta - \beta \geq a \quad \text{and} \quad (Tx)_n \leq \alpha + c_n a + \beta + \beta \leq b.$$

Therefore T maps X into itself. To show that T is a contraction mapping on X, let x and $y \in X$. Then

$$\begin{aligned} |(Tx)_n - (Ty)_n| &\leq |c_n|\,|x_{n-l_n} - y_{n-l_n}| + \left|\sum_{s=n}^{\infty}[H(s,x_{s-h_s}) - H(s,y_{s-h_s})]\right| \\ &\quad + \left|\sum_{s=n}^{\infty}[G(s,x_{s-g_s}) - G(s,y_{s-g_s})]\right| \\ &\leq |c_n|\,\|x-y\| + \frac{2\beta}{b}\|x-y\| \\ &\leq (|c| + \frac{2\beta}{b})\|x-y\| \qquad \text{for} \quad n \geq m. \end{aligned}$$

It is easy to verify that $|c| + \frac{2\beta}{b} < 1$. Clearly, the same holds for $m - r \leq n \leq m-1$. Thus T is a contraction mapping on X. By the Contraction Mapping Principle in Banach spaces, there exists a point $x \in Y$ such that $Tx = x$. But then x must be in X and is a bounded positive solution of Eq. (1.1). □

In a manner similar to the proof of Theorem 2.1 just completed, we can prove the following theorem.

Theorem 2.2 *Assume $0 < a < b$, $0 \leq c < 1$ and $a/b < (1+c)/(1-3c)$. Also assume that (1.2) and (2.1) hold and there is a $n_2 \in N$ such that $0 \leq c_n \leq c < 1$, for all $n \geq n_2 \geq n_0$. If both the series $\sum_{n=0}^{\infty} \xi_n$ and $\sum_{n=0}^{\infty} \eta_n$ are convergent, then Eq. (1.1) has a bounded eventually positive solution.*

In the cases that c_n is in $(-\infty, 1) \cup (1, \infty)$ for each n, we use $L(n) := n - l_n$. If

$$L(n) \text{ is monotone,} \tag{2.5}$$

then we have the following two theorems.

Theorem 2.3 *Assume $0 < a < b$, $c' \leq c < -1$ and $a/b < (c+1)/(4c'-3c+1)$. Also assume that (1.2), (2.1) and (2.5) hold and there is an $n_3 \in N$ such that $c' \leq c_n \leq c < -1$ for all $n \geq n_3 \geq n_0$. If both the series $\sum_{n=0}^{\infty} \xi_n$ and $\sum_{n=0}^{\infty} \eta_n$ are convergent, then Eq. (1.1) has a bounded eventually positive solution.*

Proof The idea is the same as Theorem 2.1. But now we let $\beta = (b-a)(|c|-1)/8|c|$ and let m be large enough so that

$$\sum_{n=m}^{\infty} \xi_n < \frac{|c|}{b}\beta \quad \text{and} \quad \sum_{n=m}^{\infty} \eta_n < \frac{|c|}{b}\beta.$$

We define an operator $T : X \to Y$ as follows:

$$(Tx)_n = \begin{cases} \frac{1}{c_{L^{-1}(n)}} \left[\alpha + x_{L^{-1}(n)} - \sum_{s=L^{-1}(n)}^{\infty} [H(s, x_{s-h_s}) + G(s, x_{s-g_s})] \right], & n \geq m \\ (Tx)_m, & m - r \leq n \leq m-1 \end{cases}$$

where $\alpha = (b + a)(c - 1)/2$ and $r = r_m$. Similarly it can be shown that $TX \subseteq X$ and

$$\|Tx - Ty\| \leq \left(\frac{1}{|c|} + \frac{2\beta}{b} \right) \|x - y\|.$$

It is easy to verify that $\frac{1}{|c|} + \frac{2\beta}{b} < 1$. Therefore T is a contraction mapping on X. □

In a manner similar to the proof of Theorem 2.3, we can prove the following theorem.

Theorem 2.4 *Assume $0 < a < b$, $1 < c \leq c'$ and $a/b < (c-1)/(4c' - 3c - 1)$. Also assume that (1.2), (2.1) and (2.5) hold and there is an $n_4 \in N$ such that $1 < c \leq c_n \leq c'$, for all $n \geq n_4 \geq n_0$. If both the series $\sum_{n=0}^{\infty} \xi_n$ and $\sum_{n=0}^{\infty} \eta_n$ are convergent, then Eq. (1.1) has a bounded eventually positive solution.*

3 Oscillation of solutions

Suppose that there exist nonnegative real sequences $\{p_n\}$ and $\{q_n\}$ such that for all $x \in R$ and all $n \in N$

$$\begin{cases} xH(n, x) > 0 \text{ when } x \neq 0 & \text{and} \quad |H(n, x)| \geq p_n |x|^\lambda, \\ xG(n, x) > 0 \text{ when } x \neq 0 & \text{and} \quad |G(n, x)| \geq q_n |x|^\lambda, \end{cases} \tag{3.1}$$

where $0 < \lambda = p/q < 1$ with both p and q odd integers and neither $\{p_n\}$ nor $\{q_n\}$ is eventually identically zero.

For simplicity of notation in the lemmas, theorems, and examples that follow, we use $l(n) := n - l_n$.

Lemma 3.1 *Suppose $c_n > 0$ eventually. Let x_n be an eventually positive solution of Eq. (1.1) and assume that (1.2) and (3.1) hold. Also assume that there is an increasing sequence $\{\sigma_n\}$ of positive integers with $\sigma_n \to \infty$ as $n \to \infty$ and $l(\sigma_{n+1}) = \sigma_n$ for all $n \geq n^* \geq m_0$ so that*

$$\sum_{i=1}^{n} \prod_{j=1}^{i} \frac{1}{c_{\sigma_j}} \to \infty \qquad \textit{as } n \to \infty. \tag{3.2}$$

Then $z_n := x_n - c_n x_{n-l_n}$ satisfies $\Delta z_n \leq 0$ and $z_n > 0$ eventually.

Proof Let Pick $m_0 \in$ N so that $n \geq m_0$ implies $x_n > 0$. From Eq. (1.1) and (3.1), we immediately get

$$\Delta z_n = -H(n, x_{n-h_n}) - G(n, x_{n-g_n}) < 0$$

for $n \geq m_1$ where m_1 is chosen so that $n \geq m_1$ implies $n - h_n \geq m_0$ and $n - g_n \geq m_0$. Hence z_n must have a limit as $n \to \infty$. Since neither $\{p_n\}$ nor $\{q_n\}$ is eventually identically zero, it follows that z_n is eventually of one sign. Suppose

$z_n < 0$ eventually. Then there exists an $m_2 \geq m_1$ and a $\delta < 0$ so that $n \geq m_2$ implies $z_n < \delta < 0$, i.e.,

$$x_n < \delta - c_n x_{l(n)} \quad for \quad n \geq m_2. \tag{3.3}$$

where $l(n) = n - l_n$. Next pick k so that $\sigma_k \geq m_2$. Then

$$x_{\sigma_{k+1}} < \delta + c_{\sigma_{k+1}} x_{l(\sigma_{k+1})} = \delta + c_{\sigma_{k+1}} x_{\sigma_k}.$$

Applying inequality (3.3) inductively for $n = \sigma_k, \sigma_{k+1}, \sigma_{k+2}, \cdots$ gives

$$x_{\sigma_n} < \delta \left(1 + \sum_{i=k+2}^{n} \prod_{j=i}^{n} c_{\sigma_i} \right) + x_{\sigma_k} \prod_{i=k+1}^{n} c_{\sigma_i}. \tag{3.4}$$

Now let

$$P_n = 1 + \sum_{i=k+2}^{n} \prod_{j=i}^{n} c_{\sigma_i}, \quad \Pi_n = \prod_{i=k+1}^{n} c_{\sigma_i}, \quad \text{and} \quad S_n = \sum_{i=1}^{n} \prod_{j=1}^{i} \frac{1}{c_{\sigma_j}}.$$

Then

$$S_n^* := \frac{P_n}{\Pi_n} = \left(S_n - \sum_{i=1}^{k} \prod_{j=1}^{i} \frac{1}{c_{\sigma_j}} \right) \prod_{j=1}^{k} c_{\sigma_j}$$

tends to infinity with n by condition (3.2). Using this notation, (3.4) can be rewritten as

$$x_{\sigma_n} < \delta S_n^* \Pi_n + x_{\sigma_k} \Pi_n = \left(S_n^* + \frac{x_{\sigma_k}}{\delta} \right) \delta \Pi_n.$$

Now pick $j > m_2$ so that $S_j^* + \frac{x_{\sigma_k}}{\delta} > 0$. Then $x_{\sigma_j} < \left(S_j^* + \frac{x_{\sigma_j}}{\delta} \right) \delta \Pi_j < 0$. But that is impossible since $x_{\sigma_n} > 0$ for $\sigma_n > m_0$. Hence we conclude that $z_n > 0$ eventually. □

For the case that Eq. (1.1) has an eventually negative solution, the next lemma can be proved directly following the proof of Lemma 3.1 or the eventually negative solution x_n can be transformed into a positive solution $u_n = -x_n$ with appropriate modification of function H and G so that Lemma 3.1 can be applied.

Lemma 3.2 *Suppose $c_n > 0$ eventually. Let x_n be an eventually negative solution of Eq. (1.1) and assume that (1.2) and (3.1) hold. Also assume that there is an increasing sequence σ_n of positive integers with $\sigma_n \to \infty$ as $n \to \infty$ so that $l(\sigma_{n+1}) = \sigma_n$ for all $n \geq 1$ and*

$$\sum_{i=1}^{n} \prod_{j=1}^{i} \frac{1}{c_{\sigma_j}} \to \infty \qquad as \qquad n \to \infty.$$

Then $z_n := x_n - c_n x_{n-l_n}$ satisfies $\Delta z_n > 0$ and $z_n < 0$ eventually.

Theorem 3.3 *Assume $H(n, x)$ and $G(n, x)$ are nondecreasing in x and satisfy condition (3.1). Suppose that $c_n \geq 0$ for all $n \geq m_0$, that condition (3.2) holds and that*

$$n - h_n - l_n \to \infty \quad and \quad n - g_n - l_n \to \infty \quad as \quad n \to \infty. \tag{3.5}$$

If $\sum_{n=0}^{\infty} \left(c_{n-h_n}^{\lambda} p_n + c_{n-g_n}^{\lambda} q_n \right)$ is a divergent series, then Eq. (1.1) is oscillatory.

Proof Assume that $\{x_n\}$ is an eventually negative solution of Eq. (1.1). Then there exists $m_1 \geq m_0$ such that $x_n < 0$ when $n \geq m_1 - r_m$. By Lemma 3.2, we have

$$\lim_{n\to\infty} z_n = z^* \leq 0. \tag{3.6}$$

Select $m_2 \geq m_1$ such that $z_n < 0$ when $n \geq m_2$. By condition (3.5) there is an $m_3 \geq m_2$ so that $n \geq m_3$ implies $n - h_n - l_n \geq m_2$ and $n - g_n - l_n \geq m_2$. Since $H(n,x)$ and $G(n,x)$ are nondecreasing in x, $c_n \geq 0$, and $z_n = x_n - c_n x_{n-l} \geq x_n$ when $n \geq m_3$, then we have

$$\begin{cases} H(n, z_{n-h_n}) \geq H(n, x_{n-h_n}), \\ G(n, z_{n-g_n}) \geq G(n, x_{n-g_n}). \end{cases} \tag{3.7}$$

Now $z_n \geq z_{n-h_n}$ and $z_n \geq z_{n-g_n}$ when $n \geq m_3$. Since $0 < \lambda = p/q < 1$ with both p and q odd integers, Eq. (1.1) and conditions (3.1) and (3.5) yield

$$\begin{aligned} &\Delta z_n + p_n (c_{n-h_n} z_n)^\lambda + q_n (c_{n-g_n} z_n)^\lambda \\ &\quad \geq \Delta z_n + H(n, c_{n-h_n} z_n) + G(n, c_{n-g_n} z_n) \\ &\quad \geq \Delta z_n + H(n, c_{n-h_n} z_{n-h_n-l_n}) + G(n, c_{n-g_n} z_{n-g_n-l_n}) \\ &\quad \geq \Delta z_n + H(n, c_{n-h_n} x_{n-h_n-l_n}) + G(n, c_{n-g_n} x_{n-g_n-l_n}) \\ &\quad \geq \Delta z_n + H(n, x_{n-h_n}) + G(n, x_{n-g_n}) = 0, \qquad \text{for } n \geq m_3. \end{aligned}$$

Since $z_n < 0$ when $n \geq m_3$ we get

$$\begin{aligned} \left(c^\lambda_{n-h_n} p_n + c^\lambda_{n-g_n} q_n\right) &\leq \frac{-\Delta z_n}{z_n^\lambda} = \frac{z_{n+1} - z_n}{|z_n|^\lambda} \\ &= -\frac{|\mu_n|^\lambda}{|z_n|^\lambda} \int_{z_n}^{z_{n+1}} \frac{1}{t^\lambda} dt \leq -\int_{z_n}^{z_{n+1}} \frac{1}{t^\lambda} dt, \end{aligned} \tag{3.8}$$

where $z_n \leq \mu_n \leq z_{n+1} < 0$. Summing the two sides of (3.8) from m_3 to ∞ and using $\lambda < 1$, we have

$$\begin{aligned} \sum_{n=m_3}^{\infty} \left(c^\lambda_{n-h_n} p_n + c^\lambda_{n-g_n} q_n\right) &\leq -\sum_{n=m_3}^{\infty} \int_{z_n}^{z_{n+1}} \frac{1}{t^\lambda} dt \\ &= -\lim_{n\to\infty} \int_{z_{m_3}}^{z_{n+1}} \frac{1}{t^\lambda} dt \leq -\int_{z_{m_3}}^{0} \frac{1}{t^\lambda} dt < \infty. \end{aligned}$$

But that is impossible since

$$\sum_{n=0}^{\infty} \left(c^\lambda_{n-h_n} p_n + c^\lambda_{n-g_n} q_n\right) = \infty$$

by assumption. So Eq. (1.1) has no eventually negative solution.

We can prove that Eq. (1.1) has no eventually positive solution in a similar fashion by using Lemma 3.1. Thus, Eq. (1.1) is oscillatory. □

Let us consider the following examples to better understand Theorem 3.3. We note that the conclusions do not appear to follow from the known oscillation criteria in the literature [11]–[14] since in those papers c_n must satisfy $0 \leq c_n \leq 1$ or $c_n \geq 1$.

Example 3.4 Consider neutral difference equation of the form

$$\Delta \left(x_n - \left(1 + \frac{\gamma}{n}\right) x_{n-l}\right) + \frac{1}{n} (x_{n-\gamma_1})^{\frac{1}{3}} + \frac{1}{n} (x_{n-\gamma_2})^{\frac{1}{3}} = 0, \tag{3.9}$$

where $0 < \gamma \leq l$, γ_1, $\gamma_2 > 0$.

Here $l(n) := n - l_n = n - l$. Pick $\sigma_n = nl$ for $n \geq 1$. Then the sum (3.2) in Lemma 3.1 is bounded below by $\sum_{i=1}^{n} \frac{1}{i+1}$ and so is divergent. On the other hand,

$$\begin{aligned} \sum_{\gamma_1+\gamma_2}^{\infty} (c_{n-h_n}^{\lambda} p_n + c_{n-g_n}^{\lambda} q_n) &= \sum_{\gamma_1+\gamma_2}^{\infty} \left(\frac{1}{n}\left(1+\frac{1}{n-\gamma_1}\right)^{\frac{1}{3}} + \frac{1}{n}\left(1+\frac{1}{n-\gamma_2}\right)^{\frac{1}{3}} \right) \\ &\geq \sum_{n=\gamma_1+\gamma_2}^{\infty} \left(\frac{1}{n}+\frac{1}{n}\right) = \infty. \end{aligned}$$

The other conditions of Theorem 3.3 also are satisfied. Hence Eq. (3.9) is oscillatory.

Example 3.5 Consider neutral difference equation with variable coefficients of the form

$$\Delta\left(x_n - \left(n^{\gamma/\lambda} + 1 + n^{\gamma/\lambda}\sin\frac{n\pi}{2}\right) x_{n-(8+4\times(-1)^n)}\right) + \frac{1}{n^{\gamma}}(x_{n-4})^{\lambda} + \frac{1}{n^{\gamma}}(x_{n-8})^{\lambda} = 0, \tag{3.10}$$

where $\gamma > 0$, $0 < \lambda = p/q < 1$ and both p and q are odd integers.

Let $n^* = 3$, $\sigma_j = n^* + \sum_{k=1}^{j}(8 + 4\times(-1)^k)$, where $j = 1, 2, \ldots$. Then

$$S_n = \sum_{i=1}^{n}\prod_{j=1}^{i}\frac{1}{c_{\sigma_j}} = \sum_{i=1}^{n} 1 = n \to \infty$$

as $n \to \infty$. On the other hand,

$$\begin{aligned} &\sum_{n=1}^{\infty} (c_{n-h_n}^{\lambda} p_n + c_{n-g_n}^{\lambda} q_n) \\ &= \sum_{n=1}^{\infty} \left(\frac{1}{n^{\gamma}}\left(n^{\gamma/\lambda} + 1 + n^{\gamma/\lambda}\sin\frac{(n-4)\pi}{2}\right)^{\lambda} \right. \\ &\qquad \left. + \frac{1}{n^{\gamma}}\left(n^{\gamma/\lambda} + 1 + n^{\gamma/\lambda}\sin\frac{(n-8)\pi}{2}\right)^{\lambda} \right) \\ &\geq \sum_{n=1}^{\infty} \left(\frac{1}{n^{\gamma}}\left(n^{\gamma/\lambda}\left(1+\sin\frac{n\pi}{2}\right)\right)^{\lambda} + \frac{1}{n^{\gamma}}\left(n^{\gamma/\lambda}\left(1+\sin\frac{n\pi}{2}\right)\right)^{\lambda} \right) \\ &\geq \sum_{n=1}^{\infty} \left(\left(1+\sin\frac{n\pi}{2}\right)^{\lambda} + \left(1+\sin\frac{n\pi}{2}\right)^{\lambda} \right) = \infty. \end{aligned}$$

All conditions of Theorem 3.3 are satisfied and hence Eq. (3.10) is oscillatory.

Using Lemma 3.1 and Lemma 3.2, one can prove that the following theorem in a manner similar to that of Theorem 3.3. Most of the details of the proof are therefore are omitted.

Theorem 3.6 *Suppose that $H(n, x)$ and $G(n, x)$ are nondecreasing in x and satisfy conditions* (3.1) *and* (3.2). *Also suppose that (1.2) hold and there exists $m_0 \in N$ so that $0 \leq c_n \leq 1$ for all $n \geq m_0$. If $\sum_{n=0}^{\infty}(p_n + q_n)$ is a divergent series, then Eq. (1.1) is oscillatory.*

Proof The process is the same as Theorem 3.3. Assume that x_n is an eventually negative solution of Eq. 1.1. From (3.2), (3.8) and Eq. (1.1), we get that

$$\begin{aligned}\Delta z_n &+ p_n z_n^\lambda + q_n z_n^\lambda \\ &\geq \Delta z_n + H(n, z_n) + G(n, z_n) \\ &\geq \Delta z_n + H(n, x_{n-h_n}) + G(n, x_{n-g_n}) = 0 \quad \text{for} \quad n \geq m_3.\end{aligned}$$

Since $z_n < 0$ when $n \geq m_3$, we get

$$p_n + q_n \leq \frac{-\Delta z_n}{z_n^\lambda} = \frac{z_{n+1} - z_n}{|z_n|^\lambda} = -\frac{|\mu_n|^\lambda}{|z_n|^\lambda} \int_{z_n}^{z_{n+1}} \frac{1}{t^\lambda} dt \leq -\int_{z_n}^{z_{n+1}} \frac{1}{t^\lambda} dt,$$

where $z_n \leq \gamma_n \leq z_{n+1} < 0$. The rest of the proof is similar to that of Theorem 3.3 and hence is omitted. □

Acknowledgement

Many thanks to Professor Ronald M. Mathsen for his assistance and encouragement.

References

[1] K.A. Dib and R.M. Mathsen, Oscillation of solutions of neutral delay differential equations, Math. and Computer Modelling., 32 (2000), 609–619.

[2] L.H. Erbe, Q. Kong and B.G. Zhang, Oscillation Theory for Functional Differential Equations, Marcel Dekker, New York, 1995.

[3] E.A. Grove, M.R.S. Kulenovic and G. Ladas, Sufficient conditions for oscillation and nonoscillation of neutral equations, J. Differential Equations, 68 (1987), 673–682.

[4] Y. Sahiner Yilmaz and A. Zafer, Bounded oscillation of nonlinear neutral differential equations of arbitrary order, Czechoslovak Math. J., 51(126) (2001), 185–195.

[5] A. Zafer and R.S. Dahiya, Oscillation of bounded solutions of neutral differential equations, Appl. Math. Lett., 6 (1993), 43–46.

[6] A. Zafer, Oscillation criteria for even order neutral differential equations, Appl. Math. Lett., 11 (1998), 21–25.

[7] Z. Zhou and J.S. Yu, Stability for nonautonomous neutral delay difference equations, Acta Math. Sinica., 42(6) (1999), 1093–1102.

[8] M.P. Chen, B.S. Lalli and J.S. Yu, Oscillation in neutral delay difference equations with variable coefficients, Computers Math. Applic., 29(2) (1995), 5–11.

[9] D.A. Georgiou, E.A. Grove and G. Ladas, Oscillation of neutral delay difference equations, Applic. Anal., 33 (1989), 243–253.

[10] W.G. Kelly and A.C. Peterson, Difference Equations: An Introduction with Applications, New York: Academic Press, 1991.

[11] B.S. Lalli, B.G. Zhang and L.J.Zhao, On the oscillation of solutions and existence of positive solutions of neutral difference equations, J. Math. Anal. Appl.,158 (1991), 213–233.

[12] Q.G. Tang and L. Zeng, Asymptotic behavior and Oscillation for nonlinear neutral delay difference equations of higher order, J. of Math., 20(2) (2000), 207–210. (In Chinese)

[13] Z.C. Wang and J.S. Yu, Oscillation of neutral delay difference equations with summable coefficients, Proceedings of the second international conference on difference equations, Veszpreu Hungary. Gordon and Breach Publishers, Aug. 7∼11(1995).

[14] L. Wang, Oscillation of nonlinear neutral difference equations, Appl. Math., 12(1) (1999), 97–100. (In Chinese)

Fields Institute Communications
Volume **42**, 2004

Oscillation of Higher-Order Nonlinear Delay Difference Equations

Zhou Yinggao
College of Mathematics
Central South University
Changsha, Hunan 410083, P. R. China
xhtang2003@hotmail.com

Abstract. The oscillatory behavior of higher-order nonlinear difference equation of the form

$$\triangle^l x_n + p_n \prod_{i=1}^{m} |x_{n-k_i}|^{\alpha_i} \mathrm{sgn} x_{n-k_i} = 0, n = 0, 1, 2, ... \qquad (*)$$

is investigated. A sufficient condition for oscillation of Eq.(*) is obtained. And two applications are given.

1 Introduction

Consider the following delay difference equation:

$$\triangle^l x_n + p_n \prod_{i=1}^{m} x_{n-k_i}|^{\alpha_i} sgn x_{n-k_i} = 0, \quad n = 0, 1, 2, ... \tag{1.1}$$

Its first-order form is that

$$\triangle x_n + p_n \prod_{i=1}^{m} |x_{n-k_i}|^{\alpha_i} sgn x_{n-k_i} = 0, \quad n = 0, 1, 2, ... \tag{1.2}$$

and the inequality relative to equation (1.2) is that

$$\triangle x_n + p_n \prod_{i=1}^{m} |x_{n-k_i}|^{\alpha_i} sgn x_{n-k_i} \le 0, \quad n = 0, 1, 2, ... \tag{1.3}$$

where $\{p_n\}$ is a sequence of nonnegative numbers, l is odd,and k_i are positive integers, $\alpha_i > 0$, $i = 1, 2, ..., m$; and $\triangle$ is the first-order forward difference operator, $\triangle x_n = x_{n+1} - x_n$, and $\triangle^l x_n = \triangle^{l-1}(\triangle x_n)$ for $l \ge 2$.

By a solution of equation (1.1) or (1.2) or inequality (1.3), we mean a nonconstant real sequence $\{x_n\}$ satisfying equation (1.1) or (1.2) or inequality (1.3) for $n \ge 0$. A solution $\{x_n\}$ is said to be oscillatory if it is neither eventually positive

2000 *Mathematics Subject Classification.* Primary 39A11; Secondary 92D25.
This work was Supported by NNSF of China.

nor eventually negative and nonoscillatory otherwise. An equation is said to be oscillatory if its every solution is oscillatory.

In 2001, Tang [5] obtained the sufficient and necessary conditions for oscillation of equation (1.1) with $l = 1$ and $m = 1$ in two cases where $\alpha < 1$ and $\alpha > 1$ respectively. In 2002, Zhou and Tang [1] studied the oscillatory behavior of equation (1.2) with $\sum_{i=1}^{m} \alpha_i > 1$,and got an oscillatory criterion for equation (1.2), which is stated as the following theorem.

Theorem A (see [1]) *Assume that $\sum_{i=1}^{m} \alpha_i > 1$ and there exists a $\lambda > 0$ such that*

$$\sum_{i=1}^{m} \alpha_i e^{-\lambda k_i} < 1 \tag{1.4}$$

and

$$\liminf_{n\to\infty}[p_n exp(-e^{\lambda n})] > 0 \tag{1.5}$$

then, every solution of (1.2) oscillates.

However,to the best of our knowledge,there is hardly any results on oscillation for (1.1) with $l > 1$. In this paper,we will discuss the oscillation for (1.1) and establish an oscillatory criterion for equation (1.1) with $\sum_{i=1}^{m} \alpha_i > 1$ and give their applications to the following nonlinear delay difference equations

$$\triangle^l x_n + p_n f(x_{n-k_1}, ..., x_{n-k_m}) = 0, \quad n = 01, 2, ... \tag{1.6}$$

and

$$\triangle^l x_n + \sum_{i=1}^{m} p_i(n) x_{n-k_i}^{\beta_i} = 0, n = 0, 1, 2, ... \tag{1.7}$$

where p_n, k_i and l are the same as (1.1),$\beta_i > 0$, $i = 1, ..., m$ are rational numbers with denominator being positive odd integers and

(C) $f \in (R^m, R)$, $f(x_1, ..., x_m)$ is nondecreasing on each x_i, $i = 1, ..., m$, and

$$x_i > 0 \text{ for } i = 1, ..., m \Rightarrow f(x_1, ..., x_m) > 0$$

$$x_i < 0 \text{ for } i = 1, ..., m \Rightarrow f(x_1, ..., x_m) < 0$$

and

$$\lim_{(x_1,...,x_m)\to(0,...,0)} \frac{|f(x_1, ..., x_m)|}{\prod_{j=1}^{m} |x_j|^{\alpha_j}} = M > 0$$

2 Main results

We shall need the following several lemmas in order to prove our results.

Lemma 2.1 *Consider the following inequality relative to (1.6)*

$$\triangle^l x_n + p_n f(x_{n-k_1}, ..., x_{n-k_m}) \le 0, n = 0, 1, 2, ... \tag{2.1}$$

Assume that (C) holds,and for sufficiently large n

$$(p_n, p_{n+1}, ..., p_{n+k-1}) \neq 0 \tag{2.2}$$

where $k = max\{k_1, ..., k_m\}$. Then, (1.6) has an eventually positive solution if and only if its corresponding inequality (2.1) has also an eventually positive solution.

Especially, (1.2) oscillates if and only if (1.3) oscillates.

The proof of Lemma 2.1 is similar to the proof of [3, Theorem 1] and hence we omit it here.

On the basis of Theorem A and Lemma 2.1, we can easily obtain the following Lemma 2.2.

Lemma 2.2 *If (1.4) and (1.5) hold, then (1.3) has no eventually positive solution.*

Lemma 2.3 (see [2]) *Let x_n be defined for $n \geq n_0$ and $x_n > 0$ with $\Delta^l x_n$ eventually of one sign. Then, there exists an integer $j, 0 \leq j \leq m$ with $(m+j)$ odd for $\Delta^l \leq 0$ and $(m+j)$ even for $\Delta^l \geq 0$ and an integer $N \geq n_0$, such that for all $n \geq N$*

$$j \leq l-1, \text{ implies } (-1)^{j+i}\Delta^i x_n > 0, \quad j \leq i \leq l-1$$

and

$$j \geq 1, \text{ implies } \Delta^i x_n > 0, \quad 1 \leq i \leq j-1.$$

Further [5], if $\{x_n\}$ is bounded, then we have

$$(-1)^i \Delta^i x_n > 0, \text{ for all large } n \geq n_0, \quad i = 0, 1, ..., l-1$$

Lemma 2.4 (see [2]) *Let x_n be defined for $n \geq n_0$, and $x_n > 0$ with $\Delta^l x_n \leq 0$ for $n \geq n_0$ and not identically zero. Then, there exists a large integer $n_1 \geq n_0$ such that*

$$x_n \geq \frac{2^{2-2l}}{(l-1)!} n^{(l-1)} \Delta^{l-1} x_n, \text{ for all } n \geq 2^l n_1$$

Specially, we have

$$x_n \geq \frac{\theta}{(l-1)!} n^{l-1} \Delta^{l-1} x_n, \text{ for suffiently large } n$$

where $n^{(t)} = n(n-1)...(n-t+1)$, for every $t \in R$, and $0 < \theta < 1$.

Theorem 2.5 *Assume that $\sum_{i=1}^m \alpha_i > 1$, and there exists a $\lambda > 0$ such that*

$$\sum_{i=1}^m \alpha_i e^{-\lambda k_i} < 1 \tag{2.3}$$

and

$$\liminf_{n \to \infty} \left[p_n \exp(-e^{\lambda n}) \right] > 0. \tag{2.4}$$

Then every solution of (1.1) oscillates.

Proof Without loss generality, assume that (1.1) has an eventually positive solution x_n. Then there exists an integer $N_1 > 0$ such that

$$x_{n-k} > 0, \quad \text{and} \quad \Delta^l x_n \leq 0, \quad n \geq N_1.$$

By Lemma 2.3, $\Delta^i x_n$ are eventually of one sign for each $i \in \{1, ..., m\}$, and $\Delta^{l-1} x_n > 0$ holds for large n. We prove this theorem in two cases: (A) $\Delta x_n > 0$ and (B) $\Delta x_n < 0$.

Case A: In this case, $\{x_n\}$ is increasing. In view of Lemma 2.4, there exist θ and an integer $N_2 \geq N_1$ such that

$$x_n \geq \frac{\theta}{(l-1)!} n^{l-1} \Delta^{l-1} x_n, \quad n \geq N_2 - k.$$

It follows that

$$x_{n-k_i} \geq \frac{\theta}{(l-1)!}(n-k)^{l-1}\triangle^{l-1}x_{n-k_i}, \quad i=1,...,m, \quad n \geq N_2. \tag{2.5}$$

Set $y_n = \triangle^{l-1}x_n$. Then $y_n > 0$, $n \geq N_1 - k$, and

$$x_{n-k_i} \geq \frac{\theta}{(l-1)!}(n-k)^{l-1}y_{n-k_i}, \quad i=1,...,m, \quad n \geq N_2.$$

Hence, from (1.1), we have

$$\triangle y_n + \left[\frac{(n-k)^{l-1}\theta}{(l-1)!}\right]^{\sum_{i=1}^m \alpha_i} p_n \prod_{i=1}^m y_{n-k_i}^{\alpha_i} \text{sgn } y_{n-k_1} \leq 0, \qquad n \geq N_2. \tag{2.6}$$

Set $q_n = \left[(n-k)^{l-1}\theta/(l-1)!\right]^{\sum_{i=1}^m \alpha_i} p_n$. Then we can re-write (2.6) as

$$\triangle y_n + q_n \prod_{i=1}^m y_{n-k_i}^{\alpha_i} \text{sgn } y_{n-k_1} \leq 0, \qquad n \geq N_2. \tag{2.7}$$

This shows that (2.7) has an eventually positive solution.

On the other hand, by (2.4), we have

$$\liminf_{n\to\infty} \left[q_n \exp(-e^{\lambda n})\right] \geq \frac{\theta}{(l-1)!} \liminf_{n\to\infty} \left[p_n \exp(-e^{\lambda n})\right] > 0. \tag{2.8}$$

In view of Lemma 2.2, (2.3) and (2.8) implies that inequality (15) has no eventually positive solution. This is a contradiction.

Case B: In this case, there exists an integer $N_3 > N_1$ such that

$$(-1)^i \triangle^i x_n > 0, \qquad n \geq N_3, \quad i = 0, 1, ..., l-1.$$

Summing (1.1) from $n \geq N_3$ to ∞ $l-1$ times and using the above, we have

$$\triangle x_n + \sum_{j=n}^{\infty} \frac{(j+l-2-n)^{(l-2)}}{(l-2)!} p_j \prod_{i=1}^m |x_{j-k_i}|^{\alpha_i} \text{sgn } x_{j-k_1} \leq 0, \qquad n \geq N_3,$$

which implies

$$\triangle x_n + \frac{(l-2)^{(l-2)}}{(l-2)!} p_n \prod_{i=1}^m |x_{n-k_i}|^{\alpha_i} \text{sgn } x_{n-k_1} \leq 0, \qquad n \geq N_3.$$

Similar to Case A, we can include a contradiction. Hence, The proof is complete. □

Theorem 2.6 *Assume that (C) holds and* $\sum_{i=1}^m \alpha_i > 1$. *If there exists a number* $\lambda > 0$ *such that (2.3) and (2.4) holds, then every solution of (1.6) is oscillatory.*

Proof Assume that,for the sake of contradiction,x_n is an eventually positive solution of (1.6),then (C) means that there exists an integer $N_1 > 0$ such that

$$x_n > 0, x_{n-k_i} > 0, \quad i = 1, ..., m, \text{ for } n \geq N_1$$

and

$$f(x_{n-k_1}, ..., x_{n-k_m}) \geq \frac{1}{2} M \prod_{i=1}^m x_{n-k_i}^{\alpha_i}, \text{ for } n \geq N_1 \tag{2.9}$$

Substituting (2.9) into (1.6),we get

$$\triangle^l x_n + \tilde{p}_n \prod_{i=1}^{m} x_{n-k_i}^{\alpha_i} \le 0 \quad \text{for} \quad n \ge N_1 \tag{2.10}$$

where $\tilde{p}_n = \frac{1}{2}Mp_n$, which implies that (2.10) has an eventually positive solution. According to Lemma 2.1, the corresponding equation

$$\triangle^l x_n + \tilde{p}_n \prod_{i=1}^{m} |x_{n-k_i}|^{\alpha_i} sgn x_{n-k_i} = 0, \quad \text{for } n \ge N_1 \tag{2.11}$$

has also an eventually positive solution. But, by Theorem 2.5, the every solution of (2.11) oscillates, and this is contradictory. The proof is completed. □

Theorem 2.7 *Assume that* $\sum_{i=1}^{m} \beta_i > m$, *and there exists a* $\lambda > 0$ *such that*

$$\sum_{i=1}^{m} \beta_i e^{-\lambda k_i} < m \tag{2.12}$$

and

$$\liminf_{n\to\infty} \left\{ \left[\prod_{i=1}^{m} p_i(n) \right] \exp(-me^{\lambda n}) \right\} > 0 \tag{2.13}$$

Then every solution of (1.7) is oscillatory.

Proof Suppose to the contrary that x_n is an eventually positive solution of (1.7). Then there exists an integer $N > 0$ such that

$$x_n > 0, x_{n-k_i} > 0, \quad i = 1, ..., m, \quad n \ge N.$$

It follows that

$$\triangle^l x_n + \tilde{p}_n \prod_{i=1}^{m} x_{n-k_i}^{\frac{\beta_i}{m}} \le 0, \quad n \ge N, \tag{2.14}$$

where $\tilde{p}_n = m \left[\prod_{i=1}^{m} p_i(n)\right]^{\frac{1}{m}}$. This says that (2.14) has an eventually positive solution. Applying Lemma 2.1, the corresponding equation

$$\triangle^l x_n + \tilde{p}_n \prod_{i=1}^{m} x_{n-k_i}^{\frac{\beta_i}{m}} = 0, \quad n \ge N \tag{2.15}$$

has also an eventually positive solution. But, that every solution of (2.15) is oscillatory, according to hypothesis and Theorem 2.5, results in a contradiction, and this completes the proof. □

References

[1] Y.G. Zhou and X.H. Tang (2002). Oscillation of first-order nonlinear delay difference equations (in Chinese), *Appl. Math.*, **15**, 135–138.

[2] R.P. Agarwal (1992). *Difference Equations and Inequalities*, Marcel Dekker, Inc., New York.

[3] . Zhang and Y. Gao (199). Positive solutions of higher order nonlinear difference equation, *Sys. Sci. Math. Sci.*, **19**, 157–161.

[4] Q.G. Tang and L. Zeng (2000). Asymptotic behavior and oscillation for nonlinear neutral delay difference equations of higher order (in Chinese), *J. of Math. (PRC)*, **20**, 207–210.

[5] X.H. Tang and Y.J. Liu (2001). Oscillation for nonlinear delay difference equations, *Tamkang Journal of Math.*, **329**, 394–400.

[6] G. Zhang and Y. Gao (2001). *Oscillatory Theory for Difference* (in Chinese), Higher Education Press, Beijing.

[7] M.Q. Wang (1990). *Ordinary Difference Equations* (in Chinese), Xinjiang University Press, XinJiang.

Fields Institute Communications
Volume **42**, 2004

Coexistence of Multiple Attractive Periodic Solutions in a Discrete-Time Neural Network

Zhaohui Yuan
College of Mathematics and Econometrics
Hunan University
Changsha, Hunan 410082, P. R. China
zhaohuiyuan@hotmail.com

Lihong Huang
College of Mathematics and Econometrics
Hunan University
Changsha, Hunan 410082, P. R. China
lhhuang@hnu.net.cn

Zhan Zhou
College of Mathematics and Econometrics
Hunan University
Changsha, Hunan 410082, P. R. China
zzhou@hnu.net.cn

Dedicated to Professor Zhicheng Wang on the occasion of his seventieth birthday

Abstract. In this paper, we consider a class of discrete-time systems which describing the evolution of network with a ring of three identical neurons, where each neuron receives two inputs: One from the next neuron by inhibitory interaction and another from the previous neuron by excitative interaction. By studying its limiting equation applied its symmetry and contractive map principle, we obtain sufficient condition for the coexistence of multiple attracting periodic solutions.

1 Introduction

Due to network with a ring of neurons have been found in a variety of neural structures such as neocortex [1], cerebellum [2] and hippocampus [3], much progress has been made in study of such networks since the pioneering work of Atiya and Badi [4], Campbell et al. [5], Chen et al. [6], Huang and Wu [7]. In this paper, we consider a class of discrete-time system which describing the evolution of network with a ring

2000 *Mathematics Subject Classification.* Primary 39A11, 92B20.
The second author was supported by the Foundation for University Excellent Teacher by the Ministry of Education of China and by the NNSF(10371034) of China.

of three identical neurons where each neuron receives two inputs: One from the next neuron by inhibitory interaction and another from the previous neuron by excitative interaction. The dynamic of such a network is described by the following system of discrete-time systems

$$x_i(n+1) = \lambda x_i(n) + \frac{1}{2}[f(x_{i-1}(n)) - f(x_{i+1}(n))], \tag{1.1}$$

where $i = 1, 2, 3 \pmod 3$, $\lambda \in (0, 1)$ is a given constant, $f : R \to R$ is an activation function and assumed to satisfy the following

$$\begin{cases} |f(x)+1| & \leq & \epsilon \text{ if } x \leq -r, \\ |f(x)-1| & \leq & \epsilon \text{ if } x \geq r \end{cases} \tag{1.2}$$

and

$$|f(x) - f(y)| \leq L|x-y| \text{ if } x, y \in (-\infty, -r_1) \cup (r_1, \infty) \tag{1.3}$$

for some positive constants ϵ, r, r_1 and L.

Discrete-time system (1.1) can be regarded as a discrete analogy of the differential system

$$\frac{d}{dt}x_i(t) = -\mu x_i(t) + \frac{1}{2}\{f(x_{i-1}(t)) - f(x_{i+1}(t))\}$$

or the system with a piecewise constant arguments

$$\frac{d}{dt}x_i(t) = -\mu x_i(t) + \frac{1}{2}\{f(x_{i-1}([t])) - f(x_{i+1}([t]))\}$$

where $i = 1, 2, 3 \pmod 3$, $\mu > 0$, $[.]$ denotes the greatest integer function. The method for discrete analogy, we refer to [9,10].

As an application of neural networks, the convergence of such systems have been studied in a number of areas such as content addressable memory and pattern identification, However, because of nonlinear of neural networks, they can exhibit various kinds of attractors, sometimes these attractors exists more efficient performance than conventional ones. For example, periodic sequence of neural impulses are of fundamental significance for the control of dynamic functions of the body such as heart beat which occurs with great regularity and breathing. It is of great interest to understand the mechanisms which cause and sustain periodic activities in neural networks.

In recent papers, Walther [11] and Zhou [10] constructed contracting return maps in the study of a scalar delay differential equation and discrete-time system, provided the function f is Lipschitz continuous and in a specific sense to close such as a step function

$$f_0(x) = \begin{cases} 1 & \text{if} \quad x > 0, \\ -1 & \text{if} \quad x \leq 0. \end{cases} \tag{1.4}$$

This was based on the simple fact their limiting equation have periodic solutions which can be computed explicitly and have strong attraction properties.

In this paper, by studying the limiting equation (1.1) with $f = f_0$ applied its symmetry, we use contractive map principle to obtain the coexistence of multiple attracting periodic solutions for the general nonlinear activation function $f : R \to R$ satisfying (1.2) and (1.3).

We now present some notions and notations that will be used throughout this papers. Let N denote the set of all nonnegative integers. For any $m, n \in N$, define $N(m) = \{m, m+1, \cdots\}$ and $N(m, n) = \{m, m+1, \cdots, n\}$ whenever $m \leq n$. In particular, $N = N(0)$. Let $I_- = (-\infty, 0]$ and $I_+ = (0, +\infty)$, we denote $R_{abc} = I_a \times I_b \times I_c$ for $a, b, c \in \{-, +\}$. Obviously, $R^3 = \bigcup R_{abc}$ for $a, b, c \in \{-, +\}$.

For the general system

$$\xi(n+1) = G(\xi(n)), \quad n \in N, \tag{1.5}$$

where $G : R^k \to R^k$ is a given mapping. We recall some definition in [12].

Definition 1.1 A solution $\{\xi(n)\}$ of system (1.5) is called *periodic* if there exists a positive integer such that

$$\xi(n+m) = \xi(n), \quad \text{whenever} \quad n \in N. \tag{1.6}$$

The least positive integer m such that (1.6) holds is called the *period* of the periodic solution.

Definition 1.2 The system (1.5) is γ-*symmetric* provided

$$\gamma G(\xi) = G(\gamma \xi), \quad \text{whenever} \quad \xi \in R^k,$$

where $\gamma \in R^{k \times k}$ is a nonsingular real matrix.

2 Main results

For the convenience of expression, we rewrite the system (1.1) as the following form

$$\xi(n+1) = \Phi(\xi(n)), \quad n \in N, \tag{2.1}$$

where $\xi(n) = (x_1(n), x_2(n), x_3(n))^T \in R^3$ ($\cdot^T$ is the transpose of a vector or a matrix) and the mapping Φ can be expressed as

$$\begin{pmatrix} x_1 \\ x_2 \\ x_3 \end{pmatrix} \to \begin{pmatrix} \lambda x_1 + \frac{1}{2}[f(x_3) - f(x_2)] \\ \lambda x_2 + \frac{1}{2}[f(x_1) - f(x_3)] \\ \lambda x_3 + \frac{1}{2}[f(x_2) - f(x_1)] \end{pmatrix}.$$

For the function $f = f_0$, we use the mapping $\Phi_0 : R^3 \to R^3$ to replace Φ in (2.1). If we take the composition of F with itself n times as F^n and $F^0 = Id$ (the identify mapping on the corresponding space). Then, for the function f is given by (1.2) and (1.3), the solution $\{\xi(n)\}$ of (1.1) satisfies $\xi(n) = \Phi^n(\xi(0))$ with initial value condition $\xi(0) = (x_1(0), x_2(0), x_3(0))^T$. Similarly, for the function $f = f_0$, the solution $\{\xi(n)\}$ of (1.1) satisfies $\xi(n) = \Phi_0^n(\xi(0))$ with initial value condition $\xi(0) = (x_1(0), x_2(0), x_3(0))^T$.

Lemma 2.1 *For a nonsingular matrix $\gamma \neq I$ (the identify matrix) such that $\gamma^p = I$ holds (p is a minimal positive integer). Let the system (1.5) be γ-symmetric and there exist $m \in N(1)$ and $\xi^* \in R^k$ such that*

(H_1) $G^i(\xi^*) \neq \gamma \xi^*$ *and* $G^m(\xi^*) = \gamma \xi^*$ *for* $i \in N(1, m-1)$;

(H_2) $\gamma^j G^i(\xi^*) \neq \xi^*$ *for* $i^2 + j^2 \neq 0, i \in N(0, m-1)$ *and* $j \in N(0, p-1)$.

Then, the system (1.5) has a nontrivial mp-periodic solution.

Proof Let $\{\xi(n)\}$ denotes the solution of (1.5) with initial value condition $\xi(0) = \xi^*$. Note that system (1.5) has a unique solution $\{\xi(n)\}$ starting from the initial value ξ^*, we can define $G^{-d}(\xi(n)) = G^{n-d}(\xi^*)$ for $d \in N(0, n)$. Since the mapping G is γ-symmetric, it follows from (H_1) that

$$\xi(m+d) = G^{m+d}(\xi^*) = G^d(\gamma \xi^*) = \gamma G^d(\xi^*) = \gamma \xi(d) \quad \text{for} \quad d \in N(0, m-1). \tag{2.2}$$

By induction, we get

$$\xi(lm+d) = \gamma^l \xi(d) \quad \text{for} \quad l \in N, d \in N(0, m-1). \tag{2.3}$$

Especially, we have

$$\xi(pm) = \gamma^p \xi(0) = \xi(0) = \xi^*. \tag{2.4}$$

Now, we prove $\xi(i) \neq \xi(j)$ for $i \neq j$ and $i, j \in N(0, mp-1)$.

By way of contradiction, we assume that there exists $T_1, T_2 \in N(0, mp-1)$ and $T_1 \neq T_2$ such that $\xi(T_1) = \xi(T_2)$. For $i = 1, 2$, let $T_i = l_i m + d_i$ with $l_i \in N(0, p-1)$, $d_i \in N(0, m-1)$. It follows from (H_1) and (2.3) that

$$\xi(T_i) = G^{l_i m + d_i}(\xi^*) = G^{d_i} \circ G^{l_i m}(\xi^*) = G^{d_i}(\gamma^{l_i} \xi^*) = \gamma^{l_i} G^{d_i}(\xi^*).$$

Recalling $\xi(T_1) = \xi(T_2)$, we get

$$\gamma^{l_1} G^{d_1}(\xi^*) = \gamma^{l_2} G^{d_2}(\xi^*). \tag{2.5}$$

Without loss of generality, we assume that $l_1 \geq l_2$. We distinguish the following cases

Case 1. $d_1 \geq d_2$. From (2.5), we obtain

$$\gamma^{l_1 - l_2} G^{d_1 - d_2}(\xi^*) = \xi^*. \tag{2.6}$$

Case 2. $d_1 < d_2$. We have

$$\gamma^{l_1 - l_2} \xi^* = G^{d_2 - d_1}(\xi^*). \tag{2.7}$$

If $l_1 = l_2$ and $d_1 < d_2$. It follows from (2.7) that

$$G^{d_2 - d_1}(\xi^*) = \xi^*. \tag{2.8}$$

If $l_1 > l_2$ and $d_1 < d_2$. It follows from (2.7) that

$$\gamma^{-(l_1 - l_2)} G^{d_2 - d_1}(\xi^*) = \xi^*. \tag{2.9}$$

Noticing that $\gamma^p = I$, we know that $\gamma^{-(l_1 - l_2)} = \gamma^{p-(l_1 - l_2)}$. Hence, from (2.9), we have

$$\gamma^{p-(l_1 - l_2)} G^{d_2 - d_1}(\xi^*) = \xi^*. \tag{2.10}$$

Obviously, (2.6), (2.8) and (2.10) contradict the assumption (H_2). This, we complete the proof of Lemma 2.1. □

Using the similar argument as in [10], we get the following results.

Lemma 2.2 *Let $f = f_0$ and there exists some $m \in N(1)$ such that*

$$-\lambda^{2m} + \lambda^m + \lambda^{m-1} - 1 > 0. \tag{2.11}$$

Then, the system (1.1) has at least m attractive periodic solutions, more precisely, for every $k \in N(1, m)$, (1.1) possesses a nontrivial periodic solution with period $6k$.

Proof Let

$$\gamma = \begin{pmatrix} 0 & -1 & 0 \\ 0 & 0 & -1 \\ -1 & 0 & 0 \end{pmatrix}.$$

By direct verification, for $f = f_0$, we know that the system (1.1) is γ-symmetric. Now, we compute the vector ξ^* required in Lemma 2.1. For $k \in N(1, m)$, let $\{\xi_n\}$

be a solution of the system (1.1) satisfying $\xi(i) = (x_1(i), x_2(i), x_3(i))^T \in R_{-++}$ for $i \in N(0, k-1)$. It follows from (1.1) that

$$\begin{cases} x_1(i) &= \lambda x_1(i-1), \\ x_2(i) &= \lambda x_2(i-1) - 1, \qquad i \in N(1,k), \\ x_3(i) &= \lambda x_3(i-1) + 1, \end{cases} \tag{2.12}$$

which implies that

$$\begin{cases} x_1(i) &= \lambda^i x_1(0), \\ x_2(i) &= \lambda^i x_2(0) - \frac{1-\lambda^i}{1-\lambda}, \qquad i \in N(1,k). \\ x_3(i) &= \lambda^i x_3(0) + \frac{1-\lambda^i}{1-\lambda}, \end{cases} \tag{2.13}$$

Especially

$$\begin{cases} x_1(k) &= \lambda^k x_1(0), \\ x_2(k) &= \lambda^k x_2(0) - \frac{1-\lambda^k}{1-\lambda}, \\ x_3(k) &= \lambda^k x_3(0) + \frac{1-\lambda^k}{1-\lambda}. \end{cases} \tag{2.14}$$

This is

$$\xi(k) = \lambda^k \xi(0) + \eta(k),$$

where $\eta(k) = (0, -\frac{1-\lambda^k}{1-\lambda}, \frac{1-\lambda^k}{1-\lambda})^T$.

Let $\xi(k) = \gamma\xi(0)$. From (2.14), it is easy to verify that

$$\xi^* = \xi(0) = \left(-\frac{1-\lambda^{2k}}{(1-\lambda)(1+\lambda^{3k})}, \frac{\lambda^k(1-\lambda^{2k})}{(1-\lambda)(1+\lambda^{3k})}, \frac{(1-\lambda^k)^2(1+\lambda^k)}{(1-\lambda)(1+\lambda^{3k})}\right)^T, \tag{2.15}$$

which, together with (2.13), implies that

$$\begin{array}{rcl} x_1(0) < x_1(1) < \cdots < x_1(k-1) &=& -\frac{\lambda^k(1-\lambda^{2k})}{(1-\lambda)(1+\lambda^{3k})}, \\ x_2(0) > x_2(1) > \cdots > x_2(k-1) &=& \frac{(1+\lambda^k)(-\lambda^{2k}+\lambda^k+\lambda^{k-1}-1)}{(1-\lambda)(1+\lambda^{3k})}, \\ x_3(0) < x_3(1) < \cdots < x_3(k-1) &=& \frac{(1+\lambda^k)(\lambda^{2k}-\lambda^{2k-1}-\lambda^k+1)}{(1-\lambda)(1+\lambda^{3k})}. \end{array} \tag{2.16}$$

Let $g(k) = -\lambda^{2k} + \lambda^k + \lambda^{k-1} - 1$. Clearly, $g(k)$ is monotonically decreasing, which, together with (2.11), implies that $g(k) = -\lambda^{2k} + \lambda^k + \lambda^{k-1} - 1 > 0$ for $k \in N(1, m)$. Thus, combining (2.15) with (2.16), we have

$$\xi(i) = \Phi_0^i(\xi^*) \in R_{-++} \;\text{ and }\; \xi(i) \neq \xi(j) \;\text{ for }\; i, j \in N(0, k-1), i \neq j. \tag{2.17}$$

On the other hand, it is easy to find that

$$\gamma\nu \in R_{--+} \;\text{ for any }\; \nu \in R_{-++}. \tag{2.18}$$

Noticing that $\xi^* \in R_{-++}$, we have

$$\gamma\xi^* \in R_{--+},$$

which, together with (2.17), implies that $\xi(i) \neq \gamma\xi^*$, that is, $\Phi_0^i(\xi^*) \neq \gamma\xi^*$ for $i \in N(0, k-1)$. Hence, the initial value ξ^* satisfies the assumption (H_1) in Lemma 2.1.

Now, we begin to verify ξ^* satisfies (H_2) in Lemma 2.1. Since

$$\xi(k+i) = \Phi_0^{k+i}(\xi^*) = \Phi_0^i(\xi(k)) = \Phi_0^i(\gamma\xi^*) = \gamma\Phi_0^i(\xi^*) = \gamma\xi(i) \;\text{ for }\; i \in N(0, k-1),$$

which, together with (2.17) and (2.18), implies that

$$\xi(k+i) = \gamma\Phi_0^i(\xi^*) \in B_{--+} \;\text{ for }\; i \in N(0, k-1).$$

Similarly,

$$\xi(2k+i) = \gamma^2\Phi_0^i(\xi^*) \in R_{+-+}, \quad \xi(3k+i) = \gamma^3\Phi_0^i(\xi^*) \in R_{+--},$$
$$\xi(4k+i) = \gamma^4\Phi_0^i(\xi^*) \in R_{++-}, \quad \xi(5k+i) = \gamma^5\Phi_0^i(\xi^*) \in R_{-+-}$$

and

$$\xi(6k+i) = \gamma^6\Phi_0^i(\xi^*) = \Phi_0^i(\xi^*) = \xi(i)$$

for $i \in N(0, k-1)$. Hence, from (2.16), we know that the assumption (H_2) holds. Thus, the solution $\{(x_1^*(n), x_2^*(n), x_3^*(n))^T\}$ with initial value $(x_1^*(0), x_2^*(0), x_3^*(0))^T = (a_1^*, a_2^*, a_3^*)^T$ is a $6k$-periodic solution, where $(a_1^*, a_2^*, a_3^*)^T = \xi^*$ given by (2.15).

Let

$$\begin{aligned} d &= \min\{|x_1^*(n)|, |x_2^*(n)|, |x_3^*(n)|\}_{n\in N} \\ &= \min\{x_2^*(k-1), x_3^*(0)\} \\ &= \min\{\tfrac{(1+\lambda^k)(-\lambda^{2k}+\lambda^k+\lambda^{k-1}-1)}{(1-\lambda)(1+\lambda^{3k})}, \tfrac{(1-\lambda^k)^2(1+\lambda^k)}{(1-\lambda)(1+\lambda^{3k})}\} > 0. \end{aligned}$$

For any $(a_1, a_2, a_3)^T \in R_{-++}$, if $|a_i - a_i^*| < d$ for $i \in N(1,3)$, then the solution $\{(x_1(n), x_2(n), x_3(n))^T\}$ of (1.1) with $x_i(0) = a_i$ satisfies

$$x_1(1) = \lambda a_1 \quad x_2(1) = \lambda a_2 - 1 \quad x_3(1) = \lambda a_3 + 1$$

which implies

$$x_i(1) - x_i^*(1) = \lambda(a_i - a_i^*) \quad \text{for} \quad i \in N(1,3).$$

By induction, it is easy to find that $x_i(n)x_i^*(n) > 0$ and

$$x_i(n) - x_i^*(n) = \lambda^n(a_i - a_i^*) \quad \text{for} \quad i \in N(1,3),$$

which lead to the periodic solution $\{(x_1^*(n), x_2^*(n), x_3^*(n))^T\}$ is attracting. □

Theorem 2.3 *Assume that $\lambda \in (0,1)$, f is given by (1.2) and (1.3). For some $m \in N(1)$, if the inequality (2.11) holds, then there exist positive constants ϵ, r, r_1 and L with $r > r_1$ such that system (1.1) has at least m attractive periodic solutions, more precisely, for every $k \in N(1,m)$, (1.1) possesses a nontrivial periodic solution with period $6k$.*

Proof For $f = f_0$, using Lemma 2.2, we know that the system (1.1) has at least m attractive periodic solution. Suppose $\{\xi^*(n)\}$ is such a $6k$-periodic solution, where $\xi^*(n) = (x_1^*(n), x_2^*(n), x_3^*(n))^T$ and $\xi^*(0) = \xi^*$ given by (2.15). From the proof of Lemma 2.2, we see that $x_i^*(n) \neq 0$ for $n \in N$.

For any vector $\xi = (x_1, x_2, x_3)^T$, define a **norm** as

$$||\xi|| = \max\{|x_1|, |x_2|, |x_3|\}.$$

Let $r = \delta = \frac{1}{2}\min\{|x_1^*(n)|, |x_2^*(n)|, |x_3^*(n)|\}_{n\in(1,k)}$.

$$\epsilon = (1-\lambda)\delta, \quad r_1 = \frac{1}{2}\delta, \quad L = \frac{1}{2}(1-\lambda). \tag{2.19}$$

Denote

$$U(\xi_0, \epsilon) = \{\xi : \xi \in R^3 \text{ and } ||\xi - \xi_0|| < \epsilon\}.$$

Claim: The mapping Φ maps the neighborhood $U(\xi^*(n), \delta)$ into $U(\xi^*(n+1), \delta)$ for $n \in N$. And Φ is a contractive mapping on $U(\xi^*(n), \frac{3}{2}\delta)$.

In fact, if $\xi = (x_1, x_2, x_3)^T \in U(\xi^*(n), \delta)$, then

$$|x_i - x_i^*(n)| < \delta \quad \text{for} \quad i \in N(1,3).$$

According to the definition of δ, we see that $x_j \cdot x_j^*(n) > 0$. Therefore

$$\begin{aligned}
& |\lambda x_i + \frac{1}{2}[f(x_{i-1}) - f(x_{i+1})] - x_i^*(n+1)| \\
= \ & |\lambda(x_i - x_i^*(n)) - \frac{1}{2}[f(x_{i-1}) - f_0(x_{i-1}^*(n))] - \frac{1}{2}[f(x_{i+1}) - f_0(x_{i+1}^*(n))]| \\
\leq \ & \lambda|x_i - x_i^*(n)| + \frac{1}{2}|f(x_{i-1}) - f_0(x_{i-1}^*(n))| + \frac{1}{2}|f(x_{i+1}) - f_0(x_{i+1}^*(n))| \\
\leq \ & \lambda\delta + \frac{\epsilon}{2} + \frac{\epsilon}{2} \\
= \ & \delta,
\end{aligned}$$

$i = 1, 2, 3 (\mathrm{mod}\ 3)$. This shows $\Phi(\xi) \in U(\xi^*(n+1), \delta)$ and the first part of Claim is proved.

Let $\xi_j = (x_1^j, x_2^j, x_3^j)^T \in U(\xi^*(n), \frac{3}{2}\delta)$ for $j = 1, 2$, then

$$x_i^j \cdot x_i^*(n) > 0 \quad |x_i^j| > \frac{1}{2}\delta = \gamma_1 \quad \text{for} \quad i \in N(1,3) \quad \text{and} \quad j = 1, 2.$$

Combining (1.3) with (2.19), we have

$$\begin{aligned}
& |\lambda x_i^1 + \frac{1}{2}[f(x_{i-1}^1) - f(x_{i+1}^1)] - \lambda x_i^2 - \frac{1}{2}[f(x_{i-1}^2) - f(x_{i+1}^2)]| \\
\leq \ & \lambda|x_i^1 - x_i^2| + \frac{L}{2}|x_{i-1}^1 - x_{i-1}^2| + \frac{L}{2}|x_{i+1}^1 - x_{i+1}^2| \\
\leq \ & \frac{1}{2}(1+\lambda)||\xi_1 - \xi_2||
\end{aligned}$$

$i = 1, 2, 3 (\mathrm{mod}\ 3)$, which implies

$$||\Phi(\xi_1) - \Phi(\xi_2)|| \leq \frac{1}{2}(1+\lambda)||\xi_1 - \xi_2||. \tag{2.20}$$

Since $\frac{1}{2}(1+\lambda) < 1$, we see that Φ is a contractive mapping on $U(\xi^*(n), \frac{3}{2}\delta)$. Thus, the proof of Claim is completed. From Claim, we see that the compositive mapping Φ^k is also a contractive return map on $U(\xi^*, \delta)$. Hence, Φ^k has a unique fixed point $\overline{\eta} \in U(\xi^*, \delta)$ and the solution $\{\overline{\eta}(n)\}$ of (1.1) with initial value $\overline{\eta}(0) = \overline{\eta}$ is a $6k$-periodic solution. Moreover, for any $\eta \in U(\overline{\eta}, \frac{\delta}{2}) \subset U(\xi^*, \frac{3}{2}\delta)$. From (2.20), we get

$$||\Phi^n(\eta) - \Phi^n(\overline{\eta})|| \leq (\frac{1+\lambda}{2})^n||\eta - \overline{\eta}||,$$

which implies the attractability of the solution $\{\Phi^n(\eta^*)\}$ of (1.1). The proof of Theorem 2.3 is now completed. □

References

[1] J. Szenfagothai, *The "module-concept" in cerebral cortex aschitecture*, Brain Reseach, *95*(1976), 475-495.

[2] J. C. Eccles, M. Ito and J. Szenfagogathai, *The cerebellum as neuronal machine*, New York: Springer, 1967.

[3] P. Anderson, O. Gross and T. Lomo, *Participation of inhibitory interneurons in the control of hippocampal cortial output*, in "The Interneurons" (M. Brazier ed.), Los Angeles: University of California Press, 1969.

[4] Baldi Atiya, *Oscillations and synchronizations in neural networks: an exploration of the labelling hypothesis*, International Journal of Neural Systems, **1**(1989), 103-124.

[5] S. A. Campbell et al., *Qualitative analysis of a neural network model with multiple time delays*, International Journal of Bifurcation and Chaos, *9*(1999), 1585-1595.

[6] Yuming Chen, Ying Sue Huang and Jianhong Wu, *Desynchronization of larger scale delayed neural networks*, Proceedings of the American Mathematical Society, *128*(2000), 2365-2371.
[7] L. H. Huang and J. Wu, *Nonlinear wave in networks of neurons with delayed feedback : Pattern formation and continuation*, SIAM J. Math. Anal., *34*(4) (2003), 836-860..
[8] F. C. Hoppensteadt and E. M. Izhikevich, *Weakly connected neural networks*, New York (Spring), 1997.
[9] Z. H. Yuan, L. H. Huang and Y. M. Chen, *Convergence and periodicity of solutions for a discrete-time network model of two neurons,* Mathematical and Computer Modelling, *35*(2002), 941-950.
[10] Z. Zhou, *Periodic solution of a class of difference system*, to appear.
[11] Walther, *Contracting return maps for monotone delayed feedback*, Discrete and Continuous Dynamical Systems, *7* (2001), 259-274.
[12] Yun Tang, *Elements of symmetric bifurcation theory*, Science Press, 2000(in Chinese).
[13] Y. Chen, *Global attractors of delay differential systems arising from networks of neurons*, Ph. D. Thesis, York University, Ontario, Canada, 2000.

Fields Institute Communications
Volume **42**, 2004

Necessary and Sufficient Conditions for the Existence of Positive Solutions of Nonlinear Difference Equations

R. Y. Zhang
Department of Mathematics
Northeast Normal University
Changchun 130024, P. R. China
zry@nenu.edu.cn

Z. C. Wang
College of Mathematics and Econometrics
Hunan University
Changsha 410082, P. R. China
zcwang@hnu.net.cn

J. S. Yu
College of Mathematics and Econometrics
Hunan University
Changsha 410082, P. R. China

Abstract. In this paper, we obtained necessary and sufficient conditions for the existence of strictly monotone increasing positive solutions of the equation

$$\Delta(r_k(\Delta y_{k-1})^p) + s_k y_k^p = 0.$$

Its corresponding geometric significance has been stated in terms of solution curves of a certain autonomous system. The influence of perturbation in coefficients on the existence has also been discussed.

1 Introduction

In this paper, we study the existence of strictly monotone increasing positive solutions (MIPS, for short) of the nonlinear difference equation of the form

$$\Delta(r_k(\Delta y_{k-1})^p) + s_k y_k^p = 0, \tag{1.1}$$

where $p > 0$,$\{r_k\}_{k=1}^{\infty}$ is a sequence of positive real numbers and $\{s_k\}_{k=1}^{\infty}$ means a sequence of nonnegative real numbers.

The sufficient conditions for the existence of positive solutions of equation (1.1) are obtained by Cheng [1]. For $p = 1$, the necessary and sufficient conditions for oscillation of equation (1.1) are obtained by [7]. In this paper, we obtain some

2000 *Mathematics Subject Classification.* Primary 39A11.
Project Supported by NNSF of China (No. 10271044).

necessary and sufficient conditions for the existence of MIPS of Eq. (1.1). Those conditions are more simply and intuitively easy to understood.

Applying the results obtained, some sufficient conditions and necessary conditions for existence of MIPS are also obtained, respectively.

The main objective of this paper is to present a geometric interpretation on the existence of MIPS. It is interesting to have graphic method to know the relationship between $\{r_k\}_{k=1}^{\infty}$ and $\{s_k\}_{k=1}^{\infty}$ which enables us to conclude that Eq. (1.1) has a MIPS. With the help of the relationship, we study the existence of MIPS under small perturbation in coefficients.

We end this introduction by reducing the Eq. (1.1) to its simple form. If $\{y_k\}_{k=1}^{\infty}$ is a MIPS of Eq. (1.1), set

$$u_k = r_{k+1}(y_{k+1}/y_k - 1)^p, \quad t_k = (1/r_{k+1})^{1/p} \tag{1.2}$$

for $k = 0, 1, 2, \cdots$ and

$$x(t, u) = u(1 + tu^{1/p})^{-p} \tag{1.3}$$

for $u > 0, t \geq 0$.Then $\{u_k\}_{k=0}^{\infty}$ satisfies

$$u_k = x(t_{k-1}, u_{k-1}) - s_k \tag{1.4}$$

for $k = 1, 2, 3, \cdots$. Conversely, if $\{u_k\}_{k=0}^{\infty}$ is a positive solution of Eq. (1.4), a MIPS of Eq. (1.1) can be defined by (1.2) as follows

$$y_0 = 1, \quad y_k = \prod_{j=0}^{k-1}(t_j u_j^{1/p} + 1) \tag{1.5}$$

for $k = 1, 2, 3, ...$ Therefore, in subsequent sections we do not distinguish the case where Eq. (1.1) has a MIPS from those where Eq. (1.4) has a positive solution.

2 Main results

In this section we prove the main theorem, i.e, Theorem 2.2. The proofs given here take advantage of the properties of the function $x(t; u)$ as a solution curves of autonomous system. We start by showing that.

Let $x(t; u, 0)$ denote the solution curve of following initial value problem

$$\begin{cases} \dfrac{dx}{dt} = -px^{(p+1)/p}, \\ x(0) = u. \end{cases} \tag{2.1}$$

Lemma 2.1 *The function $x(t, u)$ defined in (1.3) has following properties:*

$$\begin{aligned} x(t, u) &= x(t; u, 0), \\ x(t; x(t'; u, 0), 0) &= x(t + t'; u, 0) \ for \ \ u > 0, t, t' \geq 0, \end{aligned} \tag{2.2}$$

and $x(t; u, 0) > 0$ for $u > 0, t \geq 0$.

Proof It is easy to verify the first equality of the lemma. Equality (2.2) follows from that the differential equation (2.1) is autonomous. □

In the following we shall also consider $x(t, u)$ as a solution curve.

The main result of this paper is the following.

Theorem 2.2 *(i) If Eq. (1.1) has a MIPS, say* $\{y_k\}_{k=0}^{\infty}$ *,then*

$$s_k = x(T_{k-1} + t_{k-1}; u_0, 0) - x(T_k; u_0, 0) \tag{2.3}$$

where $u_0 = r_1(y_1/y_0 - 1)$, $x(t; u_0, 0)$ *is defined as in (2.1) with* $u = u_0$, T_k *satisfies:*

$$x(T_k; u_0, 0) = r_{k+1}(y_{k+1}/y_k - 1)^p, \quad k = 0, 1, 2, \cdots .$$

(ii) If there exist $v_0 \geq 0$ *and* $\{T_k\}_{k=0}^{\infty}$ *such that* $T_0 = 0$ *and*

$$s_k \leq x(T_{k-1} + t_{k-1}; v_0, 0) - x(T_k; v_0, 0), \tag{2.4}$$

then Eq. (1.1) has a MIPS, say $\{y_k\}_{k=0}^{\infty}$, *satisfies* $r_1(y_1/y_0 - 1) = v_0$.

Proof (i) If Eq. (1.1) has a MIPS, say $\{y_k\}_{k=0}^{\infty}$, we set

$$u_k = r_{k+1}(y_{k-1}/y_k - 1)^p, \quad k = 0, 1, 2, \cdots .$$

As has been said in Introduction, $\{u_k\}_{k=0}^{\infty}$ satisfies Eq. (1.4), the fact that $x(t, u) = x(t; u, 0)$ yields for $k = 1, 2, 3, \cdots$

$$s_k = x(t_{k-1}; u_{k-1}, 0) - u_k. \tag{2.5}$$

For $k = 1$, $s_1 = x(t_0; u_0, 0) - u_1$. So $u_1 \leq x(t_0; u_0, 0)$. From

$$x'(t; u_0, 0) = -p[x(t; u_0, 0)]^{(p+1)/p} < 0,$$

we know that $x(t; u_0, 0)$ tends to zero as t goes to infinity. Therefore, we can define $u_1 = x(T_1; u_0, 0)$ where $T_1 \geq t_0$. It follows by $T_0 = 0$ that

$$s_1 = x(T_0 + t_0; u_o, 0) - x(T_1; u_0, 0).$$

For the purpose of induction, assume that for $k \geq 1$, we have already proved

$$\begin{aligned} u_l &= x(T_{l-1}; u_0, 0), \\ s_l &= x(T_{l-1} + t_{l-1}; u_0, 0) - x(T_l; u_0, 0), \end{aligned}$$

for $l = 1, \cdots, k$. For $k + 1$, by (2.5), $s_{k+1} = x(t_k; u_k, 0) - u_{k+1}$. By the induction assumption that $u_k = x(T_k; u_0, 0)$ and (2.2), we get

$$x(t_k; u_k, 0) = x(t_k; x(T_k; u_0, 0), 0) = x(T_k + t_k; u_0, 0).$$

Thus we obtain

$$u_{k+1} = x(t_k; u_k, 0) - s_{k+1} = x(T_k + t_k; u_0, 0) - s_{k+1} \leq x(T_k + t_k; u_0, 0).$$

Similarly, we put $u_{k+1} = x(T_{k+1}; v_0, 0)$ and

$$s_{k+1} = x(T_k + t_k; u_0, 0) - x(T_{k+1}; u_0, 0).$$

Therefore, by the induction principle, we obtain $\{T_k\}_{k=0}^{\infty}$ such that

$$\begin{aligned} T_0 &= 0, \; x(T_m; u_0, 0) = r_{m+1}(y_{m+1}/y_m - 1)^p, \\ s_{m+1} &= x(T_m + t_m; u_0, 0) - x(T_{m+1}; u_0, 0) \end{aligned}$$

for $m = 0, 1, 2, \cdots$. That is, (2.3) holds, and the first part of Theorem 2.2 is proved.

(ii) Conversely, let $\{T_k\}_{k=0}^{\infty}$ be a sequence such that (2.4) holds. We define $\{v_k\}_{k=0}^{\infty}$ as follows, $v_k = x(T_k; v_0, 0)$ for $k = 0, 1, 2, \cdots$. Then from (2.2) we have $v_k > 0$ for $k = 1, 2, \cdots$, and

$$x(T_{k-1} + t_{k-1}; v_0, 0) = x(t_{k-1}; x(T_{k-1}; v_0, 0), 0) = x(t_{k-1}; v_{k-1}, 0).$$

This, together with (2.4), shows that

$$s_k \leq x(t_{k-1}; v_{k-1}, 0) - v_k \tag{2.6}$$

for $k = 1, 2, 3, \cdots$.

Now, we define $\{u_k\}_{k=0}^{\infty}$ as follows

$$u_0 = v_0, \quad u_k = x(t_{k-1}; u_{k-1}, 0) - s_k \tag{2.7}$$

for $k = 1, 2, \cdots$.

As has been said, it is enough to show that $u_k \geq v_k$, for $k = 1, 2, \cdots$. For $k = 1$, by (2.6) and (2.7), $v_1 \leq x(t_0; u_0, 0) - s_1 = u_1$. Similarly to the proof of the first part, assume that for $k \geq 1$, we already proved $u_l \geq v_l$, for $l = 1, \cdots, k$. For $k+1$, from the induction assumption that $v_k \leq u_k$ and noting that

$$\frac{\partial x(t; u, 0)}{\partial u} > 0 \text{ for } \quad u > 0, \quad t > 0,$$

we get $x(t_k; v_k, 0) \leq x(t_k; u_k, 0)$. Finally, we have by this and (2.6)–(2.7) that

$$v_{k+1} \leq x(t_k; v_k, 0) - s_{k+1} \leq x(t_k; u_k, 0) - s_{k+1} = u_{k+1}.$$

Thus, we obtain $u_k \geq v_k > 0$ for $k = 0, 1, 2, \cdots$. Therefore, Eq. (1.4) has a positive solution $\{u_k\}_{k=0}^{\infty}$. That is, Eq. (1.1) has a MIPS. This accomplishes the proof of Theorem 2.2. □

Consider the solution curve $x(t; u', 0)$ of initial value problem (2.1) with $u' > u_0$. $\frac{\partial x(t;u,0)}{\partial u} > 0$ implies that

$$\begin{aligned} s_k &= x(T_{k-1} + t_{k-1}; u_0, 0) - x(T_k; u_0, 0) \\ &= \int_{T_{k-1}+t_{k-1}}^{T_k} p[x(t; u_0, 0)]^{(p+1)/p} dt \\ &< \int_{T_{k-1}+t_{k-1}}^{T_k} p[x(t; u', 0)]^{(p+1)/p} dt \\ &= x(T_{k-1} + t_{k-1}; u', 0) - x(T_k; u', 0). \end{aligned}$$

Theorem 2.2 tell us that Eq. (1.1) has a MIPS , say $\{\tilde{y}_k\}_{k=0}^{\infty}$, satisfies

$$r_1(\tilde{y}_1/\tilde{y}_0 - 1) = u'.$$

So, the following assertion holds.

Corollary 2.3 *If Eq. (1.1) has a MIPS , say* $\{y_k\}_{k=0}^{\infty}$, *satisfies* $r_1(y_1/y_0 - 1) = u_0$, *then for any* $u' > u_0$, *Eq.(1) has a MIPS , say* $\{\tilde{y}_k\}_{k=0}^{\infty}$, *satisfies* $r_1(\tilde{y}_1/\tilde{y}_0 - 1) = u'$.

As a consequence of Theorem 2.2, we have the following theorem:

Theorem 2.4 *Eq. (1.1) has a MIPS if and only if there exist* $c_0 > 0$ *and a sequence* $\{T_k\}_{k=0}^{\infty}$ *such that* $T_0 = 0$,

$$s_k \leq [c_0 + T_{k-1} + (\frac{1}{r_k})^{1/p}]^{-p} - [c_0 + T_k]^{-p} \quad \text{for} \quad k = 1, 2, \cdots. \tag{2.8}$$

Proof Let $v_0 > 0$ satisfy $(1/v_0)^{1/p} = c_0$, then for any $t > 0$, we get

$$x(t; v_0, 0) = \left(\frac{v_0^{1/p}}{1 + t v_0^{1/p}} \right)^p = (c_0 + t)^{-p}.$$

Noting that $t_{k-1} = (1/r_k)^{1/p}$, for $k = 1, 2, \cdots$, the inequality (2.8) is equivalent to the inequality (2.4) . Now, we conclude by Theorem 2.2 (ii) that Eq. (1.1) has a MIPS .

Conversely, if Eq. (1.1) has a MIPS, inequality (13) can be derived immediately from inequality (2.3) in Theorem 2.2 with $c_0 = (1/u_0)^{1/p}$. This completes the proof of Theorem 2.4. □

3 Some necessary conditions and sufficient conditions

In this section, applying Theorem 2.2 we obtain some necessary conditions and sufficient conditions for the existence of MIPS of Eq. (1.1), respectively.

First, if Eq. (1.1) has a MIPS, by inequality (2.3) , we have

$$\begin{aligned}\sum_{k=m}^{\infty} s_k &= \sum_{k=m}^{\infty}[x(T_{k-1}+t_{k-1};u_0,0)-x(T_k;u_0,0)]\\ &= x(T_{m-1}+t_{m-1};u_0,0)-\sum_{k=m}^{\infty}[x(T_k;u_0,0)-x(T_k+t_k;u_0,0)]\\ &< x(T_{m-1}+t_{m-1};u_0,0).\end{aligned}$$

Noting that $T_0 = 0, T_k \geq T_{k-1}+t_{k-1}$, $for\ k = 1,2,\cdots$, it is easy to verify that $T_1 \geq t_0 = (1/r_1)^{1/p}$ and

$$T_{m-1}+t_{m-1} \geq \sum_{j=0}^{m-2} t_j + t_{m-1} = \sum_{j=1}^{m-1} t_j = \sum_{j=1}^{m}(\frac{1}{r_j})^{1/p}$$

for $m = 2,3,\cdots$. So we obtain

$$\begin{aligned}\sum_{k=m}^{\infty} s_k &\leq x(T_{m-1}+t_{m-1};u_0,0) \leq x(\sum_{j=1}^{m}(\frac{1}{r_j})^{1/p},u_0,0)\\ &= \left[(1/u_0)^{1/p}+\sum_{j=1}^{m}(\frac{1}{r_j})^{1/p}\right]^{-p} < \left[\sum_{j=1}^{m}(\frac{1}{r_j})^{1/p}\right]^{-p}.\end{aligned}$$

In brief, we have following theorem:

Theorem 3.1 *If Eq. (1.1) has a MIPS, then for each positive integer m,*

$$\left(\sum_{j=1}^{m}(\frac{1}{r_j})^{1/p}\right)^{p}\left(\sum_{k=m}^{\infty} s_k\right) < 1. \tag{3.1}$$

It is not hard to prove that $\int_0^\infty x(t;u_0,0)dt < \infty$, for $p > 1$. As has been said that

$$\begin{aligned}\left(\frac{1}{r_m}\right)^{1/p}\left(\sum_{k=m}^{\infty} s_k\right) &= t_{m-1}\left(\sum_{k=m}^{\infty} s_k\right) < x(T_{m-1}+t_{m-1};u_0,0)t_{m-1}\\ &< \int_{T_{m-1}}^{T_{m-1}+t_{m-1}} x(t;u_0,0)dt.\end{aligned}$$

Combining the previous discussion we may obtain

Theorem 3.2 *If Eq. (1.1) has a MIPS , $p > 1$, then*

$$\sum_{m=1}^{\infty}(1/r_m)^{1/p}(\sum_{k=m}^{\infty} s_k) < \infty. \tag{3.2}$$

By our previous inequalities, we also obtain that

$$\left(\sum_{k=m}^{\infty} s_k\right)^{1/p} < [x(T_{m-1}+t_{m-1};u_0,0)]^{1/p} < [x(T_{m-1};u_0,0)]^{1/p} = u_{m-1}^{1/p}.$$

On the other hand, if Eq. (1.1) has a bounded MIPS , say $\{y_k\}_{k=0}^{\infty}$, then

$$\begin{aligned} 1+\sum_{m=1}^{\infty}\left(\frac{1}{r_m}\right)^{1/p} u_{m-1}^{1/p} &= \lim_{n\to\infty}\left[1+\sum_{m=1}^{n}(\frac{1}{r_m})^{1/p}u_{m-1}^{1/p}\right] \\ &\le \lim_{n\to\infty}\prod_{j=0}^{n-1}\left(1+\left(\frac{1}{r_{j+1}}\right)^{1/p} u_j^{1/p}\right) \\ &= \lim_{n\to\infty}\prod_{j=0}^{n-1}(t_j u_j^{1/p}+1) \\ &= \lim_{n\to\infty} y_n < \infty. \end{aligned}$$

We are thus led to the following conclusion:

Theorem 3.3 *If Eq. (1.1) has a bounded MIPS , then*

$$\sum_{m=1}^{\infty}\left(\frac{1}{r_m}\sum_{k=m}^{\infty} s_k\right)^{1/p} < \infty. \tag{3.3}$$

From Theorem 3.1, we see that condition (3.1) is necessary for the existence of MIPS of Eq. (1.1). It is interesting to ask if the left-hand side of inequality (3.1) can be arbitrary close to one under the condition that Eq. (1.1) has a MIPS? For each given positive integer m , we have following result:

Theorem 3.4 *For each given positive integer m and arbitrary positive integer n, there exists a sequence $\{s_k\}$ which guarantees that Eq. (1.1) has a MIPS satisfies:*

$$\left(\sum_{j=1}^{m}\left(\frac{1}{r_j}\right)^{1/p}\right)^{p}\left(\sum_{k=m}^{\infty} s_k\right) = \frac{n-1}{n}. \tag{3.4}$$

Proof Since m and n have been chosen, c can be chosen sufficiently small and then $n' \in N$ can be chosen sufficiently large so that

$$\left(\frac{n'-1}{n'}\right)^{1/p}\sum_{j=1}^{m}\left(\frac{1}{r_j}\right)^{1/p} = \left(\frac{n-1}{n}\right)^{1/p}\left[\sum_{j=1}^{m}\left(\frac{1}{r_j}\right)^{1/p}+c\right],$$

in other words,

$$\frac{n'-1}{n'}\left[c+\sum\left(\frac{1}{r_j}\right)^{1/p}\right]^{-p} = \frac{n-1}{n}\left[\sum\left(\frac{1}{r_j}\right)^{1/p}\right]^{-p}.$$

For $m \ge 2,\ 1 \le k \le m-1$, choose $u_o > 0$ so that $(1/u_0)^{1/p} = c/2$ and let

$$T_0 = 0, \quad T_k = \sum_{j=0}^{k-1} t_j + \frac{k}{2(m-1)}c.$$

If $m = 1$, we choose u_0 such that $(1/u_0)^{1/p} = c$. Then we have

$$\left[c + \sum_{j=1}^{m} \left(\frac{1}{r_j}\right)^{1/p}\right]^{-p} = \left[\left(\frac{1}{u_0}\right)^{1/p} + T_{m-1} + t_{m-1}\right]^{-p} = x(T_{m-1} + t_{m-1}; u_0, 0).$$

In sum, we have

$$\frac{n'-1}{n'} x(T_{m-1} + t_{m-1}; u_0, 0) = \frac{n-1}{n} \left[\sum_{j=1}^{m} \left(\frac{1}{r_j}\right)^{1/p}\right]^{-p}. \tag{3.5}$$

If $m \geq 2$, we set that for $k = 1, 2, \cdots, m-1$,

$$s_k = x(T_{k-1} + t_{k-1}; u_0, 0) - x(T_k; u_0, 0). \tag{3.6}$$

Because $x(t; u_0, 0)$ is monotone decreasing continuous function with respect to t and

$$\lim_{t \to \infty} x(t; u_0, 0) = 0,$$

we can find an unique T_m such that

$$x(T_{m-1} + t_{m-1}; u_0, 0) - x(T_m; u_0, 0) = (n'-1)[x(T_m; u_0, 0) - x(T_m + t_m; u_0, 0)].$$

We set that

$$s_m = x(T_{m-1} + t_{m-1}; u_0, 0) - x(T_m; u_0, 0).$$

It is clear that

$$s_m = \frac{n'-1}{n'}[x(T_{m-1} + t_{m-1}; u_0, 0) - x(T_m + t_m; u_0, 0)].$$

For the purpose of inductive definition, assume that for $k \geq 1$, we have already defined $\{s_{m+l}\}$, $\{T_{m+l}\}$ for $l = 0, 1, \cdots, k-1$ very well such that

$$\begin{aligned} s_{m+l} &= x(T_{m+l-1} + t_{m+l-1}; u_0, 0) - x(T_{m+l}; u_0, 0) \\ &= \frac{n'-1}{n'}[x(T_{m+l-1} + t_{m+l-1}; u_0, 0) - x(T_{m+l} + t_{m+l}; u_0, 0)]. \end{aligned}$$

For $k+1$, similarly, we can find an unique T_{m+k} such that

$$\begin{aligned} x(T_{m+k-1} + t_{m+k-1}; u_0, 0) - x(T_{m+k}; u_0, 0) = (n'-1)[x(T_{m+k}; u_0, 0) \\ - x(T_{m+k} + t_{m+k}; u_0, 0)]. \end{aligned}$$

Therefore, we can set that

$$s_{m+k} = x(T_{m+k-1} + t_{m+k-1}; u_0, 0) - x(T_{m+k}; u_0, 0).$$

Then we have

$$s_{m+k} = \frac{n'-1}{n'}[x(T_{m+k-1} + t_{m+k-1}; u_0, 0) - x(T_{m+k} + t_{m+k}; u_0, 0)].$$

Thus, by the induction principle, we obtain $\{T_k\}_{k=m}^{\infty}$ and $\{s_k\}_{k=m}^{\infty}$ such that for $l = 0, 1, 2, \cdots$

$$s_{m+l} = x(T_{m+l-1} + t_{m+l-1}; u_0, 0) - x(T_{m+l}; u_0, 0), \tag{3.7}$$

$$s_{m+l} = \frac{n'-1}{n'}[x(T_{m+l-1} + t_{m+l-1}; u_0, 0) - x(T_{m+l} + t_{m+l}; u_0, 0)]. \tag{3.8}$$

This, together with (3.6), we conclude by Theorem 2.2 that Eq. (1.1) has a MIPS. Combining (3.8) and (3.5) yields

$$\sum_{k=m}^{\infty} s_k = \frac{n'-1}{n'}\sum_{l=0}^{\infty}[x(T_{m+l-1}+t_{m+l-1};u_0,0)-x(T_{m+l}+t_{m+l};u_0,0)]$$

$$= \frac{n'-1}{n'}x(T_{m-1}+t_{m-1};u_0,0) = \frac{n-1}{n}\left[\sum_{j=1}^{m}\left(\frac{1}{r_j}\right)^{1/p}\right]^{-p}.$$

Thus the proof is complete. □

Note that various sufficient conditions can be obtained by making use of Theorem 2.2(ii). The following result is a simply case.

Theorem 3.5 *Eq. (1.1) has a MIPS provided that*

$$s_k \le (\frac{1}{2})^{p+1}\left\{\left[\sum_{j=1}^{k}(\frac{1}{r_j})^{1/p}\right]^{-p} - \left[\sum_{j=1}^{k+1}(\frac{1}{r_j})^{1/p}\right]^{-p}\right\} \tag{3.9}$$

for $k = 1, 2, \cdots$.

Proof For each positive integer k , let C_k be the right-hand side of (3.9). Choosing $v_0 = r_1$, noting that $t_j = (1/r_{j+1})^{1/p}$ for $j = 0, 1, \cdots$, and according to (1.2), we may rewrite C_k as:

$$C_k = \frac{1}{2}\left[x(t_0+2\sum_{j=1}^{k-1}t_j;\ v_0,0) - x(t_0+2\sum_{j=1}^{k}t_j;\ v_0,0)\right],\quad k=2,3,\cdots,$$

and

$$C_1 = \frac{1}{2}\left[x(t_0;v_0,0)-x(t_0+2t_1;v_0,0)\right].$$

Since $d^2x/dt^2 \ge 0$, we conclude that $x(t;u_0,0)$ is a convex function with respect to t, hence for $t' > t'' > 0$,

$$\frac{1}{2}\left[x(t';v_o,0)-x(t'';v_0,0)\right] \le x(t';v_0,0) - x(\frac{1}{2}(t'+t'');\ v_0,0).$$

It follows from above inequality with $t' = t_0 + 2\sum_{j=1}^{k-1}t_j$ and $t'' = t_0 + 2\sum_{j=1}^{k}t_j$ that

$$C_k \le x(t_0+2\sum_{j=1}^{k-1}t_j;\ v_0,0) - x(t_0+2\sum_{j=1}^{k-1}t_j+t_k;\ v_0,0),\quad k=2,3,\cdots. \tag{3.10}$$

Similarly,

$$C_1 \le x(t_0;v_o,0) - x(t_0+t_1;v_0,0).$$

We set $T_0 = 0,\ T_1 = t_0 + t_1,\ T_k = t_0 + 2\sum_{j=1}^{k-1}t_j + t_k,\ \ k = 2, 3, \cdots$. It is easy to see that

$$s_k \le C_k \le x(T_{k-1}+t_{k-1};v_0,0) - x(T_k;v_0,0).$$

The Theorem 3.5 now follows by Theorem 2.2(ii). □

Remark If we choose

$$s_k = \alpha(\frac{1}{2})^{p+1} \left\{ \left[\sum_{j=1}^{k} (\frac{1}{r_j})^{1/p} \right]^{-p} - \left[\sum_{j=1}^{k+1} (\frac{1}{r_j})^{1/p} \right]^{-p} \right\}, \quad 0 < \alpha \leq 1,$$

then Theorem 3.5 shows that Eq. (1.1) has a MIPS. In this case we get

$$\left(\frac{1}{r_m} \sum_{k=m}^{\infty} s_k \right)^{1/p} = \alpha^{1/p} (\frac{1}{2})^{1+1/p} \left\{ (\frac{1}{r_m})^{1/p} \left[\sum_{j=1}^{m} (\frac{1}{r_j})^{1/p} \right]^{-1} \right\}.$$

If $\sum_{m=1}^{\infty} (1/r_j)^{1/p} = \infty$, we have by N. H. Abel and U. Dini criterion that

$$\sum_{m=1}^{\infty} \left(\frac{1}{r_m} \sum_{k=m}^{\infty} s_k \right)^{1/p} = \infty,$$

from this, we conclude by Theorem 3.3 that the MIPS is unbounded.

4 Perturbed problem

We assume that $\{s_k\}_{k=0}^{\infty}$ is a sequence of positive real numbers through this section. In many cases, it is necessary to know if the existence of MIPS is influenced by perturbation in coefficients. In other words, if Eq. (1.1) with given coefficients $\{r_k\}_{k=0}^{\infty}$ and $\{s_k\}_{k=0}^{\infty}$ has a MIPS , the sequences $\{r'_k\}_{k=0}^{\infty}$ and $\{s'_k\}_{k=0}^{\infty}$ is closed to the sequences $\{r_k\}_{k=0}^{\infty}$ and $\{s_k\}_{k=0}^{\infty}$, respectively, if Eq. (1.1) with coefficients $\{r'_k\}_{k=0}^{\infty}$ and $\{s'_k\}_{k=0}^{\infty}$ has also a MIPS?

Theorem 4.1 *If Eq. (1.1) with given coefficients $\{r_k\}_{k=0}^{\infty}$ and $\{s_k\}_{k=0}^{\infty}$ has a MIPS , then there exists a sequence of positive real numbers $\{\delta_k\}_{k=0}^{\infty}$ such that*

$$|r'_k - r_k| \leq \delta_k, \quad |s'_k - s_k| \leq \delta_k, \quad for\ k = 1, 2, \cdots \tag{4.1}$$

imply that Eq. (1.1) with coefficients $\{r'_k\}_{k=0}^{\infty}$ and $\{s'_k\}_{k=0}^{\infty}$ has a MIPS.

To prove Theorem 4.1, we need following result:

Lemma 4.2 *Let $u > 0$, $0 < t_1 < t_2$, $0 < u_1 < u_2$. Then we have*

$$\begin{aligned} &[x(t_1; u_2, 0) - x(t_2; u_2, 0)] - [x(t_1; u_1, 0) - x(t_2; u_1, 0)] \\ &\geq \ p(p+1) \left[(\frac{1}{u_1})^{1/p} - (\frac{1}{u_2})^{1/p} \right] \left[(\frac{1}{u_1})^{1/p} + t_2 \right]^{-(p+2)} (t_2 - t_1), \end{aligned} \tag{4.2}$$

and

$$x(t_1; u, 0) - x(t_2; u, 0) \leq p \left[(\frac{1}{u})^{1/p} + t_1 \right]^{-(p+1)} (t_2 - t_1). \tag{4.3}$$

Proof With the help of properties of the function $x(t; u, 0)$, it is not hard to obtain following estimations:

$$\begin{aligned} &[x(t_1; u_2, 0) - x(t_2; u_2, 0)] - [x(t_1; u_1, 0) - x(t_2; u_1, 0)] \\ &= \int_{t_1}^{t_2} [-x'(t; u_2, 0) + x'(t; u_1, 0)] dt \\ &= p \int_{t_1}^{t_2} \left\{ \left[(\frac{1}{u_2})^{1/p} + t \right]^{-(p+1)} - \left[(\frac{1}{u_1})^{1/p} + t \right]^{-(p+1)} \right\} dt \end{aligned}$$

$$
\begin{aligned}
&= \int_{t_1}^{t_2} p(p+1)\left\{(\frac{1}{u_2})^{1/p}+\theta\left[(\frac{1}{u_1})^{1/p}-(\frac{1}{u_2})^{1/p}\right]\right. \\
&\quad \left.+t\right\}^{-(p+2)}\left[(\frac{1}{u_1})^{1/p}-(\frac{1}{u_2})^{1/p}\right]dt \\
&\geq p(p+1)\left[(\frac{1}{u_1})^{1/p}-(\frac{1}{u_2})^{1/p}\right]\int_{t_1}^{t_2}\left[(\frac{1}{u_1})^{1/p}+t\right]^{-(p+2)}dt \\
&= p\left[(\frac{1}{u_1})^{1/p}-(\frac{1}{u_2})^{1/p}\right]\left\{-\left[(\frac{1}{u_1})^{1/p}+t\right]^{-(p+1)}\Bigg|_{t_1}^{t_2}\right\} \\
&\geq p\left\{\left[(\frac{1}{u_1})^{1/p}+t_1\right]^{-(p+1)}-\left[(\frac{1}{u_1})^{1/p}+t_2\right]^{-(p+1)}\right\}\left[(\frac{1}{u_1})^{1/p}-(\frac{1}{u_2})^{1/p}\right] \\
&= -p(p+1)\left[(\frac{1}{u_1})^{1/p}+t_1+\tau(t_2-t_1)\right]^{-(p+2)}(t_1-t_2)\left[(\frac{1}{u_1})^{1/p}-(\frac{1}{u_2})^{1/p}\right] \\
&\geq \left[(\frac{1}{u_1})^{1/p}+t_2\right]^{-(p+2)}p(p+1)\left[(\frac{1}{u_1})^{1/p}-(\frac{1}{u_2})^{1/p}\right](t_2-t_1),
\end{aligned}
$$

the proof of part one is complete. To prove second part, we have

$$
\begin{aligned}
x(t_1;u,0)-x(t_2;u,0) &= \int_{t_1}^{t_2} -x'(t;u,0)dt \\
&= \int_{t_1}^{t_2} p[x(t;u,0)]^{(p+1)/p}dt \\
&\leq p[x(t_1;u,0)]^{(p+1)/p}(t_2-t_1) \\
&= p\left[(\frac{1}{u})^{1/p}+t_1\right]^{-(p+1)}(t_2-t_1).
\end{aligned}
$$

This completes the proof of second part. □

We are now ready to prove Theorem 4.1.

Proof of Theorem 4.1 If Eq. (1.1) has a MIPS for given coefficients $\{r_k\}_{k=0}^{\infty}$ and $\{s_k\}_{k=0}^{\infty}$, say $\{y_k\}_{k=0}^{\infty}$, letting $u_0 = r_1(y_1/y_0 - 1)$ and $\{I_k\}_{k=0}^{\infty}$ be an order distribution of $\{\mho_k\}_{k=0}^{\infty}$, then Theorem 2.2 tell us that

$$
s_k = x(T_{k-1}+t_{k-1};u_0,0)-x(T_k;u_0,0) \quad k=1,2,\cdots, \tag{4.4}
$$

where $T_k = L_{I_k}$. For each positive integer k, we now define

$$
\begin{aligned}
\omega_k &= p(p+1)\left[(\frac{1}{u_0})^{1/p}-(\frac{1}{u_0+1})^{1/p}\right] \\
&\quad \times\left[(\frac{1}{u_0})^{1/p}+T_k\right]^{-(p+2)}(T_k-T_{k-1}-t_{k-1}).
\end{aligned} \tag{4.5}
$$

It is easy to know by $s_k > 0$ that $\omega_k > 0$. For each ω_k, we can find $\delta_k : 0 < \delta_k < min\{\frac{1}{2}\omega_k,\ s_k,\ r_k\}$ such that $|r'_k - r_k| \leq \delta_k$ implies that

$$
\left|(\frac{1}{r'_k})^{1/p}-(\frac{1}{r_k})^{1/p}\right| \leq \frac{1}{2p}\left[(\frac{1}{u_0+1})^{1/p}+T_{k-1}+t_k-1\right]^{(p+1)}\omega_k, \tag{4.6}
$$

for $k=1,2,\cdots$. Now we already defined the sequence $\{\delta_k\}_{k=1}^{\infty}$.

In view of Theorem 2.2 (ii), it is sufficient to show that for each positive integer k, $|r'_k - r_k| \le \delta_k$ and $|s'_k - s_k| \le \delta_k$ imply that

$$s'_k \le x(T_{k-1} + t'_{k-1}; u_0 + 1, 0) - x(T_k; u_0 + 1, 0). \tag{4.7}$$

In fact, we note that

$$\begin{aligned} &x(T_{k-1} + t'_{k-1}; u_0 + 1, 0) - x(T_k; u_0 + 1, 0) \\ &= \{x(T_{k-1} + t'_{k-1}; u_0 - 1, 0) - x(T_{k-1} + t_{k-1}; u_0 + 1, , 0)\} \\ &\quad + \{[x(T_{k-1} + t_{k-1}; u_0 + 1, 0) - x(T_k; u_0 + 1, 0)] \\ &\quad - [x(T_{k-1} + t_{k-1}; u_0, 0) - x(T_k; u_0, 0)]\} \\ &\quad + \{x(T_{k-1} + t_{k-1}; u_0, 0) - x(T_k; u_0, 0)\} \\ &= \{\ II_1\ \} + \{\ II_2\ \} + \{\ II_3\ \}. \end{aligned}$$

Now, we estimate each term. For II_1, if $t'_{k-1} \le t_{k-1}$, it follows from $x'(t; u, 0) \le 0$ that $II_1 \ge 0$. Otherwise, by (4.3) with $u = u_0 + 1$, $t_1 = T_{k-1} + t_{k-1}$ and $t_2 = T_{k-1} + t'_{k-1}$, we obtain

$$II_1 \ge -p \left[(\frac{1}{u_0 + 1})^{1/p} + T_{k-1} + t_{k-1} \right]^{-(p+1)} (t'_{k-1} - t_{k-1}).$$

Noting that (4.6), we get $II_1 \ge -\frac{1}{2}\omega_k$. In sum we have

$$II_1 \ge -\frac{1}{2}\omega_k. \tag{4.8}$$

Applying (4.2) with $u_2 = u_0 + 1$, $u_1 = u_0$, $t_1 = T_{k-1} + t_{k-1}$, and $t_2 = T_k$, we get

$$II_2 \ge p(p+1) \left[(\frac{1}{u_0})^{1/p} - (\frac{1}{u_0 + 1})^{1/p} \right] \left[(\frac{1}{u_0})^{1/p} + T_k \right]^{-(p+2)} (T_k - T_{k-1} - t_{k-1}).$$

It follows immediately from (4.5) that

$$II_2 \ge \omega_k. \tag{4.9}$$

At the same time, we know by (4.4) that

$$II_3 = s_k. \tag{4.10}$$

Together (4.8)–(4.10), we obtain

$$x(T_{k-1} + t'_{k-1};\ u_0 + 1,\ 0) - x(T_k; u_o + 1, 0) \ge s_k + \frac{1}{2}\omega_k.$$

On the other hand, (4.1) gives us that $|s'_k - s_k| \le \delta_k \le \frac{1}{2}\omega_k$. Thus, we have $s_k + \frac{1}{2}\omega_k \ge s_k + s'_k - s_k = s'_k$, that is (4.7) holds. The proof of Theorem 4.1 is complete. □

References

[1] Chen, M. P. and Yu, J. S. [1995] *Oscillations for delay difference equations with variable coefficients*, Proceedings of the first International Conference on Difference Equations, Edited by S. N. Elaydi, et al.

[2] Cheng, S. S. and Patula, W. T. [1993] *An existence theorem for a nonlinear difference equation*, Nonlinear Analysis **20**, 193-203.

[3] Cheng, S. S. and Zhang, B. G. [1995] *Nonexistence of positive nondecreasing solutions of a nonlinear difference equation*, Proceedings of the First International Conference on Difference Equations, Edited by S. N. Elaydi, et al.

[4] Cheng, S. S. [1994] *Hille-Wintner type comparison theorems for nonlinear difference equations*, Funkcialaj Ekvacioj **37**, 531-535.

[5] Cheng, S. S. and Zhang, B. G. [1994] *Monotone solutions of a class of nonlinear difference equations*, Computers Math. Appl. **28**, 71-79.

[6] Cheng, S. S., Horng, J. L. and William, T. P. [1989] *Bounded and zero convergent solutions of second-order difference equations*, J. Math. Anal. Appl. **141**, 463-483.

[7] Chen, Shaozhu [1994] *Disconjugacy, disfocality, and oscillation of second order difference equation*, J. Differential Equations **107**, 383-394.

Fields Institute Communications
Volume **42**, 2004

Periodic Solutions of a Class of Difference Systems

Zhan Zhou
Department of Applied Mathematics, Hunan University
Changsha, Hunan 410082, P. R. China
zzhou@hnu.net.cn.com

Qinqin Zhang
Department of Applied Mathematics, Hunan University
Changsha, Hunan 410082, P. R. China
13975104698@hnmcc.com

Dedicated to Professor Zhicheng Wang on the occasion of his seventieth birthday

Abstract. Consider the difference system

$$x_n = \beta x_{n-1} + f(y_{n-1}),\ y_n = \beta y_{n-1} - f(x_{n-1}),\ n = 1, 2, \cdots \qquad (*)$$

which is a discrete-time neural network describing the dynamic interaction of two neurons, where $\beta \in (0,1)$ is the internal decay rate, g is a signal transmission function. The emphasis is put on the case where β is near to 1. We construct explicitly a $4m$-periodic solution when f is a step function (McCulloch-Pitts Model), we also obtain the necessary conditions for the periods of the periodic solutions of (*). For the general nonlinear signal transmission functions, we apply the contractive map principle to obtain the existence and attractivity of the $4m$-periodic solutions.

1 Introduction

We consider the following nonlinear difference system

$$\begin{cases} x_n = \beta x_{n-1} + f(y_{n-1}), \\ y_n = \beta y_{n-1} - f(x_{n-1}), \end{cases} \quad n \in N \tag{1.1}$$

where N is the set of all positive integers, $\beta \in (0,1)$, f is a nonlinear function. System (1.1) describes the evolution of a discrete-time network of two neurons with inhibitory interations, where β is the internal decay rate, f is the signal transmission function. See [6] for general backgrounds on delayed neural networks, and [7, 8] for related discrete-time networks.

2000 *Mathematics Subject Classification.* Primary 39A11, 39A12; Secondary 37E15.

This work was supported by the Natural Science Foundation of Hunan Province of China.

A basic and important signal transmission function is the so-called McCulloch-Pitts nonlinearity given by

$$f_0(x) = \begin{cases} -1 & x > 0, \\ 1 & x \leq 0. \end{cases} \tag{1.2}$$

This function describes the simple on-or-off characteristic of a single neuron. For more general case, we consider a signal transmission function satisfying the following conditions:

$$\begin{cases} |f(x)+1| \leq \epsilon & \text{if} \quad x \geq r \\ |f(x)-1| \leq \epsilon & \text{if} \quad x \leq -r \end{cases} \tag{1.3}$$

and

$$|f(x) - f(y)| \leq L|x-y| \text{ if } x, y \in (-\infty, -r_1) \text{ or } x, y \in (r_1, \infty). \tag{1.4}$$

for some positive constants ϵ, r, r_1 and L. This type of nonlinearity satisfying (1.3) and (1.4) was also used in [4, 5] in the study of a scalar delay differential equation with negative feedback, where the author successfully constructed an attractive returning map whose fixed point gives an asymptotically stable periodic solution.

Difference system (1.1) can be regarded as a discrete analogy of the differential system

$$\begin{cases} \frac{d}{dt}x(t) = -\mu x(t) + f(y(t)) \\ \frac{d}{dt}y(t) = -\mu y(t) - f(x(t)), \end{cases} \quad t \geq t_0 \tag{1.5}$$

with $\mu > 0$, which was extensively studied in the literature. In (1.5), replace $\frac{d}{dt}x(t)$ and $\frac{d}{dt}y(t)$ by $x(t+h) - x(t)/h$ and $y(t+h) - y(t)/h$ respectively, here h is a small positive constants, and let $x_n = x(t_0 + nh)/h$, $y_n = y(t_0 + nh)/h$ and $\beta = 1 - \mu h$, then (1.5) reduces to (1.1). We notice that $\beta \to 1$ as $h \to 0$. This is why we pay attention to the case where β near to 1 for (1.1).

Recall that by a **solution** of (1.1), we mean a sequence $\{(x_n, y_n)\}$ of points in R^2 that is defined for every nonnegative integers. Clearly, for every $z = (x, y) \in R^2$, system (1.1) has a unique solution (x_n^z, y_n^z) satisfying the initial condition

$$(x_0^z, y_0^z) = (x, y). \tag{1.6}$$

A solution $\{(x_n, y_n)\}$ of (1.1) is called **periodic** if there exists a positive integer m such that

$$(x_{n+m}, y_{n+m}) = (x_n, y_n), \quad n = 0, 1, 2, \cdots. \tag{1.7}$$

The least positive integer m such that (1.7) holds is called the **period** of the periodic solution.

For general background, we refer to [1, 2, 3].

This paper serves two goals. One is that most papers deal with the attractivity of scalar difference equation, less is known for the existence and attractivity of periodic solutions of the difference system due to the lack of effective methods. The other is that the difference systems studied here have potential applications in the realization of neural networks.

The remaining part of this paper is organized as follows. In section 2, we consider the case where $f = f_0$ (which is defined in (1.2)). We will construct explicitly $4m$-periodic solutions of (1.1) for every $m \in N$. When β is near to 1, we find that there is no odd periodic solution for system (1.1). In section 3, we use the contractive map principle to obtain the existence and attractivity of $4m$-periodic solutions.

2 McCulloch-Pitts signal transmission function

In this section, we consider system (1.1) where the signal transmission function is with McCulloch-Pitts type, that is $f = f_0$. Let

$$\begin{array}{ll} R_{11} = \{(x,y)|x > 0, y > 0\}, & R_{12} = \{(x,y)|x > 0, y \leq 0\}, \\ R_{21} = \{(x,y)|x \leq 0, y > 0\}, & R_{22} = \{(x,y)|x \leq 0, y \leq 0\}. \end{array} \tag{2.1}$$

Then $R^2 = R_{11} \cup R_{12} \cup R_{21} \cup R_{22}$. We notice that the solution $\{(x_n, y_n)\}$ of (1.1) moves according to the sequence $R_{11} \to R_{21} \to R_{22} \to R_{12} \to R_{11} \to \cdots$.

Theorem 2.1 *Let $f = f_0$, $\beta \in (0,1)$. For any $m \in N$, if*

$$2\beta^{m-1} - 1 - \beta^{2m} > 0. \tag{2.2}$$

Then system (1.1) has an attractive periodic solution of period $4m$.

Proof Let $\{(x_n, y_n)\}$ be a solution of (1.1) with the initial condition $x_0 = a > 0, y_0 = b > 0$, then

$$x_1 = \beta a - 1, \quad y_1 = \beta b + 1. \tag{2.3}$$

Assume that $(x_n, y_n) \in R_{21}$ for $n = 1, \cdots, m_1$; $(x_n, y_n) \in R_{22}$ for $n = m_1 + 1, \cdots, m_1 + m_2$; $(x_n, y_n) \in R_{12}$ for $n = m_1 + m_2 + 1, \cdots, m_1 + m_2 + m_3$; $(x_n, y_n) \in R_{11}$ for $n = m_1 + m_2 + m_3 + 1, \cdots, m_1 + m_2 + m_3 + m_4; \cdots$. Then

$$\left\{ \begin{array}{l} x_n = \beta x_{n-1} - 1 \\ y_n = \beta y_{n-1} - 1 \end{array} \right. \quad \text{for} \quad 2 \leq n \leq m_1 + 1. \tag{2.4}$$

Thus,

$$\left\{ \begin{array}{l} x_n = x_1 \beta^{n-1} + \frac{\beta^{n-1}-1}{1-\beta} \\ y_n = y_1 \beta^{n-1} + \frac{\beta^{n-1}-1}{1-\beta} \end{array} \right. \quad \text{for} \quad 2 \leq n \leq m_1 + 1. \tag{2.5}$$

Especially, we get

$$\left\{ \begin{array}{l} x_{m_1+1} = x_1 \beta^{m_1} + \frac{\beta^{m_1}-1}{1-\beta} \\ y_{m_1+1} = y_1 \beta^{m_1} + \frac{\beta^{m_1}-1}{1-\beta}. \end{array} \right. \tag{2.6}$$

In general, we can obtain

$$\left\{ \begin{array}{l} x_{m_1+m_2+m_3+m_4+1} = x_1 \beta^{m_1+m_2+m_3+m_4} + \frac{\beta^{m_1+m_2+m_3+m_4} - 2\beta^{m_2+m_3+m_4} + 2\beta^{m_4} - 1}{1-\beta} \\ y_{m_1+m_2+m_3+m_4+1} = y_1 \beta^{m_1+m_2+m_3+m_4} + \frac{\beta^{m_1+m_2+m_3+m_4} - 2\beta^{m_3+m_4} + 1}{1-\beta}. \end{array} \right. \tag{2.7}$$

Let $m_1 = m_2 = m_3 = m_4 = m$, and $x_1 = x_{4m+1}$, from (2.7), we get

$$\left\{ \begin{array}{l} x_1 = \frac{\beta^{4m} - 2\beta^{3m} + 2\beta^m - 1}{(1-\beta^{4m})(1-\beta)} = -\frac{(1-\beta^m)^2}{(1+\beta^{2m})(1-\beta)} \\ y_1 = \frac{\beta^{4m} - 2\beta^{2m} + 1}{(1-\beta^{4m})(1-\beta)} = \frac{(1-\beta^{2m})}{(1+\beta^{2m})(1-\beta)}. \end{array} \right. \tag{2.8}$$

This leads to

$$x_0 = a^* = \frac{2\beta^{m-1} - 1 - \beta^{2m}}{(1+\beta^{2m})(1-\beta)}, \quad y_0 = b^* = \frac{1 + \beta^{2m} - 2\beta^{2m-1}}{(1+\beta^{2m})(1-\beta)}. \tag{2.9}$$

Clearly, the solution $\{(x_n^*, y_n^*)\}$ of (1.1) with initial condition $x_0^* = a^*, y_0^* = b^*$ is a $4m$-periodic solution.

Let

$$\begin{array}{ll} d & = \min\{|x_n^*|, |y_n^*|\}_{n \in N} = \min\{x_0^*, x_{2m+1}^*\} \\ & = \min\{\frac{2\beta^{m-1}-1-\beta^{2m}}{(1+\beta^{2m})(1-\beta)}, \frac{(1-\beta^m)^2}{(1+\beta^{2m})(1-\beta)}\} > 0. \end{array} \tag{2.10}$$

For any $(a,b) \in R_{11}$, if $|a-a^*|<d, |b-b^*|<d$, then the solution $\{(x_n,y_n)\}$ of (1.1) with $x_0=a, y_0=b$ satisfies

$$x_1=\beta a-1, \quad y_1=\beta b+1,$$

which implies

$$x_1-x_1^*=\beta(a-a^*), \quad y_1-y_1^*=\beta(b-b^*).$$

By induction, it is easy to find that $x_n x_n^*>0, y_n y_n^*>0$ and

$$x_n-x_n^*=\beta^n(a-a^*), \quad y_n-y_n^*=\beta^n(b-b^*)$$

which leads to the attractivity of the periodic solution $\{(x_n^*,y_n^*)\}$. □

Remark 2.2 If $m=1$, then (2.2) holds for any $\beta\in(0,1)$. If $m\geq 2$, then for $\beta\in(0,1)$,

$$2\beta^{m-1}-1-\beta^{2m}=0$$

has a unique solution $\beta_0(m)$. Thus (2.2) is equivalent to $\beta\in(\beta_0(m),1)$. And Theorem 2.1 implies for any fixed $m\geq 2$, (1.1) has always $4m$ periodic solution when β is near to 1.

For the sake of convenience, we will use the convention

$$\sum_{i=1}^{j} z_j \equiv 0 \quad \text{whenever} \quad j<1.$$

Now we turn to the condition of periods for the periodic solutions of (1.1). Before giving the theorem, we establish a lemma which is useful in its own right.

Lemma 2.3 *Assume $k\in N$, $m_i\in N$ for $i=1,2,\cdots,2k$, c is a constant in $(0,1)$, and $\{z_n\}$ is a sequence. If, for any $\beta\in(c,1)$, $\{z_n\}$ satisfy*

$$\begin{cases} z_n=\beta z_{n-1}-1, & \sum_{i=1}^{2j-2} m_i+2\leq n\leq \sum_{i=1}^{2j-1} m_i+1 \\ z_n=\beta z_{n-1}+1, & \sum_{i=1}^{2j-1} m_i+2\leq n\leq \sum_{i=1}^{2j} m_i+1 \end{cases} \tag{2.11}$$

for $j=1,\cdots,k$;

$$z_{\sum_{i=1}^{2j-1} m_i+1}\leq 0, \quad z_{\sum_{i=1}^{2j} m_i+1}>0, \quad \text{for} \quad j=1,\cdots,k. \tag{2.12}$$

and

$$z_1=z_{\sum_{i=1}^{2k} m_i+1}. \tag{2.13}$$

Then, $\sum_{i=1}^k m_{2i-1}=\sum_{i=1}^k m_{2i}$.

Proof From (2.11), we have

$$\begin{aligned} z_{m_1+1} &= \beta^{m_1} z_1+\tfrac{\beta^{m_1}-1}{1-\beta} \\ z_{m_1+m_2+1} &= \beta^{m_2} z_{m_1+1}-\tfrac{\beta^{m_2}-1}{1-\beta} \\ &\vdots \\ z_{\sum_{i=1}^{2k-1} m_i+1} &= \beta^{m_{2k-1}} z_{\sum_{i=1}^{2k-2} m_i+1}+\tfrac{\beta^{m_{2k-1}}-1}{1-\beta} \\ z_{\sum_{i=1}^{2k} m_i+1} &= \beta^{m_{2k}} z_{\sum_{i=1}^{2k-1} m_i+1}-\tfrac{\beta^{m_{2k}}-1}{1-\beta}. \end{aligned}$$

In view of (2.13), this implies

$$z_1=\frac{\beta^{m_1+m_2+\cdots+m_{2k}}-2\beta^{m_2+\cdots+m_{2k}}+2\beta^{m_3+\cdots+m_{2k}}-\cdots+2\beta^{m_{2k-1}+m_{2k}}-2\beta^{m_{2k}}+1}{(1-\beta^{m_1+m_2+\cdots+m_{2k}})(1-\beta)}.$$

Let

$$g(\beta)=\beta^{m_1+m_2+\cdots+m_{2k}}-2\beta^{m_2+\cdots+m_{2k}}+2\beta^{m_3+\cdots+m_{2k}}-\cdots+2\beta^{m_{2k-1}+m_{2k}}-2\beta^{m_{2k}}+1. \tag{2.14}$$

Clearly, $g(1)=1$, and that $z_1>0$ for $\beta\in(c,1)$ is equivalent to $g(\beta)>0$ for $\beta\in(c,1)$. Thus

$$\begin{aligned} g'(1) &= \textstyle\sum_{i=1}^{2k} m_i - 2\sum_{i=2}^{2k} m_i + 2\sum_{i=3}^{2k} m_i - \cdots + 2\sum_{i=2k-1}^{2k} m_i - 2m_{2k} \\ &= \textstyle\sum_{i=1}^{k} m_{2i-1} - \sum_{i=1}^{k} m_{2i} \le 0. \end{aligned} \tag{2.15}$$

Similarly, that $z_{m_1+1}\le 0$ for $\beta\in(c,1)$ implies

$$\sum_{i=1}^{k} m_{2i-1}-\sum_{i=1}^{k} m_{2i}\ge 0. \tag{2.16}$$

Combining (2.15) with (2.16), we know

$$\sum_{i=1}^{k} m_{2i-1}=\sum_{i=1}^{k} m_{2i}.$$

The proof is complete. □

Theorem 2.4 *System (1.1) has no periodic solution of odd period.*

Proof Assume that (1.1) has a periodic solution $\{(x_n,y_n)\}$ of period p. Since $\{(x_n,y_n)\}$ moves according to the sequence $R_{11}\to R_{21}\to R_{22}\to R_{12}\to R_{11}$, there is no harm in assuming that $(x_0,y_0)\in R_{11}$; and

$$\begin{array}{ll} (x_n,y_n)\in R_{21} & \text{for } 1+\sum_{i=1}^{j-1}v_i\le n\le r_j+\sum_{i=1}^{j-1}v_i; \\ (x_n,y_n)\in R_{22} & \text{for } r_j+1+\sum_{i=1}^{j-1}v_i\le n\le r_j+s_j+\sum_{i=1}^{j-1}v_i; \\ (x_n,y_n)\in R_{12} & \text{for } r_j+s_j+1+\sum_{i=1}^{j-1}v_i\le n\le r_j+s_j+t_j+\sum_{i=1}^{j-1}v_i; \\ (x_n,y_n)\in R_{11} & \text{for } r_j+s_j+t_j+1+\sum_{i=1}^{j-1}v_i\le n\le \sum_{i=1}^{j}v_i \end{array}$$

where $v_i=r_i+s_i+t_i+u_i$, $j=1,\cdots,m$ and

$$p=\sum_{i=1}^{m}(r_i+s_i+t_i+u_i). \tag{2.17}$$

From $y_{p+1}=y_1$, (1.1) and Lemma 2.3, we know that

$$\sum_{i=1}^{m}(r_i+s_i)=\sum_{i=1}^{m}(t_i+u_i). \tag{2.18}$$

Similarly, from $x_{p+1}=x_1$, (1.1) and Lemma 2.3, we get

$$\sum_{i=1}^{m}(s_i+t_i)=\sum_{i=1}^{m}(r_i+u_i). \tag{2.19}$$

Combining (2.18) with (2.19), we get

$$\sum_{i=1}^{m} s_i=\sum_{i=1}^{n} u_i,\quad \sum_{i=1}^{m} r_i=\sum_{i=1}^{m} t_i, \tag{2.20}$$

which implies that p is even, and the proof is complete. □

The following corollary is a partial converse of Theorem 2.1.

Corollary 2.5 Assume that $c \in (0,1)$. If $\{(x_n, y_n)\}$ is a p-periodic solution of (1.1) satisfying $(x_0, y_0) \in R_{11}$; $(x_n, y_n) \in R_{21}$ for $1 \le n \le r$; $(x_n, y_n) \in R_{22}$ for $r+1 \le n \le r+s$; $(x_n, y_n) \in R_{12}$ for $r+s+1 \le n \le r+s+t$; $(x_n, y_n) \in R_{11}$ for $r+s+t+1 \le n \le r+s+t+u$; and $p = r+s+t+u$. Then $r = s = t = u$ and $x = x_n^*$, $y_n = y_n^*$ when $\{(x_n^*, y_n^*)\}$ is defined in the proof of Theorem 2.1 for $m = r$.

Proof According to the proof of Theorem 2.4, we see that

$$r = t, \quad s = u. \tag{2.21}$$

And

$$\begin{cases} x_{r+s+t+u+1} &= x_1\beta^{r+s+t+u} + \frac{\beta^{r+s+t+u} - 2\beta^{s+t+u} + 2\beta^u - 1}{1-\beta} \\ y_{r+s+t+u+1} &= y_1\beta^{r+s+t+u} + \frac{\beta^{r+s+t+u} - 2\beta^{t+u} + 1}{1-\beta}. \end{cases}$$

Noticing (2.21) and $p = r+s+t+u$, we see

$$\begin{cases} x_1 = -\frac{\beta^{r+s} - 2\beta^s + 1}{(1+\beta^{r+s})(1-\beta)} \\ y_1 = -\frac{1-\beta^{r+s}}{(1+\beta^{r+s})(1-\beta)}. \end{cases} \tag{2.22}$$

$x_1 \le 0$ for $\beta \in (c, 1)$ implies that

$$r \le s. \tag{2.23}$$

Next, we have

$$\begin{aligned} y_{r+1} &= (y_1 + \tfrac{1}{1-\beta})\beta^r - \tfrac{1}{1-\beta} \\ &= \tfrac{2\beta^r - 1 - \beta^{r+s}}{(1+\beta^{r+s})(1-\beta)}. \end{aligned} \tag{2.24}$$

$y_{r+1} \le 0$ for $\beta \in (c, 1)$ leads to

$$r \ge s. \tag{2.25}$$

In view of (2.21), (2.23) and (2.25), we see that

$$r = s = t = u.$$

The second part of Corollary 2.5 follows the proof of Theorem 2.1. □

3 General nonlinear signal transmission function

In this section, we consider the general case where the signal transmission function satisfying (1.3) and (1.4).

First, we establish

Theorem 3.1 *Assume that $\beta \in (0,1)$, $k \in N$. And for $f = f_0$, (1.1) has a k-periodic solution $\{\bar{x}_n, \bar{y}_n\}$ with $\bar{x}_n \ne 0, \bar{y}_n \ne 0$ for $n \in N$. Then there exist positive constants ϵ, r, r_1 and L with $r > r_1$, for any f satisfying (1.3) and (1.4), (1.1) has an attractive k-periodic solution.*

Proof For any $(x,y) \in R^2$, define an operator $F: R^2 \to R^2$ by

$$F(x,y) = (\beta x + f(y), \beta y - f(x)). \tag{3.1}$$

Then the solution of (1.1) satisfies

$$(x_n, y_n) = F(x_{n-1}, y_{n-1}).$$

For any $(x,y) \in R^2$, define a **norm** as

$$||(x,y)|| = \max\{|x|, |y|\}. \tag{3.2}$$

Let

$$r = \delta = \frac{1}{2}\min\{|\bar{x}_n|, |\bar{y}_n|; n = 1, \cdots, k\},\ \epsilon = (1-\beta)\delta,\ r_1 = \frac{1}{2}\delta,\ L = \frac{1}{2}(1-\beta). \tag{3.3}$$

Denote

$$N((x_0, y_0), \epsilon) = \{(x, y) \in R^2; ||(x, y) - (x_0, y_0)|| \leq \epsilon\}.$$

We can get the following lemma.

Lemma 3.2 *The operator F maps $N((\bar{x}_i, \bar{y}_i), \delta)$ into $N((\bar{x}_{i+1}, \bar{y}_{i+1}), \delta)$ for $i \in N$. And F is a contractive operator on $N((\bar{x}_i, \bar{y}_i), \frac{3}{2}\delta)$.*

Proof of Lemma 3.2 In fact, if $(x, y) \in N((\bar{x}_i, \bar{y}_i), \delta)$, then $|x - \bar{x}_i| \leq \delta$, $|y - \bar{y}_i| \leq \delta$. According to the definition of δ, we see that $x\bar{x}_i > 0, y\bar{y}_i > 0$. Therefore,

$$\begin{aligned} |\beta x + f(y) - \bar{x}_{i+1}| &= |\beta(x - \bar{x}_i) + f(y) - f_0(\bar{y}_i)| \\ &\leq \beta|x - \bar{x}_i| + |f(y) - f_0(\bar{y}_i)| \\ &\leq \beta\delta + \epsilon = \delta. \end{aligned}$$

Similarly, we can show that

$$|\beta y - f(x) - \bar{y}_{i+1}| \leq \delta.$$

This shows that $F(x, y) \in N((\bar{x}_{i+1}, \bar{y}_{i+1}), \delta)$ and proves the first part of the Lemma.

Next, if $(x_i, y_i) \in N((\bar{x}_i, \bar{y}_i), \frac{3}{2}\delta)$ for $j = 1, 2$. Then,

$$x\bar{x}_i > 0, \ y\bar{y}_i > 0, \ |x| \geq \frac{1}{2}\delta, \ |y| \geq \frac{1}{2}\delta.$$

In view of (1.4) and (3.3), we have

$$\begin{aligned} &|\beta x_1 + f(y_1) - (\beta x_2 + f(y_2))| \\ \leq\ & \beta|x_1 - x_2| + L|y_1 - y_2| \\ \leq\ & \tfrac{1}{2}(1 + \beta)||(x_1, y_1) - (x_2, y_2)||. \end{aligned}$$

Similarly,

$$|\beta y_1 - f(x_1) - (\beta y_2 - f(x_2))| \leq \frac{1}{2}(1 + \beta)||(x_1, y_1) - (x_2, y_2)||.$$

This shows that

$$||F(x_1, y_1) - F(x_2, y_2)|| \leq \frac{1}{2}(1 + \beta)||(x_1, y_1) - (x_2, y_2)||. \tag{3.4}$$

Since $\frac{1}{2}(1 + \beta) < 1$, we see that F is a contractive operator on $N((\bar{x}_i, \bar{y}_i), \frac{3}{2}\delta)$. The proof of the lemma is complete. □

Define

$$F^2 = F \circ F, \ F^{i+1} = F^i \circ F \text{ for } i \in N.$$

In view of Lemma 3.2, we see that F^m is a contractive return map on $N((\bar{x}_0, \bar{y}_0), \delta)$. Thus F has a unique fixed point $(x^*, y^*) \in N((\bar{x}_0, \bar{y}_0), \delta)$. The solution of (1.1) with the initial condition $x_0 = x^*, y_0 = y^*$ is a k-periodic solution. Moreover, for any $(a, b) \in N((x^*, y^*), \frac{1}{2}\delta) \subset N((\bar{x}_0, \bar{y}_0), \frac{3}{2}\delta)$, from (3.4), we get

$$||F^n(a, b) - F^n(x^*, y^*)|| \leq (\frac{1 + \beta}{2})^n||(a, b) - (x^*, y^*)||.$$

which implies the attractivity of the solution $\{F^n(x^*, y^*)\}$ of (1.1). The proof of Theorem 3.1 is now complete. □

Combining Theorem 2.1 and Theorem 3.1, we have

Corollary 3.3 For any fixed $m \in N$, if $\beta \in (0,1)$ and (2.2) holds for m. Then there exist positive constants ϵ, r, r_1 and L with $r > r_1$ and for f satisfying (1.3) and (1.4), (1.1) has an attractive $4m$-periodic solution.

References

[1] W. G. Kelly and A. C. Peterson, *Difference Equations, An Introduction with Applications*, Academic Press, New York, 1991.

[2] V. LJ. Kocic and G. Ladas, *Global Behavior of Nonlinear Difference Equations of Higher Order with Applications*, Kluwer Academic Publishers, 1993.

[3] M. Martelli, *Introduction to Discrete Dynamical Systems and Chaos*, John Wiley and Sons, Inc, 1990.

[4] H. O. Walther, Contracting return maps for monotone delayed feedback, *Discrete and Continuous Dynamical Systems*, **7**(2001), 259-274.

[5] H. O. Walther, Contracting return maps for some delay differential equations, in "Topics in Functional Differential and Difference Equations" (T. Faria, P. Freitas, eds), *Fields Inst. Communications*, **29**(2001), 349-360.

[6] J. Wu, *Introduction to Neural Dynamics and Signal Transmission Delay*, De-Gruyter, Berlin, 2001.

[7] Z. Zhou and J. Wu, Attractive periodic orbits in nonlinear discrete-time neural networks with delayed feedback, *J. Diff. Eqns. Appl.* **8**(5) (2002), 467-483.

[8] Z. Zhou, J. S. Yu and L. H. Huang, Asymptotic behavior of delay difference systems, *Computers Math. Applic.* **42** (2001), 283–290.

Fields Institute Communications
Volume **42**, 2004

Dynamic Analysis of a Discrete-Time Network of Two Neurons with Delayed Feedback

Huiyan Zhu
College of Mathematics-Physics
Nanhua University
Hengyang, Hunan 421001, P. R. China
lyzn@21cn.com

Lihong Huang
College of Mathematics and Econometrics
Hunan University
Changsha, Hunan 410082, P. R. China
lhhuang@hnu.net.cn

Dedicated to Professor Zhicheng Wang on the occasion of his seventieth birthday

Abstract. We consider a nonlinear discrete-time system describing the dynamic interaction of two identical neurons, which includes the discrete version of an artificial neural network model of two neurons with delayed feedback. A detailed analysis is presented for the dynamics of the system, and some interesting results are obtained for the periodicity and convergence of solutions of the system.

1 Introduction

Two-neuron networks have played an important role in the analysis of delay-induced changes in the dynamics of neural networks. Despite the small number of neurons in the system, in many instances, two-neuron networks with delay display the same behavior as larger networks and many techniques developed to deal with two-neuron networks can be carried over to large size networks such as those used in practical applications. Therefore, a two-neuron net can be used as prototypes to improve our understanding of the influence of delay. Much has been done for

2000 *Mathematics Subject Classification.* Primary 92B20; Secondary 39A11.
Research supported by National Natural Science Foundation of China (10071016), the Key Research Program of Science and Technology of the Ministry of Education of China, the Doctor Program Foundation of the Ministry of Education of China (20010532002) and Foundation for University Excellent Teacher by the Ministry of Education.

the dynamic analysis of two-neuron networks described by delay differential equations(see, for example, [3–7, 9–12]). As we know, the system

$$\begin{cases} \dot{x} = -x + f(y(t-\tau)), \\ \dot{y} = -y + f(x(t-\tau)) \end{cases} \tag{1.1}$$

has been used as the model for an artificial neural network of two neurons, and such a model has attracted the attention of many researchers (see, for example, [3–6, 9, 10]), where x and y denote the activations of the neurons, $f : R \to R$ denotes the signal function, τ denotes the signal delay related to each connection. In particular, Huang and Wu [6] gave a detailed dynamic analysis for the system (1.1) when f is of the form

$$f(u) = \begin{cases} -1, & u > \sigma^*, \\ 1, & u \leq \sigma^* \end{cases} \tag{1.2}$$

for some constant $\sigma^* \in R$.

Recently, there has been increasing interest in the study of dynamics for nonlinear discrete-time neural networks with delayed feedback. But there are only a limited number of works about it, for some results, we refer to [13, 14, 15]. For the general background and results on discrete-time systems, we refer to [1, 2, 8] and references therein. In a recent paper [14], Zhou, Yu and Huang replaced $\dot{x}$ and $\dot{y}$ in (1.1) by the backward difference $x_n - x_{n-1}$ and $y_n - y_{n-1}$ respectively to get the following discrete analog of the model (1.1).

$$\begin{cases} 2x_n - x_{n-1} = f(y_{n-k}), \\ 2y_n - y_{n-1} = f(x_{n-k}), \end{cases} \quad n = 0, 1, 2, \cdots, \tag{1.3}$$

where k is a positive integer. They studied the asymptotic behavior of the system (1.3) when f satisfies (1.2), and obtained some interesting results.

The purpose of this paper is to give a dynamic analysis for the discrete-time system (1.3) when the signal function f is of the following form:

$$f(\xi) = \begin{cases} 1, & \xi \in (0, \sigma], \\ 0, & \xi \in (-\infty, 0] \bigcup (\sigma, \infty) \end{cases} \tag{1.4}$$

for some constant $\sigma \in (0, \infty)$. Such a signal function shows that if the activation of one neuron is between 0 and σ, then it has a constant active affection to the other one; otherwise, it has no affection.

To simplify our presentation, we introduce some notations and notions first. Let Z denote the set of all integers. For any $a, b \in Z$ with $a \leq b$, define $N(a) = \{a, a+1, \cdots\}$ and $N(a, b) = \{a, a+1, \cdots b\}$. In particular, let $N = N(0)$. Moreover, let X denote the set of mappings from $N(-k, -1)$ to R^2 and define

$$\begin{aligned} R_1 &= \{\varphi; \quad \varphi : N(-k, -1) \to R \quad \text{and } \varphi(i) \leq 0 \text{ for } i \in N(-k, -1)\}, \\ R_2 &= \{\varphi; \quad \varphi : N(-k, -1) \to R \quad \text{and } 0 < \varphi(i) \leq \sigma \text{ for } i \in N(-k, -1)\}, \\ R_3 &= \{\varphi; \quad \varphi : N(-k, -1) \to R \quad \text{and } \varphi(i) > \sigma \text{ for } i \in N(-k, -1)\}, \\ X_{ij} &= \{\Phi; \quad \Phi = (\varphi, \psi) \in X, \varphi \in R_i \quad \text{and } \psi \in R_j\}, \quad i, j = 1, 2, 3 \end{aligned}$$

and

$$X_\sigma = \bigcup_{i,j=1}^{3} X_{ij}.$$

A below semicycle of a sequence $\{x_n\}_{n=-k}^{\infty}$ relative to zero consists of a string of terms $\{x_l, x_{l+1}, \cdots, x_m\}$, all less than or equal to zero, with $l \geq -k$ and $m \leq \infty$

and such that

$$\text{either } l = -k \text{ or } l > -k \text{ and } x_{l-1} > 0,$$

and

$$\text{either } m = \infty \text{ or } m < \infty \text{ and } x_{m+1} > 0.$$

A middle semicycle of a sequence $\{x_n\}_{n=-k}^{\infty}$ relative to zero and σ consists of a string of terms $\{x_l, x_{l+1}, \cdots, x_m\}$, all greater than zero and less than or equal to σ, with $l \geq -k$ and $m \leq \infty$ and such that

$$\text{either } l = -k \text{ or } l > -k \text{ and either } x_{l-1} \leq 0 \text{ or } x_{l-1} > \sigma,$$

and

$$\text{either } m = \infty \text{ or } m < \infty \text{ and either } x_{m+1} \leq 0 \text{ or } x_{m+1} > \sigma.$$

An above semicycle of a sequence $\{x_n\}_{n=-k}^{\infty}$ relative to σ consists of a string of terms $\{x_l, x_{l+1}, \cdots, x_m\}$, all greater than zero, with $l \geq -k$ and $m \leq \infty$ and such that

$$\text{either } l = -k \text{ or } l > -k \text{ and } x_{l-1} \leq 0,$$

and

$$\text{either } m = \infty \text{ or } m < \infty \text{ and } x_{m+1} \leq 0.$$

By a solution of (1.3), we mean a sequence $\{(x_n, y_n)\}$ of points in R^2 that is defined for all $n \in N(-k)$ and satisfies (1.3) for $n \in N$. Clearly, for any $\Phi = (\varphi, \psi) \in X$, (1.3) has a unique solution (x_n^{Φ}, y_n^{Φ}) satisfying the initial condition:

$$(x_i^{\Phi}, y_i^{\Phi}) = (\varphi(i), \psi(i)) \quad \text{for } i \in N(-k, -1). \tag{1.5}$$

Our main goal is to determine the large time behavior of (x_n^{Φ}, y_n^{Φ}). In particular, we concentrate on the case where $\varphi, \varphi - \sigma, \psi$ and $\psi - \sigma$ have no sign changes on $N(-k, -1)$. Namely, we consider those $\Phi \in X_\sigma$.

The main results of this paper are as follows.

Theorem 1.1 *Assume that $0 < \sigma < 1$ and $\Phi = (\varphi, \psi) \in X_{22} \bigcup X_{33}$ and $\varphi(-1) = \psi(-1)$. Then the solution $\{(x_n^{\Phi}, y_n^{\Phi})\}_{n=-k}^{\infty}$ of (1.3) with the initial Φ and f satisfying (1.4) satisfies $x_n^{\Phi} = y_n^{\Phi}$ for $n \in N$, and the following statements are true.*

- **(i):** *Every above semicycle of $\{x_n^{\Phi}\}$ relative to σ except perhaps for the first one, has at least k terms, and the number of terms is less than $\log_2 \frac{2^k+\sigma-1}{\sigma}$.*
- **(ii):** *Every middle semicycle of $\{x_n^{\Phi}\}$ relative to σ except perhaps for the first one, has at least k terms, and the number of terms is less than $\log_2 \frac{2^k-\sigma}{1-\sigma}$.*

Corollary 1.2 *Assume that $\Phi = (\varphi, \psi) \in X_{22} \bigcup X_{33}$ and $\varphi(-1) = \psi(-1)$, and that (x_n^{Φ}, y_n^{Φ}) is the solution of (1.3) with the initial value Φ and f satisfying (1.4). Then $x_n^{\Phi} = y_n^{\Phi}$ for all $n \in N$, furthermore, we have the following.*

- **(i):** *For $0 < \sigma < \frac{2^k}{2^{k+1}-1}$, every middle semicycle of $\{x_n^{\Phi}\}$, except perhaps for the first one, has just k terms.*
- **(ii):** *For $\frac{2^k-1}{2^{k+1}-1} < \sigma < 1$, every above semicycle of $\{x_n^{\Phi}\}$, except perhaps for the first one, has just k terms.*

(iii): *For* $\frac{2^k-1}{2^{k+1}-1} < \sigma < \frac{2^k}{2^{k+1}-1}$, *there exists* $\Phi_0 = (\varphi_0, \psi_0) \in X_{33}$ *with* $\varphi_0(-1) = \psi_0(-1)$ *such that the solution* $(x_n^{\Phi_0}, y_n^{\Phi_0})$ *of (1.3) with initial value* Φ_0 *is periodic with the minimal period* $2k$. *Moreover, there is an integer* $m \in N(0, 2k-1)$ *such that*

$$\lim_{n\to\infty} (x_n^{\Phi} - x_{n-m}^{\Phi_0}) = 0.$$

Theorem 1.3 *Let* $\sigma > 1$ *and* $\Phi = (\varphi, \psi) \in X_\sigma$. *Then each solution* (x_n^Φ, y_n^Φ) *of (1.3) with the initial value* Φ *and* f *satisfying (1.4) is convergent as* $n \to \infty$. *More precisely, we have*

(i): $(x_n^\Phi, y_n^\Phi) \to (0,0)$ *as* $n \to \infty$ *for* $\Phi \in X_{11}$.
(ii): $(x_n^\Phi, y_n^\Phi) \to (1,1)$ *as* $n \to \infty$ *for* $\Phi \in X_\sigma - X_{11}$.

Theorem 1.4 *Let* $\sigma = 1$. *Then every solution* (x_n^Φ, y_n^Φ) *of (1.3) with the initial value* Φ *and* f *satisfying (1.4) is convergent as* $n \to \infty$. *More precisely, we have the following conclusions:*

(i): *If* $\Phi \in X_{11}$, *then* $(x_n^\Phi, y_n^\Phi) \to (0,0)$ *as* $n \to \infty$;
(ii): *If* $\Phi \in X_{22} \bigcup X_{12} \bigcup X_{13} \bigcup X_{21} \bigcup X_{31}$, *then* $(x_n^\Phi, y_n^\Phi) \to (1,1)$ *as* $n \to \infty$;
(iii): *If* $\Phi \in X_{23}$, *then* $(x_n^\Phi, y_n^\Phi) \to (0,1)$ *as* $n \to \infty$;
(iv): *If* $\Phi \in X_{32}$, *then* $(x_n^\Phi, y_n^\Phi) \to (1,0)$ *as* $n \to \infty$;
(v): *If* $\Phi \in X_{33}$, *then*
  (a): $(x_n^\Phi, y_n^\Phi) \to (1,0)$ *as* $n \to \infty$ *and* $\varphi(-1) \geq \psi(-1)2^k$,
  (b): $(x_n^\Phi, y_n^\Phi) \to (1,1)$ *as* $n \to \infty$ *and* $\psi(-1) < \varphi(-1) \leq \psi(-1)2^{k-1}$,
  (c): $(x_n^\Phi, y_n^\Phi) \to (0,1)$ *as* $n \to \infty$ *and* $\psi(-1) \geq \varphi(-1)2^k$,
  (d): $(x_n^\Phi, y_n^\Phi) \to (1,1)$ *as* $n \to \infty$ *and* $\varphi(-1) < \psi(-1) \leq \varphi(-1)2^{k-1}$,
  (e): $(x_n^\Phi, y_n^\Phi) \to (1,1)$ *as* $n \to \infty$ *and* $\varphi(-1) = \psi(-1)$.

2 Proofs of main results

Assuming $n_0 \in N$, we first note that the solution of the difference equation

$$2x_n - x_{n-1} = 0, \quad n \in N(n_0), \tag{2.1}$$

with initial condition $x_{n_0-1} = a$, is

$$x_n = a(\frac{1}{2})^{n+1-n_0}, \quad n \in N(n_0). \tag{2.2}$$

The solution of the difference equation

$$2x_n - x_{n-1} = 1, \quad n \in N(n_0), \tag{2.3}$$

with initial condition $x_{n_0-1} = a$, is

$$x_n = (a-1)(\frac{1}{2})^{n+1-n_0} + 1, \quad n \in N(n_0). \tag{2.4}$$

For the sake of simplicity, in the sequel, we always denote x_n^Φ and y_n^Φ by x_n and y_n, respectively.

Proof of Theorem 1.1 We only consider the case where $\Phi = (\varphi, \psi) \in X_{33}$. The case where $\Phi = (\varphi, \psi) \in X_{22}$ can be dealt with analogously.

By (1.3), we can easily obtain that $x_n = y_n$ for all $n \in N$. Therefore, it suffices to show that the solution $\{x_n\}$ of the equation

$$2x_n - x_{n-1} = f(x_{n-k}), \quad n = 0, 1, 2, \cdots, \tag{2.5}$$

with initial condition

$$x_i \in R_3 \text{ for } i \in N(-k,-1),$$

satisfies (i) and (ii).

Assume that the first semicycle of $\{x_n\}$ relative to σ is $\{x_{-k}, \cdots, x_{m_1-1}\}$, which is an above semicycle. This implies that

$$x_{m_1-1} > \sigma \quad \text{and } x_{m_1} \leq \sigma. \tag{2.6}$$

Using (2.5) and (1.4), we see that

$$2x_n - x_{n-1} = 0, \quad n \in N(0, m_1 + k - 1).$$

In view of (2.2), we have

$$x_n = x_{m_1-1}(\frac{1}{2})^{n+1-m_1}, \quad n \in N(m_1, m_1 + k - 1). \tag{2.7}$$

Assume that the second semicycle of $\{x_n\}$ relative to zero and σ is $\{x_{m_1}, \cdots, x_{m_2-1}\}$. This is a middle semicycle, which implies that

$$0 < x_{m_2-1} \leq \sigma \quad \text{and } x_{m_2} > \sigma. \tag{2.8}$$

By (2.7), it is easy to see that

$$0 < x_{m_1+k-1} \leq x_{m_1+k-2} \cdots \leq x_{m_1} \leq \sigma.$$

Thus, $m_2 \geq m_1 + k$, and

$$2x_n - x_{n-1} = 1, \quad n \in N(m_1 + k, m_2 + k - 1).$$

By (2.4), we get

$$\begin{aligned} x_n &= (x_{m_1+k-1} - 1)(\tfrac{1}{2})^{n-m_1-k+1} + 1 \\ &= x_{m_1-1}(\tfrac{1}{2})^{n-m_1+1} - (\tfrac{1}{2})^{n-m_1-k+1} + 1. \end{aligned}$$

In particular, we have

$$x_{m_2-1} = x_{m_1}(\frac{1}{2})^{m_2-m_1} - (\frac{1}{2})^{m_2-m_1-k} + 1.$$

From (2.6) and (2.8), it follows that

$$\sigma(\frac{1}{2})^{m_2-m_1} - (\frac{1}{2})^{m_2-m_1} 2^k + 1 < x_{m_2-1} \leq \sigma,$$

which implies

$$m_2 - m_1 < \log_2 \frac{2^k - \sigma}{1 - \sigma}.$$

Assume that the third semicycle of $\{x_n\}$ relative to σ is $\{x_{m_2}, \cdots, x_{m_3-1}\}$. Clearly, this is an above semicycle. By a similar way to the second semicycle, we can prove

$$k < m_3 - m_2 < \log_2 \frac{2^k + \sigma - 1}{\sigma}.$$

By induction, it is easy to finish the proof of Theorem 1.1. □

Proof of Corollary 1.2 If $0 < \sigma < \frac{2^k}{2^{k+1}-1}$, since

$$\log_2 \frac{2^k - \sigma}{1 - \sigma} < k + 1,$$

we know that the conclusion (i) holds

Similarly, we see that the conclusion (ii) holds.

If $\frac{2^k-1}{2^{k+1}-1} < \sigma < \frac{2^k}{2^{k+1}-1}$, then

$$\log_2 \frac{2^k+\sigma-1}{\sigma} < k+1 \text{ and } \log_2 \frac{2^k-\sigma}{1-\sigma} < k+1.$$

By Theorem 1.1, we see that every semicycle of $\{x_n\}$, except perhaps for the first one, has k terms. We consider the case where $\Phi = (\varphi, \psi) \in X_{33}$, the case where $\Phi = (\varphi, \psi) \in X_{22}$ is similar and the proof is omitted. Assume that m_1 is the least nonnegative integer such that (2.6) holds. Then the i-th semicycle of $\{x_n\}$ is $\{x_{m_1+(i-2)k}, x_{m_1+(i-2)k+1}, \cdots, x_{m_1+(i-1)k-1}\}$ for $i \in N(2)$. From (2.5), we have, for $i \in N(1)$,

$$2x_n - x_{n-1} = 0, \quad n \in N(m_1+(2i-2)k, m_1+(2i-1)k-1)$$

and

$$2x_n - x_{n-1} = 1, \quad n \in N(m_1+(2i-1)k, m_1+2ik-1).$$

Thus, we have

$$x_n = x_{m_1-1}(\frac{1}{2})^{n+1-m_1}, \quad n \in N(m_1, m_1+k-1)$$

and

$$x_n = x_{m_1-1}(\frac{1}{2})^{n-m_1+1} - (\frac{1}{2})^{n-m_1-k+1} + 1, \\ n \in N(m_1+k, m_1+2k-1).$$

In general, for $i \in N(1)$, we have

$$x_n = \begin{cases} (\frac{1}{2})^{n-m_1-k+1}\frac{2^{(2i-2)k}-1}{2^k+1} + x_{m_1-1}(\frac{1}{2})^{n-m_1+1} \\ \quad \text{for } n \in N(m_1+(2i-2)k, m_1+(2i-1)k-1); \\ -(\frac{1}{2})^{n-m_1-k+1}\frac{2^{(2i-1)k}+1}{2^k+1} + x_{m_1-1}(\frac{1}{2})^{n-m_1+1} + 1 \\ \quad \text{for } n \in N(m_1+(2i-1)k, m_1+2ik-1). \end{cases} \tag{2.9}$$

Let $\Phi_0 = (\varphi_0, \psi_0) \in X_{33}$ with $\varphi_0(-1) = \psi_0(-1) = \frac{2^k}{1+2^k}$. Then, from (2.9) we get, for $i \in N(1)$,

$$x_n^{\Phi_0} = \begin{cases} \frac{1}{2^k+1}(\frac{1}{2})^{n-(2i-1)k+1} \text{ for } n \in N((2i-2)k, (2i-1)k-1), \\ -\frac{1}{2^k+1}(\frac{1}{2})^{n-2ik+1} + 1\text{for } n \in N((2i-1)k, 2ik-1). \end{cases}$$

Clearly, $\{x_n^{\Phi_0}\}_{n=0}^{\infty}$ is periodic with minimal period $2k$. In view of (2.9), it is easy to see that

$$x_{m_1+j} - x_j^{\Phi_0} = (x_{m_1-1} - \frac{2^k}{1+2^k})(\frac{1}{2})^{j+1} \to 0 \text{ as } j \to \infty. \tag{2.10}$$

Let $m = m_1 - 2k[\frac{m_1}{2k}]$, where $[\cdot]$ denotes the greatest function. Then $m \in N(0, 2k-1)$, and it follows from (2.10) that

$$\lim_{n\to\infty}(x_n - x_{n-m}^{\Phi_0}) = 0.$$

This completes the proof of Corollary 1.2. □

Proof of Theorem 1.3 We first show the conclusion (i). Since $\Phi = (\varphi, \psi)^T \in X_{11}$, from (1.3) and (1.4), we see that

$$\begin{cases} 2x_n - x_{n-1} = 0, \\ 2y_n - y_{n-1} = 0 \end{cases} \tag{2.11}$$

for $n \in N(0, k-1)$.Therefore,

$$\begin{cases} x_n = \varphi(-1)(\frac{1}{2})^{n+1}, \\ y_n = \psi(-1)(\frac{1}{2})^{n+1} \end{cases} \tag{2.12}$$

for $n \in N(0, k-1)$. This implies that $x_n < 0$ and $y_n < 0$ for $n \in N(0, k-1)$, and (2.11) is satisfied for $n \in N(k, 2k-1)$. Thus, $x_n < 0$ and $y_n < 0$ for $n \in N(k, 2k-1)$. Repeating this procedure, we can obtain that (x_n, y_n) satisfies (2.12) for all $n \in N$, from which it follows that $(x_n, y_n) \to (0, 0)$ as $n \to \infty$.

We next let $\Phi \in X_\sigma - X_{11}$. To show the conclusion (ii), we distinguish several cases.

Case 1. $\Phi \in X_{22}$. In view of (1.3) and (1.4), we see that

$$\begin{cases} 2x_n - x_{n-1} = 1, \\ 2y_n - y_{n-1} = 1 \end{cases} \tag{2.13}$$

for $n \in N(0, k-1)$. It follows that

$$\begin{cases} x_n = (\varphi(-1) - 1)(\frac{1}{2})^{n+1} + 1, \\ y_n = (\psi(-1) - 1)(\frac{1}{2})^{n+1} + 1 \end{cases} \tag{2.14}$$

for $n \in N(0, k-1)$.

Note that

$$0 < -(\frac{1}{2})^{n+1} + 1 < x_n \leq (\sigma - 1)(\frac{1}{2})^{n+1} + 1 < \sigma$$

and

$$0 < -(\frac{1}{2})^{n+1} + 1 < y_n \leq (\sigma - 1)(\frac{1}{2})^{n+1} + 1 < \sigma$$

for $n \in N(0, k-1)$, it follows that (2.13) is satisfied for $n \in N(k, 2k-1)$. Thus, $0 < x_n \leq \sigma$ and $0 < y_n \leq \sigma$ for $n \in N(k, 2k-1)$. Repeating this procedure, we can obtain that (x_n, y_n) satisfies (2.14) for all $n \in N$, from which we know that $(x_n, y_n) \to (1, 1)$ as $n \to \infty$.

Case 2. $\Phi \in X_{33}$. From (1.3) and (1.4), we see that

$$\begin{cases} 0 \leq 2x_n - x_{n-1} \leq 1, \\ 0 \leq 2y_n - y_{n-1} \leq 1 \end{cases} \tag{2.15}$$

for $n \in N$. Using induction, it is easy to show that

$$\begin{cases} 0 < \varphi(-1)(\frac{1}{2})^{n+1} \leq x_n \leq (\varphi(-1) - 1)(\frac{1}{2})^{n+1} + 1, \\ 0 < \psi(-1)(\frac{1}{2})^{n+1} \leq y_n \leq (\psi(-1) - 1)(\frac{1}{2})^{n+1} + 1, \end{cases} \quad n \in N. \tag{2.16}$$

From (2.16), we see that there exists a positive integer m such that $0 < x_n \leq \sigma$ and $0 < y_n \leq \sigma$ for $n \in N(m_1)$. This, together with (1.3) and (1.4), implies that (2.13) holds for $n \in N(m_1 + k)$. Therefore, we have

$$\begin{cases} x_n = (x_{m_1+k-1} - 1)(\frac{1}{2})^{n-m_1-k+1} + 1, \\ y_n = (y_{m_1+k-1} - 1)(\frac{1}{2})^{n-m_1-k+1} + 1, \end{cases} \quad n \in N(m_1 + k).$$

It follows that $(x_n, y_n) \to (1, 1)$ as $n \to \infty$.

Case 3. $\Phi \in X_{12} \cup X_{13} \cup X_{23}$. In view of (1.3) and (1.4), it is easy to see that

$$\begin{cases} x_n \leq (\varphi(-1) - 1)(\frac{1}{2})^{n+1} + 1, \\ 0 < \psi(-1)(\frac{1}{2})^{n+1} \leq y_n \leq (\psi(-1) - 1)(\frac{1}{2})^{n+1} + 1 \end{cases} \tag{2.17}$$

for $n \in N$. This implies that there exists a positive integer m_1 such that $0 < y_n \leq \sigma$ for $n \in N(m_1)$. Thus, by (1.3) and (1.4), we can obtain

$$2x_n - x_{n-1} = 1, \quad n \in N(m_1 + k).$$

Therefore, it follows from (2.3) and (2.4) that $x_n = (x_{m_1+k-1}-1)(\frac{1}{2})^{n-m_1-k+1}+1$ for $n \in N(m_1+k)$, which implies

$$\lim_{n\to\infty} x_n = 1 > 0.$$

Hence, by (2.15), it is certain that there exists a positive integer $m_2 > m_1 + k$ such that $0 < x_n \le \sigma$ for $n \in N(m_2)$. Therefore,

$$\begin{cases} x_n = (x_{m_2+k-1}-1)(\frac{1}{2})^{n-m_2-k+1}+1, \\ y_n = (y_{m_2+k-1}-1)(\frac{1}{2})^{n-m_2-k+1}+1, \end{cases} \quad n \in N(m_2+k),$$

which implies that $(x_n, y_n) \to (1,1)$ as $n \to \infty$.

Case 4. $\Phi \in X_{21} \cup X_{31} \cup X_{32}$. For this case, the conclusion follows from Case 3 and the symmetry of (1.3). This completes the proof of Theorem 1.3. □

Proof of Theorem 1.4 The proof of the conclusion (i) is the same as that of the conclusion (i) of Theorem 1.3.

To prove the conclusion (ii), we distinguish several cases.

Case 1. $\Phi \in X_{22}$. For this case, the conclusion can be obtained by a similar argument as that in Case 1 of the conclusion (ii) for the proof of Theorem 1.3.

Case 2. $\Phi \in X_{21}$. From (1.3) and (1.4), we see that

$$\begin{cases} 2x_n - x_{n-1} = 0, \\ 2y_n - y_{n-1} = 1 \end{cases} \tag{2.18}$$

for $n \in N(0, k-1)$.Therefore,

$$\begin{cases} x_n = \varphi(-1)(\frac{1}{2})^{n+1}, \\ y_n = [\psi(-1)-1](\frac{1}{2})^{n+1}+1 \end{cases} \tag{2.19}$$

for $n \in N(0, k-1)$.

Let m_1 be the least nonnegative integer such that

$$y_{m_1} - 1 < 0 \quad \text{and} \quad y_{m_1} \ge 0. \tag{2.20}$$

Then (2.18) and so (2.19) hold for $n \in N(0, m_1+k-1)$. By (2.19) and (2.20), we have

$$\begin{cases} 1 > y_{m_1+i} = [\psi(-1)-1](\frac{1}{2})^{m_1+i+1}+1 \ge 0, \\ 1 \ge x_{m_1+i} = \varphi(-1)(\frac{1}{2})^{m_1+i+1} > 0, \end{cases} \quad i \in N(0, k-1).$$

Similar to Case 1, we know the conclusion holds.

Case 3. $\Phi \in X_{12}$. For this case, the conclusion follows from Case 2 and the symmetry of (1.3).

Case 4. $\Phi \in X_{31}$. From (1.3) and (1.4), we see that (2.11) holds for $n \in N(0, k-1)$. It follows that (2.12) is satisfied for $n \in N(0, k-1)$.

Assume that m_1 is the least nonnegative integer such that

$$x_{m_1-1} > 1 \quad \text{and} \quad x_{m_1} \le 1, \tag{2.21}$$

then (2.12) holds for $n \in N(0, m_1+k-1)$. By (2.12), we obtain

$$\begin{aligned} &0 < x_{m_1+i} = \varphi(-1)(\tfrac{1}{2})^{m_1+i+1} \le x_{m_1} \le 1, \\ &y_{m_1+i} = \psi(-1)(\tfrac{1}{2})^{m_1+i+1} \le 0, \end{aligned} \quad i \in N(0, k-1).$$

Similar to the Case 2, we know the conclusion holds.

Case 5. $\Phi \in X_{13}$.

For this case, the conclusion follows from Case 4 and the symmetry of (1.3).

Now we show the conclusion (iii). Let $\Phi \in X_{23}$. From (1.3) and (1.4), we see that (2.18) and (2.19) hold for $n \in N(0, k-1)$. This implies that $0 < x_n \leq 1$ and $y_n > 1$ for $n \in N(0, k-1)$, and that (2.19) is satisfied for $n \in N(k, 2k-1)$. Thus, $0 < x_n \leq 1$ and $y_n > 1$ for $n \in N(k, k-1)$. Repeating this procedure, we can obtain that (x_n, y_n) satisfies (2.19) for all $n \in N$, from which we know that $(x_n, y_n) \to (0, 1)$ as $n \to \infty$.

We next show the conclusion (iv).

Let $\Phi \in X_{32}$. For this case, (iv) follows from (iii) and the symmetry of (1.3).

Finally, we prove the conclusion (v). Let $\Phi \in X_{33}$ and $\varphi(-1) \geq \psi(-1)2^k$. From (1.3) and (1.4), we see that (2.11) and (2.12) hold for $n \in N(0, k-1)$. Let m_1 be the least nonnegative integer such that

$$y_{m_1-1} > 1 \quad \text{and} \quad y_{m_1} \leq 1. \tag{2.22}$$

Then (2.12) holds for $n \in N(0, m_1 + k - 1)$. By (2.12), we obtain

$$0 < y_{m_1+i} = \psi(-1)(\frac{1}{2})^{m_1+i+1} \leq y_{m_1} \leq 1$$

and

$$\begin{aligned} x_{m_1+i} &= \varphi(\tfrac{1}{2})^{m_1+i+1} \geq \psi(-1)(\tfrac{1}{2})^{m_1+i+1-k} \\ &\geq \varphi(-1)(\tfrac{1}{2})^{m_1} = y_{m_1-1} > 1, \;\; i \in N(0, k-1). \end{aligned}$$

Thus, $x_{m_1+i} > 1$ and $0 < y_{m_1+i} \leq 1$ for $i \in N(0, k-1)$. By the conclusion (iv), the case (a) can be proved.

Case (b). $\Phi \in X_{33}$ and $1 < \frac{\varphi(-1)}{\psi(-1)} \leq 2^{k-1}$. By the same argument as in that of Case (a), we have

$$0 < x_{m_1+k-1} = \varphi(-1)(\frac{1}{2})^{m_1+k} \leq \psi(-1)(\frac{1}{2})^{m_1+k+1-k} = y_{m_1} \leq 1$$

and

$$0 < y_{m_1+k-1} = \psi(-1)(\frac{1}{2})^{m_1+k} \leq 1,$$

which implies that

$$0 < x_n \leq 1 \quad \text{and} \quad 0 \leq y_n \leq 1 \quad \text{for} \quad n \in N(m_1 + 2k - 1).$$

Therefore, (2.13) holds for $n \in N(m_1 + 2k - 1)$. It follows that (2.14) holds for all $n \in N(m_1 + 2k - 1)$, and hence $(x_n, y_n) \to (1, 1)$ as $n \to \infty$.

For the case (c) where $\psi(-1) \geq \varphi(-1)2^k$, the proof follows from Case (a) and the symmetry of (1.3). For the case (d) where $\varphi(-1) < \psi(-1) \leq \varphi(-1)2^{k-1}$, the proof follows from Case (b) and the symmetry of (1.3).

Case (e). $\Phi \in X_{33}$ and $\varphi(-1) = \psi(-1)$. Using (1.3), we can easily obtain that $x_n = y_n$ for all $n \in N$. Clearly, x_n satisfies equation

$$2x_n - x_{n-1} = f(x_{n-k}) \tag{2.23}$$

with initial conditions $x_i \in R$ for $i \in N(-k, -1)$. By (1.4) and (2.2), we have

$$x_n = \varphi(-1)(\frac{1}{2})^{n+1} \quad \text{for} \quad n \in N(0, k-1). \tag{2.24}$$

Assume that m_1 is the least nonnegative integer such that

$$x_{m_1-1} > 1 \quad \text{and} \quad x_{m_1} \leq 1, \tag{2.25}$$

then (2.24) holds for $n \in N(0, m_1 + k - 1)$. By (2.24), we have

$$0 < x_{m_1+k-1} = \varphi(-1)(\frac{1}{2})^{m_1+k} \leq \varphi(-1)(\frac{1}{2})^{m_1+1} = x_{m_1} \leq 1$$

which implies that $0 < x_n \leq 1$ for $n \in N(m_1 + k - 1)$. Therefore

$$2x_n - x_{n-1} = 1, \ n \in N(m_1 + 2k - 1).$$

In view of (2.4), we see that $x_n \to 1$ as $n \to \infty$, and thus $(x_n, y_n) \to (1, 1)$ as $n \to \infty$. The proof is complete. □

References

[1] R. P. Agarwal and P. J. Y. Wong, *Advanced Topics in Difference Equations*, Kluwer, Dordrecht, 1997.

[2] R. P. Agarwal, *Difference Equations and Inequalities: Theory, Method and Applications*, Marcel Dekker, New York, 1992.

[3] Y. Chen and J. Wu, *Minimal instability and unstable set of a phase-locked periodic orbit in a delayed neural network*, Physica D, **134** (1999), 185-199.

[4] Y. Chen and J. Wu, *Existence and attraction of a phase-locked oscillation in a delayed network of two neurons*, Integral and Differential Equations, **14** (2001), 1181-1238.

[5] Y. Chen, J. Wu and Krisztin, *Connection orbits from synchronous periodic solutions to phase-locked periodic solutions in a delay differential system*, J. Differential Equations **163** (2000),130-173.

[6] L. H. Huang and J. Wu, *Dynamics of inhibitory artificial neural network with threshold nonlinearity*, Fields Institute Communications, **29** (2001), 235-243.

[7] L. H. Huang and J. Wu, *The role of threshold in preventing delay-induced oscillations of frustrated neural networks with McCulloch-Pitts nonlinearity*, Int. J. Math. Game Theory and Algebra, **11** (2001), no.6, 71-100.

[8] V. L. J. Kocic and G. Ladas, *Global Behavior of Nonlinear Difference Equations of Higher Order with Applications*, Kluwer Academic, 1993.

[9] K. Pakdaman, C. Grotta-Ragazzo and C. P. Malta, *Transient regime duration in continuous-time neural networks delays*, Phys. Rev. E, **58** (1998), 3623-3627.

[10] K. Pakdaman, C. Grotta-Ragazzo, C. P. Malta, O. Arino and J. F. Vibert, *Effect of delay on the boundary of the basin attraction in a system of two neurons, Neural Networks*, **11** (1998), 509-519.

[11] T. Roska and L. O. Chua, *Cellular neural networks with nonlinear and delay-type template elements*, Int. J. Circuit Theory Appl. **20** (1992),469-481.

[12] J. Wu, *Introduction to Neural Dynamics and Signal Transmission Delay*, Berlin: De-Gruyter, 2001.

[13] Z. H. Yuan, L. H. Huang and Y. M. Chen, *Convergence and periodicity of solutions for a discrete-time network model of two neurons*, Mathematical and Computer Modelling, **35** (2002), 941-950.

[14] Z. Zhou, J. S. Yu and L. H. Huang, *Asymptotic behavior of delay difference systems*, Computers Math. Applic., **42** (3-5)(2001), 283-290.

[15] Z. Zhou and J. Wu, *Attractive periodic orbits in nonlinear discrete-time neural networks with delayed feedback*, J. Difference Equations and Applications, **8(5)** (2001), 467-483.

OPEN PROBLEMS AND CONJECTURES

Introduction
Open Problems and Conjectures

An open problem session was held during the annual meeting of the International Society of Difference Equations at Changsha. We include five papers by Professors Herman Brunner, Saber Elaydi, Ed Grove and Gerry Ladas, Robert Sacker and Hubertus von Bremen, and Jianhong Wu. Professor Brunner poses two open problems on the uniform boundedness and the asymptotic stability of solutions to systems of difference equations arising in the discretization of linear Volterra difference equations. Professor Elaydi poses several problems concerning the global stability of cycles and the extension of Sharkovsky's theorem in nonautonomous difference equations. Professors Grove and Ladas introduce six open problems and six conjectures. The authors consider a second order rational difference equation and the question of the existence of 2-cycles and their local and global stability. Professors Sacker and von Bremen pose three open problems on cycling in a 4-dimensional genetic model that has been developed by the authors (see this proceedings). Professor Wu considers a system of two delay difference equations that arises in the modeling of a network with two neurons. He poses four open problems concerning the number of periodic and quasi periodic orbits and their stability.

Saber Elaydi

Fields Institute Communications
Volume **42**, 2004

Discretization of Volterra Integral Equations with Weakly Singular Kernels

Hermann Brunner
Department of Mathematics and Statistics
Memorial University of Newfoundland
St. John's, NL Canada A1C 5S7
hbrunner@mun.ca

We present two open problems on the uniform boundedness and the asymptotic stability of solutions to systems of difference equations arising in the discretization of linear first- and second-kind Volterra integral equations with weakly singular kernels.

1 Description of the problems

Let $\alpha \in (0,1)$, $g \in C^1(I)$ $(I := [0,T])$, with $g(0) = 0$. Consider the (scalar) difference equation

$$a_0(\alpha; c_1)y_{n+1} + \sum_{\ell=0}^{n-1} a_{n-\ell}(\alpha; c_1)y_{\ell+1} = b_n(\alpha; c_1) \quad (n = 0, 1, \ldots, N-1), \tag{1.1}$$

where the parameter c_1 lies in $(0,1]$ and where the coefficients are given by

$$a_0(\alpha; c_1) := \int_0^{c_1} (t-s)^{-\alpha} ds, \tag{1.2}$$

$$a_{n-\ell}(\alpha; c_1) := \int_0^1 (n + c_1 - \ell - s)^{-\alpha} ds \quad (\ell < n), \tag{1.3}$$

$$b_0(\alpha; c_1) := h^{\alpha-1} g(t_n + c_1 h). \tag{1.4}$$

The points t_n belong to a uniform mesh $I_h := \{t_n := nh : \ n = 0, 1, \ldots, N; \ t_N = T\}$ on I.

Problem 1.1 Given $\alpha \in (0,1)$, for which values of $c_1 = c_1(\alpha) \in (0,1]$ do the solutions $\{y_{n+1}\}$ of the difference equation (1.1) remain uniformly bounded as $N \to \infty$, $h \to 0$ (with $Nh = T$) ?

As we shall see in Section 2, this difference equation arises in the discretization of the first-kind Volterra integral equation (first studied by N.H. Abel in 1823),

$$(\mathcal{V}_\alpha y)(t) := \int_0^t (t-s)^{-\alpha} y(s) ds = g(t), \quad t \in I, \tag{1.5}$$

2000 *Mathematics Subject Classification.* Primary 65R20; Secondary 39A12, 45E10.

by collocation using piecewise constant functions. Its (unique) solution $y \in C(I)$ is given by

$$y(t) = \frac{1}{\gamma_\alpha} \int_0^t (t-s)^{\alpha-1} g'(s)ds, \quad t \in I \quad (\gamma_\alpha := \Gamma(\alpha)\Gamma(1-\alpha)).$$

We note that if $\alpha = 0$ in (1.1), then the complete answer to Problem 1.1 is known: the solutions of the difference equation (1.1) remain uniformly bounded if, and only if, $c_1 \geq 1/2$ (compare, e.g., [2, 3]). For $0 < \alpha < 1$ a *sufficient condition* for uniform boundedness is $c_1 \geq c_1^*(\alpha) := (1/2)[\alpha(1-\alpha)\gamma_\alpha]^{1/(1-\alpha}$ (see [2]). For example, $c_1^*(\alpha) \doteq 0.3084$ when $\alpha = 1/2$.

In order to state Problem 1.2, let $\lambda < 0$, and assume that α and the coefficients $a_0(\alpha; c_1)$, $a_{n-\ell}(\alpha; c_1)$ and $b_n(\alpha; c_1)$ are as in Problem 1.1. The difference equation

$$y_{n+1} = 1 + \lambda h^{1-\alpha}\{a_0(\alpha; c_1)y_{n+1} + \sum_{\ell=0}^{n-1} a_{n-\ell}(\alpha; c_1)y_{\ell+1}\} \quad (n \geq 0) \tag{1.6}$$

results from the approximation of the linear second-kind Volterra integral equation,

$$y(t) = 1 + \lambda(\mathcal{V}_\alpha y)(t), \quad t \geq 0 \tag{1.7}$$

by collocation in the space of piecewise constant functions (compare Section 2). Its (unique) continuous solution is given by

$$y(t) = E_{1-\alpha}(\lambda\Gamma(1-\alpha)t^{1-\alpha}), \quad t \geq 0,$$

and it has the property that, for any $\alpha \in (0,1)$, $\lim_{t\to\infty} y(t) = 0$ whenever $\lambda < 0$. This follows from the definition of the *Mittag-Leffler function* E_β $(\beta > 0)$:

$$E_\beta(z) := \sum_{j=0}^{\infty} \frac{z^j}{\Gamma(1+j\beta)}.$$

Observe that for $\beta = 1$ we have $E_1(z) = \exp(z)$.

Problem 1.2 Given $\alpha \in (0,1)$, for which values of $c_1 = c_1(\alpha) \in (0,1]$ are the solutions of the difference equation (1.6) asymptotically stable for any fixed $h > 0$; i.e. when does

$$\lim_{n\to\infty} y_n = 0 \qquad (h > 0 \text{ fixed}) \tag{1.8}$$

hold ?

A *sufficient condition* was derived in [4] (see also [1]): asymptotic stability holds for (6) if $c_1 \geq \hat{c}_1(\alpha) := [2^{1/(1-\alpha)} - 1]^{-1}$. For $\alpha = 1/2$ we have $\hat{c}_1(\alpha) = 1/3$.

2 Background and generalization

Let

$$(\mathcal{V}_\alpha \phi)(t) := \int_0^t p_\alpha(t-s)\phi(s)ds, \quad t \in I := [0,T], \tag{2.1}$$

with $p_\alpha(z) := t^{-\alpha}$ if $0 < \alpha < 1$, and $p_\alpha(t) := \log(t)$ if $\alpha = 1$. Suppose that the Volterra integral equations (1.5) and (1.7), with the weakly singular integral operator $\mathcal{V}_\alpha$ $(0 < \alpha \leq 1)$ given by (2.1), are solved by collocation in the piecewise polynomial space

$$S_{m-1}^{(-1)}(I_h) := \{u_h : u_h|_{\sigma_n} \in \pi_{m-1} \; (0 \leq n \leq N-1)\},$$

where $I_h := \{t_n := nh : \ 0 \le n \le N;\ t_N = T\}$, $\sigma_n := (t_n, t_{n+1}]$, and where π_{m-1} $(m \ge 1)$ denotes the space of real polynomials of degree not exceeding $m-1$. Let the set of collocation points with respect to the given mesh I_h be

$$X_h := \{t_{n,i} := t_n + c_i h :\ 0 < c_1 < \cdots < c_m \le 1\ (0 \le n \le N-1)\}.$$

The collocation solutions $u_h \in S_{m-1}^{(-1)}(I_h)$ are determined by the collocation equations

$$(\mathcal{V}_\alpha u_h)(t) = g(t), \quad t \in X_h$$

(for the first-kind equation (1.5)) and

$$u_h(t) = 1 + \lambda(\mathcal{V}_\alpha u_h)(t), \quad t \in X_h$$

(for the second-kind equation (1.7)). For the computation of u_h we will employ the local Lagrange representation of u_h on σ_n, namely

$$u_h(t_n + vh) = \sum_{j=1}^{m} L_j(v) U_{n,j}, \quad v \in (0,1], \quad U_{n,j} := u_h(t_{n,j}), \tag{2.2}$$

where $L_j(v) := \prod_{k\ne j}^{m}(v - c_k)/(c_j - c_k)$. Let first $0 < \alpha < 1$, and define the $m \times m$ matrices

$$\begin{aligned} A_0(\alpha;\mathbf{c}) &:= \left(\begin{array}{c} \int_0^{c_i} (c_i - s)^{-\alpha} L_j(s)ds \\ (i,j = 1,\dots,m) \end{array} \right), \\ A_{n-\ell}(\alpha;\mathbf{c}) &:= \left(\begin{array}{c} \int_0^{1} (n + c_i - \ell - s)^{-\alpha} L_j(s)ds \\ (i,j = 1,\dots,m) \end{array} \right) \quad (\ell < n), \end{aligned}$$

with $\mathbf{c} := (c_1,\dots,c_m)$. The vectors $\mathbf{U}_n := (\,U_{n,1},\dots,U_{n,m}\,)^T \in \mathbb{R}^m$ defining u_h via (2.2) are then given, respectively, by the solutions of the linear difference equations

$$A_0(\alpha;\mathbf{c})\mathbf{U}_n + \sum_{\ell=0}^{n-1} A_{n-\ell}(\alpha;\mathbf{c})\mathbf{U}_\ell = \mathbf{b}_n(\alpha;\mathbf{c}) \quad (n = 0,1,\dots,N-1), \tag{2.3}$$

where $\mathbf{b}_n(\alpha;\mathbf{c}) := h^{\alpha-1}(\,g(t_n + c_1 h),\dots,g(t_n + c_m h)\,)^T \in \mathbb{R}^m$, and

$$\mathbf{U}_n = \mathbf{e} + \lambda\{A_0(\alpha;\mathbf{c})\mathbf{U}_n + \sum_{\ell=0}^{n-1} A_{n-\ell}(\alpha;\mathbf{c})\mathbf{U}_\ell\},$$

corresponding to the integral equations (1.5) and (1.7) with $0 < \alpha < 1$. Here, $\mathbf{e} := (\,1,\dots,1\,)^T \in \mathbb{R}^m$. For $m = 1$ these difference equations reduce to the scalar equations (1.1) and (1.6), where $\mathbf{U}_n = y_{n+1} = u_h(t_n + vh)$, $v \in (0,1]$.

Note that the solutions of the difference equation (2.3) with $\alpha = 0$ remain uniformly bounded as $N \to \infty$, $h \to 0$ $(Nh = T)$ if and only if the collocation parameters $\{c_i\}$ are such that

$$\rho_m := (-1)^m \prod_{i=1}^{m} \frac{1 - c_i}{c_i} \in [-1,1]$$

holds (see [2, 3] for details and original references).

For $0 < \alpha < 1$, the analogues of Problem 1.1 and Problem 1.2 for the vectors $\mathbf{U}_n \in \mathbb{R}^m$, and hence for the corresponding collocation solutions $u_h \in S_{m-1}^{(-1)}(I_h)$, remain completely open when $m \ge 2$. The same is true when $\alpha = 1$ in (2.1): in the case of a logarithmic kernel singularity even partial answers (e.g. sufficient conditions) to the two problems remain unknown when $m = 1$.

References

[1] L. Blank, *Stability of collocation for weakly singular Volterra equations,* IMA J. Numer. Anal., **15** (1995), 357-375.

[2] H. Brunner, *On systems of Volterra difference equations associated with collocation methods for weakly singular Volterra integral equations,* New Developments in Difference Equations and Applications (Taipei 1997) (S.S. Cheng, S. Elaydi and G. Ladas, eds.), pp. 75-92, Gordon and Breach, Amsterdam, 1999.

[3] H. Brunner, *Collocation Methods for Volterra Integral and Related Functional Differential Equations*, Cambridge University Press, Cambridge, 2004 (to appear).

[4] H. Brunner, R. Crisci, E. Russo and A. Vecchio, *A family of methods for Abel integral equations of the second kind,* J. Comput. Appl. Math., **34** (1991), 211-219.

Fields Institute Communications
Volume **42**, 2004

Nonautonomous Difference Equations: Open Problems and Conjectures

Saber N. Elaydi
Department of Mathematics
Trinity University
San Antonio, TX 78212 U.S.A.
selaydi@trinity.edu

1 Introduction

Autonomous difference equations of the form $x_{n+1} = f(x_n)$ may model populations of species with nonoverlapping generations such as fish, orchard pests, etc. The drawback of such models is that they do not account for environmental fluctuations or seasonal changes. Hence we are led to nonautonomous difference equations of the form $x_{n+1} = f_n(x_n)$, $n \in \mathbb{Z}^+$. Our main focus in this note will be on periodic difference equations in which the sequence f_n is periodic. Most of the open problems and conjectures in this part are motivated by recent work by Elaydi and Sacker [3], Elaydi and Yakubu [4] [5], and Elaydi [2]. The second part of the paper discusses the connection between a nonautonomous difference equation and its limiting equation. We present here several conjectures and open problems pertaining to the properties of omega limit sets (see Kempf [7]) and the question of lifting properties from the limiting equation to the original equation. For the convenience of the reader we introduce in Section 4 some rudiments of the theory of skew-product dynamical systems [8].

2 Periodic difference equations

Consider the nonautonomous difference equation

$$x_{n+1} = f(n, x_n) \tag{2.1}$$

which may be written in the form

$$x_{n+1} = f_n(x_n), \quad n \in \mathbb{Z}^+ \tag{2.2}$$

where $\mathbb{Z}^+$ is the set of nonnegative integers and $f_n : X \to X$ is a continuous function on a metric space X. If $f_{n+p} = f_n$ for all $n \in \mathbb{Z}^+$, then Eq. (2.1) is said to be periodic of period p or p-periodic. On the other hand, if $f_n = f$ for all $n \in \mathbb{Z}^+$, then Eq. (2.1) reduces to the autonomous difference equation

$$x_{n+1} = f(x_n) \tag{2.3}$$

2000 *Mathematics Subject Classification.* Primary 39A11; Secondary 57B55.

The "semi flow" of equation (2.2) is defined by the double sequence

$$\Phi(n, n_0) = f_{n+n_0-1} \circ f_{n+n_0-2} \circ \cdots \circ f_{n_0}, \quad \Phi(n_0, n_0) = id \tag{2.4}$$

If $n_0 = 0$, then we write $\Phi(n, 0) = \Phi_n$.

The orbit of x_0 is given by

$$\mathcal{O}(x_0) = \{\Phi_n(x_0) : n \in \mathbb{Z}^+\}.$$

Elaydi and Yakubu [4] showed that a globally asymptotically stable periodic cycle of Eq. (2.3) must be a fixed point if the space X is connected. A natural question is whether or not this theorem extends to nonautonomous difference equations. A definitive answer was recently provided by Elaydi and Sacker [3].

Theorem 2.1 [3] *Suppose that $f_{n+p} = f_n$ for all $n \in \mathbb{Z}^+$, for some $p \in \mathbb{Z}^+$. Then a globally asymptotically stable periodic cycle of Eq.* (2.1) *must be of period p or a divisor of p.*

Theorem 2.2 [3] *Given any two positive integers r and k such that r is a divisor of k. Then there exists a nonautonomous periodic difference equation with minimum period k and which has a globally asymptotically stable periodic orbit with minimal period r*

The following example illustrates the above theorems.

Example 2.3 [3] Consider the equation

$$\begin{aligned} x_{n+1} &= f_n(x_n) \\ \text{with } f_0(x) &= 1 + \alpha x \\ f_1(x) &= -\alpha + \alpha x \\ f_2(x) &= 1 + \beta x \\ f_3(x) &= -\beta + \beta x \\ f_n &= f \mod 4 \end{aligned}$$

The 2-cycle $\{0, 1\}$ is globally asymptotically stable, while the difference equation is 4-periodic.

A periodic solution of (2.1) may have a periodic solution whose period is unrelated to p as demonstrated by the following example.

Example 2.4 Consider the system of difference equations.

$$x_{n+1} = x_n - y_n + (x_n^2 + y_n^2 - 1) \cos \frac{2\pi}{3} n$$

$$y_{n+1} = x_n + y_n + (x_n^2 + y_n^2 - 1) \sin \frac{2\pi}{3} n$$

This is a periodic system with period $p = 3$. However, the solution $x_n = \cos \frac{n\pi}{4}$, $y_n = \sin \frac{n\pi}{4}$ is of period 8. Notice that, by virtue of Theorem 2.1 this 8-periodic orbit cannot be globally asymptotically stable.

Such periodic orbits will not be considered and we will be interested in periodic orbits of period p or in "superharmonics" whose minimal period is a multiple of p or in "subharmonics" whose period is a divisor of p.

Lemma 2.5 *A solution of* $\varphi(n) = \Phi_n(x_0)$ *is p-periodic if and only if*

$$\Phi_p(x_0) = x_0.$$

Proof Clearly if $\varphi(n)$ is p-periodic, then $\Phi_p(x_0) = x_0$. Conversely, define $\psi(n) = \varphi(n+p)$.
Then

$$\begin{aligned}\psi(n+1) &= \varphi(n+1+p) \\ &= f_{n+p}(\varphi(n+p)) \\ &= f_n(\psi(n))\end{aligned}$$

Consequently, $\psi(n)$ is a solution of equation Eq. (2.1). Moreover, $\psi(0) = \varphi(p) = \varphi(0)$ which implies by uniqueness of solutions that $\psi(n) = \varphi(n)$. Hence $\varphi(n+p) = \varphi(n)$ for all $n \in \mathbb{Z}^+$. □

We now turn our attention to scalar periodic difference equations.

Theorem 2.6 *Eq.* (2.1) *has a p-periodic solution if and only if the operator* $\Phi_p : X \to X$ *has a fixed point* x_o. *If* x_0 *is a fixed point of* Φ_p, *then* $\varphi(n, x_0)$ *is a periodic solution of Eq.*(2.1).

Theorem 2.7 *Suppose that Eq.* (2.1) *has a bounded solution. Then it has a p-periodic solution.*

Proof Let $x_n = \Phi_n(x_0)$ be a bounded solution of Eq. (2.1). Consider the point $x_p = \Phi_p(x_0)$. Then either $x_p > x_0$ or $x_p < x_0$. For if $x_p = x_0$ we then would have a p-periodic solution. Assume that $x_p > x_0$. If $x_{2p} = \Phi_p(x_p) < x_p$, then Eq. (2.1) has a p-periodic solution. So assume that the sequence $\Phi_{np}(x_0)$ is monotonically increasing. And since it is bounded it converges to a point $\bar{z}$ which is clearly a p-periodic solution of Eq. (2.1).

□

Remark 2.8 The main difficulty in corollary 2.7 is that it does not distinguish non-constant periodic solutions from equilibria (fixed points). One exception is the periodically forced autonomous systems of the form

$$x_{n+1} = f(x_n) + b(n) \tag{2.5}$$

where $b(n+p) = b(n)$ for all $n \in \mathbb{Z}^+$. If b is nonconstant, then equation (2.5) has no equilibrium points.

Consider the logistic map

$$x_{n+1} = \mu x_n(1 - x_n) \tag{2.6}$$

It is known that the positive fixed point of Eq. (2.6) $x^* = \frac{\mu-1}{\mu}$ is globally asymptotically stable if $1 < \mu \leq 3$.

Suppose now we replace μ with a periodic sequence μ_n with period p, i.e.,

$$\mu_{n+p} = \mu_n \text{ for all } n \in \mathbb{Z}^+.$$

We than have

$$x_{n+1} = \mu_n x_n(1 - x_n) \tag{2.7}$$

Open Problem 2.9 Find conditions on μ_n under which Eq. (2.7) has a globally asymptotically stable p-cycle

The Ricker's map given by $R(x) = xe^{\beta - x}$ has a globally asymptotically stable fixed β if $0 < \beta \leq 2$. The corresponding p-periodic difference equation is given by

$$x_{n+1} = x_n e^{\beta_n - x_n} \tag{2.8}$$

Open Problem 2.10 Find conditions under which Eq. (2.8) has a globally asymptotically p-cycle.

A more general open problem concerns itself with difference equations with a parameter such as

$$x_{n+1} = f(x_n, \alpha) \tag{2.9}$$

Open Problem 2.11 Suppose that Eq. (2.9) has a globally asymptotically stable fixed point in a positive cone in $\mathbb{R}^m$ for $c < \alpha < d$. Find conditions under which the equation

$$x_{n+1} = f(x_n, \alpha_n) \tag{2.10}$$

where $\alpha_{n+p} = \alpha_n$. has a globally asymptotically stable r-cycle where r is a divisor of p .

3 Extension of Sharkovsky's Theorem to periodic difference equations

Consider the following ordering of the natural numbers:

$3 \rhd 5 \rhd 7 \rhd \ldots$ odd integers	$2 \times 3 \rhd 2 \times 5 \rhd 2 \times 7 \rhd \ldots$ $2 \times$ odd integers
$2^2 \times 3 \rhd 2^2 \times 5 \rhd 2^2 \times 7 \rhd \ldots$ $2^2 \times$ odd integers	$2^n \times 3 \rhd 2^n \times 5 \rhd 2^n \times 7 \rhd \ldots$ $2^n \times$ odd integers
$\ldots \ldots \ldots$	$\cdots \rhd 2^n \rhd \cdots \rhd 2^3 \rhd 2^2 \rhd 2 \rhd 1$ powers of 2.

This exhausts all the natural numbers. This is called the Sharkovsky's ordering of the natural numbers. Sharkovsky's Theorem (f. Sharkovsky [9], Elaydi [1]) is

Theorem 3.1 *Suppose $f : \mathbb{R} \to \mathbb{R}$ is continuous and f has a periodic point of minimal period k. If $k \rhd r$ in the Sharkovsky's ordering, then f also has a periodic point of period r.*

It is of great interest to extend this remarkable result to periodic difference equations.

Conjecture 3.2 *Suppose that $f_{n+p} = f_n$ in Eq. (2.1) and $f_n : \mathbb{R} \to \mathbb{R}$ is continuous, $n \in \mathbb{Z}^+$. Suppose that Eq. (2.1) has a periodic point of minimal period k. Let $s = [p, k]$, be the least common divisor and let $\ell = s/k$. If $\ell \rhd r$ in the Sharkovsky's ordering, then f also has a periodic point of period r. In particular, if p and k are relatively prime, and k rhdr, then f has a point of period r.*

A converse of Sharkovsky's theorem was proved in Elaydi [2]

Theorem 3.3 *Given any two positive integers r and k such as $r \rhd k$ in the Sharkovsky's order. Then there exists a continuous function $f : \mathbb{R} \to \mathbb{R}$ which has a point of period k but no points of period r.*

Open Problem 3.4 Extend the above theorem to periodic difference equations.

In the next section we introduce the notion of a skew-product dynamical system which is, in our view, the main tool in studying nonautonomous difference equations. Notice that, unlike autonomous difference equations, nonautonomous difference equations do not give rise to a dynamical system. Such defect is remedied by defining a dynamical system on an enlarged space. This is in essence the main idea behind the notion of skew-product dynamical systems. In [8], Sacker an Sell developed the theory of skew-product dynamical systems and applied it to the study of nonautonomous differential equations. In [3], Elaydi and Sacker is developing the theory of discrete skew-product semidynamical systems and applying it to the study of nonautonomous difference equations. We now introduce some rudiments of the theory of skew-product dynamical systems.

4 Skew-product dynamical system

Definition 4.1 Let X and Y be two metric spaces. A dynamical system π on a product space $X \times Y$ is said to be a skew-product dynamical system if there exist continuous mappings $\varphi : X \times Y \times \mathbb{Z} \to X$ and $\sigma : Y \times \mathbb{Z} \to Y$ such that

$$\pi(x, y, n) = (\varphi(x, y, n), \sigma(y, n)).$$

If $\mathbb{Z}$ is replaced by $\mathbb{Z}^+$, then π is called a skew-product semi-dynamical system.

Now let us construct a skew-product semi-dynamical system for the periodic nonautonomous difference equation Eq. (2.1)

We let $Y = \{f_n : n \in \mathbb{Z}^+\}$. Now for f_i, we define $\sigma^r(f_i) = f_{i+r}$ as the shift map.

The map φ is defined as

$$\varphi(x_0, f_i, n) = \Phi(n, i)x_0$$

Hence

$$\pi((x_0, f_i), n) = (\Phi(n, i)x_0, f_{i+n})$$

For the general construction of skew-product dynamical systems we refer the reader to Elaydi and Sacker[3]. If p is the projection map, then the following diagram commutes.

$$\begin{array}{ccc} X \times Y \times \mathbb{Z}^+ & \xrightarrow{\pi} & X \times Y \\ \Big\downarrow{\scriptstyle p\times id} & & \Big\downarrow{\scriptstyle p} \\ Y \times \mathbb{Z}^+ & \xrightarrow{\sigma} & Y \end{array}$$

Hence π is "fiber preserving": For every $(y, n) \in Y \times \mathbb{Z}^+$, the fiber over (y, n) is defined to be $(p^{-1} \times id)(y, n)$ and fiber preserving means that $\pi((p^{-1} \times id)(y, n)) \subset p^{-1}(\sigma(y, n))$.

Open Problem 4.2 Suppose that Y is p-periodic. Under what conditions does Eq.(2.1) have a r-periodic solution, where r divides p?

5 Limiting equations

Suppose that $f_n : D \to D$, where D is a compact subset of $\mathbb{R}^m$, is a sequence of continuous functions that converges uniformly to a continuous function $f_\infty : D \to D$. Then the equation

$$x_{n+1} = f_\infty(x_n) \tag{5.1}$$

is called the limiting equation of the equation

$$x_{n+1} = f_n(x_n) \tag{5.2}$$

Open Problem 5.1 Suppose that $D = [a, b]$, where $[a, b] \subset \mathbb{R}$ is a compact interval, and that f_∞ has a 2^k-cycle but no 2^{k+1}-cycles. Find conditions under which the omega-limit set of any point in Eq. (5.2) is a 2^r-cycle of f_∞, with $0 \leq r < k$.

Under a summability condition on the sequence $\{f_n\}$, Kempf [7] proved the conjecture for $k = 1$. Hirsch, Smith, and Zhao [6] proved that the omega-limit of any point in Eq. (5.2) is internally chain transitive.

Open Problem 5.2 1. Under what conditions does stability of cycles under f_∞ can be lifted to the nonautonomous system?
2. Under what conditions does periodicity and almost periodicity can be lifted from f_∞ to the nonautonomous system?

References

[1] S. Elaydi. *Discrete Chaos,* Chapman & Hall, Boca Raton, 2000.

[2] S. Elaydi, *On a converse of Sharkovsky's theorem,* American Math. Monthly, **103** (1996) 386-392.

[3] S. Elaydi and R. Sacker, *Global stability of periodic orbits of nonautonomous difference equations,* Reprint.

[4] S. Elaydi and A. Yakubu, *Global stability of cycles: Lotka-Volterra competition model with stocking,* J. Diff. Equations and Appl. **8(6)** (2002), 537-549.

[5] S. Elaydi and A. Yakubu, *Basins of attraction of stable cycles,* J. Diff. Equations Appl. **8(8)** (2002), 755-760.

[6] M.W. Hirsch, H.L. Smith, X-Q. Zhao, *Chain transitivity, attractivity and strong repellors for semidynamical systems,* J. Dynamics and Diff. Equations **13** (2001).

[7] R. Kempf, *On Ω-limit sets of discrete-time dynamical systems,* J. Diff. Equations and Appl. **8(12)** (2002), 1121-1131.

[8] S. Sacker and G. Sell, *Lifting properties in skew-product flows with applications to differential equations,* Memoirs Amer. Math. Soc. **11, No. 190** (1977).

[9] A. Sharkovsky, *Coexistence of cycles of a continuous mapping of the line into itself (in Russian),* Ukrain. Mat. Ž. **16** (1964), 61-71.

Fields Institute Communications
Volume **42**, 2004

On Period-Two Solutions of $x_{n+1} = \frac{\alpha+\beta x_n+\gamma x_{n-1}}{A+Bx_n+Cx_{n-1}}$

E. A. Grove
Department of Mathematics
University of Rhode Island
Kingston, RI 02881-0816 U.S.A.
grove@math.uri.edu

G. Ladas
Department of Mathematics
University of Rhode Island
Kingston, RI 02881-0816 U.S.A.
gladas@math.uri.edu

Dedicated to Professor Zhicheng Wang on his 70th Birthday

Abstract. We state some facts, and we pose several open problems and conjectures about the period-two solutions of the equation in the title.

1 Introduction

Consider the non-linear, second order, rational difference equation

$$x_{n+1} = \frac{\alpha + \beta x_n + \gamma x_{n-1}}{A + Bx_n + Cx_{n-1}} \quad , \qquad n = 0, 1, \ldots \tag{1.1}$$

with non-negative parameters and non-negative initial conditions chosen such that the denominator is always positive.

Our goal is to state some facts and to pose several open problems and conjectures regarding period-two solutions of Eq. (1.1).

Eq. (1.1) has been the subject of several investigations. See the monograph [4] and the references cited therein.

In order to avoid degenerate cases, we shall assume that

$$\alpha + \beta + \gamma > 0 \quad , \quad B + C > 0 \quad , \quad \beta + B > 0 \quad \text{and} \quad \gamma + C > 0. \tag{1.2}$$

2000 *Mathematics Subject Classification.* Primary 39A10, 39A11.
Key words and phrases. Difference equations, period-two solution.

2 The case $C = 0$

Suppose $C = 0$. Then Eq. (1.1) reduces to the equation

$$x_{n+1} = \frac{\alpha + \beta x_n + \gamma x_{n-1}}{A + B x_n}, \qquad n = 0, 1, \ldots. \tag{2.1}$$

The character of the solutions of Eq. (2.1) is described by the following trichotomy result. See [1], [2] and [4].

Theorem 2.1 *The following three statements are true:*

(a) *Every solution of Eq. (2.1) converges if and only if*

$$\gamma < \beta + A. \tag{2.2}$$

(b) *Every solution of Eq. (2.1) converges to a period-two solution of Eq. (2.1) if and only if*

$$\gamma = \beta + A. \tag{2.3}$$

(c) *There exist unbounded solutions of Eq. (2.1) if and only if*

$$\gamma > \beta + A. \tag{2.4}$$

Although the "trichotomy" Theorem 2.1 gives a detailed description of the solutions of Eq. 2.1), it leaves many open problems which require further study. Here are some.

Open Problem 2.2 Assume (2.3) holds, and let

$$\phi, \psi, \phi, \psi, \ldots \tag{2.5}$$

be a period-two solution of Eq. (2.1). Determine the set of all initial conditions (x_{-1}, x_0) of the solutions of Eq. (2.1) which converge to (2.5).

Open Problem 2.3 Assume (2.3) holds, and let $\{x_n\}_{n=-1}^{\infty}$ be a positive solution of Eq. (2.1). Determine the period-two solution

$$\phi, \psi, \phi, \psi, \ldots$$

to which $\{x_n\}_{n=-1}^{\infty}$ converges.

Conjecture 2.4 *Assume (2.4) holds. Show that every bounded solution of Eq. (2.1) converges.*

3 The case $C > 0$

Suppose $C > 0$. Then one can see that in order for Eq. (1.1) to have a periodic solution with prime period-two, it is necessary to have $\gamma > 0$. In this case, we may write Eq. (1.1) in the normalized form

$$x_{n+1} = \frac{\alpha + \beta x_n + x_{n-1}}{A + B x_n + x_{n-1}}, \qquad n = 0, 1, \ldots \tag{3.1}$$

with non-negative parameters and non-negative initial conditions. (Recall that we are also assuming that $\beta + B > 0$.)

3.1 The case $\alpha = \beta = 0$. Suppose $\alpha = \beta = 0$. Then Eq. (3.1) reduces to the equation

$$x_{n+1} = \frac{x_{n-1}}{A + Bx_n + x_{n-1}} \quad , \qquad n = 0, 1, \ldots \tag{3.2}$$

with

$$A \in [0, \infty) \quad \text{and} \quad B > 0. \tag{3.3}$$

Eq. (3.2) has periodic solutions with prime period-two if and only if

$$A < 1. \tag{3.4}$$

Furthermore, when (3.3) and (3.4) hold, the following statements are true. See [4].

(i) When $B = 1$, the periodic solutions of Eq. (3.2) with prime period-two are given by

$$\ldots, \phi, \psi, \phi, \psi, \ldots$$

where

$$\phi + \psi = 1 - A \quad \text{with} \quad \phi, \psi \in [0, 1 - A] \text{ and } \phi \neq \psi.$$

(ii) When $B \neq 1$, Eq. (3.2) has the unique periodic solution with prime period-two

$$\ldots, 0, 1 - A, 0, 1 - A, \ldots \tag{3.5}$$

and the solution (3.5) of Eq. (3.2) is locally asymptotically stable when $B > 1$ and is unstable when $0 < B < 1$.

3.2 The case $\alpha + \beta > 0$. Suppose $\alpha + \beta > 0$. Then Eq. (3.1) has periodic solutions with prime period-two if and only if

$$\beta + A < 1\,, \; B > 1\,, \; \text{and } 4\alpha < (1 - \beta - A)[B(1 - \beta - A) - (1 + 3\beta - A)]. \tag{3.6}$$

Furthermore when (3.6) holds, Eq. (3.1) has the unique period-two solution

$$\ldots, \phi, \psi, \phi, \psi, \ldots \tag{3.7}$$

where the values of ϕ and ψ are the two positive and distinct roots of the quadratic equation

$$t^2 - (1 - \beta - A)t + \frac{\alpha + \beta(1 - \beta - A)}{B - 1} = 0. \tag{3.8}$$

Conjecture 3.1 *Assume that (3.6) holds. Show that the following statements are true:*

1. *The positive equilibrium of Eq. (3.1) is an unstable saddle-point equilibrium.*
2. *The period-two solution (3.7) of Eq. (3.1) is locally asymptotically stable.*

Open Problem 3.2 Assume that (3.6) holds. Investigate the basin of attraction of the period-two solution (3.7) of Eq. (3.1).

Conjecture 3.3 *Assume that one of the inequalities in (3.6) is reversed. Show that every solution of Eq. (3.1) has a finite limit. More precisely, show that the following is true. If $\tilde{x} = 0$ is the only equilibrium of Eq. (3.1), then $\tilde{x} = 0$ is globally asymptotically stable. If Eq. (3.1) has a positive equilibrium $\bar{x} > 0$, then $\bar{x} > 0$ is locally asymptotically stable and is a global attractor of all nonzero solutions of Eq. (3.1).*

Open Problem 3.4 Investigate the global stability and the periodic nature of the solutions of Eq. (3.1) with positive parameters and non-negative initial conditions.

4 Equations with a unique prime period-two solution

It follows from Sections 2 and 3 that after some obvious renormalizations, the only equations of the type of Eq. (1.1) with a unique prime period-two solution are the following seven equations with positive parameters.

Case 1 [See [3] and ([4], p. 92).]

$$y_{n+1} = \frac{p + y_{n-1}}{qy_n + y_{n-1}} \quad , \quad n = 0, 1, \ldots \tag{4.1}$$

Case 2 [See [5] and ([4], p. 113).]

$$y_{n+1} = \frac{py_n + y_{n-1}}{qy_n + y_{n-1}} \quad , \quad n = 0, 1, \ldots \tag{4.2}$$

Case 3 [See [4], p. 133]

$$y_{n+1} = \frac{y_{n-1}}{p + qy_n + y_{n-1}} \quad , \quad n = 0, 1, \ldots \tag{4.3}$$

Case 4 [See [4], p. 149.]

$$y_{n+1} = \frac{p + qy_{n-1}}{1 + y_n + ry_{n-1}} \quad , \quad n = 0, 1, \ldots \tag{4.4}$$

Case 5 [See [4], p. 158.]

$$y_{n+1} = \frac{py_n + y_{n-1}}{r + qy_n + y_{n-1}} \quad , \quad n = 0, 1, \ldots \tag{4.5}$$

Case 6 [See [4], p. 175.]

$$y_{n+1} = \frac{r + py_n + y_{n-1}}{qy_n + y_{n-1}} \quad , \quad n = 0, 1, \ldots \tag{4.6}$$

Case 7 [See [4], p. 183.]

$$y_{n+1} = \frac{p + qy_n + y_{n-1}}{r + sy_n + y_{n-1}} \quad , \quad n = 0, 1, \ldots \tag{4.7}$$

The local asymptotic stability of the prime period-two solutions of Eqs. (4.5), (4.6), and (4.7) has not yet been investigated.

Conjecture 4.1 *Assume that*

$$p + 1 > r \quad \textit{and} \quad q + r > 3p + 1 + qr + pq$$

and let ϕ *and* ψ *be the two positive and distinct roots of the quadratic equation*

$$t^2 - (1 - p - r)t + \frac{p(1 - p - r)}{q - 1} = 0.$$

Show that

$$\ldots, \phi, \psi, \phi, \psi, \ldots$$

is a locally asymptotically stable prime period-two solution of Eq. (4.5).

Conjecture 4.2 *Assume that*

$$p < 1 \quad , \quad q > 1 \quad , \quad \text{and} \quad 4r < (1-p)(q-pq-3p-1)$$

and let ϕ and ψ be the two positive and distinct roots of the quadratic equation

$$t^2 - (1-p)t + \frac{r+p(1-p)}{q-1} = 0.$$

Show that

$$\ldots, \phi, \psi, \phi, \psi, \ldots$$

is a locally asymptotically stable prime period-two solution of Eq. (4.6).

Conjecture 4.3 *Assume that*

$$q+r < 1 \quad , \quad s > 1 \quad , \quad \text{and} \quad p < \frac{(1-q-r)[s(1-q-r)-(1+3q-r)]}{4}$$

and let ϕ and ψ be the two positive and distinct roots of the quadratic equation

$$t^2 + (1-q-r)t + \frac{p+q(1-q-r)}{s-1} = 0.$$

Show that

$$\ldots, \phi, \psi, \phi, \psi, \ldots$$

is a locally asymptotically stable prime period-two solution of Eq. (4.7).

Open Problem 4.4 For each of the Equations (4.1)–(4.7), determine the basin of attraction of their respective unique prime period-two solution.

Open Problem 4.5 Assume that $\{p_n\}_{n=0}^{\infty}$, $\{q_n\}_{n=0}^{\infty}$, $\{r_n\}_{n=0}^{\infty}$, and $\{s_n\}_{n=0}^{\infty}$ are period-two sequences with non-negative values. Investigate the local stability character of the non-negative period-two solutions of the equation

$$y_{n+1} = \frac{p_n + q_n y_n + y_{n-1}}{r_n + s_n y_n + y_{n-1}} \quad , \quad n = 0, 1, \ldots \tag{4.8}$$

References

[1] C.H. Gibbons, M.R.S. Kulenović, and G. Ladas, *On the recursive sequence $x_{n+1} = (\alpha + \beta x_{n-1})/(\gamma + x_n)$*, Math. Sci. Res. Hot-Line, **4**(2) (2000), 1-11.

[2] C.H. Gibbons, M.R.S. Kulenović, G. Ladas, and H.D. Voulov, *On the trichotomy character of $x_{n+1} = (\alpha + \beta x_n + \gamma x_{n-1})/(A + x_n)$*, J. Diff. Equations Appl., **8**(1) (2002), 75-92.

[3] W.A. Kosmala, M.R.S. Kulenović, G. Ladas, and C.T. Teixeira, *On the recursive sequence $y_{n+1} = (p + y_{n-1})/(qy_n + y_{n-1})$*, J. Math. Anal. Appl., **251** (2000), 571-586.

[4] M.R.S. Kulenović and G. Ladas, *Dynamics of Second Order Rational Difference Equations; With Open Problems and Conjectures*, Chapman & Hall/CRC Press, 2001.

[5] M.R.S. Kulenović, G. Ladas, and W.S. Sizer, *On therecursive sequence $x_{n+1} = (\alpha x_n + \beta x_{n-1})/(\gamma x_n + C x_{n-1})$*, Math. Sci. Res. Hot-Line 2, (1998), no. 5, 1-16.

Fields Institute Communications
Volume **42**, 2004

Open Problems on Cycling in Genetic Systems

Robert J. Sacker
Department of Mathematics
University of Southern California
Los Angeles, CA 90089-1113 U.S.A.
rsacker@math.usc.edu

Hubertus F. von Bremen
Department of Mathematics
University of Southern California
Los Angeles, CA 90089-1113 U.S.A.
vonbrem@usc.edu

In [1], we developed the following difference equation

$$x^* = G(x, r) \qquad x = (x_1, x_2, x_3, x_4)^T$$

where x^* represents the image of x. The quantities x_i represent frequencies of certain gametic types in an infinite population and therefore $\sum_{i=1}^{4} x_1 = 1$. The mapping G, the *generational map* is given by

$$G = S \circ R$$

where S is the *selection mapping* given by

$$x \;\rightarrow\; x^* = S(x) \qquad x = (x_1, x_2, x_3, x_4)^T$$

where

$$x_i^* \;=\; x_i\left(\frac{F_i}{\overline{F}}\right) \tag{1.1}$$

and

$$F_i = \sum_{j=1}^{4} F_{ij}x_j \qquad \textit{and} \qquad \overline{F} = \sum_{i,j=1}^{4} F_{ij}x_ix_j \;=\; \sum_{i=1}^{4} F_ix_i$$

The quantity F_i is the *marginal mean fitness* of x_i and $\overline{F}$ is the *mean fitness* of the population. Dividing by the quantity $\overline{F}$ in (1.1) preserves the relation $\sum x_i = 1$.

The mapping R is given by

$$x \;\rightarrow\; x^* = R(x) \qquad x = (x_1, x_2, x_3, x_4)^T$$

2000 *Mathematics Subject Classification.* Primary 39A11; Secondary 92O10, 92O25.

where

$$\begin{aligned} x_1^* &= x_1 - r(x_1x_4 - x_2x_3) \\ x_2^* &= x_2 + r(x_1x_4 - x_2x_3) \\ x_3^* &= x_3 + r(x_1x_4 - x_2x_3) \\ x_4^* &= x_4 - r(x_1x_4 - x_2x_3) \end{aligned} \tag{1.2}$$

and r, the *recombination coefficient*, satisfies $0 \leq r \leq 1/2$. The coefficient r is a measure of the degree to which crossover takes place between the two genetic loci: $r \approx 0$ represents *tight linkage* whereas $r \approx 1/2$ represents *loose linkage* between the loci.

It was shown in Sacker and von Bremen [1, 2] that for certain values of the fitnesses F_{ij} there is a critical value r_0 and a stable fixed point x_0 such that as r decreases through r_0, x_0 loses its stability and bifurcates into a stable invariant curve, Γ.

Open Problem 1.1 The motion on Γ is such that it takes 217 iterations of G for a point to return approximately to where it began. This is due to the fact that in the Neimark-Sacker bifurcation which takes place, the imaginary part of the complex eigenvalue λ which exits the unit circle is very small.

Question 1.2 *Can fitnesses F_{ij} be found such that this "return time" of* 217 *is reduced to something in the order of* 10 *to* 20*?*

Question 1.3 *Can fitnesses F_{ij} be found such that the "non resonance" condition of the bifurcation theorem is violated, i.e., λ passes through a 3rd or 4th root of unity as it exits the unit circle?*

Open Problem 1.4 Can conditions on the fitnesses F_{ij} be found which guarantee that cycling cannot take place for any value of the recombination coefficient?

Open Problem 1.5 The fitnesses F_{ij} can vary in time due to periodic environmental influences. Does this enhance or hinder the onset of cycling and if cycling does occur are the iterates of the basic map confined to a curve or some other regular set, e.g. a torus?

References

[1] Robert J. Sacker and Hubertus F. von Bremen. Bifurcation of maps and cycling in genetic systems, *in Proceedings of 7th Intl. Conf. on Difference Eq. and Appl., Changsha, China, 2002.*

[2] Robert J. Sacker and Hubertus F. von Bremen. A new approach to cycling in a 2-locus 2-allele genetic model, *J. Difference Eq. and Appl., 9(5)2003, pp.441-448*

Fields Institute Communications
Volume **42**, 2004

On Periodic Solutions of Systems of Two Difference Equations

Jianhong Wu
Department of Mathematics and Statistics
York University
Toronto, ON, Canada, M3J 1P3
wujh@yorku.ca

In the design of a neural network architecture for the purpose of content-addressable memory of prescribed spatio-temporal patterns, it is important that the corresponding dynamical model describing the computational performance of the network has the co-existence of multiple stable solutions since each memory has to be associated with a stable and attractive solution. In Wu and Zhang [1], it was shown that a very simple and small network of neurons with delayed excitation/inhibition exhibits multistability and possesses a very large number of stable attractive periodic orbits with prescribed periods if the delay is sufficiently large and the updating is discrete.

More precisely, Wu and Zhang considered the following system of two difference equations coupled through an excitatory feedback with an integer delay given below

$$\begin{cases} x(n) = & \beta x(n-1) + \alpha f(y(n-k)), \\ y(n) = & \beta y(n-1) + \alpha f(x(n-k)), \end{cases} \tag{1.1}$$

where $n \in \mathbf{N}$ (the set of all nonnegative integers), $\alpha > 0$, $\beta \in (0,1)$, $k \geq 1$ is a fixed integer, $f : \mathbf{R} \to \mathbf{R}$ is a nonlinear function satisfying standard conditions of the usual signal function such as McCulloch-Pitts step-function and sigmoid function with large gain $f'(0)$. Such a system describes the dynamics, *updating discretely*, of a network of two identical neurons where the information processing of a neuron involves the internal decay and feedback from another with a delay.

It was shown that, under certain technical conditions, for any positive integer p with $p|2k$ system (1.1) has exactly $N(p)/p$ distinct asymptotically stable p-periodic solutions. Here $N(p)$ is related to the well-known Möbius inversion formula described below: Let $p = \prod_1^l p_i^{m_i}$, where p_i, $i = 1, 2, \cdots, l$ are primes. For every subset I of the set $\{1, \cdots, l\}$, let $p_I = \prod_{i \in I} p_i$. $N(p) = \sum_{I \subset \{1,\cdots,l\}} (-1)^{|I|} 2^{p/p_I}$.

Therefore, a network of two neurons, when updated discretely, allows the coexistence of an amazingly large number of stable periodic orbits. However, it is still a remaining open problem how this can be used as a device for memory storage.

[1] *A Simple Delayed Neural Network with Large Capacity for Associative Memory, Discrete and Continuous Dynamical Systems, Series B, in press, 2003.*

The problem is that the model involves very limited number of parameters (β: the internal decay rate; α: the synaptic weight; and k the delay) and thus it is difficult to train such a network so that a large number of memories can be stored as stable periodic orbits. One way seems to replace the coupling term $\alpha f(y(n-k))$ by a weighted sum involving multiple delays

$$\sum_{j=1}^{M} \alpha_j f(y(n-k_j))$$

and then to determine the coefficients α_j from the training data sets. This is equivalent to the training the density of the distribution of delay in a model of difference equations with distributed delays.

In summary, we wish to consider the following system

$$\begin{cases} x(n) = & \beta x(n-1) + \sum_{j=1}^{M} \alpha_j f(y(n-k_j)), \\ y(n) = & \beta y(n-1) + \sum_{j=1}^{M} \alpha_j f(x(n-k_j)). \end{cases} \tag{1.2}$$

We would like to know some answers, for a given set of integers $(k_j)_{j=1}^{M}$, for each of the following questions:

(1) How many periodic solutions can system (1.2) possess?
(2) Which of the above periodic solutions are asymptotically stable?
(3) How these periodic solutions are related to the parameters α_j and the integers k_j?
(4) Are there any quasi-periodic solutions and if yes, are they stable?